Fundamentals of Mechatronics

Fundamentals of Mechatronics

Second Edition

Musa Jouaneh
Department of Mechanical,
Industrial, and Systems Engineering
University of Rhode Island

Cengage

Australia • Brazil • Canada • Mexico • Singapore • United Kingdom • United States

Fundamentals of Mechatronics,
Second Edition
Musa Jouaneh

SVP, Product Management: Cheryl Costantini

VP, Product Management: Heather Bradley Cole

Portfolio Product Director: Colin Grover

Portfolio Product Manager: Timothy Anderson

Product Assistant: Lea Sanchez

Learning Designer: MariCarmen Constable

Content Manager: Sean Campbell

Digital Project Manager: Nikkita Kendrick

Director, Product Marketing: Danae April

Product Marketing Manager: Chris Walz

Content Acquisition Analyst: Deanna Ettinger

Production Service: MPS Limited

Designer: Chris Doughman

Cover Image Source: jntvisual/istock photo

Previous edition: © 2013

Library of Congress Control Number: 2024908308

ISBN: 978-0-357-68487-0

Cengage
5191 Natorp Boulevard
Mason, OH 45040
USA

Printed in the United States of America
Print Number: 01   Print Year: 2024

To the LORD who has done wonderful things in my life
and to my lovely wife for her encouragement and support

Contents

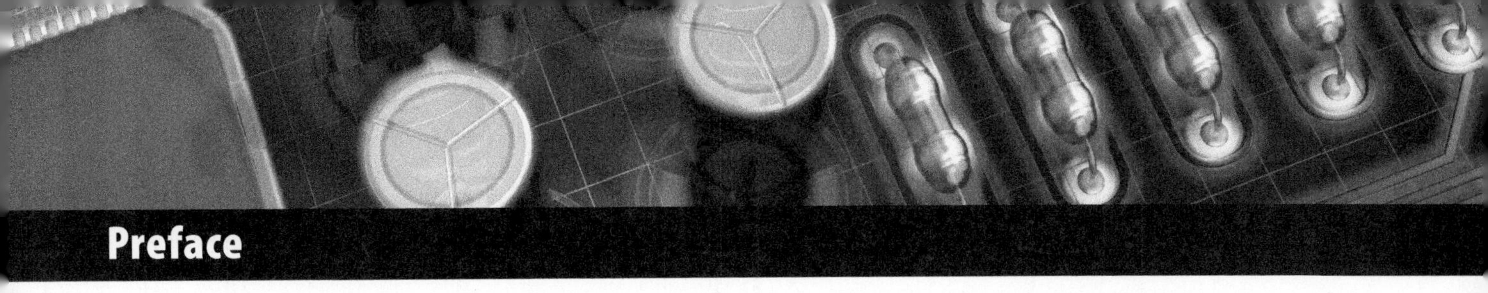

Preface

Fundamentals of Mechatronics is designed to serve as a textbook for an undergraduate course in Mechatronics Systems Design. The book focuses on the fundamental concepts, operating principles, application considerations, and relevant practical issues that arise in the selection and design of mechatronics components and systems. The text gives complete treatment to the subject matter by covering both hardware and software aspects of mechatronics systems in a single text, The book is comprehensive in its coverage covering all the important topics such as analog and digital circuits, basic electronics, microcontrollers, data acquisition, control software, system response, sensors, actuators, and control systems. The book covers the Arduino platform and has code examples in both C and MATLAB. The book has over 90 worked examples and hundreds of end-of-chapter questions and problems. It also includes four integrated case studies. A separate laboratory book, with additional exercises, is provided to give guided hands-on experience with many of the topics covered in the text.

Content and Organization

Fundamentals of Mechatronics focuses on the fundamental concepts, applications, modeling considerations, and relevant practical issues that arise in the selection and design of mechatronics components and systems. The text has comprehensive coverage of all the essential topics in mechatronics including hardware and software and provides an in-depth exploration of key mechatronics concepts. This comprehensive coverage ensures that instructors have a single, cohesive resource for teaching a wide range of topics. The textbook includes numerous real-world examples, leveraging current manufacturers' data sheets to illustrate practical applications of theoretical concepts. The text also includes an in-depth exploration of Arduino microcontrollers, focusing on their role in mechatronics. Detailed guidance on interfacing microcontrollers and PCs with mechatronic components using both asynchronous and synchronous serial communication techniques is also provided. Additionally, comprehensive coverage of structured control software development, including timing, event-driven programming, and state-transition diagrams is included in the text. The text incorporates software coding examples in both C and MATLAB. Furthermore, nearly every topic in the textbook is supplemented with video links, offering students additional resources for understanding and visualizing the concepts discussed. Finally, four detailed case studies are interwoven throughout the book, encompassing design, modeling, simulation, control, and implementation aspects of mechatronic systems. This offers a unique teaching tool that demonstrates the interconnectedness of mechatronics concepts in real-world scenarios, aiding in the delivery of more interactive and impactful lessons.

The book is organized into 11 chapters and several appendices. Chapter 1 is an introductory chapter. Chapters 2 and 3 focus on circuits and electronics. Chapters 4 through 6 focus on microcontrollers, interfacing, and control software development. Chapter 7 focuses on system response, while Chapters 8 and 9 focus on sensors and actuators. Chapter 10 focuses on the basics of feedback control. The final Chapter 11 provides additional details on the four mechatronics integrated cases studies that are discussed throughout the text. The text also has several appendices. Appendix A lists standard resistor values, while Appendix B has a list of 7-bit ASCII codes. Appendix C gives a brief review of the Laplace Transform, and Appendix D gives an overview of the dynamic modeling of mechanical systems.

Course Outlines

Although the intended market for this text is junior/senior-level undergraduate students taking a course in mechatronics, the text is also suitable for other courses. It can serve as a textbook for a junior-level undergraduate introductory course on circuits, measurements and sensing, or an advanced undergraduate/beginning graduate-level course with focus on control software. For a junior/senior-level mechatronics undergraduate course, some of the sections in Chapter 5 (Data Acquisition and Microcontroller/PC Interfacing), Chapter 6 (Control Software), Chapter 7 (System Response), and Chapter 10 (Feedback Control) could be skipped. For a junior-level undergraduate introductory course on circuits, measurements and sensing, the course could just focus on the contents of Chapters 2, 3, 5, 7-8, and selected topics from other chapters. For an advanced undergraduate level/beginning graduate-level course with focus on control software, the course could just focus on the contents of Chapters 4–7, 10–11, and selected topics from the remaining chapters. In the undergraduate mechatronics course (MCE433) that the author teaches at the University of Rhode Island (URI), he covers Chapter 1, most of the material in Chapters 2, 3, and 4, selected topics from Chapter 5, most of Chapter 6, and selected topics from Chapters 7–11.

What's New in This Edition

The Second Edition of *Fundamentals of Mechatronics* has been extensively revised and updated to offer students broader and deeper coverage of topics, enhanced relevance to current technology, and a substantial increase in the number of worked examples, as well as end-of-chapter problems and questions.

The main changes from the first edition include:

- The introduction of coverage on Arduino microcontrollers, a new chapter on System Response, a detailed appendix on modeling mechanical systems with worked examples, and two new appendices on the Laplace Transform and standard resistor values.
- Expanded and revised content across numerous topics, particularly in chapters covering microcontrollers, data acquisition, control software, sensors, actuators, and control systems.
- More emphasis on operating principles and application considerations and less emphasis on coding. Also, coverage of Visual Basic was removed from text.
- A significant increase in end-of-chapter problems, tripling the number to over 320, a doubling of worked examples from 46 to 92, and an increase in end-of-chapter questions from 135 to 215.
- The addition of four integrated case studies, woven throughout the book, that address the design, modeling, simulation, control, and implementation of mechatronic systems.
- The incorporation of video links for nearly every topic covered in the text. The ▶ symbol indicates that a specific video is linked to a topic. To access these videos, please visit www.cengage.com and access the book's companion site. The list of videos will be present as a downloadable file.

Below is a list of the specific changes made in each chapter.

Chapter 1 (Introduction to Mechatronics): A new example of mechatronic systems (Airbag Safety System) was introduced, and two other examples were revised. Also, a new section on the Integrated Mechatronics Cases Studies was added in this chapter.

Chapter 2 (Analog Circuits and Components): A new section on power supplies and batteries was added, as well as the coverage of instrumentation amplifiers in the op-amps section. The coverage of analog circuit elements and electromechanical relays was revised.

Chapter 3 (Semiconductor Electronic Devices and Digital Circuits): The coverage of diodes, bipolar junction transistors, and H-bridges was revised.

Chapter 4 (Microcontrollers): This chapter was completely rewritten to replace the coverage of PIC microcontrollers with Arduino platform microcontrollers.

Chapter 5 (Data Acquisition and Microcontroller/PC Interfacing): The section on data acquisition was completely revised and new material on MATLAB/Simulink data acquisition was added. Also, the coverage of the sampling was substantially revised.

Chapter 6 (Control Software): New sections on timing in AVR microcontrollers and events and event-driven programming were added. The presentation and organization of the task/state control software structure was substantially revised.

Chapter 7 (System Response): This chapter is a new one that focuses on system response. Material from two of the appendices in the first edition was used in this chapter. The chapter covers time and frequency response of first and second-order systems including topics such as resonance, bandwidth, and various methods of simulating of the response of dynamic systems in MATLAB. Coverage of various types of filters were moved from the Sensors chapter to this chapter, and their coverage was expanded.

Chapter 8 (Sensors): New sections that discuss capacitive proximity sensors, multi-axis force/torque sensors, and IMU sensors were added. The coverage of thermistors, thermocouples, and 4-20 mA sensor output was substantially revised. Also, many sections in this chapter were updated and revised.

Chapter 9 (Actuators): New sections that discuss pneumatic and hydraulic actuators, piezoelectric, and shape memory alloy actuators were added. The coverage of solenoids moved from Chapter 2 to this chapter. Also, many sections in this chapter were updated and revised.

Chapter 10 (Feedback Control): This entire chapter was completely revised, and the discussion on PID-control was substantially expanded and split into two sections, one that focuses on P-, PI-, PD-, and PID-control of first-order systems, and the other on P-, PI-, PD-, and PID-control of second-order systems. Also new sections on cascaded control loop structure, controller bandwidth, tuning (via the Ziegler-Nichols method), and control robustness were added.

Chapter 11 (Mechatronics Cases Studies): The previous chapter on Mechatronics projects was replaced by a chapter that focuses on the four mechatronics case studies. Two of the case studies (Linear Motion Slide and Mobile Robot) were newly added in this edition.

Appendices: Three new appendices were added that includes one that reviews the dynamic modeling of mechanical systems which has several worked examples.

Supplements and Ancillaries

A separate laboratory book called *Laboratory Exercises in Mechatronics* is available for purchase through Cengage Learning. *Laboratory Exercises in Mechatronics* details several laboratory exercises and projects to facilitate guided hands-on experience with many of the topics covered in this text.

Additional instructor resources for this book are available online. Instructor assets include a Solution and Answer Guide and PowerPoint® slides. Sign up or sign in at www.cengage.com to search for and access this product and its associated resources.

Acknowledgements

I would like to acknowledge the many students who were enrolled in the mechatronics class at the University of Rhode Island and who had provided useful suggestions and comments which shaped the current manuscript.

I would also like to acknowledge James Byrnes, the electronic technician in the Mechanical, Industrial, and Systems Engineering Department at URI who has helped in building and wiring some of the circuits used in this book. I am also thankful to several of my colleagues at the University of Rhode Island who provided valuable comments. These include Professors Godi Fisher, William Palm, and Richard Vaccaro.

I also wish to acknowledge the valuable comments and suggestions for the first or second edition from the manuscript reviewers:

Amit Banerjee, Penn State Harrisburg

Alan A. Barhorst, Texas Tech University

Jordan M. Berg, Texas Tech University

William W. Clark, University of Pittsburgh

Burford Furman, San Jose State University

Hector Gutierrez, Florida Institute of Technology

Md Ahasan Habib, Rochester Institute of Technology

Shouling He, Vaughn College

Steve Hung, Clemson University

Marcia K. O'Malley, Rice University

Horacio Vasquez, University of Texas Rio Grande Valley

Thanks to Tim Anderson, portfolio product manager, at Cengage Learning, for his help in bringing this book to fruition.

The author and the publisher are grateful for the support extended by the instructors at Penn State. We also thank the many students who used the manuscript and provided useful comments.

Musa Jouaneh

Kingston, Rhode Island

About the Author

Musa Jouaneh is a Professor of Mechanical Engineering at the University of Rhode Island (URI). His research interests include mechatronics, manufacturing automation, robotics, and engineering education, and he has worked on many industrially funded projects in these areas. Dr. Jouaneh received his B.S. in Mechanical Engineering from the University of Louisiana, Lafayette, and his M. Eng. and Ph.D. degrees in Mechanical Engineering from the University of California at Berkeley. He has been teaching the senior-level Mechatronics course (MCE430) at URI since 2003. In addition to his Fundamentals of Mechatronics and Laboratory Exercises in Mechatronics textbooks, he is also the author of the book *Arduino-Based Introductory Guided Exercises in Mechatronics*. Professor Jouaneh has been the recipient of several awards including two URI College of Engineering (COE) Faculty Excellence Awards, two URI Outstanding Contribution to Intellectual Property Awards, the URI Foundation Teaching Excellence Award, and the URI COE Frank M. White Award for Excellence in Engineering Teaching. Dr. Jouaneh is a Fellow member of the American Society of Mechanical Engineers (ASME), a senior member of the Institute of Electrical and Electronics Engineering (IEEE), and a member of the American Society of Engineering Education (ASEE).

Chapter 1

Introduction to Mechatronics

Chapter Objectives:

When you have finished this chapter, you should be able to:

- Explain what a mechatronics system is
- List the components of a mechatronics system
- Provide examples of real-world mechatronics systems

1.1 Introduction

Mechatronics (▶ available—V1.1) is the field of study concerned with the design, selection, analysis, and control of systems that combine mechanical elements with electronic components, including computers and/or microcontrollers. Mechatronics topics involve elements from mechanical engineering, electrical engineering, and computer science, and the subject matter is directly related to advancements in computer technology. Yasakawa Electric Company coined the term **'mechatronics'** [1] to refer to the use of electronics in mechanical control (i.e., **'mecha'** from mechanical engineering and **'tronics'** from electrical or electronic engineering). Auslander et al. [2] defined mechatronics as the application of complex decision-making to the operation of physical systems. This definition eliminates the reference to the particular technology used when performing the operation.

A block diagram of a typical mechatronic system is shown in Figure 1.1. A mechatronic system has at its core a **mechanical system** that needs to be commanded or controlled. Such a system could be a vehicle braking system, a positioning table, an oven, or an assembly machine. The controller needs information about the state of the system. This information is obtained from a variety of **sensors,** such as those that give proximity, velocity, temperature, or displacement information. In many cases, the signals produced by the sensors are not in a form readable by the controller and need some **signal conditioning operations** performed on them. The conditioned, sensed signals are then converted to a digital form (if not already in that form) and presented to the controller.

The **controller** is the 'mind' of the mechatronic system. It processes **user commands** and sensed signals to generate command signals to be sent to the actuators in the system. The user commands are obtained from a variety of devices, including command buttons, graphical user interfaces (GUIs), touch screens, or pads. In some cases, the command signals are sent to the actuators without utilizing any feedback information from the sensors. This is called **open-loop operation**, and for it to work, it requires a good calibration between the input and output of the system with minimal disturbances. The more common mode of operation is the **closed-loop mode** in which the command signals sent to the actuators utilize the feedback information from the sensors. This mode of operation does not require calibration information, and it is much better suited for handling disturbances and noise.

Figure 1.1

Typical components of a mechatronic system

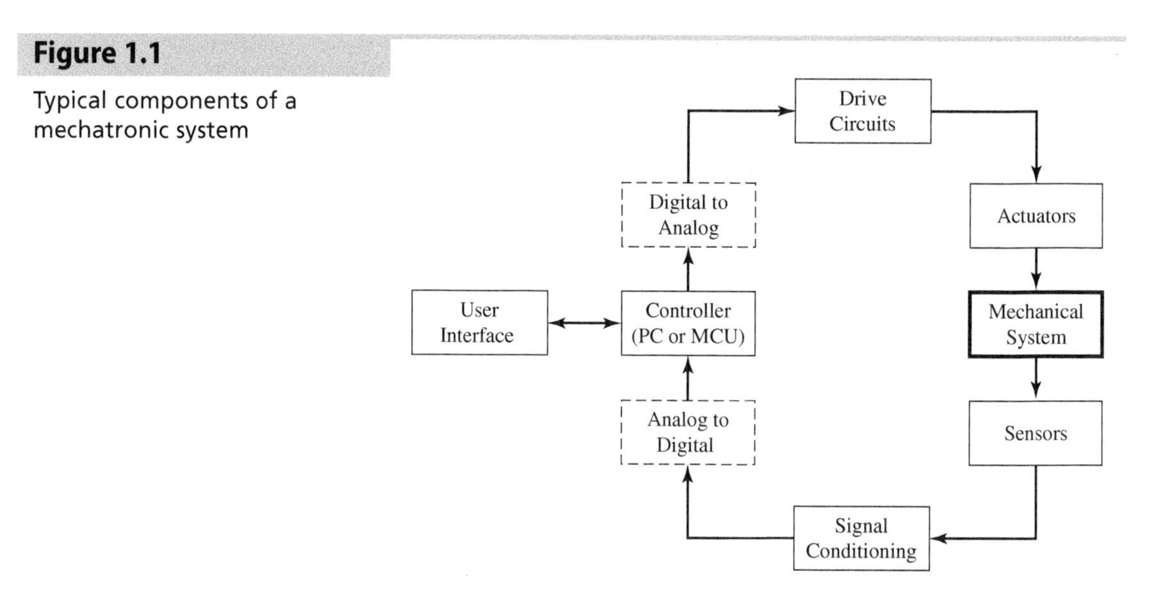

In many cases, the command signals to the actuators are first converted from a digital to an analog form. Amplifiers implemented in the form of **drive circuits** also can amplify the command signals sent to the **actuators**. The actuator is the mechanism that converts electrical signals into useful mechanical motion or action. The choice of the controller for the mechatronic system depends on many factors, including cost, size, ease of development, and transportability. Many mechatronic systems use personal computers (PCs) with data acquisition capabilities for implementation. Examples include control of manufacturing processes such as welding, cutting, and assembly. A significant number of controllers for a mechatronic system are implemented using a **microcontroller unit** (MCU), which is a single-chip device that includes a processor, memory, and input-output devices on the same chip. Microcontrollers often control many consumer devices, including toys, hand-held electronic devices, and vehicle safety systems. Control systems that use MCUs often are referred to as **embedded control systems**.

The control system for a mechatronic system can be classified as either a discrete-event control system or a feedback control system. In a **discrete-event system,** the controller controls the execution of a sequence of events, while in a **feedback control system,** the controller controls one or more variables using feedback sensors and feedback control laws. Almost all realistic systems involve a combination of the two. This textbook will discuss these two classes in detail.

A mechatronic system integrates mechanical components, electronic components, and software implemented either on a PC or MCU to produce a flexible and intelligent system that performs the complex processing of signals and data. In many cases, a mechatronic system improves the performance of a system beyond what can be achieved using manual means. An example is the speed control of rotating equipment. In some cases, a mechatronic system is the only means by which that system can operate (such as the control of magnetic bearings and in nano-positioning control applications).

1.2 Examples of Mechatronic Systems

Modern society depends on mechatronic-based systems for its conveniences and luxurious standard of living. From intelligent appliances to safety features in cars (such as airbags and anti-lock brakes), mechatronic systems are widely used in everyday life. The availability of low-cost, compact, and powerful processors in the form of MCUs accelerated the widespread use of mechatronic systems. An example is using embedded controllers to control many of the devices in a vehicle (▶ available—V1.2). A list of such applications is shown in Table 1.1. Note that in a vehicle, an MCU is housed in a unit called an **electronic control unit** (or **ECU**), which holds the MCU and some peripherals. Each ECU is configured for a specific function as the ECU to control anti-lock braking.

Table 1.1

Listing of sample applications of mechatronic systems in vehicles

Application Area		
Safety	**Comfort**	**Power Train**
• Airbag system	• Door locks	• Engine controls
• Anti-lock braking system	• Keyless entry system	• Fuel pump controls
• Daytime running light	• Heating system controls	• Fuel sensing controls
• Electronic stability controls	• Seat positioning controls	• Gearbox controls

To further illustrate mechatronic systems, we will discuss four systems: the airbag safety system in a vehicle, an industrial robot, a mobile robot, and a parking garage gate.

1.2.1 Airbag Safety System

The airbag safety system (▶ available—V1.3) is designed to protect the occupants of a vehicle in the case of a collision, and it consists of three main components: sensors to detect a crash, an ECU to process the sensor signals, and actuators to deploy the airbags. For frontal collisions (see Figure 1.2), most airbag safety systems use two sensors. A 'crash' sensor is located in the crush zone of the vehicle to detect the collision, and a 'safety' or 'arming' sensor is located in the passenger compartment area to prevent false deployment of the airbag. The signals from these two sensors are fed into the airbag ECU which is typically located in the center console area of the vehicle. The software in the ECU processes the sensor signals and employs logic to decide if the airbags need to be deployed. If that is the case, then an electrical signal is sent to the airbag canister or inflator which detonates a small amount of igniter compound. The ignition causes a chemical reaction that causes the nylon fabric airbag to be rapidly filled with Nitrogen from the chemicals that are placed in the canister. Airbag systems also use seat occupancy sensors to detect the presence and weight of the occupant in the front passenger seat. If the seat is not occupied or a low-weight occupant is identified, then the passenger airbag will not be deployed (such as to protect a small person sitting in the front seat from the force of the airbag). Advanced airbag ECUs can also activate seat belt pretensioners, which are devices that quickly retract and tighten the seatbelts, to restrain the vehicle occupants in their seats in the case of a collision. It is important to note that it takes less than 50 milliseconds (ms) from the time the collision is detected by the airbag sensors to the time the airbag is fully inflated, so it is of utmost importance that the control software for an airbag system operates with no delay. As explained previously, the airbag safety system is a sophisticated mechatronic system that uses various sensors in combination with a control unit and actuators that need to work reliably and rapidly.

Figure 1.2

Frontal collision sensors in an airbag system

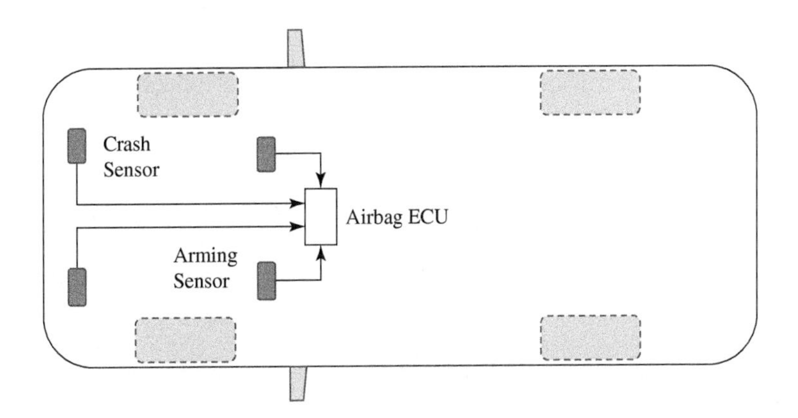

1.2.2 Industrial Robots

Industrial robots (▶ available—V1.4) are a prime example of mechatronic systems, whether they are fixed in one location or mobile. Figure 1.3 is an illustration of an industrial robot arm, which embodies the core principles of mechatronics. A **robot**, in essence, is a mechanical apparatus that can be programmed to execute a wide range of tasks and applications. The fundamental components of a robot system consist of the controller and the mechanical arm. The controller assumes several crucial functions within the robotic system, including managing the user interface, programming operations, and controlling the movements of the arm. Its responsibilities encompass facilitating communication between the user and the robot, enabling the programming of various tasks, and overseeing the precise execution of these tasks by the mechanical arm. The mechanical arm itself is typically composed of multiple interconnected mechanical links, joined together by joints. Each of these links is driven by an actuator, which imparts motion to the arm. Furthermore, each actuator is equipped with a feedback sensor that provides valuable information regarding the position of its corresponding link. This sensor-based feedback mechanism allows for accurate and precise control of the robotic arm's movements. A multi-link robot, due to its intricate design and structure, necessitates a sophisticated level of coordination among the individual links. This coordination is primarily facilitated by the control software, which operates by processing input data derived from the desired motion of the arm, as well as the feedback received from the sensors. To enhance a robot's ability to adapt to variations within its operating environment, additional sensors are typically incorporated into the system. Vision sensors, for example, enable the robot to perceive and interpret visual information, while proximity sensors aid in detecting the presence or proximity of objects. By utilizing such additional sensors, the mechatronic system enhances the robot's perception and understanding of its surroundings, allowing it to effectively navigate and interact within its environment. In summary, the mechatronic system of an industrial robot encompasses most, if not all, of the elements depicted in Figure 1.1. Through the seamless integration of mechanical components, actuators, sensors, control software, and additional sensing capabilities, industrial robots exemplify the synergy between mechanical engineering, electronics, and computer science.

Figure 1.3

Industrial robot

(Baloncici/Shutterstock.com)

1.2.3 Mobile Robots

Mobile robots are currently being used in a wide diversity of applications. Whether vacuum cleaning (▶ available—V1.5), assisting soldiers in combat operations, or delivering food and medicine in hospitals, their use is increasing. Similar to their fixed counterparts, a mobile robot consists of several modules that are commanded by a controller. Due to their operation in unstructured environments, mobile robots rely heavily on sensors to guide them in navigation and to avoid obstacles. Examples of sensors used by mobile robots include ultrasonic proximity sensors, vision sensors, and global positioning system sensors. An example of a mobile robot is the Roomba® vacuum-cleaning robot (see Figure 1.4) made by iRobot® Corporation. The Roomba has a cylindrical shape, two-wheel modules, and several sensors (see Figure 1.5) including cliff sensors, a wall-following sensor, a floor tracking sensor, and a camera. The infrared or cliff sensors are located at the edge of the robot that let the robot know when it is close to a 'cliff' such as stairs. These cliff sensors measure the distance between the robot base and the floor by sending an infrared light that is reflected from the floor surface and detected by receivers on the robot. If the receiver does not receive any signal, then the robot figures out that it is close to the ledge, and backs away from that position.

The wall-following sensor also uses infrared lighting to maintain a given distance from a wall. The floor tracking sensor measures the distance and direction that the robot has traveled. The control system on the Roomba uses advanced algorithms to combine data from several sensors to create a map of the room that is being cleaned and to navigate around obstacles. The Roomba has all the main components of a mechatronic system: actuators (wheel modules), sensors, and a controller.

Figure 1.4

Roomba® vacuum cleaning robot

(Courtesy of iRobot Corporation, Bedford, MA)

Figure 1.5

Sensors on Roomba® vacuum cleaning robot

(Courtesy of iRobot Corporation, Bedford, MA)

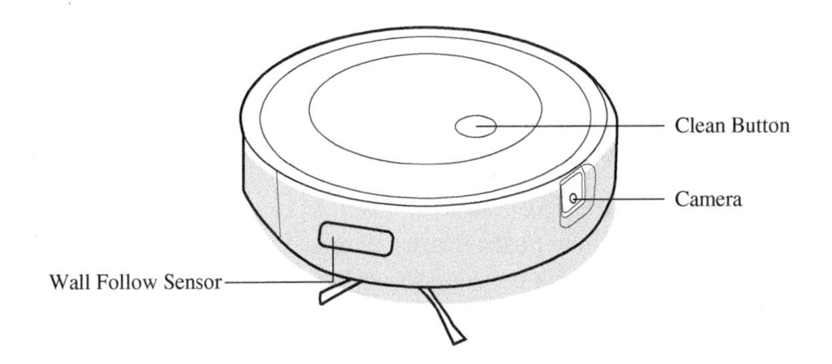

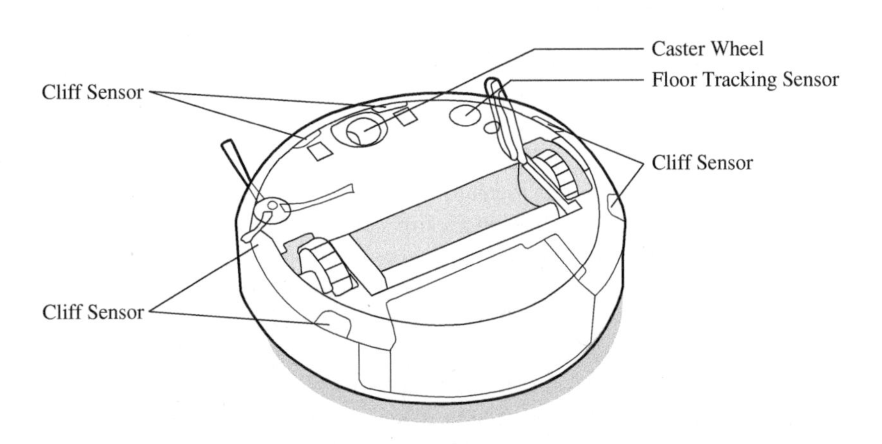

1.2.4 Parking Gate

A **parking garage gate** (▶ available—V1.6) is another example of a mechatronic system that involves several elements (see Figure 1.6). The system has an electric motor to raise and lower the gate arm. It also has a proximity sensor to prevent the gate from striking people and vehicles. In addition, it has a controller in which software is used to run the gate in different operating modes. Typically, a parking garage gate operates as follows: The user presses a button to get a ticket or swipes a card in a card scanner. Once the ticket is picked up by the user or the card is validated, the gate arm rotates upward. The gate arm remains in a raised position until the vehicle has completely cleared the gate, at which point the gate drops down. The operation of each stage of this system is dependent on sensor feedback and timing information. The controller for this system cycles between the different operating stages each time a vehicle needs to enter the parking garage.

Figure 1.6

Parking gate
(BigPileStock/Alamy Stock Photo)

The previous examples illustrate a wide range of mechatronic systems. Mechatronics, the interdisciplinary field that combines mechanical engineering, electrical engineering, and computer science encompasses numerous **enabling technologies** crucial to the design, operation, and control of modern smart systems. These technologies include signal processing, system interfacing, sensor integration, drive technology, actuation systems, software programming, and motion-control systems. The future of mechatronics is quite promising. Mechatronics will continue to play a vital role in the development of advanced robotics and automation systems. These systems will become more sophisticated, and capable of performing complex tasks with greater precision, agility, and adaptability with applications in different industries, including manufacturing, healthcare, agriculture, and logistics. The integration of mechatronics with the Internet of Things (IoT) will enable the development of smart and interconnected devices that allow them to collect and exchange data, perform real-time analysis, and respond intelligently to their environments. This integration will lead to advancements in areas such as smart homes, autonomous vehicles, and wearable technology.

Overall, mechatronics will continue to drive innovation and contribute to the advancement of numerous fields. Its interdisciplinary nature and integration with emerging technologies will shape the future of robotics, automation, and smart systems, leading to improved efficiency, convenience, and functionality across various sectors.

1.3 Overview of Text

This book covers topics that are needed to design, analyze, and implement a complete mechatronic system. The text is organized into eleven chapters and several appendices and covers all the areas related to the mechatronic components shown in Figure 1.1. **Chapter 2** covers the basics of analog circuits and components. Virtually every mechatronic device has some form of an electric circuit, and thus understanding and analyzing electrical circuits is important in mechatronics. **Chapter 3** discusses the operation of semiconductor electronic devices (such as diodes, thyristors, and transistors) that are used in many circuits and devices for switching or amplification purposes. It also covers digital circuits. In many situations, the current and voltage capabilities of interface devices available on PCs and MCUs are not adequate to operate real devices (such as motors and heaters), and semiconductor electronic components (such as transistors) are needed. **Chapter 4** discusses the use and programming of microcontrollers in detail. The objective of this chapter is to give the reader complete coverage of the features and capabilities of a typical microcontroller. Unlike combinational and sequential circuits, microcontrollers offer a flexible but complex method to implement control logic.

Chapter 5 discusses techniques to interface a processor to the outside world using different interface devices (such as analog-to-digital converters, digital-to-analog converters, digital input/output ports, and asynchronous and synchronous serial ports). **Chapter 6** focuses on software development issues when using a microcontroller (and to a less extent, a PC) as the controller in a mechatronic system. Some of these issues include how to incorporate time into a control program, how to structure the operation and control of physical systems into tasks and states, and how to write control code that is responsive and not blocking. A software-based control system offers flexibility over a hardware-based one since the controller structure and control logic can be changed by simply changing the code in the program. **Chapter 7** focuses on system response in both the time and frequency domains. In mechatronic systems design, it is important to study the response of the system as this will provide evaluation data on the performance of the system. Understanding system response is also important in the selection of sensors and actuators as well as in control system design. **Chapter 8** focuses on sensors. Sensors are vital components of mechatronic systems since they provide the feedback information that enables automated systems to function. A sensor is an element that produces an output in response to changes in physical quantity (such as temperature, force, or displacement). **Chapter 9** discusses actuators, which are the key components of all mechanized equipment. An actuator is a device that converts energy to mechanical motion. This chapter primarily focuses on electrically powered actuators, which are commonly used in mechatronic systems. **Chapter 10** covers the basics of feedback control systems. The objective is to illustrate to the reader the design, simulation, and implementation of basic feedback control systems. **Chapter 11** provides detailed setup information, code listings, and additional modeling, simulation, and experimental results about the four integrated case studies that are discussed throughout the text. The text also has several appendices. **Appendix A** lists standard resistor values, while **Appendix B** has a list of 7-bit ASCII codes. **Appendix C** gives a brief review of the Laplace Transform, and **Appendix D** gives an overview of the dynamic modeling of mechanical systems.

1.4 Integrated Mechatronics Case Studies

This second edition of the text includes four integrated mechatronics **case studies** that are covered throughout the text. The purpose of these case studies is to illustrate the integration of the many topics covered in this book in the form of experimental systems as well as to provide practical information on the design and control of the mechatronic systems used in these case studies. A picture and a detailed description of the hardware for each case are provided in Chapter 11 along with additional material. The cases, the subtopics covered in the text for each case, and where that subtopic is covered in the text are listed below. Table 1.2 shows the distribution of the subtopics in the text.

Case Study I: Stepper Motor-Driven Rotary Table

- I.A Photo Interrupter as a Homing Sensor for the Table (Chapter 3)
- I.B A State-Transition Diagram for Table Operation (Chapter 6)
- I.C Stepper Motor and Driver for the System (Chapter 9)

Case Study II: DC Motor-Driven Linear Motion Slide

- II.A Power Supply and Buck Converter for the System (Chapter 2)
- II.B Data Acquisition using MATLAB/Simulink (Chapter 5)
- II.C Brushless DC Motor and Servo Drive (Chapter 9)
- II.D Closed-Loop Position Controller (Chapter 10)

Case Study III: Temperature-Controlled Heating System

- III.A Transistor for Controlling the Voltage (Chapter 3)
- III.B Control Software for System Operation (Chapter 6)
- III.C System Model Parameters from Step Response Data (Chapter 7)

Case Study IV: Mobile Robot

- IV.A H-Bridge Driver for the Motor (Chapter 3)
- IV.B Microcontroller to Control the Robot (Chapter 4)
- IV.C State-Transition Diagram for Robot Motion (Chapter 6)
- IV.D IMU Sensor (Chapter 8)

Table 1.2

Distribution of the Integrated Mechatronics Case Studies subtopics

Chapter	Case Study I: Stepper Motor-Driven Rotary Table	Case Study II: DC Motor-Driven Linear Motion Slide	Case Study III: Temperature-Controlled Heating System	Case Study IV: Mobile Robot
Two (Analog Circuits and Components)		II.A Power Supply and Buck Converter for the System (Section 2.11)		
Three (Semiconductor Electronic Devices and Digital Circuits)	I.A Photo Interrupter as a Homing Sensor for the Table (Section 3.4.4)		III.A Transistor for Controlling the Voltage (Section 3.5)	IV.A H-Bridge Driver for the Motor (Section 3.10)
Four (Microcontrollers)				IV.B Microcontroller to Control the Robot (Section 4.4.5)
Five (Data Acquisition and Microcontroller/PC Interfacing)		II.B Data Acquisition using MATLAB/Simulink (Section 5.5.2)		
Six (Control Software)	I.B A State-Transition Diagram for Table Operation (Section 6.5.3)		III.B Control Software for System Operation (Section 6.5.3)	IV.C State-Transition Diagram for Robot Motion (Section 6.5.3)
Seven (System Response)			III.C System Model Parameters from Step Response Data (Section 7.2)	
Eight (Sensors)				IV.D IMU Sensor (Section 8.10.3)
Nine (Actuators)	I.C Stepper Motor and Driver for the System (Section 9.5.2)	II.C Brushless DC Motor and Servo Drive (Section 9.2.3)		
Ten (Feedback Control)		II.D Closed-Loop Position Controller (Section 10.8.5)		

Questions

1.1 What is mechatronics?

1.2 What are the elements of a mechatronic system?

1.3 How are mechatronic systems implemented?

Problems

P1.1 Research mechatronic systems that are used in vehicles. Identify the type of sensors and/or actuators that are used in the following systems.

a. Anti-lock braking

b. Door locks

c. Powered side mirrors

P1.2 Research and identify all the mechatronic components used in the following devices.

a. Modern washing machine

b. Servo-driven industrial robot

c. Automated entry door

P1.3 Research and identify several sensors that are used in autonomous vehicles.

Chapter 2

Analog Circuits and Components

Chapter Objectives:

When you have finished this chapter, you should be able to:

- Explain the characteristics of basic circuit components
- Explain the different types of switches
- Use a relay as a switch
- Perform circuit analysis using Kirchhoff's voltage law (KVL) and Kirchhoff's current law (KCL)
- Determine an equivalent circuit for a given two-terminal circuit
- Explain the concept of impedance and loading effects
- Determine the RMS current and voltage in an AC circuit
- Determine the components of power in an AC circuit
- Analyze different op-amp circuits
- Explain proper grounding techniques
- Explain the different types of power supplies
- Compare batteries to power supplies

2.1 Introduction

Virtually every mechatronic device has some form of an electric circuit, and thus, understanding and analyzing electrical circuits is important in mechatronics. A **circuit** is a closed path through a series of electronic components in which a current flows through. Electric circuits are classified into analog and digital types. In **analog circuits**, the voltage is continuous and can have any value within a specified range. On the other hand, **digital circuits** represent voltage signals using two distinct levels, commonly 0 volts and a higher voltage level (such as 5 volts). Analog circuits are generally more susceptible to noise and disturbances compared to digital circuits. In an analog circuit, any noise present in the system can lead to changes in the analog signal or result in a loss of information. These variations can impact the accuracy and reliability of the circuit's output. In contrast, digital circuits are more robust against small disturbances or noise. As long as the voltage signal stays within the specified range for a digital circuit, it represents the same information. This property of digital circuits makes them less susceptible to the effects of noise or small disruptions. Digital circuits are a tool to perform logic operations using hardware instead of software.

Two basic quantities in electrical circuits are voltage, or electric potential, and current. **Voltage** refers to the capability of driving a stream of electrons through a circuit, similar to the concept of a force in a mechanical system, while **current** is a measure of the flow of charge in the circuit. The unit of measurement for current is Amperes (A). The time integral of the current is defined as an **electrical charge** and has units of Coulombs (C). When the voltage or current does not change its value with respect to time in a circuit, it is called a direct current (DC) circuit. On the other hand, when the voltage or current periodically changes in magnitude and direction, it is called an alternating current (AC) circuit.

This chapter covers the basics of analog circuits and components. Digital circuits are covered in the next chapter. For further reading, see [3-6].

2.2 Analog Circuit Elements

Circuit elements include power sources (such as a power supply or a battery), switches to open and close the circuit, and circuit components (such as resistors and capacitors). A schematic of an electrical circuit is shown in Figure 2.1. The figure shows a closed loop where a conducting element (such as a copper wire) connects a voltage source (or a current source) to load elements on the circuit.

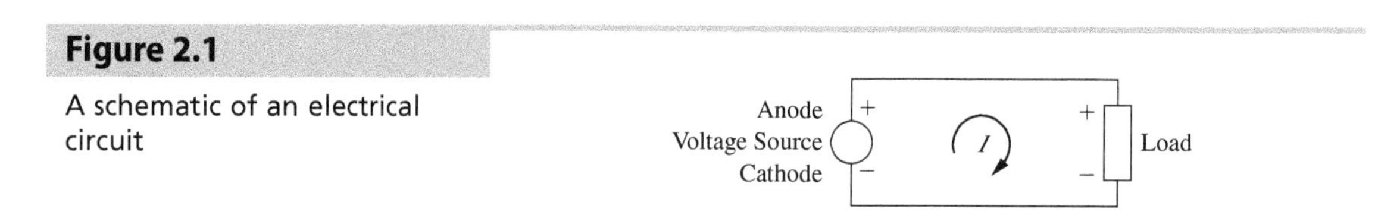

Figure 2.1

A schematic of an electrical circuit

Figure 2.1 also shows the direction of conventional current flow from the **anode** (positive terminal) of the power source to the **cathode** (negative terminal). The conventional current flow direction is the opposite of the direction of actual electron flow in the circuit but was kept due to Benjamin Franklin, who thought that electrical current is due to the motion of positively charged particles.

Circuit components can be of the passive type, which require no external power to operate (such as resistors and capacitors), or active components which require power to operate (such as operational amplifiers). Three basic passive circuit elements are the resistor, the capacitor, and the inductor. Table 2.1 shows the electrical symbols for these elements. The table also shows the symbols for two energy sources that are normally represented in circuits. These include an ideal voltage source and an ideal current source. The sources are considered ideal because they do not have any internal resistance, capacitance, or inductance.

Table 2.1

Symbols of basic circuit elements

Element	Reference	Circuit Symbol
Resistor	R	
Capacitor	C	
Inductor	L	
Ideal Voltage Source	V	
Ideal Current Source	I	

The **resistor** is an element that dissipates energy. The constitutive relation for an ideal resistor is given by **Ohm's law**, in which the voltage drop across the resistor is linearly related to the current through the resistor, or

(2.1)
$$V = IR$$

The resistance is measured in units of ohms (Ω). Resistors (▶ available—V2.1) can be either fixed type or variable. Fixed-type resistors are made in a variety of forms including surface mount, wire wound, thick film, and carbon composition (see Figure 2.2).

Figure 2.2

Resistor types
(a) surface mount, (b) wire wound, (c) thick film, and (d) carbon composition

(Courtesy of Ohmite Mfg. Co., Arlington Heights, IL)

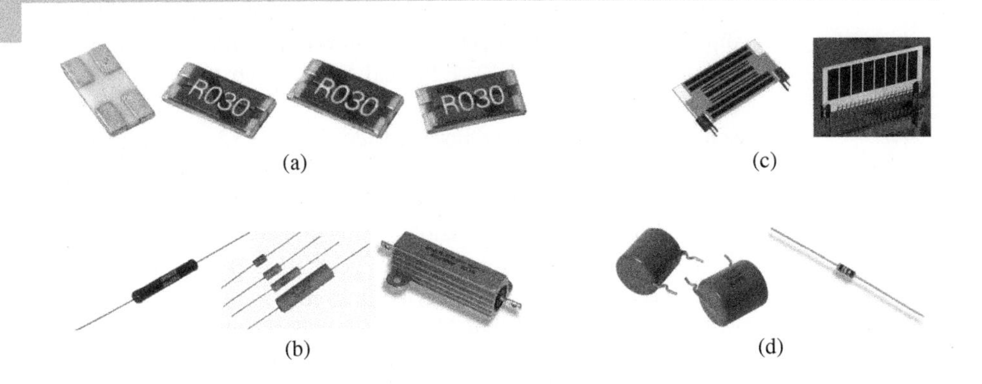

Typical fixed-type, low-wattage resistors are made of molded carbon composition and have a resistance that ranges from a few ohms to about 20 MΩ. These resistors have a cylindrical shape and have sizes that increase with the power rating of the resistor. The typical power rating is 1/4 to 1 watt (W). The resistance can be read from the color code printed on the resistors. Typically, four color bands are shown on the resistor, as shown in Figure 2.3, but resistors with five or six color bands are also available. For four color bands, the left three bands give the resistor value, while the fourth band gives the resistance tolerance (Tol). The resistance is given by the formula

$$R = ab \times 10^{c}(\mp\%\text{Tol}) \tag{2.2}$$

where the a band is the value of the tens digit, the b band is the value of the ones digit, the c band is the base-10 exponent power value, and the Tol band gives the tolerance or expected percentage variation in the resistor value. Table 2.2 gives these values.

Figure 2.3

Resistor color bands

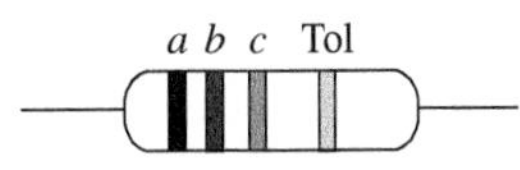

Table 2.2

Resistor bands color code

a, b, c Band Color	Value	Tol Band Color	Value (+/−)
Black	0	Silver	10%
Brown	1	Gold	5%
Red	2	Brown	1%
Orange	3	Red	2%
Yellow	4	Green	0.5%
Green	5	Blue	0.25%
Blue	6	Violet	0.1%
Violet	7	Gray	0.05%
Gray	8		
White	9		

As an example, a resistor whose bands are colored brown, black, orange, and silver has a resistance of 10k +/−10% ohms. The resistance of a real resistor is not constant, but it increases with temperature.

Commercial resistors are typically available in a set of **standard values** known as the **E-series** (see Appendix A). These values are organized according to the tolerance of the resistors. The three most common E-series are E12, E24, and E96, which are used for resistors with tolerance values of $\pm 10\%$, $\pm 5\%$, and $\pm 1\%$, respectively. For each series, the number represents the number of distinct values within each decade of resistance. For instance, in the E12 series, there are 12 distinct values ranging from 10 to 100 ohms (10, 12, 15, 18, 22, 27, 33, 39, 47, 56, 68, and 82), and the pattern repeats for each subsequent decade (100 to 1000, 1000 to 10,000, etc.). Similarly, E24 has 24 distinct values per decade, providing a greater choice for more precision applications.

Variable-type resistors include rheostats and potentiometers. **Rheostats** are two-terminal resistors, while **potentiometers** are three-terminal resistors. They can be of the linear or rotary type, and the resistance between the terminals is changed as the position of the wiper terminal is changed.

Unlike a resistor, a **capacitor** is an energy storage element. The constitutive relation for a capacitor is

(2.3) $$\frac{dV}{dt} = \frac{1}{C}I$$

where C is the capacitance in Farads (F). Small capacitors are typically of the ceramic type (▶ available—V2.2), which can be used in both AC and DC circuits. These capacitors have a capacitance that is less than 0.1 micro-Farads (μF). **Ceramic capacitors often have a numeric code** printed on them to indicate their capacitance value. This code typically consists of two or three digits, followed by a letter. The first two digits represent the base value of the capacitance, and the third digit represents the multiplier or the number of zeros to follow the base value. This code represents the capacitance value in picofarads (pF). For instance, a ceramic capacitor with the code '104' would have a capacitance of 100,000 pF or 0.1 μF, because the '4' adds four zeroes to the base value '10'. Sometimes, there may only be two digits, which means there are no additional zeros. For example, a capacitor labeled '47' would have a capacitance of 47 pF. The letter following the number code often indicates the tolerance of the capacitor. Common tolerance letters are 'M' ($\pm 20\%$), 'K' ($\pm 10\%$), and 'J' ($\pm 5\%$), among others. For example, a capacitor labeled '104J' would have a capacitance of 0.1 μF with a tolerance of $\pm 5\%$.

Capacitors with large capacitance (up to several thousand micro Farads) are of the **electrolytic type**. These are used only in DC circuits, and their leads are polarized. One characteristic of capacitors is the leakage current, which is the current that flows between the capacitor plates when a voltage is applied across the plates of the capacitor. This current leads to the loss of charge over time from the capacitor. This current, however, is typically small, unless the capacitor is of the electrolytic type. Similar to resistors, capacitors are also available as fixed or variable types.

An **inductor** is also an energy storage element. Inductive elements in practice include solenoids and motors. The constitutive relation for an inductor is

(2.4) $$\frac{dI}{dt} = \frac{1}{L}V$$

where L is the inductance, and it is measured in units of Henry (H). Small-sized inductors are of the molded type, and they have inductance that varies from sub-micro to several thousand micro-Henry (μH).

2.3 Switches

2.3.1 Mechanical Switches

Mechanical switches are devices that make-or-break contact in electrical circuits. There are a variety of mechanical switches available (▶ available—V2.3), including toggle, push-button, rocker, slide, and others (see Figure 2.4).

Toggle switches are specified in terms of their number of poles and throws. **Poles** refer to the number of circuits that can be completed by the same switching action, while **throws** refer to the number of individual contacts for each pole. Figure 2.5 shows four different configurations of toggle switches. In Figure 2.5(a), a single-pole, single-throw (SPST) switch is shown, which is the configuration of basic switches (such as on-off switches and mechanical contact limit switches). In Figure 2.5(b), a single-pole, double-throw (SPDT) switch is shown. Figure 2.5(b) configuration is commonly used in the residential wiring of rooms that have two switches to operate a light fixture, and Figure 2.6 shows an example of such a circuit that uses two SPDT switches. Note that the SPDT switch is commonly known as a 'three-way switch.' Figure 2.5(c) shows a double-pole, single-throw (DPST) switch, which is equivalent to two SPST switches controlled by a single mechanism. Figure 2.5(d) shows a double-pole, double-throw (DPDT) switch configuration. This configuration is commonly used in the construction of electromechanical relays (to be discussed in the next section).

Figure 2.4

Mechanical switches (a) toggle, (b) pushbutton, (c) rocker, and (d) slide

(Courtesy of CIT Relay & Switch, Minneapolis, MN)

CIT
(a)
(b)
(c)
(d)

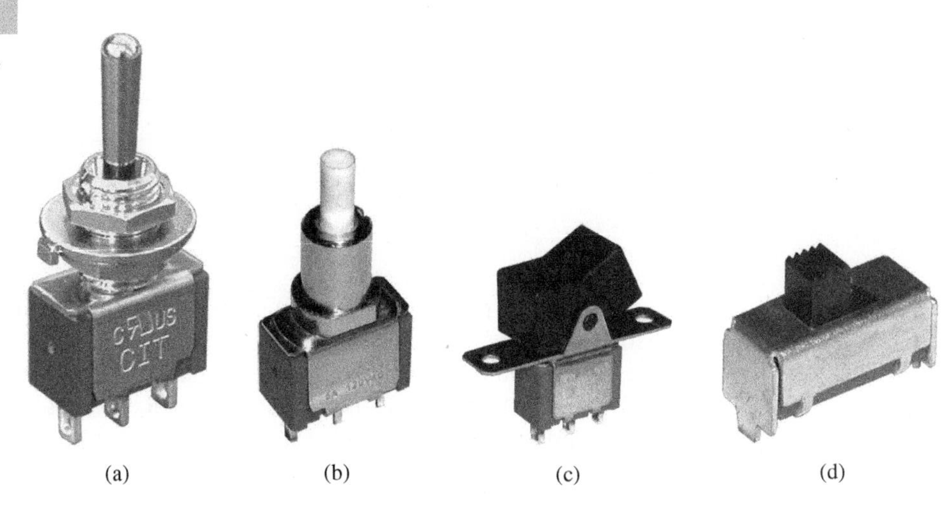

Figure 2.5

Different configurations of toggle switches

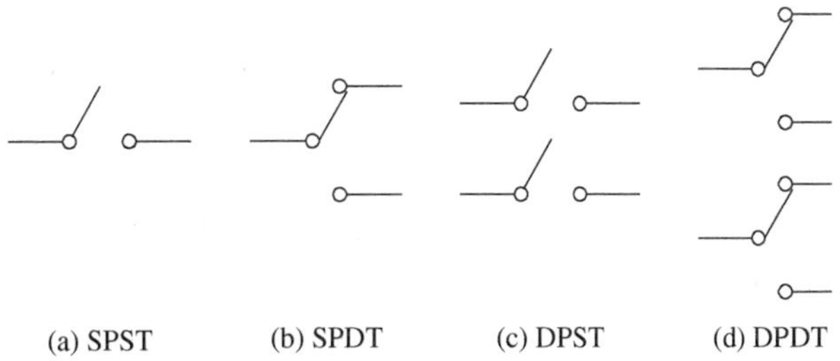

(a) SPST
(b) SPDT
(c) DPST
(d) DPDT

Figure 2.6

Wiring circuit for a light bulb using two SPDT switches

Bulb
Neutral
Live

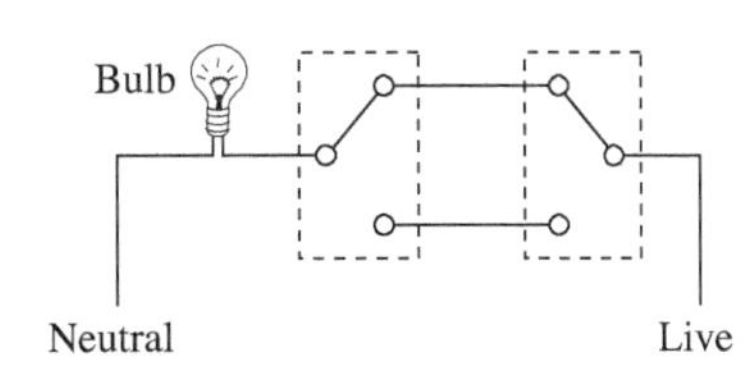

A DPDT switch that is internally wired for polarity reversal applications is commonly called a four-way switch (see Figure 2.7). Such a switch has only four wires coming out of it (instead of six) and can be inserted between two SPDT switches to enable the wiring of a single light bulb using three switches (see Problem P2.4).

Figure 2.7

DPDT switch wired as 'a four-way switch'

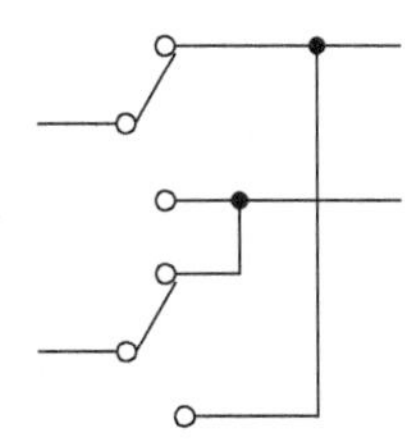

Toggle switches are known as 'break before make' type, which means that the switch pole never connects to both terminals in SPDT or DPDT switch configuration. **Push-button switches** have the symbol shown in Figure 2.8. They can be either of two types: normally open (NO) or normally closed (NC). Normally open or normally closed refer to the state of the switch before it is activated. Pushbutton switches are widely used as reset switches and doorbell switches.

Figure 2.8

Push-button switch (a) normally open and (b) normally closed

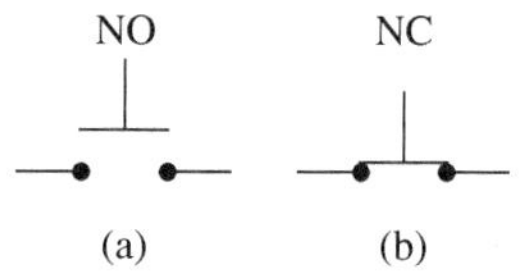

NO NC

(a) (b)

One possible disadvantage of mechanical switches is **switch bouncing**. Since the switch arm is typically a small flexible element, the opening and closing of mechanical switches cause the switch to bounce several times before settling at its desired state. Figure 2.9 shows a typical pattern in closing a switch. Note that each of the contacts during the bouncing interval, which is typically about 15 to 25 ms long, may register by a processor as separate switch action unless means were incorporated to address this issue. The most common approach to solve this problem is to provide for each switch a debouncing circuit that makes use of flip-flop circuit elements (to be discussed in Chapter 3).

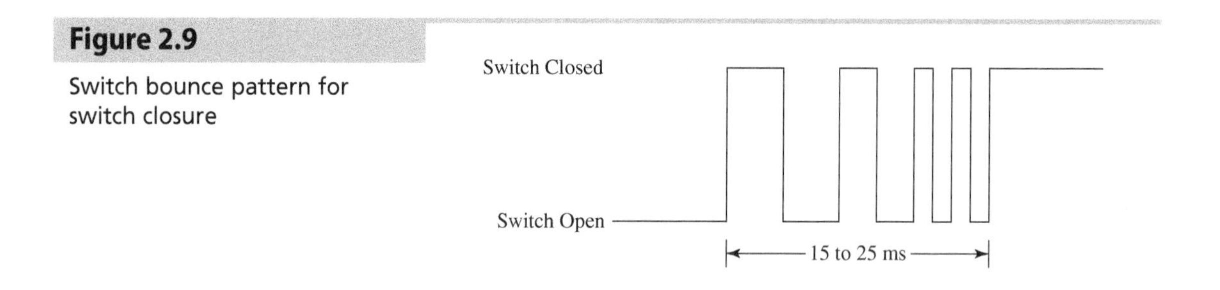

Figure 2.9

Switch bounce pattern for switch closure

2.3.2 Electromechanical Relays

Many computer interfacing applications rely on electromechanical relays, which are electrically actuated switches utilizing a solenoid (refer to Section 9.3) to establish or disrupt the mechanical contact between electrical leads. Figure 2.10 illustrates the schematic of a single-channel relay (available—V2.4). In addition to the coil leads responsible for actuating the solenoid, a relay typically features three other leads. One of these leads is called COM (or COMMON), and it serves as the connection point for the power to the load. Another lead is designated as NO (or normally open), while the third lead is labeled NC (or normally closed). When the coil of the relay is energized, it causes the pole connected to the common lead to switch from the NC contact to the NO contact. In this energized state, the NO lead establishes electrical continuity, while the NC lead becomes disconnected. Conversely, when the coil is de-energized, the NC lead restores its contact, and the NO lead breaks the connection.

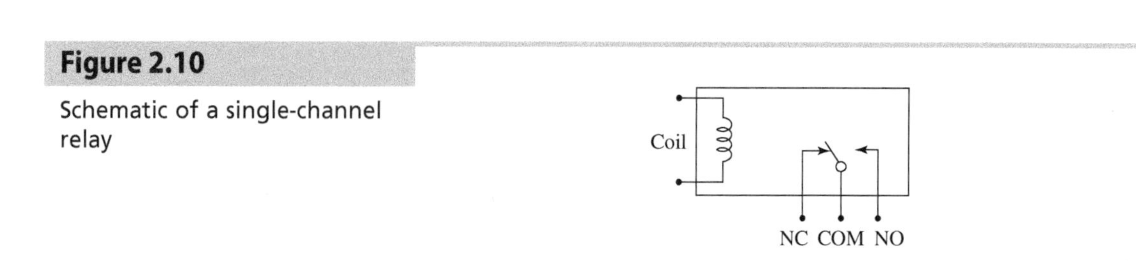

Figure 2.10

Schematic of a single-channel relay

Figure 2.11(a) shows the connection diagram of a typical small power relay. An example of using this relay to switch ON and OFF a 24 VDC motor is shown in Figure 2.11(b). When the coil circuit is closed (terminals 1 and 16), the solenoid will move the two poles that contact terminals 6 and 11 (NC terminals) to contact terminals 8 and 9 (NO terminals), respectively. For the motor example, closing the coil circuit will cause the 24 V to be applied to one lead of the motor and the motor will start rotating. The switch configuration in this relay is an example of the double-pole, double-throw switch configuration that was previously discussed. Some of the important characteristics of this relay are listed in Table 2.3. This relay can be used to switch up to 60 W (or 62.5 VA) using a coil (solenoid) that requires 5 VDC at 100 mA to operate. This current input value is beyond the current output limits of digital output ports (discussed in Chapters 4 and 5), so a current amplifying component (such as a transistor) is normally used to interface the digital output port to the coil terminals of the relay. The advantage of a relay is that the input circuit is electrically isolated from the output circuit. So, any noise-induced voltages in the output circuit have a minimal impact on the input circuit. The second advantage is that a small coil current can be used to switch a much larger load current.

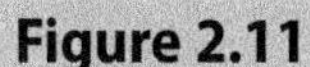

Figure 2.11

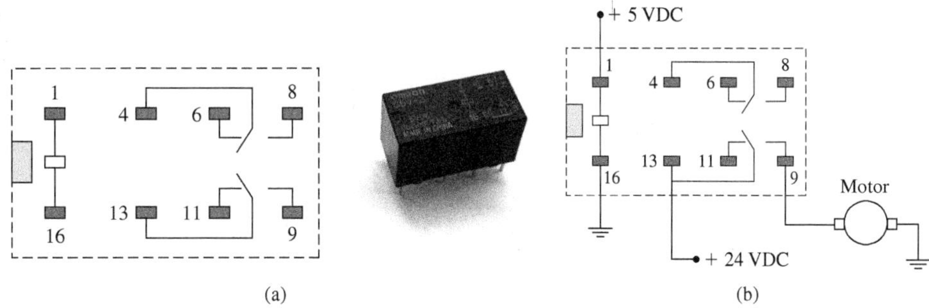

(a) G5V-2, a typical small power relay and (b) example wiring diagram

(Courtesy of Omron Corporation)

Table 2.3

Characteristics of the G5V-2 OMRON relay

Coil Rating	Rated Voltage	5 VDC
	Rated Current	100 mA
	Coil Resistance	50 Ω
Contact Rating	Rated Load	0.5 A at 125 VAC, 2 A at 30 VDC
	Max. Switching Voltage	125 VAC, 125 VDC
	Max. Switching Current	2 A
Operating Characteristics	Operate Time	7 ms max
	Release Time	3 ms max
	Max. Operating Frequency	Mechanical: 36,000 operations/hr Electrical: 18,000 operations/hr

One disadvantage of electromechanical relays is their relatively long switching time. For the previous relay, the maximum operate-release cycle time is 10 ms, and the maximum mechanical switching frequency is 10 Hz. This is in contrast to solid-state transistors (discussed in Chapter 3), which have nanoseconds switching time.

Example 2.1 illustrates the use of relays.

Example 2.1 Relays

A circuit consists of two SPST switches, A and B, and a load. Design a circuit using two single-channel relays to control the load so that it receives power when either switch A or switch B is closed and turns off only when both switches A and B are open.

Solution:

The circuit to solve this example is shown in Figure 2.12. Notice how the load voltage is applied to the COM lead on both relays. Also, notice that the NO leads on both relays are connected with the load. If either switch is closed, power is delivered to the load through the NO lead. If neither switch is pressed, no power is delivered to the load since the NO open leads will be disconnected from the load voltage.

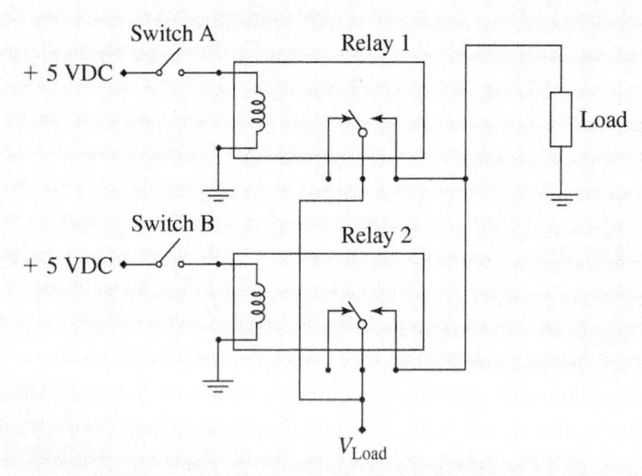

Figure 2.12 Circuit for Example 2.1

2.4 Circuit Analysis

A typical analog circuit is shown in Figure 2.13. The objective of circuit analysis is to determine the voltage and current at any point in the circuit. This is done with the aid of two laws: Kirchhoff's voltage law (KVL) and Kirchhoff's current law (KCL).

Figure 2.13

A typical electric circuit

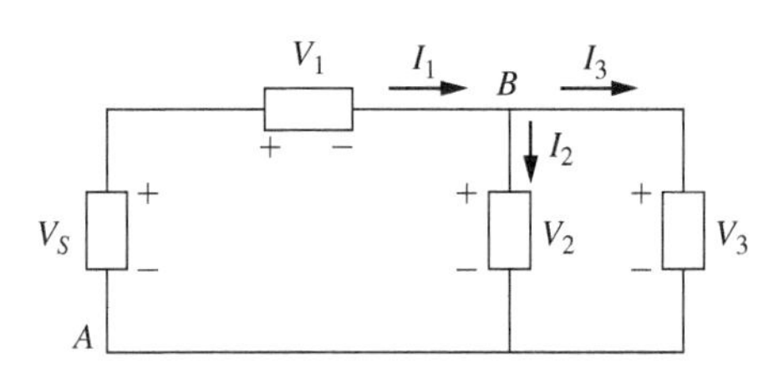

Kirchhoff's voltage law states that the algebraic sum of the voltage drops and rises around any closed path in a circuit is zero. In equation form, it is stated as

(2.5)
$$\sum_{i=1}^{i=N} V_i = 0$$

where N is the number of elements in the selected path. To illustrate this, consider the left loop of the circuit in Figure 2.13. Starting at any point in the circuit (such as point A), and going clockwise, we get

(2.6)
$$V_s - V_1 - V_2 = 0$$

In getting Equation (2.6), potential rises are considered positive, and potential drops are considered negative. As we go around the loop, the potential of the voltage source V_s rises (goes from $-$ to $+$), and the potential across the second and third elements drops (goes from $+$ to $-$). This would result in an equivalent expression if going counterclockwise around the loop. Here the sign of the potential drops and rises for the three elements would be opposite, but the final form is equivalent.

Kirchhoff's current law states that the sum of the current into a node is zero or in equation form

$$\sum_{i=1}^{i=N} I_i = 0 \tag{2.7}$$

With reference to Figure 2.13, KCL gives the following relationship between the currents at node B in the circuit.

$$I_1 - I_2 - I_3 = 0 \tag{2.8}$$

where the current is considered positive if it goes into the node and is negative if it leaves the node. Example 2.2 illustrates the use of KVL and KCL.

When two resistors are connected in series in a circuit, as in Figure 2.14(a), it is called a **voltage dividing circuit** because the voltage is divided among the two resistors, with the voltage drop across each resistor being proportional to the resistance of each resistor. For example, the output voltage across resistor R_2 is given by

$$V_2 = \frac{R_2}{R_1 + R_2} V_{\text{in}} \tag{2.9}$$

Similarly, when two resistors are connected in parallel in a circuit, as in Figure 2.14(b), it is called a **current dividing circuit**. The current through resistor R_2 is given by

$$I_2 = \frac{R_1}{R_1 + R_2} I \tag{2.10}$$

Note that the current through R_2 is the product of the input current I and the resistance of the other resistor R_1 divided by the sum of the resistance of the two resistors in the circuit.

Figure 2.14

(a) Voltage dividing circuit and
(b) current dividing circuit

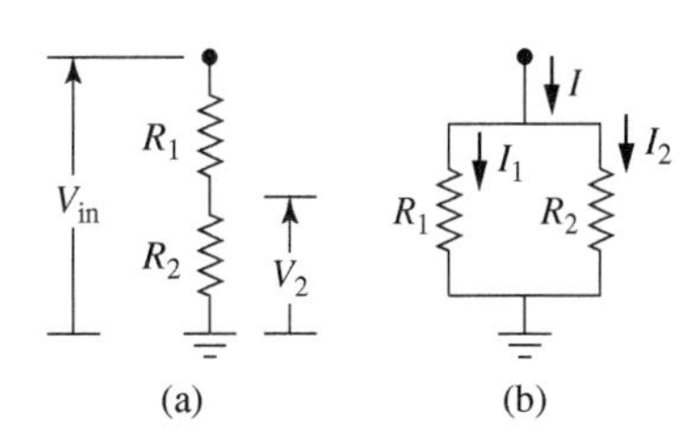

In many circuits, there could be several elements in a serial or parallel configuration. Table 2.4 gives the total resistance, capacitance, and inductance for serial and parallel combinations of these elements. The

Table 2.4

Total resistance, capacitance, and inductance

Element	Series Connection	Parallel Connection
Resistor	R_1 R_2	R_1 R_2
	$R_T = R_1 + R_1$	$1/R_T = 1/R_1 + 1/R_2$
Capacitor	C_1 C_2	C_1 C_2
	$1/C_T = 1/C_1 + 1/C_2$	$C_T = C_1 + C_2$
Inductor	L_1 L_2	L_1 L_2
	$L_T = L_1 + L_2$	$L_T = 1/L_1 + 1/L_2$

total value expressions can be derived by using the constitutive relation for the element in question and by applying KVL or KCL.

Example 2.2 Application of KVL and KCL

For the circuit shown in Figure 2.15, determine the voltage drop across each of the three resistors as well as the current through each one of them.

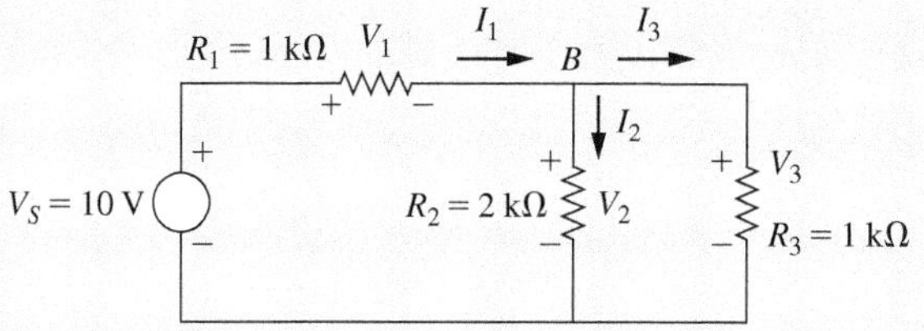

Figure 2.15 Circuit for Example 2.2

Solution:

Applying KCL at node B, results in

$$I_1 = I_2 + I_3 \quad (1)$$

Applying KVL to the left loop gives

$$10 = V_1 + V_2 \quad (2)$$

From Ohm's law applied to each resistor and using Equation (2), we get

$$I_1 = (V_S - V_2)/R_1,\ I_2 = V_2/R_2,\ \text{and}\ I_3 = V_3/R_3 \quad (3)$$

From KVL applied to the right loop, we get

$$V_3 - V_2 = 0 \quad (4)$$

Substituting the expressions in Equations (3) and (4) into Equation (1), we get

$$(10 - V_2)/R_1 = V_2/R_2 + V_2/R_3$$

Solving for V_2, we get $V_2 = 4$ V. Substituting this result into Equations (2), (3) and (4) gives $V_1 = 6$ V, $V_2 = V_3 = 4$ V, $I_1 = 6$ mA, $I_2 = 2$ mA, and $I_3 = 4$ mA.

2.5 Equivalent Circuits

For any two-terminal circuit or network that has only resistive elements (or any type of elements if impedance (see next section) is used instead of resistance), the circuit is simplified into one of two forms. These are the Thevenin equivalent circuit or the Norton equivalent circuit. These simplified forms allow us to focus on a specific portion of a network by replacing the remaining network with an equivalent circuit. The **Thevenin equivalent circuit**, as in Figure 2.16(a) consists of an ideal voltage source (V_{TH}) connected

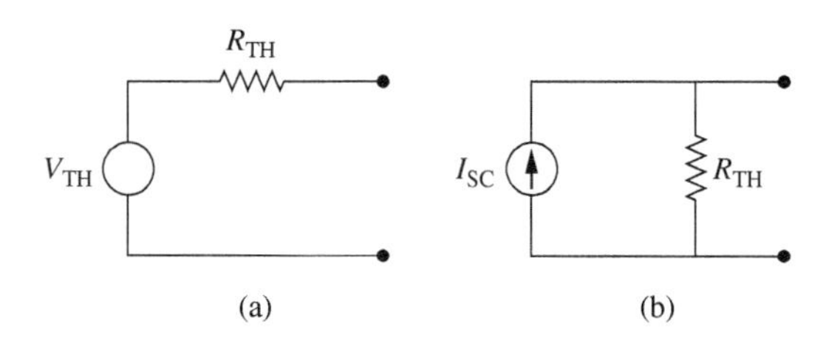

Figure 2.16

(a) Thevenin equivalent circuit and (b) Norton equivalent circuit in series with a resister (R_{TH}). The value of V_{TH} is the open-circuit voltage of the original circuit at the terminals. R_{TH} is defined as the ratio of the open-circuit voltage V_{TH} to the short-circuit current (I_{SC}), where the short-circuit current is the current that would flow through the terminals if the terminals were short-circuited. R_{TH} alternatively can be found by determining the equivalent resistance at the terminals when the voltage sources are shorted, and the current sources are replaced with open ones.

The **Norton equivalent circuit**, as in Figure 2.16(b), consists of an ideal current source I_{SC} connected in parallel with a resistor R_{TH}.

As an illustration, consider the circuit shown in Figure 2.17. Replace the circuit to the left of nodes *a* and *b* with an equivalent Thevenin circuit. The open-circuit voltage of the left circuit at terminals *a* and *b* is simply the voltage across the R_2 resistor when the load is disconnected. This voltage is given by

$$V_{TH} = \frac{R_2}{R_1 + R_2} V_S \tag{2.11}$$

Figure 2.17

Circuit to be replaced with a Thevenin equivalent circuit

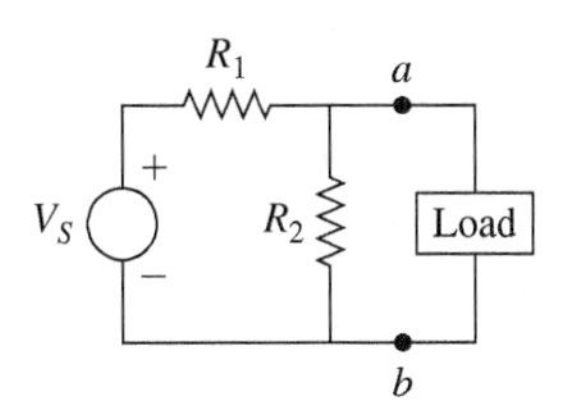

To determine R_{TH}, we need first to find the short-circuit current I_{SC}. This is the current that would flow through the terminals when the terminals *a* and *b* are shorted with the portion of the circuit to the right of the *a* and *b* terminals removed. This current is simply

$$I_{SC} = \frac{V_S}{R_1} \tag{2.12}$$

since no current will flow through R_2. Thus, R_{TH} is

$$R_{TH} = \frac{V_{TH}}{I_{SC}} = \frac{R_1 R_2}{R_1 + R_2} \tag{2.13}$$

Alternatively, R_{TH} is found by determining the resistance at the *a* and *b* terminals when the supply voltage is short-circuited. In this case, the two resistors R_1 and R_2 act as two resistors in parallel, and the resistance at the terminals is the effective resistance of these two resistors. Thus,

$$R_{TH} = R_{Eff} = \frac{R_1 R_2}{R_1 + R_2} \tag{2.14}$$

This concept is further illustrated in Example 2.3.

Example 2.3 Thevenin Equivalent Circuit

Determine the Thevenin equivalent circuit for the circuit shown in Figure 2.18. Let $V_S = 8$ V, $R_1 = 2$, and $R_2 = R_3 = 4$.

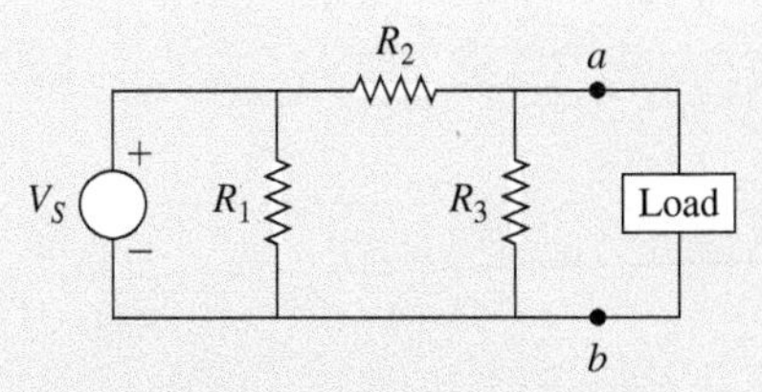

Figure 2.18 Circuit for Example 2.3

Solution:

We start by determining the open-circuit voltage at the terminals *a* and *b* when the load is removed. This is the same as the voltage drop across the R_3 resistor. Note that with no connection between terminals *a* and *b*, the resistors R_2 and R_3 are in series, and the two of them are in parallel with the R_1 resistor. The voltage drop across R_2 and R_3 is the same as the voltage across the R_1 resistor or 8 V. Using the voltage dividing rule, the voltage across R_3 is

$$V_{TH} = V_{R_3} = \frac{4}{4+4}8 = 4\text{ V}$$

To determine R_{TH}, we find the total resistance of this circuit at the terminals *a* and *b* when V_S is short circuited. In this case, the R_1 resistance does not come to play, and the resistance of this network is the parallel combination of the resistors R_2 and R_3, which is 2 Ω. Thus, R_{TH} is 2 Ω. The Thevenin equivalent circuit is shown in Figure 2.19.

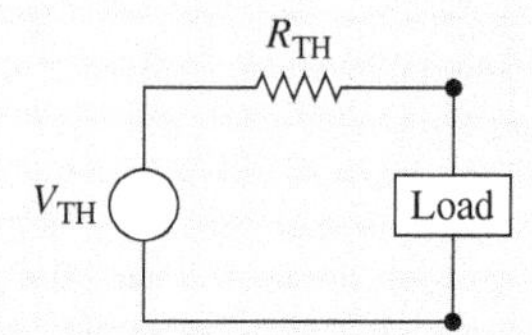

Figure 2.19 Thevenin equivalent circuit

Note if the load resistance is 8 Ω, then the current through the load is simply

$$I_L = \frac{V_{TH}}{R_{TH} + R_L} = \frac{4}{2+8} = 0.4\text{ A}$$

The reader should verify that the same value of current would have been obtained if the original circuit was analyzed.

2.6 Impedance

The concept of impedance is useful in analyzing loading effects that occur when measurement devices are connected to circuits. Impedance is a generalization of the concept of resistance. The impedance of a circuit that has only resistive elements is simply the resistance of the circuit. We define impedance for a two-terminal element as the ratio of the voltage to the current in that element or

(2.15)

$$Z = \frac{V}{I}$$

For circuits that contain other than resistive elements, the impedance is a function of the AC excitation frequency. We will obtain the impedance of different types of circuits under sinusoidal AC excitation. This allows us to see the effect of frequency on impedance. We will start with the RC circuit shown in Figure 2.20(a).

Applying KVL, the relationship between the supply voltage and the current through the circuit is given by

(2.16)

$$V_S(t) = Ri(t) + \frac{1}{C}\int i(t)\,dt$$

Figure 2.20

(a) RC circuit and (b) RL circuit

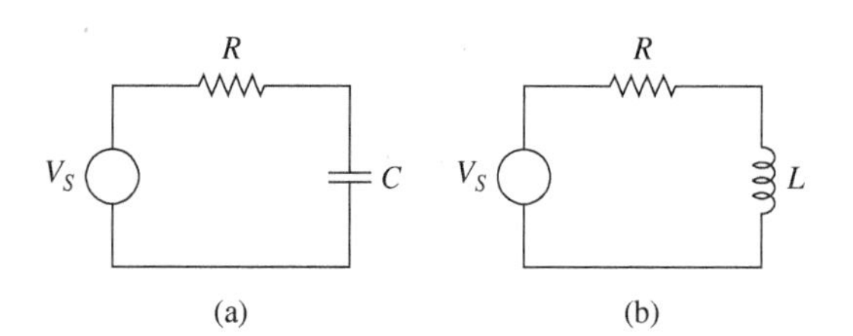

Using the Laplace Transform (see Appendix C), we can convert Equation (2.16) to an algebraic equation in the Laplace variable s as

$$V_S(s) = RI(s) + \frac{1}{Cs}I(s) \tag{2.17}$$

Solving for the ratio of $V_S(s)$ to $I(s)$, we obtain

$$\frac{V_S(s)}{I(s)} = \left[R + \frac{1}{Cs}\right] \tag{2.18}$$

Substituting $s = j\omega$ where ω is the angular frequency and $j = \sqrt{-1}$ to obtain this ratio as a function of ω, we obtain

$$Z = \frac{V_S(j\omega)}{I(j\omega)} = \left[R + \frac{1}{j\omega C}\right] \tag{2.19}$$

Equation (2.19) shows that the impedance of an RC circuit is a complex quantity that is a function of the excitation frequency ω. If the resistor in the above circuit is removed, then the impedance due to the capacitor is

$$Z = \frac{1}{j\omega C} = -\frac{1}{\omega C}j \tag{2.20}$$

Equation (2.20) shows that the impedance of a capacitor is infinite with DC excitation, and approaches zero as the frequency goes to infinity. In the same fashion, we can see that the impedance of a resistor is independent of the excitation frequency and just equal to R. Using a similar approach to that used with the RC circuit, the impedance of the RL circuit shown in Figure 2.20(b) can be obtained as

$$Z = \frac{V_S(j\omega)}{I(j\omega)} = [R + j\omega L] \tag{2.21}$$

The impedance due to the inductor is simply given by

$$Z = \frac{V_S(j\omega)}{I(j\omega)} = j\omega L \tag{2.22}$$

Similar to the capacitor, Equation (2.22) shows that impedance of an inductor is a complex quantity that depends on the frequency ω. For DC excitation, an inductor has zero impedance and acts as a short circuit, but as the frequency approaches infinity, its impedance goes to infinity and acts as an open circuit.

Because impedance can be a complex quantity, we can express impedance as

$$Z = R + jX \tag{2.23}$$

where the real part of Z is defined as the resistance R and the imaginary part of Z is defined as the **reactance** X. Thus, the reactance of a capacitor is

$$X_C = -\frac{1}{\omega C} \tag{2.24}$$

and that of an inductor is

$$X_L = \omega L \tag{2.25}$$

Note that similar to resistors, the total impedance of several elements arranged in series is the sum of the individual impedances of the elements.

Measurement devices such as voltmeters and oscilloscopes are not ideal. They have finite input impedances that could affect the value of the measured quantity. Similarly, amplifiers are not ideal devices but have finite input impedances that could affect the output of the amplifier. Also, power supply sources are not ideal and have small output impedances. When any of these devices are interfaced with a circuit, they create **loading effects** which are explained below. In general, a voltage source should have a very small output impedance, and a measurement device or amplifier should have a very large input impedance.

To illustrate loading effects, assume we have a voltage source V_S connected in series with a resistance R_S, as shown in Figure 2.21. This voltage source can be the output of a sensor or the output of a real power supply. Now assume that the value of this voltage will be measured by a multimeter or an oscilloscope. If we assume an ideal meter, then the meter has an infinite input impedance and will draw no current. Such an arrangement is shown in Figure 2.22. Because the 'ideal' voltmeter draws no current, the voltage measured by the ideal voltmeter will be the open-circuit voltage of the voltage source or V_S.

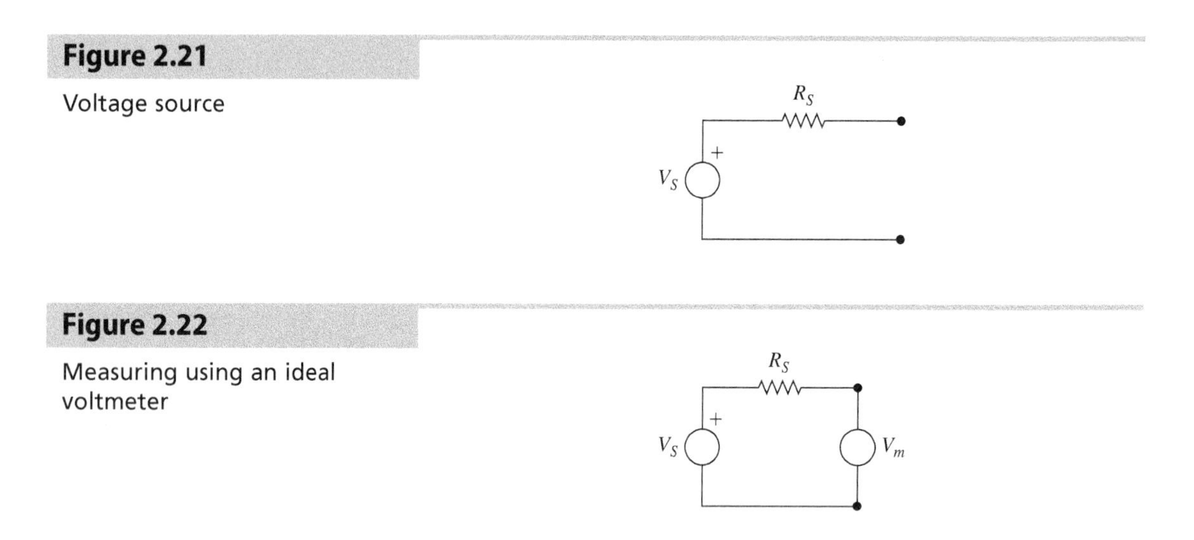

Figure 2.21

Voltage source

Figure 2.22

Measuring using an ideal voltmeter

Now assume that the ideal voltmeter is replaced with a real voltmeter that has finite impedance. Such a meter can be represented as an ideal voltmeter in parallel with the voltmeter finite impedance (R_m). When this meter is connected to the voltage source (see Figure 2.23), the output of the voltmeter will be the voltage drop across the R_m resistor.

Figure 2.23

Measuring using a real voltmeter

This value is given by the expression

(2.26)

$$V_m = \frac{R_m}{R_m + R_S} V_S$$

If R_m is much larger than R_S, then the ratio is almost equal to 1, and V_m will be close to V_S. If however, R_m is comparable to R_S, then the measured voltage by the voltmeter will be significantly different from V_S. Multimeters and oscilloscopes have large impedance (1 MΩ or higher), and power supplies have small R_S resistance (less than 1Ω), so the loading effect is negligible. However, if a voltmeter is used to measure the voltage in a circuit with large resistance, then loading effect errors could be significant. Loading effects are illustrated in Examples 2.4 and 2.5.

Example 2.4 Loading Effects in an Amplifier

A sensor with an output voltage of 1 V and a series internal resistance of 1 kΩ is connected to an amplifier that has a gain of 10 V/V and an internal resistance of 5 kΩ. Determine the output of the amplifier due to this sensor input.

Solution:

If the amplifier is an ideal amplifier, then the amplifier output is simply the voltage applied to the amplifier times the amplifier gain or $1 \times 10 = 10$ V.

However, the amplifier internal resistance acts as a voltage divider with the sensor internal resistance (a circuit similar to that shown in Figure 2.23). The voltage drop across the amplifier internal resistance is

$$V_{in} = \frac{R_A}{R_A + R_S} V_S = \frac{5}{5+1} 1 = 0.83 \text{ V}$$

$$\text{Thus, } V_{out} = \text{Gain} \times V_{in} = 8.3 \text{ V}$$

Hence, due to overloading, the amplifier output deviates from the ideal output by more than 16%. Obviously, this error can be minimized by using an amplifier with high input impedance. Operational amplifiers are such type of amplifiers. Note that if the ratio of the amplifier impedance to the sensor impedance is more than 100:1, then the error due to loading effects is less than 1%.

Example 2.5 Loading Effects in a Measurement Probe

Commonly used passive probes for oscilloscopes are available as X1 or X10 types. Explain how the X10 probe reduces the loading effects for measurements taken by the oscilloscope.

Solution:

The X1 or X10 type refers to the multiplication factor by which the impedance of the scope is multiplied by the probe. The input impedance of the oscilloscope is normally 1 MΩ. When a 10X probe is connected to the scope, the scope becomes a measurement device with an effective input impedance of 10 MΩ instead of just 1 MΩ as seen by the voltage signal to be measured. This happens because the 10X probe has an internal resistance (R_P) of 9 MΩ which combines in series with the 1 MΩ input resistance of the scope (R_S). Figure 2.24 illustrates these resistances. The higher input impedance thus will reduce any loading effects that occur in the measurement. Note the X10 probe attenuates the measured signal by a factor of ten, and this attenuation is a result of the voltage dividing circuit formed by the probe's impedance and the oscilloscope's internal impedance. For example, a 5-volt signal will be read as 0.5 volts. The above explanation applies to DC or low-frequency signal measurements. At high input signal frequencies, additional factors such as the probe's capacitance (C_P) and the oscilloscope's internal capacitance (C_S) come into play and affect the measurement. At DC or low frequencies, the reactance of these capacitors is infinite (see Equation (2.24)), and the impedance of the probe or the oscilloscope input is just determined by the values of their resistance. At high frequencies, these capacitors affect the impedance, and the attenuation of the probe depends on the values of the capacitors' reactance. However, X10 probes are designed with an adjustable capacitor (C_A) that can be adjusted to maintain the same voltage dividing ratio in the circuit. This adjustment compensates for the impact of capacitance at high frequencies, ensuring accurate measurements across a wide range of frequencies. Problem P2.22 examines the resultant impedance of the resistor R_S and the capacitor C_S at different frequencies.

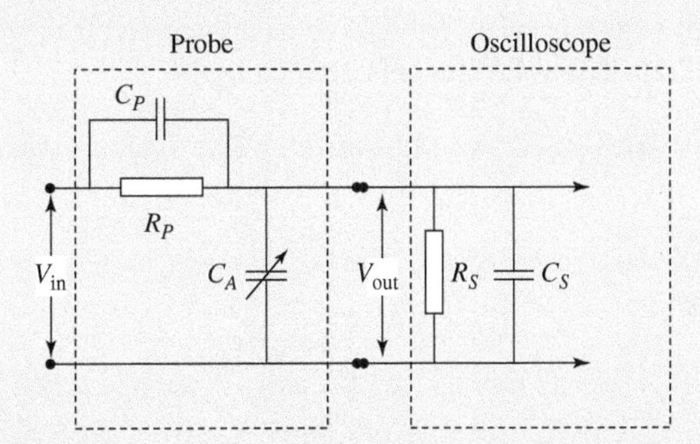

Figure 2.24 X10 probe connected to an oscilloscope

When signals are transmitted between devices that are interfaced together, it is important to ensure that the impedances of the different devices are properly matched (▶ available—V2.5). If the impedances are not matched, then a high-impedance input device can reflect some of the input signal contents that the low-impedance output device produces. As an example, a function generator is a low-impedance device that has a 50 Ω output impedance. If a function generator needs to be interfaced with a high-impedance device, then we can match the output impedance of the function generator by inserting a 50 Ω resistor in parallel with the high-impedance device. This is shown in Figure 2.25. The effective resistance of the inserted resistor and the high-impedance circuit is almost equal to the output impedance of the function generator. **Impedance matching** is important in many applications including the transmission of audio signals between the audio amplifier and the speakers, and in transmitting signals from ultrasonic transducers through cables.

Figure 2.25

Signal connection for impedance matching

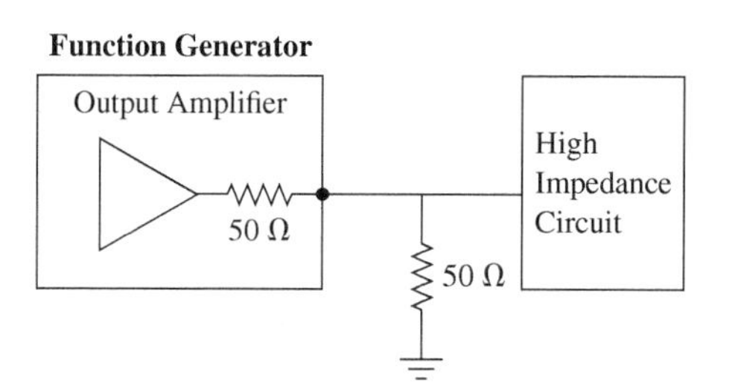

In many cases, it is desirable to deliver the maximum power to a circuit from a supply source. To achieve this, the impedances of the supply circuit should match the load circuit. As an illustration, the power delivered to a resistive load with a resistance R_L from a power supply with an output voltage of V_O, and an output impedance R_S is given by the following expression (see Figure 2.23 for a similar circuit if R_m is replaced with R_L).

(2.27)
$$\text{Power} = \frac{V_L^2}{R_L} = \frac{R_L^2}{(R_L + R_S)^2} V_O^2 \times \frac{1}{R_L} = \frac{R_L}{(R_L + R_S)^2} V_0^2$$

If Equation (2.27) was differentiated with respect to R_L and set equal to zero, then the maximum power is obtained when

(2.28)
$$R_L = R_S$$

2.7 AC Signals

While DC voltages are common in battery-powered devices and laboratory setups, power transmission and operation of industrial and residential equipment such as compressors and kitchen appliances use AC voltages. AC voltage signals have the advantage that they are more efficient to transmit over long distances. When an AC voltage signal, such as a sinusoidal voltage signal, is applied to a circuit, the voltage in the circuit will also be sinusoidal with a frequency the same as the applied frequency. Two sinusoidal voltage signals are shown in Figure 2.26. The solid signal is defined by

(2.29)
$$V = V_O \sin(\omega t + \theta)$$

Figure 2.26

Two sinusoidal voltage signals

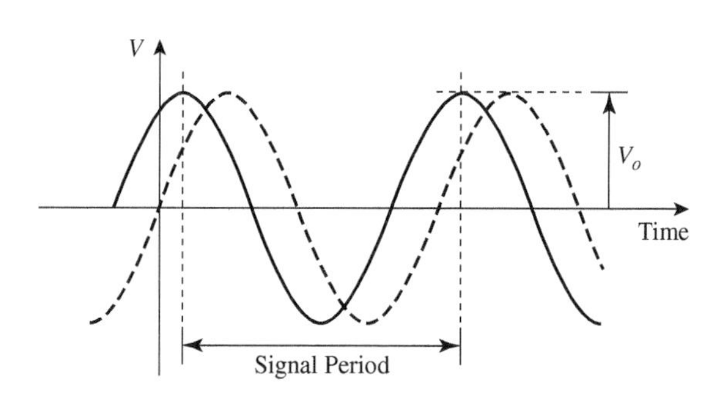

while the dashed signal is defined by

$$V = V_O \sin(\omega t) \tag{2.30}$$

where V_O is the amplitude of the signal, and ω is the angular frequency in units of radians/second. Theta (θ) is defined as the phase angle and is a measure of the lead/lag in the signal. A positive phase angle (such as that for the solid signal) means that the signal is ahead or leads the dashed signal. The lead time is given by (θ/ω).

Note that the circular frequency ω in radians/second (rad/s) and the cycle frequency f in Hertz (Hz) are related by $\omega = 2\pi f$ and that the cycle frequency is the inverse of the signal period T.

For AC circuits, the magnitude of the voltage or current is specified by using the amplitude, the peak-to-peak value, or the root mean square (RMS) value. The **RMS voltage and current** are defined by

$$V_{\text{RMS}} = \sqrt{1/T \int_0^T v^2\,dt} \tag{2.31}$$

and

$$I_{\text{RMS}} = \sqrt{1/T \int_0^T i^2\,dt} \tag{2.32}$$

Note that multimeters measure the RMS and not the amplitude value when they measure AC voltages and currents. Note also that 110 or 220 V supply is the RMS value of the voltage signal.

In resistive networks, both the current and the voltage will have the same phase angle. However, in circuits that have capacitive and inductive elements, the voltage and current will be out of phase. This is because the impedance of these elements has an imaginary or j-component. In an inductor, the voltage will lead the current by 90° (the reactance is positive for an inductor), while in a capacitor; the voltage lags the current by 90° (the reactance is negative for a capacitor).

2.8 Power in Circuits

Power is defined as the rate of doing work. Power is an important specification for electrical components as it defines the rated capability for these components. In electrical circuits, the instantaneous power in an element at any point in time is defined as the product of the voltage and current through that element or

$$P(t) = V(t)\, I(t) \tag{2.33}$$

If the voltage and current do not change with time as in DC circuits, then the instantaneous power and average power are the same. For a resistor in a DC circuit, the power can also be written as

$$P(t) = P_{\text{avg}} = VI = \frac{V^2}{R} = I^2 R \tag{2.34}$$

However, in AC circuits the voltage and current vary with time, and we need to distinguish between instantaneous power and average power. For an AC circuit with voltage $V(t) = V_O \sin(\omega t + \theta_V)$ and current $I(t) = I_O \sin(\omega t + \theta_I)$, the **instantaneous power** is given as

$$P(t) = V_O I_O \sin(\omega t + \theta_V)\sin(\omega t + \theta_I) \tag{2.35}$$

And the **average power** over one period is obtained as

$$P_{\text{avg}} = V_{\text{RMS}} I_{\text{RMS}} \cos(\theta) = \frac{V_O}{\sqrt{2}} \frac{I_O}{\sqrt{2}} \cos(\theta) \tag{2.36}$$

where θ is the difference between the voltage and current phase angles, i.e., ($\theta = \theta_V - \theta_I$). The term cos ($\theta$) in Equation (2.36) is called the **power factor** and is a measure of the presence of reactive components (capacitors and inductors in the circuit). For a purely resistive network, the power factor is 1, and for a purely inductive network, the power factor is 0. The power factor is also defined as the ratio of the real (or useful) power to the apparent power (see Figure 2.27). A component of the apparent power is the reactive (or nonworking) power that is exchanged in capacitive and inductive components. Power factor is an important specification for devices with reactive components (such as AC-powered motors), as it defines how much of the supplied power is converted to real or useful power. Power supply companies may charge industrial and commercial customers fees for operating devices with a low-power factor since a low-power factor means that these devices draw larger currents that result in bigger power distribution lines. Example 2.6 illustrates the computation of power in AC circuits.

Figure 2.27

Real and apparent power

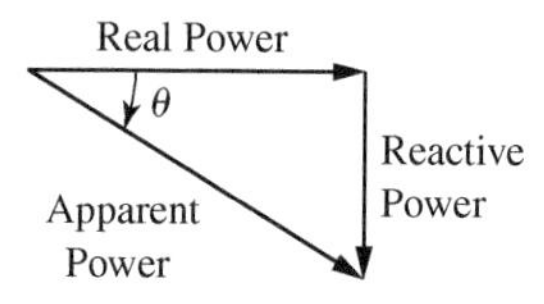

Example 2.6 Power in AC Circuits

A load with an impedance of 500 + 600j Ω is connected to a 110 V, 60 Hz source. Determine the power factor and the power absorbed by the load.

Solution:

The power factor can be determined from the angle that the load reactance makes with the load resistance. Using Equation (2.23), θ is given by

$$\theta = \tan^{-1}(X/R) = \tan^{-1}(600/500) = 50.2°$$

But the power factor is $\cos(\theta)$ or $\cos(50.2°) = 0.640$

From Equation (2.15), the current through the load is given by

$$I = \frac{V}{Z} = \frac{110}{\sqrt{500^2 + 600^2}} = 0.141 \text{ A}$$

Using Equation (2.36), the absorbed or real power is

$$\text{Power} = V_{RMS}\, I_{RMS} \cos(\theta) = 110 \times 0.141 \times 0.640 = 9.93 \text{ W}$$

2.9 Operational Amplifiers

Operational amplifiers (op-amps) are analog circuit components that require power to operate. They are widely used in amplification and signal-conditioning circuits. The symbol for an op-amp (▶ available—V2.6) is shown in Figure 2.28. The symbol is a triangle, with two leads drawn on one side of the triangle, and the third lead is drawn at the apex opposite to that side. One lead is defined as the inverting input (−), the other lead is defined as the non-inverting input (+), and the third lead is the output. The voltages at these two inputs and the op-amp output are referenced to the ground. Figure 2.28 also shows the connections for the positive and negative supply voltages, although these connections are normally omitted when an op-amp is drawn in a circuit. The supply voltage is typically ±15 V. There are two other connections to the op-amp (called the balance or null offset) that permit adjustment of the op-amp output, but they are typically not shown.

Figure 2.28

Symbol and connections for an op-amp

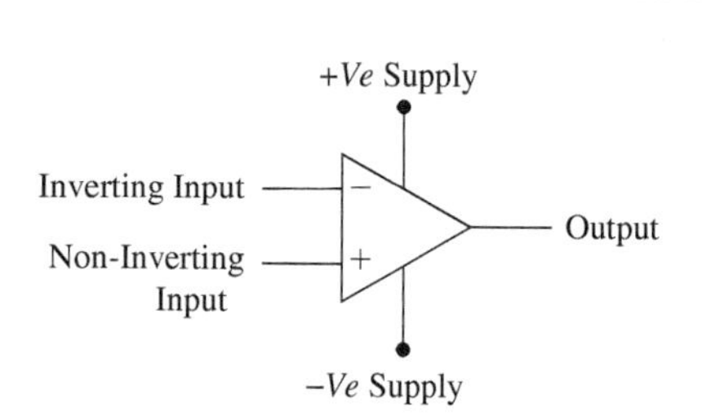

Commercially, op-amps are available in a variety of forms. A common form is a single op-amp in the form of an 8-pin integrated circuit (IC), an example of which is the LM741 chip from Texas Instruments. The pin layout of this chip is shown in Figure 2.29(a). Note that there is no connection to pin 8, and the positive and negative supply voltages are connected at pins 7 and 4, respectively. Another form is the dual op-amps on a single 8-pin package, and the pin layout for this form is different from that of a single op-amp IC. Many vendors manufacture op-amp ICs, and they are available in other chip numbers such as the LF411 chip that is also available from Texas Instruments. An op-amp is constructed from several components including transistors, diodes, capacitors, and resistors.

Figure 2.29

(a) Pin layout for the LM741 and (b) model of an ideal op-amp

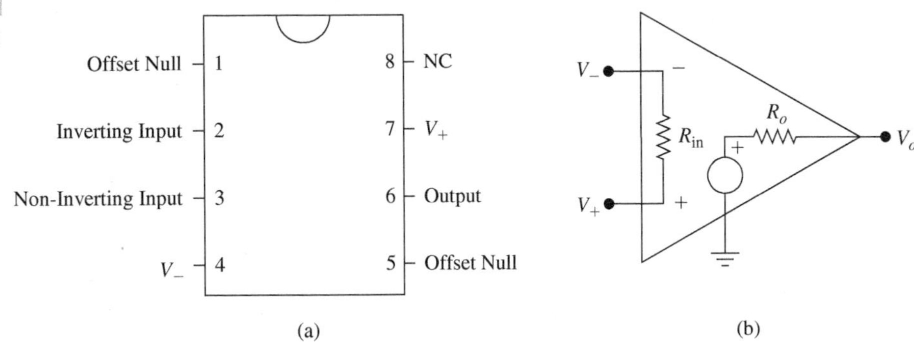

An **ideal op-amp** can be modeled as shown in Figure 2.29(b). The inputs to the op-amp can be thought to be connected internally by a high-impedance resistor R_{in}. The value of this resistance is high enough (more than 1 MΩ), such that for ideal behavior, we can assume that no current flows between the input terminals. The output of the op-amp is modeled as a voltage source connected to a low impedance resistor R_o (less than 100 Ω) in series. The voltage output is proportional to the difference between the input voltages, i.e.,

$$V_o = K_{OL} \times (V_+ - V_-) \tag{2.37}$$

where K_{OL} is the open-loop gain of the op-amp. The open-loop gain of the op-amp is usually very high (10^5 to 10^6), so a very small voltage difference between the two inputs results in a saturation of the output. For example, if the gain is 10^6, and the saturation voltage is 10 V, then the op-amp will saturate if the voltage difference between the input leads exceeds 10 μV. Since the op-amp output is finite, but the op-amp has a very large gain, we assume that $V_+ = V_-$. The assumption that $V_+ = V_-$ along with the assumption that no current flows into the input terminals are the **two basic rules** that are used to analyze ideal op-amp circuits.

The saturation voltage of an op-amp is a function of the supply voltage for the op-amp and it is slightly smaller than it. For example, at a supply voltage of ±15 V, the saturation voltage is about ±13 V. The open-loop input-output relationship for an op-amp is shown in Figure 2.30. In most cases, however, op-amps are not used in open-loop configurations but are used with a feedback loop between the output

Figure 2.30

Open-loop input/output relationship for an op-amp

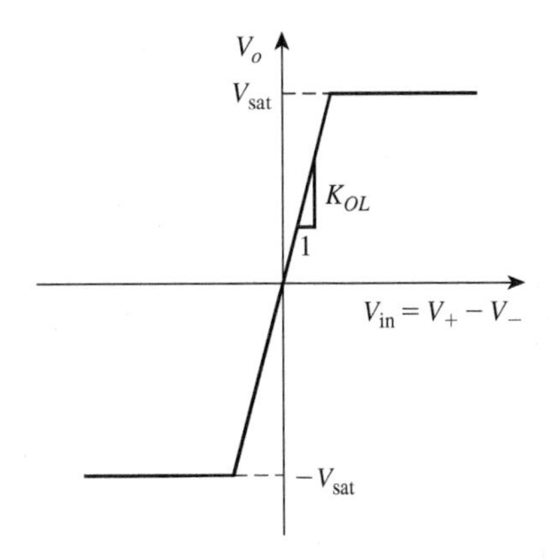

voltage lead and the inverting input lead. The closed-loop gain is much smaller than the open-loop gain, but the feedback provides more stable operating characteristics.

An op-amp gives a zero output if the two input voltages are the same. This is called the **common-mode rejection property** of the op-amp. In reality, the output will not be exactly zero, but one can use the null offset terminals on the op-amp to adjust this output. Op-amps have good frequency response characteristics, and their bandwidth (see Chapter 7) exceeds 1 MHz.

Op-amps can perform various operations such as comparison, amplification, inversion, summation, integration, differentiation, or filtering (▶ available—V2.7). The particular operation depends on how the op-amp is wired and what external components are connected to the op-amp. We will discuss below some of these operations assuming ideal behavior. In most cases, the real behavior closely follows the ideal behavior.

2.9.1 Comparator Op-Amp

A comparator is used to compare two voltage signals and switch the output to $+V_{sat}$ if one of the signals is larger than the other, and to $-V_{sat}$ otherwise, where V_{sat} is the saturated output of the op-amp. The circuit for an op-amp operating as a comparator is shown in Figure 2.31. Here the op-amp is operating in an open loop, which means there is no feedback from the op-amp output to the input. The input voltage V_i is connected to the non-inverting input (+), and the reference voltage V_{ref} is connected to the inverting input (−). The comparator output V_o is then

(2.38)
$$V_o = \begin{cases} V_{sat,} \; V_i > V_{ref} \\ -V_{sat,} \; V_i < V_{ref} \end{cases}$$

A comparator can be used, as an example, in situations where it is needed to set an output on if a sensor input exceeds a certain value. Microcontrollers such as the Atmega328P (discussed in Chapter 4) have a built-in comparator feature.

Figure 2.31

Comparator op-amp circuit

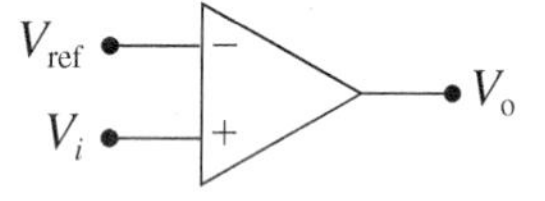

2.9.2 Inverting Op-Amp

The inverting op-amp circuit is shown in Figure 2.32 which has a feedback loop between the op-amp output and the inverting input (−). The naming of this circuit is because the output signal has a sign that is inverse of the sign of the input signal. In this circuit, an input voltage V_i is applied to the inverting input (−) through a resistor R_1, and the non-inverting input (+) is grounded. Since the non-inverting input is connected to the ground,

(2.39)
$$V_- = V_+ = 0$$

The current I_1 is equal to I_2 because virtually no current flows between the inverting and the non-inverting inputs. The current I_1 is equal to

(2.40)
$$I_1 = \frac{V_i - V_-}{R_1} = \frac{V_i}{R_1}$$

and the current I_2 is equal to

(2.41)
$$I_2 = \frac{V_- - V_o}{R_2} = -\frac{V_o}{R_2}$$

Equating I_1 to I_2, and solving for the op-amp output V_o gives

(2.42)
$$V_o = -\frac{R_2}{R_1} V_i$$

Figure 2.32

Inverting op-amp circuit

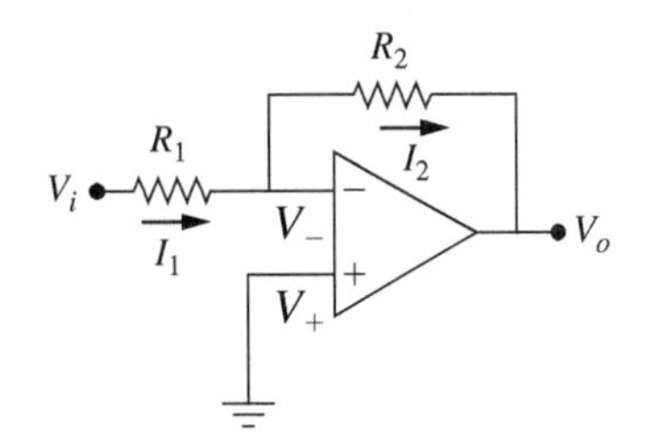

Thus, in this circuit the op-amp inverts the input voltage and amplifies it by a factor equal to the ratio of the resistance of R_2 to R_1. An application of this circuit is to perform signal inversion where the output will have a 180° phase shift with the input. Example 2.7 illustrates the inverting op-amp circuit.

Example 2.7 Summing Circuit

Draw a circuit that shows how op-amps can be used to perform summation of three analog voltage signals.

Solution:

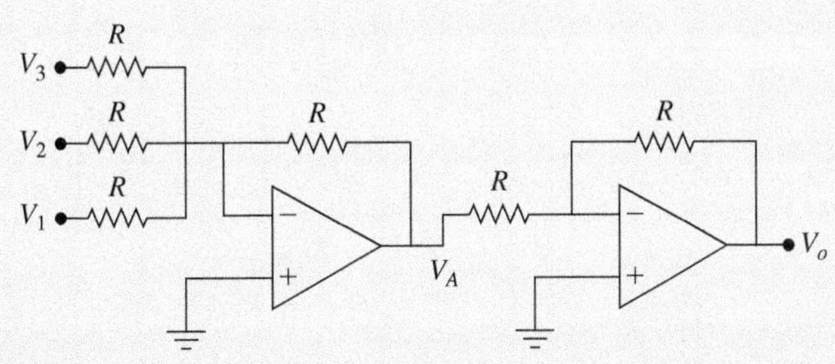

Figure 2.33 Circuit to perform summation of three analog voltage signals

The circuit that performs this operation is shown in Figure 2.33. It basically consists of two cascaded op-amps each wired as an inverting amplifier. The three voltage signals are connected to the left op-amp. Note that the sum of the currents through the three resistors that are connected to the input voltages V_1, V_2, and V_3 is the same as the current that goes through the resistor in the feedback loop in the left op-amp circuit:

$$\frac{V_1 - V_-}{R} + \frac{V_2 - V_-}{R} + \frac{V_3 - V_-}{R} = \frac{V_- - V_A}{R} \quad (1)$$

Since $V_- = 0$, and cancelling R from each term in Equation (1), this gives

$$V_A = -(V_1 + V_2 + V_3) \quad (2)$$

From the second op-amp circuit, we get

$$V_o = -V_A = V_1 + V_2 + V_3 \quad (3)$$

2.9.3 Non-Inverting Op-Amp

The non-inverting op-amp circuit is shown in Figure 2.34. The naming of this circuit is because the output has the same sign as the input with no inversion. Here the non-inverting input (+) is connected to an input

Figure 2.34

Non-Inverting op-amp circuit

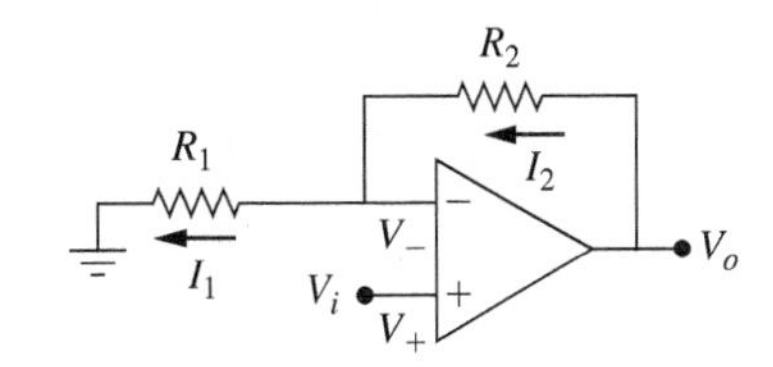

voltage V_i, and the inverting input (−) is connected to the ground through a resistor R_1. There is also a feedback loop between the op-amp output and the inverting input.

The voltage V_+ is equal to V_- and is also equal to V_i in this case. But the voltage at the inverting input is also given by

(2.43)
$$V_- = \frac{R_1}{R_1 + R_2} V_o$$

since R_1 and R_2 act as a voltage-dividing circuit between V_o and ground. Thus, the output V_o of the op-amp is given by

(2.44)
$$V_o = \frac{R_1 + R_2}{R_1} V_i = \left(1 + \frac{R_2}{R_1}\right) V_i$$

Notice how the gain of the op-amp in this case is always greater than 1. Now if we let R_2 be zero and R_1 be infinite, this gives the circuit shown in Figure 2.35. This circuit is known as a **voltage follower** (▶ available—V2.8) or buffer, and $V_o = V_i$ in this case. Because the op-amp has a low output impedance (about 75 Ω), and a high input impedance (about 2 MΩ), the voltage follower circuit can be used in a variety of ways to reduce loading effects. The voltage output from a circuit with large resistance can be connected to a voltmeter through a buffer to reduce loading effects on the voltmeter, or the buffer output can be connected to a high-impedance circuit, so the buffer output appears as a low-impedance source.

Figure 2.35

Voltage follower

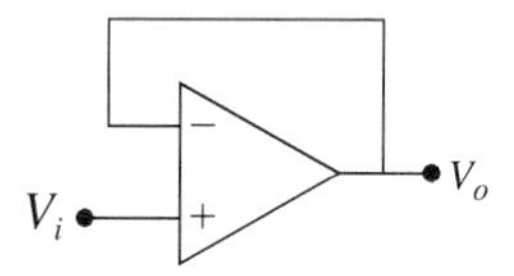

In both the inverting and the non-inverting op-amp circuits shown above, the feedback between the output voltage and the inverting input is known as negative feedback. Negative feedback results in a linear relationship between the output and input voltages. If the feedback loop was between the output voltage and the non-inverting input, then the output-input relationship is nonlinear. The nonlinearity is a hysteresis where the input has to change by a certain amount before the output changes state. Non linear op-amp circuits are utilized in the design of **Schmitt triggers**, which are IC circuits that are used for converting slowly changing or noisy analog signals into two-level digital signals (see Section 8.4.1 which discusses their use in the wiring circuit for a Hall-effect proximity sensor). The symbol for a standard (non-inverting) Schmitt trigger and the input and output voltages from a Schmitt trigger are shown in Figure 2.36. Note how the output of the Schmitt trigger goes to V_{max} when the input signal voltage exceeds the positive-going

Figure 2.36

Schmitt trigger: (a) symbol and (b) input/output relationship

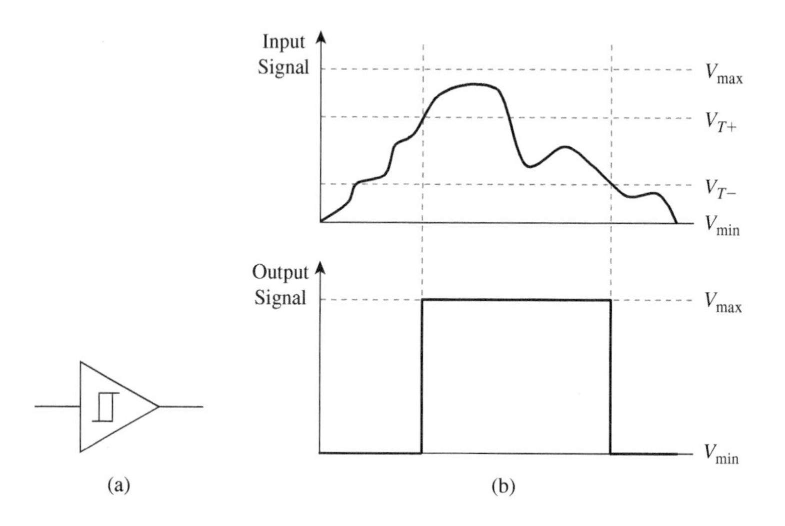

threshold voltage (V_{T+}). The output signal stays at V_{max} until the input signal drops below the negative going threshold voltage (V_{T-}), at which point the output goes to V_{min}. In Figure 2.36, V_{max} and V_{min} are the positive (typically 5 VDC) and the negative (typically 0 VDC) supply voltage, respectively, for the Schmitt trigger device. The 74HC7014 IC has six non-inverting Schmitt triggers with $V_{T+} = 3.1$ V and $V_{T-} = 2.9$ V when used with a 5 VDC supply voltage.

2.9.4 Differential Op-Amp

The differential op-amp circuit gives an output that is proportional to the difference between the two voltages applied to its inputs. A differential op-amp circuit with two voltages (V_1 and V_2) applied to its inputs is shown in Figure 2.37. Two inputs (differential input) are used to reduce the circuit sensitivity to noise since any noise applied to the circuit will be most probably the same on each of the inputs.

Figure 2.37

Differential input op-amp circuit

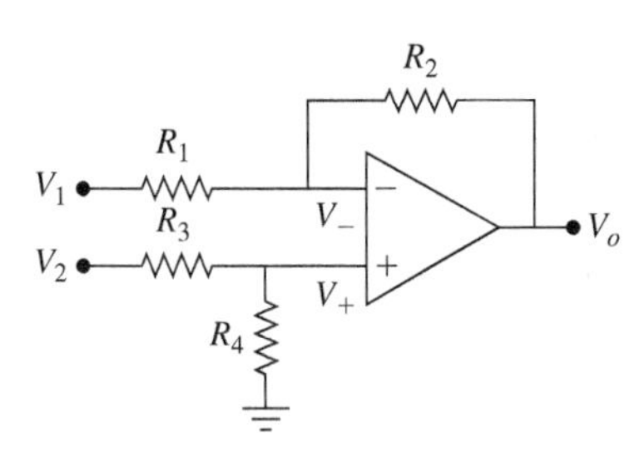

For this circuit, the current through the R_1 and the R_2 resistors is the same since no current goes through the inverting input. This current is given by

$$I_{R_1} = \frac{V_1 - V_-}{R_1} = I_{R_2} = \frac{V_- - V_o}{R_2} \tag{2.45}$$

The voltage at the non-inverting input is given by

$$V_+ = \frac{R_4}{R_3 + R_4} V_2 \tag{2.46}$$

But $V_+ = V_-$. Substituting the expression for V_+ in Equation (2.45) and solving for V_o gives

$$V_o = \frac{R_4}{R_3 + R_4}\left(1 + \frac{R_2}{R_1}\right) V_2 - \frac{R_2}{R_1} V_1 \tag{2.47}$$

If $R_3 = R_1$ and $R_4 = R_2$, this expression simplifies to

$$V_o = \frac{R_2}{R_1}(V_2 - V_1) \tag{2.48}$$

and shows that the output of this op-amp circuit is proportional to the voltage difference between the inputs V_2 and V_1. A differential amplifier circuit can be used, for example, to implement an analog proportional control feedback loop (see Figure 2.38). Feedback control is covered in Chapter 10.

Figure 2.38

Proportional control feedback loop

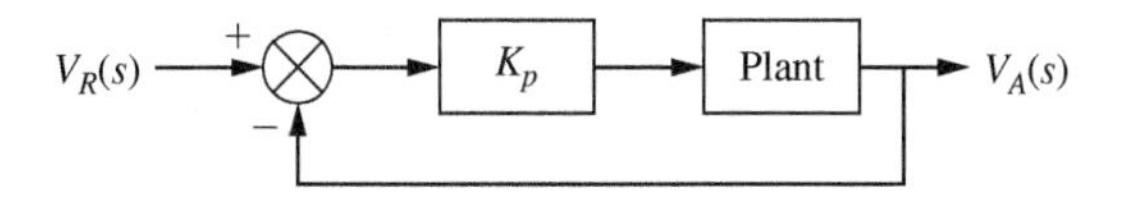

If the reference signal V_R is the V_2 voltage, the actual or measured signal V_A is the V_1 voltage, and the ratio R_2/R_1 is the proportional gain K_p, then the output of the differential amplifier will be

$$V_o = K_P(V_R - V_A) \tag{2.49}$$

The differential amplifier circuit can also amplify the difference between the voltage outputs from the arms of a Wheatstone bridge used to measure strain (see Section 8.11.3).

A limitation of the differential amplifier circuit is that its input impedances are lower than that of other operational amplifier circuits, such as the single input non-inverting amplifier circuit. This is because each input voltage is connected to the circuit through a resistor which has a lower impedance than the op-amp alone. Even if the R_1 and R_3 resistors are selected to have a high resistance (such as 1 MΩ) to improve the input impedance, this requires that resistors R_2 and R_4 need to be at least 100 MΩ to achieve a gain of at least 100 or higher that is needed to amplify small signals. Due to resistor non-idealities, it is difficult to match such large resistors with good accuracy. To improve on these limitations, another op-amp circuit configuration called the instrumentation amplifier op-amp is used. This is discussed in the next section.

2.9.5 Instrumentation Amplifier Op-Amp

The instrumentation amplifier op-amp or ia op-amp circuit (▶ available—V2.9) uses three op-amps and seven resistors and is shown in Figure 2.39. The circuit combines a differential op-amp circuit with two voltage follower circuits to create a circuit that has high input impendence and a single output that is controlled by the value of just one resistor. The right part of the circuit is the differential op-amp circuit that was discussed in the previous section, while the left part of the circuit has two voltage follower circuits that were discussed in Section 2.9.3.

Figure 2.39

Instrumentation op-amp circuit

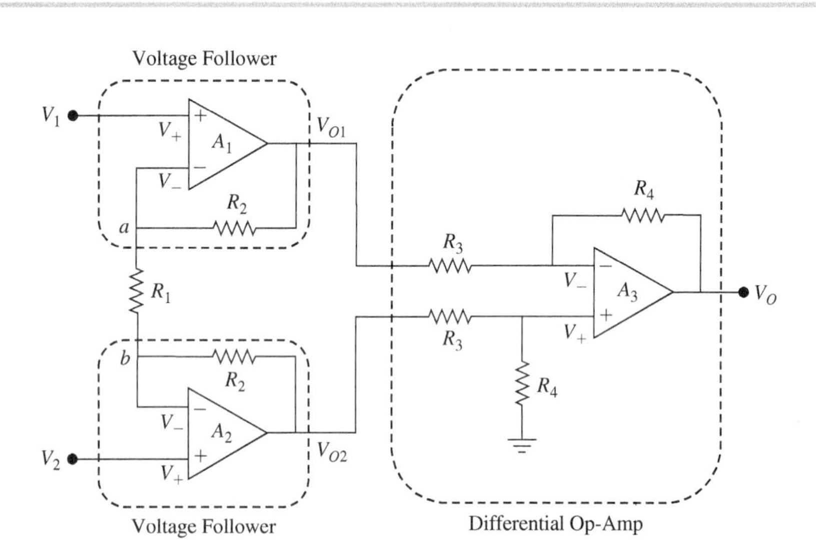

Using Equation (2.48), the output V_O of the instrumentation amplifier can be written in terms of the voltages V_{O_1} and V_{O_2} shown in Figure 2.39 as

(2.50)
$$V_O = \frac{R_4}{R_3}(V_{O_2} - V_{O_1})$$

Due to the presence of the resistor R_1 which joins the two voltage follower circuits, each of the voltages V_{O_1} and V_{O_2} depends on the two input voltages V_1 and V_2 applied to the amplifier. To obtain expressions for V_{O_1} and V_{O_2}, the principle of superposition will be used where one of the input voltages (V_2) will be first assumed to be zero and the output due to the other input voltage (V_1) will be obtained. Then V_1 will be assumed to be zero, and the output due to V_2 will be obtained. The two outputs will then be combined to give us the output due to V_1 and V_2. Letting $V_2 = 0$, the voltage $V_+ = V_- = 0$ for op-amp A_2, and the voltage at node b will also be zero. Thus op-amp A_1 will act as a non-inverting op-amp similar to Figure 2.34, and its output voltage V_{O_1} will be

(2.51)
$$V_{O_1} = \left(\frac{R_1 + R_2}{R_1}\right) V_1$$

Now if $V_1 = 0$, op-amp A_1 will function as an inverting op-amp (see Figure 2.32), and its output voltage for a non-zero V_2 will be

$$V_{O_1} = -\frac{R_2}{R_1} V_2 \tag{2.52}$$

because the voltage at node b will be V_2. Combining the outputs from Equations (2.51) and (2.52), the output from op-amp A_1 will be

$$V_{O_1} = \left(\frac{R_1 + R_2}{R_1}\right) V_1 - \frac{R_2}{R_1} V_2 \tag{2.53}$$

Similarly, the output from op-amp A_2 will be

$$V_{O_2} = \left(\frac{R_1 + R_2}{R_1}\right) V_2 - \frac{R_2}{R_1} V_1 \tag{2.54}$$

Substituting the expressions from Equations (2.53) and (2.54) into Equation (2.50) and simplifying, we obtain

$$V_O = \frac{R_4}{R_3}\left(1 + \frac{2R_2}{R_1}\right)(V_2 - V_1) \tag{2.55}$$

If $R_2 = R_3 = R_4 = R$, then the voltage gain of the instrumentation amplifier is

$$1 + \frac{2R}{R_1} \tag{2.56}$$

which can be adjusted by changing the value of the resistor R_1. The ia op-amp is available commercially as a single IC. One example of such an ia op-amp is the IA128 IC available from Texas Instruments. A pin layout and a schematic of this IC is shown in Figure 2.40.

Figure 2.40

(a) Pin layout and (b) simplified schematic of the IA128 instrumentation amplifier IC

(Courtesy of Texas Instruments Incorporated, Dallas, Texas)

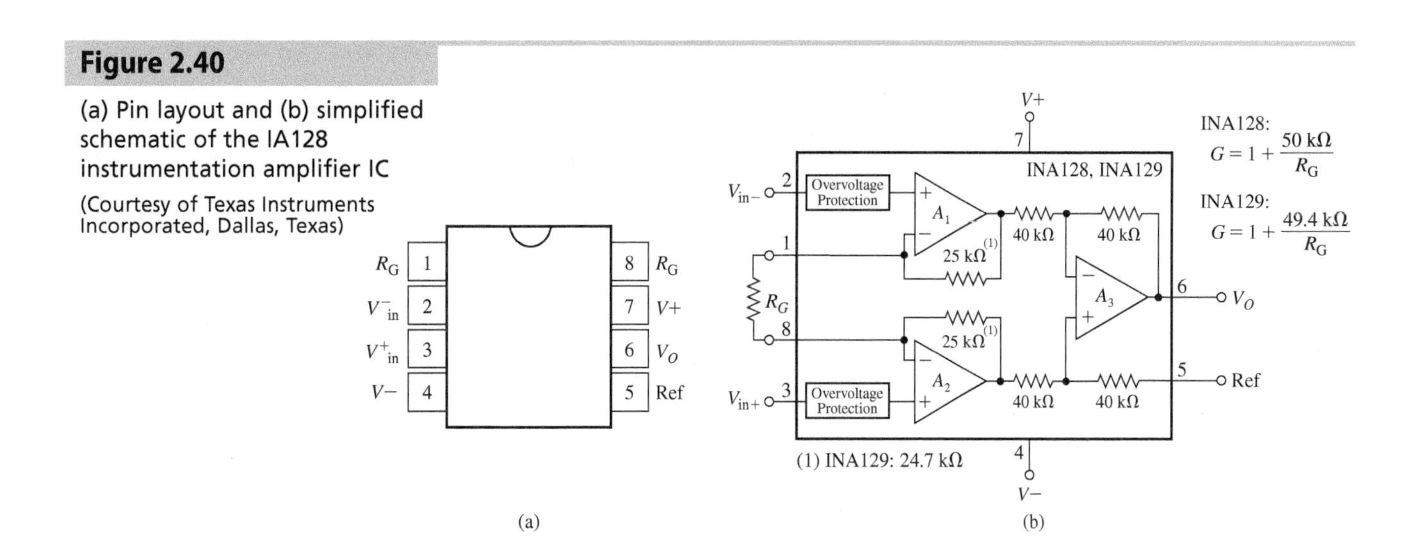

The differential input is connected to pins 2 and 3, while the positive supply voltage is connected to pin 7 and the negative supply voltage is connected to pin 4. The gain of this amplifier is $1 + 50/R_G$ where R_G (value in kΩ) is the external resistor (like resistor R_1 in Figure 2.39) that is connected across pins 1 and 8 on the IC. The 50 kΩ is the sum of the two 25 kΩ internal resistors in the chip. The gain can be set to any value between 1 and 10,000. If there is no external resistor connected, then the voltage gain is 1. The advantage of the ia op-amp is that all the resistors in its circuit except for R_1 or R_G are internal, and these internal resistors do not need to have a high resistance to achieve a high voltage gain. Furthermore, these internal resistances are typically selected to have a very good resistance matching, which improves the gain accuracy of the op-amp. Instrumentation amplifiers are widely used to process measurements from sensors and transducers. They are well suited to handle very low voltage signals such as those produced by thermocouples (see Chapter 8). Since the ia op-amp amplifies the difference between the two voltages that are applied to its input terminals, it works well with floating ground signals which have no direct connection to ground such as those produced by a load cell interfaced to a Wheatstone bridge (See Chapter 8).

In a differential op-amp or an ia op-amp as well as in all op-amps in general, since op-amps are differential amplifiers by their design, the output of the op-amp is a function of the common input applied to both inputs as well as the difference between the two inputs. In equation form, this can be written as

(2.57) $$V_O = G_c V_c + G_d V_d$$

where G_c is the **common input gain** and G_d is the **differential input gain**. The **differential voltage** V_d is defined as

(2.58) $$V_d = V_2 - V_1$$

while the **common voltage** V_c is defined as

(2.59) $$V_c = \frac{1}{2}(V_1 + V_2)$$

If $V_1 = V_2 = V$, then $V_d = 0$, and $V_c = V$. In op-amp circuits, ideally, you want the circuit to produce no output due to the common input which could be just a noise signal or an interference, but a high value due to the difference in the two inputs. This capability is quantified by what is called the **common mode rejection ratio** or CMRR. The CMRR is defined as the ratio of G_d/G_c expressed in dB units. Thus, the expression for the CMRR is

(2.60) $$CMRR = 20 \log\left(\frac{G_d}{G_c}\right)$$

Instrumentation amplifiers typically have a high CMRR that exceeds 100 dB. The IA128 has a minimum CMRR of 80 dB at a differential gain of 1 and increases to 120 dB at a gain of 1000. Example 2.8 illustrates the use of the CMRR in determining the output from an instrumentation amplifier op-amp circuit.

Example 2.8 Output of an IA OP-AMP

An ia op-amp with a CMRR value of 100 dB has a common voltage input of 1 volt. Compute the output voltage of this ia op-amp for a differential input gain of (a) 100 and (b) 1000.

Solution:

(a) For $G_d = 100$, the common input gain G_c is obtained as follows

$$CMRR = 20 \log\left(\frac{G_d}{G_c}\right) = 100 \Rightarrow \log\left(\frac{G_d}{G_c}\right) = 5 \Rightarrow \frac{G_d}{G_c} = 100{,}000 \text{ which gives } G_c \text{ as } 10^{-3}$$

Thus $V_O = 10^{-3}$ (1) = 0.001 V = 1 mV

(b) For $G_d = 1000$, the common input gain G_c is obtained similarly as 10^{-2}, and the output voltage will be $V_O = 10^{-2}$ (1) = 0.01 V = 10 mV

From this example, we can see that when the differential input gain was increased by a factor of 10, it also increased the output due to the common input by a factor of 10, which means the output noise increased. If the CMRR is 120 dB, then V_O will still be 1 mV at $G_d = 1000$. Thus, having a higher CMRR for a given differential input gain factor reduces the output due to the common input voltage.

2.9.6 Integrating Op-Amp

The circuit for an integrating op-amp is shown in Figure 2.41, which has a capacitor C in the feedback loop.

Figure 2.41

Integrating op-amp circuit

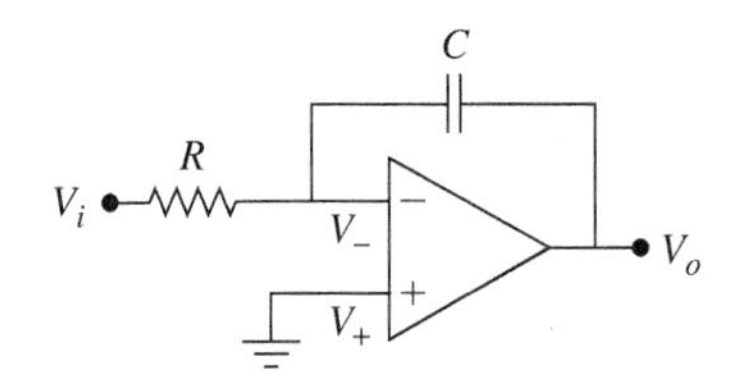

From Equation (2.3), the current through a capacitor is given by

$$I_C = C\frac{dV}{dt} \tag{2.61}$$

For this capacitor, $V = V_- - V_o = -V_o$, since $V_- = 0$. But the current through this capacitor is the same as the current that passes through the resistor R since no current flows through V_-. This current is given by

$$I_R = \frac{V_i - V_-}{R} = \frac{V_i}{R} \tag{2.62}$$

Thus,

$$\frac{dV_o}{dt} = -\frac{V_i}{RC} \tag{2.63}$$

Integrating Equation (2.63) from time $t = 0$ to time $t = t_1$ gives

$$V_o(t) = -\frac{1}{RC}\int_0^{t_1} V_i(t)\,dt + V_o(0) \tag{2.64}$$

where $V_o(0)$ is the initial condition for the capacitor voltage. Thus, in this circuit, the op-amp produces an inverted output of the integral of the applied input voltage. Note that if the capacitor and the resistor were interchanged in this circuit, the op-amp will act as a **differentiator** of the input signal. The op-amp output in this case will be

$$V_o = -RC\frac{dV_i(t)}{dt} \tag{2.65}$$

Note that any noise in the input signal will be amplified by differentiation.

2.9.7 Power Amplifier

A standard op-amp (such as the LM741) has a current output rating of about 25 mA. This is not sufficient to meet the current needs of driving loads (such as valve actuators, servo motors, and audio amplifiers). Commercial op-amps with a higher current output rating are available. These op-amps are called power op-amps, an example of which is the OPA547 chip from Texas Instruments. The OPA547 can provide a continuous output current of 500 mA with the ability to control the output current limit. Power op-amps can be conveniently used to interface a digital-to-analog (D/A) converter (see Chapter 5) that needs to drive a DC motor. Table 2.5 gives a sampling of power op-amp devices.

Table 2.5

Power op-amp devices

Device	Power Supply Range	Continuous Output Current (A)	Peak Current (A)	Slew Rate (V/μs)	Adjustable Current Limit (Y/N)
OPA547	+8V to +60V ±4V to ±30V	0.5	0.75	6	Y
LM675	±8V to ±30V	3	4	8	N
OPA541	+20V to +80V ±10V to ±40V	5	10	10	Y
OPA549	±4V to ±30V	8	10	9	Y

In Table 2.5, the *Power Supply Range* column defines the allowable voltage levels that can be applied to the positive and negative supply inputs of the op-amp. The power supply range affects the op-amp **output voltage swing**, which is the maximum voltage that the op-amp can produce without saturation for a given load. Note that the output voltage swing is proportional to the power supply range. The *Slew Rate* column defines the rate at which the op-amp output voltage will change when the op-amp gain is set to unity. Several of the power op-amps listed in Table 2.5 allow adjustment of the maximum output current of the op-amp. Due to their large output current, power op-amps are available in packages with a built-in copper tab to allow easy mounting to a heat sink for good thermal performance.

To show further the application of op-amps, Example 2.9 illustrates the use of op-amps in analog feedback control loops.

Example 2.9 PI Analog Feedback Loop

Illustrate how op-amps can be used to implement an analog proportional integral (PI) feedback control loop.

Solution:

A PI controller has the following relationship between the control output $V_o(t)$ and the error signal $e(t)$:

$$V_o(t) = K_P\, e(t) + K_I \int e(t)\, dt \tag{1}$$

where the error signal is defined as

$$e(t) = V_{ref}(t) - V_A(t) \tag{2}$$

with $V_{ref}(t)$ as the reference or desired value and $V_A(t)$ as the actual or measured value. The PI controller can be implemented as the cascade of three op-amp circuits (Figure 2.42). The first circuit is a differential op-amp circuit to compute the error signal. The second circuit, which actually consists of two op-amps, one an inverting op-amp and other an integrating op-amp, implements the P and I actions, respectively. The last stage sums and inverts the outputs from the P and I action circuits. Note that the K_P gain is the ratio of R_4 to R_3 and the K_I gain is equal to $1/(R_2C)$. To allow for variable gains, separate potentiometers can be used to replace the R_4 and R_2 resistors, respectively.

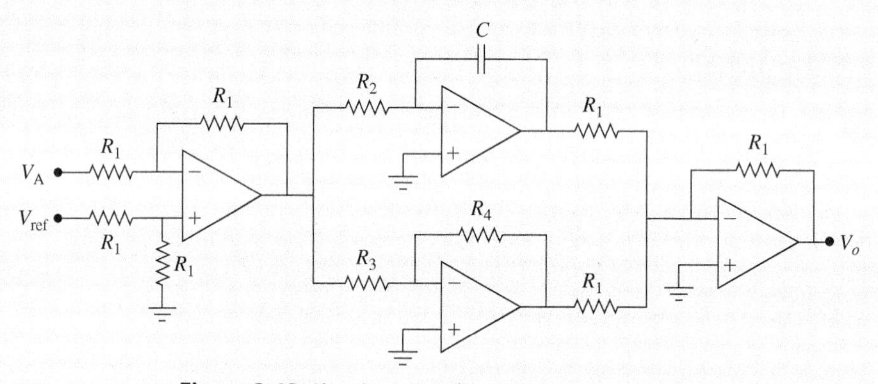

Figure 2.42 Circuit to implement a PI controller

2.10 Grounding

When we talk about voltage or potential difference, we always refer to the value of one voltage level with respect to another level. A **ground voltage** or zero voltage is commonly used as a reference (video available—V2.10). It is indicated in circuit diagrams by the symbol shown in Figure 2.43(a). Using that voltage as a reference, one can measure the other voltages in the circuit. Technically, a true ground voltage refers to the earth's ground voltage which is obtained by connecting a wire to a metal pole that is inserted into the earth's surface. However, in many circuits, a ground symbol does not mean a connection to an earth ground but to a current return path to the negative terminal of the power supply. Another type of ground reference is

Figure 2.43

(a) Ground return symbol and
(b) Chassis return symbol

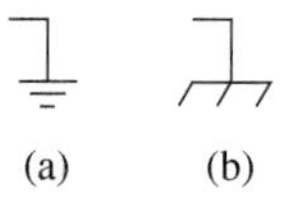

(a) (b)

called the **Chassis return**. This is indicated by the symbol shown in Figure 2.43(b). A chassis return refers to the connection between a device housing or casing and the earth line in the power cord.

It is important in circuits to have a common ground to avoid the problem that arises from ground loops. **Ground loops** form when there is more than one path to connect a circuit or system to the ground. An example of ground loop wiring is shown in Figure 2.44 where two separate ground points are used. Due to inherent impedance in the connection between the ground points [7], ground loops lead to voltage differences between the two ground points, which results in noise in the circuit. A way to eliminate ground loops is to connect all of the return paths in a circuit to a common ground point. If this is not practically possible, then all of the return paths should be connected to a common ground bus which is itself connected to the ground. In addition, if the circuit has analog and digital elements, then the analog ground and digital ground should be connected at one point.

Figure 2.44

Illustration of wiring that leads to ground loops

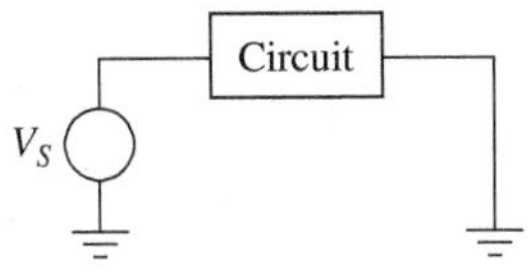

In discussing grounding, it is important to understand how this is done in AC circuits. In AC circuits (such as those in a typical home), the electrical wall socket outlet has three slots (see Figure 2.45). One slot is called live or hot, the other is neutral or return, and the third is earth or ground. The live slot carries the alternating current to the load (such as a small appliance) that is plugged into the wall socket. The neutral or return slot provides a return path of the current back to the source. The ground line, which is connected to the earth's ground, normally does not carry any current but is provided as a safety feature in case of an electrical fault within the connected equipment. The metal case or covering of the electrical device, where humans typically contact, is connected to the ground line to provide a path for the current in fault situations. In the USA and many countries, polarized plugs are used in which there is only one way of connecting the plug to the wall socket to prevent the interchange of the live and neutral lines. This is done as a safety feature to prevent, for example, a switch to open a circuit using the neutral line. While such an arrangement interrupts the current flow through the device, the live line is still connected to the device, which would cause a safety hazard if one serviced the equipment while the plug is still connected.

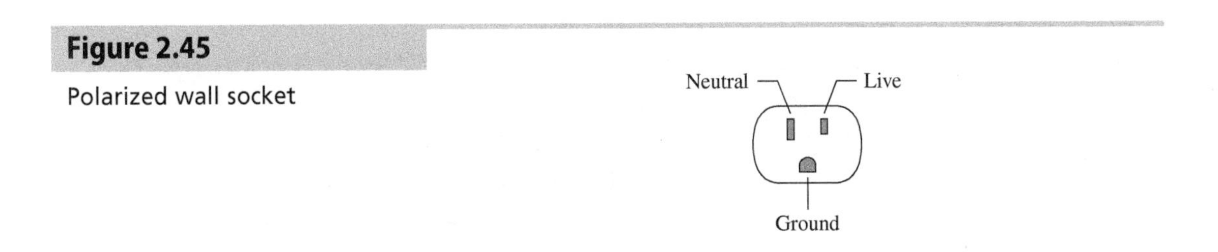

Figure 2.45

Polarized wall socket

2.11 Power Supplies and Batteries

Mechatronics systems require power to operate. Many motors as well as sensors, control boards, and microcontrollers require DC power for operation. This power can be provided by DC power supplies (PS) as well as from batteries. This section discusses these two sources of power. DC PS convert AC voltage coming from a wall outlet into DC power. There are two different types of DC PS: **linear-regulated AC/DC PS** and **switched-mode PS** (▶ available—V2.11). Table 2.6 compares the characteristics of these two kinds of power supplies. In general, both PS types provide stable regulated DC output voltage. Linear-regulated PS tend to be more precise and quieter but less efficient and bulkier, while switched-mode PS are less precise and noisy, but compact and more efficient. Linear-regulated PS use a transformer to first reduce the AC line voltage to a lower AC voltage, and then use filtering, linear regulators, and additional circuitry to rectify the AC voltage (see Section 3.2) into a very clean and stable constant DC voltage at their output. They are typically used in applications where there is a need for a stable precise DC voltage source such as low-noise amplifiers, data acquisition, signal processing, and test equipment. Switched-mode PS on the other hand do not use transformers to reduce the input voltage. Instead, the AC input voltage is first rectified and filtered at the input, and then transistors (see Chapter 3) that switch on and off at a high frequency through a process called pulse width modulation (see Chapter 4) are used to convert the voltage signal into a high-frequency pulse train. Further rectification, filtering, and regulation are then performed before the output is made available. The switching action dissipates little power but generates noise that could be reduced through filtering. Switched mode PS are typically used to provide power for telecommunication devices, computing equipment, audio equipment, mobile phone chargers, and manufacturing operations such as electroplating and anodizing, as well as to provide power to DC motor drives. A feature of switched power supplies is that they can work with different input voltage levels such as 110 or 220 V and with different AC input frequencies such as 50 or 60 Hz, so they can be used in any country without any changes to the power supply which is not the case with linear regulated PS.

Table 2.6

Characteristics of DC power supplies

Characteristics	Linear-Regulated AC/DC PS	Switched-Mode PS
Quietness	Quiet since there is no high-frequency switching	Noisy due to the high-frequency noise
Electric Noise	Small	High
Efficiency	Not very efficient and generates a lot of heat	More efficient and generates less heat
Precision	More precise than switched mode	Less precise than linear regulated
Size	Can be large and heavy	Small and compact
Output Power	Limited	Can produce large power
Cost	More expensive	Less expensive

DC PS are distinguished by their power rating as well as the DC voltages that they provide. In some mechatronics applications, there is a need to have two DC voltage levels, such as 24 V to power a motor and 5 V to power a sensor. While power supplies intended for laboratory use come with multiple DC voltage output levels, such as 5 V, 12 V, and variable 0-30 volts, other power supplies come only with one output voltage level such as 24 or 48 volts. If one needs to get a 5 V output from such a power supply, an efficient way to do this is using a **buck DC/DC converter** (see Figure 2.46).

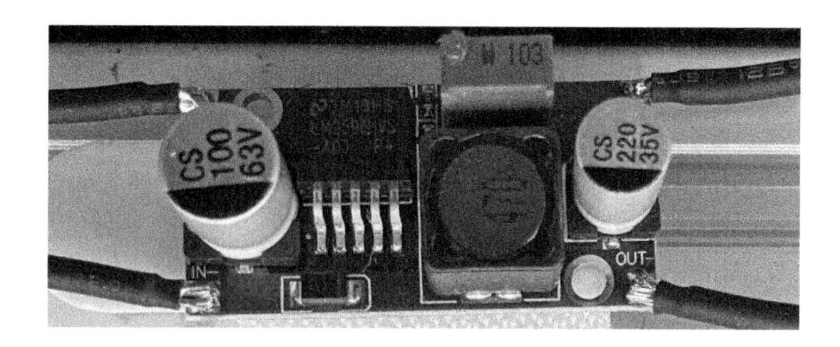

Figure 2.46

Buck DC-DC converter

(Jouaneh, University of Rhode Island)

A buck converter (available—V2.12), also known as a step-down converter, is a simple type of DC-DC converter that produces an output voltage level of the same polarity that is smaller than the input value. Figure 2.47 shows the principle of operation of a buck converter. A power switch (usually a MOSFET transistor—see Chapter 3) is turned on and off at a high frequency by a control circuit. When the switch is turned on, the input voltage is connected to the inductor, storing energy in the inductor, and causing a current to flow through it into the load and the output capacitor. In this state, the inductor acts against or 'bucks' the input voltage, thus lowering the output voltage. When the switch is turned off, the stored energy in the inductor is transferred to the output load through the diode which allows the current to flow through it, creating a closed loop path for the inductor's energy. This energy transfer results in a lower output voltage since some energy is consumed in the process. The output capacitor helps to smooth out any ripples or fluctuations in the output voltage. By controlling the duty cycle (the ratio of time the switch is on to the time it is off) of the power switch, the output voltage of the buck converter can be regulated. The efficiency of the buck converter is high because the power is regulated through switching action reducing the amount of energy lost as heat, which is not the case if the input voltage was reduced through a voltage dividing circuit. Integrated Case Study II.A discusses the power supplies and the wiring used for the DC Motor-Driven Linear Motion Slide Case Study.

Figure 2.47

Buck DC-DC converter operation

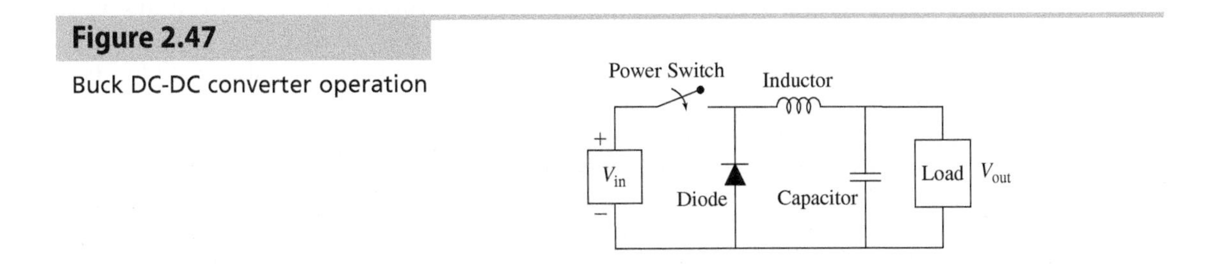

Integrated Case Study II.A: Power Supply and Buck Converter for the System

For Integrated Case Study II (DC Motor-Driven Linear Motion Slide—see Section 11.3), 24- and 5-volt DC supply voltage levels are needed to power the different components of the system. A Drok 480-watt, 0-48 V adjustable DC output with a 10 A current rating was used as the power supply for the system, and the power wiring diagram is shown in Figure 2.48. Using the provided knob on the power supply and the LED digital readout on the top of the power supply, the output voltage was adjusted to 24 volts. The 24 volts was used to power the servo drive (see Section 9.2.3) that powers up a Lin Engineering brushless DC motor (Model BL23E22-01D-06RO, see Section 9.2.2) that drives the belt-driven stage. The motor is designed to operate at 24 volts and has a rated current of 3 A. To provide means to automatically shut off the power to the servo drive in case of slide overtravel, the 24 V to the servo drive was routed through two 5 volt DC relay modules (HiLetgo 5V One Channel Relay Module from Amazon). The *IN* signal trigger input on the relay module was set to low level via jumper, which causes the relay to toggle open in the event of a low-level trigger. Each relay is activated by the output from a photo interrupter limit switch module (Teyleten Robot IR Infrared Slotted Optical Optocoupler Module from Amazon) that is placed at each end of the travel. When the limit switch is not activated, the photo interrupter module provides 5 volts output signal (or the *IN* signal to be high) which causes the COM and NC leads on the relay to be in contact and power to be conducted through the relay. When the limit switch is activated, the *IN* signal goes low, which causes the COM and NO leads on the relay to be in contact, and the 24 V transmission through the relay is broken. To provide the 5 VDC power needed to drive the relays and the photo interrupter sensors, a 24-volts to 5-volts DC-DC buck converter was also used.

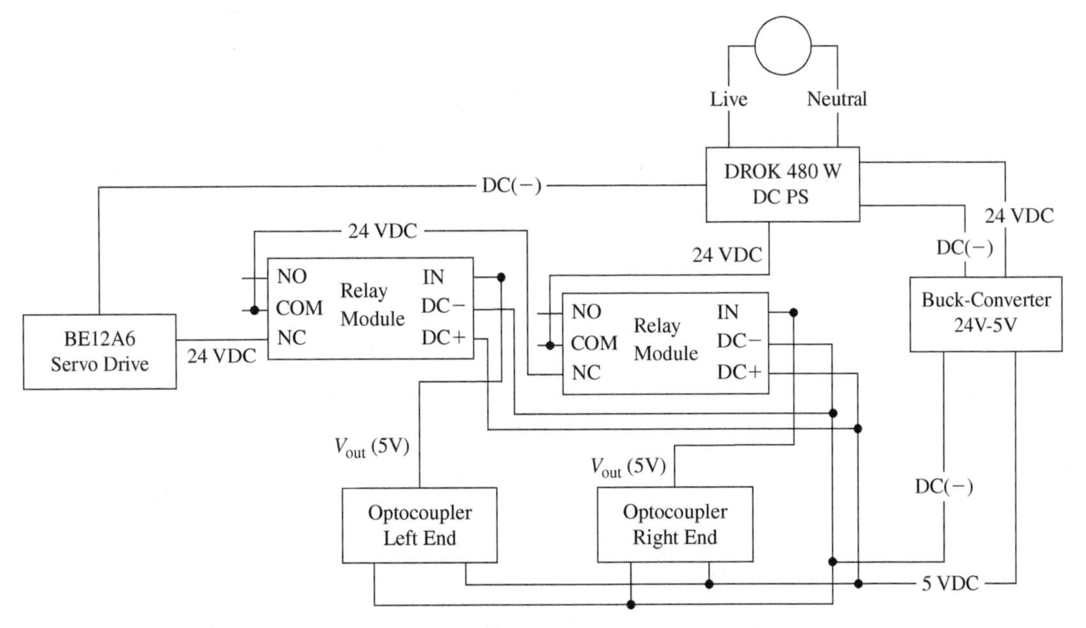

Figure 2.48 Power connection diagram for the DC-motor driven linear motion slide

Batteries offer a convenient, readily available DC power source for portable applications such as mobile robots, drones, and portable electronics. Batteries can be rechargeable or non-rechargeable kind; are available in different types or chemistry such as lithium, alkaline, or lithium-ion; and come in different sizes and forms. In addition to the voltage rating of a battery such as 1.5 V or 9 V, the capacity of a battery in milliamp-hours (mAh) or amp-hours (aH) is an important consideration when choosing a battery. **Battery capacity** refers to the amount of charge a battery holds. As an example, a battery with a 3000 mAh capacity means that the battery will provide a 100 mA current for 30 hours, or 1000 mA for 3 hours. The amount of current that the battery provides is a function of the load (or resistance) in the circuit to which the battery is connected. One disadvantage of battery power sources is the need to replace or recharge the battery source as the charge in the battery is drained. Another disadvantage is the rated voltage of the battery does not remain constant as the battery is used but decreases in a non linear fashion. Figure 2.49 shows a typical discharge plot (▶ available—V2.13) for a lithium-ion battery, but different battery types have different discharge plots, and even for the same battery type, the discharge plot differs for different rates of discharging the battery. Microcontrollers have a built-in feature that is useful when the microcontroller is powered by a battery called a brown-out detection circuit that monitors the voltage level supplied to the microcontroller and causes a reset of the microcontroller if the voltage drops below a specified level (see Section 4.7.4). Example 2.10 illustrates the computation of battery life.

Figure 2.49

Typical battery discharge plot for a lithium-ion battery

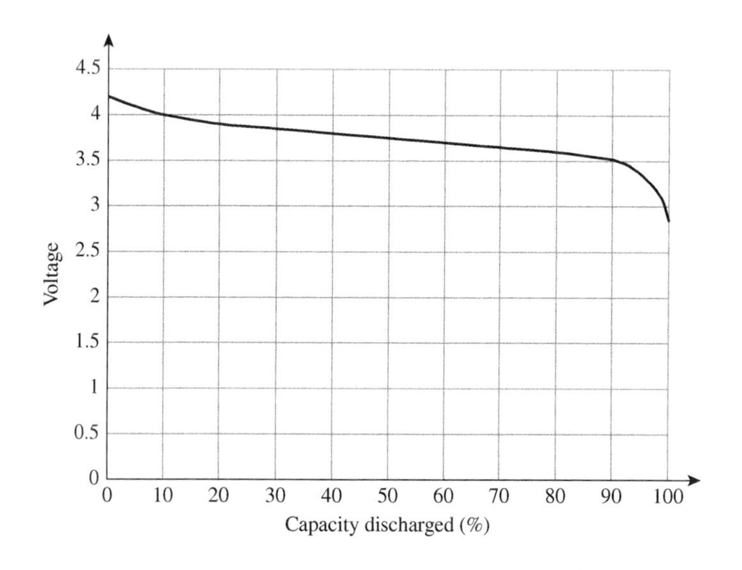

Example 2.10 Computation of Battery Life

An application uses four batteries, each with a voltage output of 1.5 volts and an initial charge capacity of 2700 mAh. Determine when the battery charge will be depleted if the batteries are used to drive a resistive load of 500 ohms, and they are wired in two different configurations: (a) in series fashion and (b) in parallel fashion. For the calculation, assume that the voltage remains constant throughout this operation.

Solution:

(a) For series wiring, the voltage produced by the four batteries will be 6 volts. The current through the 500 Ω load will be 12 mA (6/500) which is also the current that is provided by each battery. For an initial charge of 2700 mAh, each battery will run for 225 hours (2700/12) or 9.4 days if the voltage remains constant during the operation.

(b) For parallel wiring, the voltage produced by the four batteries will be 1.5 volts. The current through the 500 Ω load will be 3 mA (1.5/500). Since the batteries are connected in parallel, each battery will need to provide only one-fourth of that current or 0.75 mA. For an initial charge of 2700 mAh, each battery will run for 3600 hours (2700/0.75) or 150 days, significantly more than case *a*, if the voltage remains constant during the operation.

Problem P2.40 considers similar configurations but does not assume that the battery voltage remains constant during operation.

Power supplies and many other industrial power devices such as signal conditioners are available in two form factors or mounting options, tabletop, and DIN rail. **Tabletop** PS, also known as benchtop or desktop PS, are designed to be placed on a table or workbench. **DIN rail** PS are designed to be mounted on a standard DIN rail (see Figure 2.50), which is a metal rail used for mounting various industrial control equipment in control panels or cabinets. DIN rail PS have a more compact form factor and are intended for permanent installation in electrical enclosures or control systems. They typically have screw terminals or plug-in connectors for easy integration into the system.

Figure 2.50

Standard DIN rail

(Courtesy of Altech Corp., Flemington, NJ)

2.12 Chapter Summary

This chapter discussed analog circuits and components. The chapter started by discussing basic circuit elements (such as resistors, capacitors, and inductors) and the laws for their combination in series or parallel form. It then discussed various types of switches, including mechanical push-button and toggle switches, and electromechanical relays. Circuit analysis was then covered. Kirchhoff's voltage law (KVL) and Kirchhoff's current law (KCL) are the two basic laws that are used to perform circuit analysis. KVL states that the algebraic sum of the voltage drops and rises around any closed path in a circuit is zero, while KCL states that the sum of the current into a node is zero. To simplify certain circuits, equivalent circuits are used. The equivalent circuit can be in one of two forms: the Thevenin equivalent circuit or the Norton equivalent circuit. The concept of impedance, which is a generalization of the concept of resistance, was then covered. Impedance concepts are useful in analyzing loading effects that occur when measurement devices are connected to circuits. Alternating current circuits were also discussed, including computing the power in these circuits. Operational amplifiers or op-amps are analog circuit components that require power to operate. Op-amps can perform various operations (such as comparison, amplification, inversion, summation, integration, differentiation, or filtering). The particular operation depends on how the op-amp is wired and what external components are connected to the op-amp. Several op-amps circuits were discussed. The concepts of ground loops and proper grounding techniques were also presented. The last section discussed power supplies and batteries.

Questions

2.1 Define what is meant by an analog circuit.

2.2 List the different types of toggle switches.

2.3 What is a relay?

2.4 Name the two laws that are used to analyze electrical circuits.

2.5 Define impedance.

2.6 What impedance characteristic is desirable in measuring devices, and why?

2.7 What device has a very large impedance at low frequencies?

2.8 List an advantage of AC signals.

2.9 Define the real power of an AC circuit.

2.10 What characteristics of an op-amp make it useful to use it as an interface?

2.11 List the two rules that are used to analyze ideal op-amp circuits.

2.12 List some of the different types of op-amp circuits.

2.13 For what purpose is a voltage follower circuit used?

2.14 What type of op-amp circuit is used in the implementation of an analog proportional control feedback loop?

2.15 Can the output voltage of an op-amp circuit exceed the supply voltage?

2.16 List an advantage that an ia op-amp has over a differential op-amp.

2.17 What is the significance of a high CMRR value?

2.18 Name one way to avoid a ground loop.

2.19 List several advantages of switched power supplies.

2.20 Name some limitations of battery-powered sources.

Problems

Section 2.3 Switches

P2.1 Illustrate how a single SPST switch can be used for wiring three light bulbs that are turned on or off at the same time.

P2.2 Illustrate how an SPDT switch can be used for wiring the low/high beam for car headlights using the car battery as the voltage source.

P2.3 Identify three household/consumer applications where push-button switches are used and identify whether the switch is NO or NC type.

P2.4 Draw the wiring circuit for a light bulb that can be turned on/off from three switch locations.

P2.5 Draw the switching circuit for a three-position light bulb. A three-position light bulb has two filaments. At the low switch position, the low-intensity filament is turned on. At the medium setting, the medium-intensity filament is turned on. At the high position, both filaments are turned on.

P2.6 Draw a wiring diagram to show how a single-channel relay can be used with an NC emergency switch to shut off the power to a load when the switch is activated.

P2.7 A circuit has two switches, A and B, and a load. Design a circuit using two relays to control the load so that it turns on only when switch A and switch B are closed and turns off when either switch is open.

P2.8 Draw a circuit that uses two relays (similar to the one shown in Figure 2.11(a)) to switch the direction of rotation of a DC motor. The circuit has three inputs, the supply voltage for the motor and two control inputs *A* and *B*, as shown in Figure P2.8. When *A* is 5 V and *B* is 0 V, the motor rotates in one direction, and when *A* is 0 V and *B* is 5 V, the motor rotates in the opposite direction.

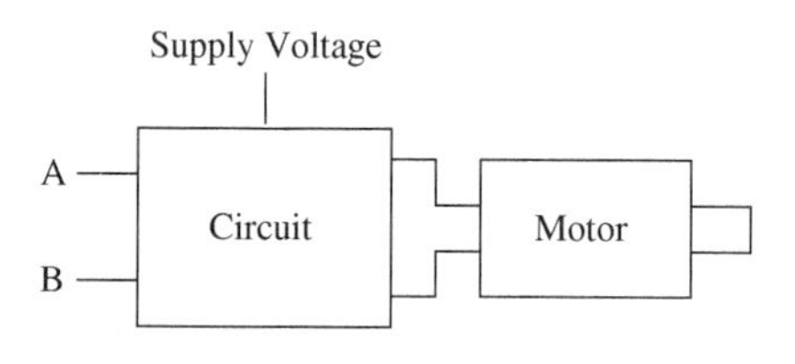

Figure P2.8

Section 2.4 Circuit Analysis

P2.9 Consider the voltage dividing circuit shown in Figure P2.9. Provide an expression for the output voltage V_{out} of the circuit.

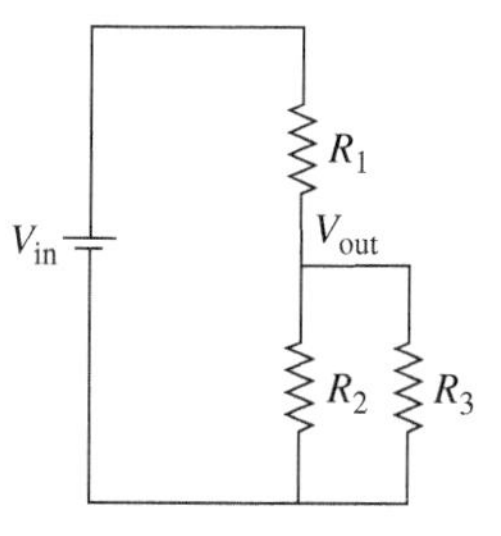

Figure P2.9

P2.10 A voltage divider circuit is constructed using a 10 kΩ fixed resistor and a 0-10 kΩ potentiometer. The output voltage is measured across the potentiometer. When the potentiometer is set to 5 kΩ, the output voltage is 2 V. However, when a 10 kΩ load is connected across the potentiometer (similar to resistor R_3 in Figure P2.9), the output voltage drops to 1.5 V. Calculate the output voltage of the voltage divider circuit when the potentiometer is set to 7 kΩ and the 10 kΩ load is connected.

P2.11 Determine the unknown currents through the resistor network shown in Figure P2.11.

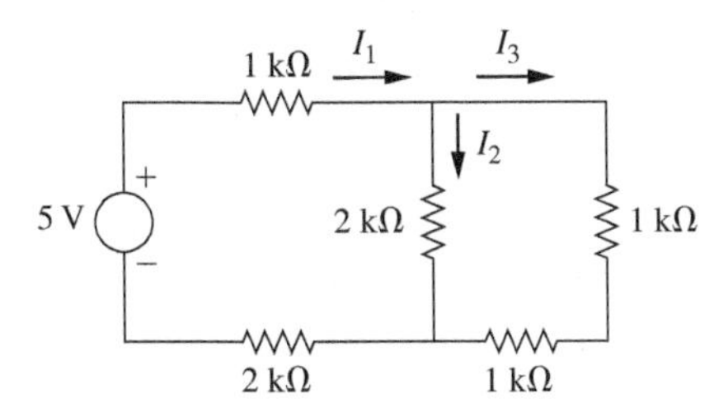

Figure P2.11

P2.12 Determine the unknown currents through the resistor network shown in Figure P2.12.

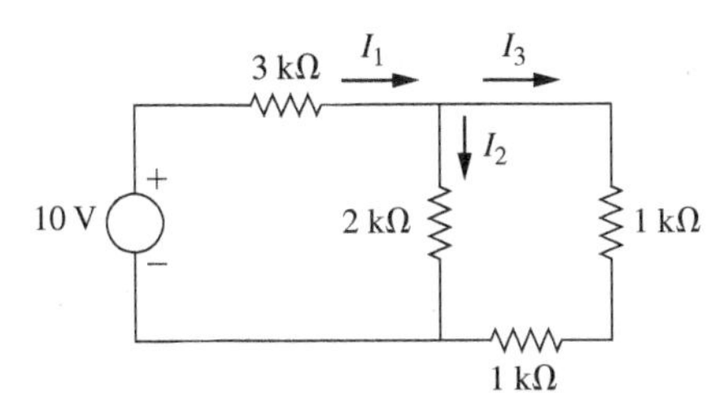

Figure P2.12

P2.13 Determine the unknown currents through the resistor network shown in Figure P2.13.

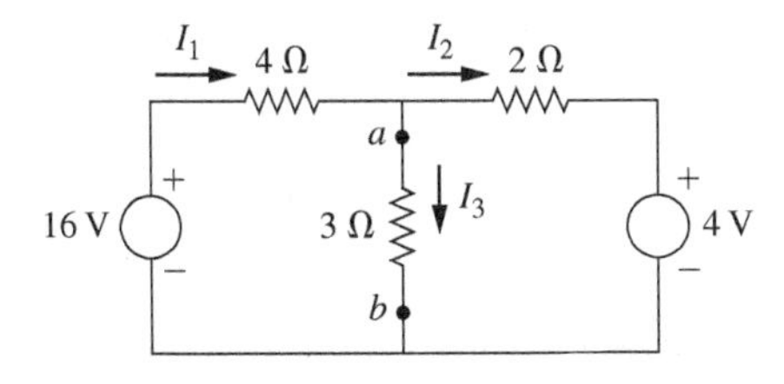

Figure P2.13

P2.14 Determine the unknown currents through the resistor network shown in Figure P2.14.

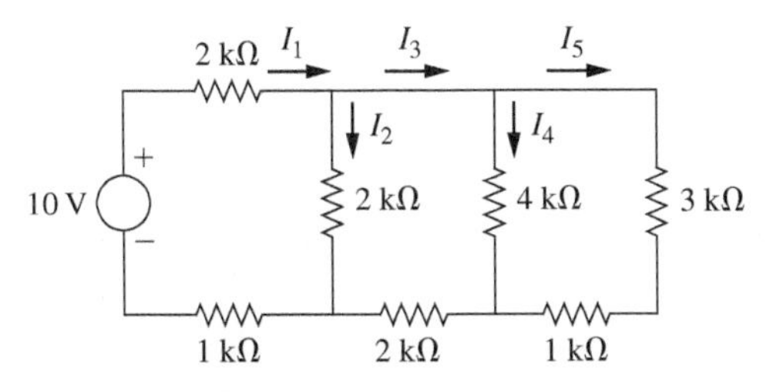

Figure P2.14

P2.15 Determine the unknown currents through the resistor network shown in Figure P2.15.

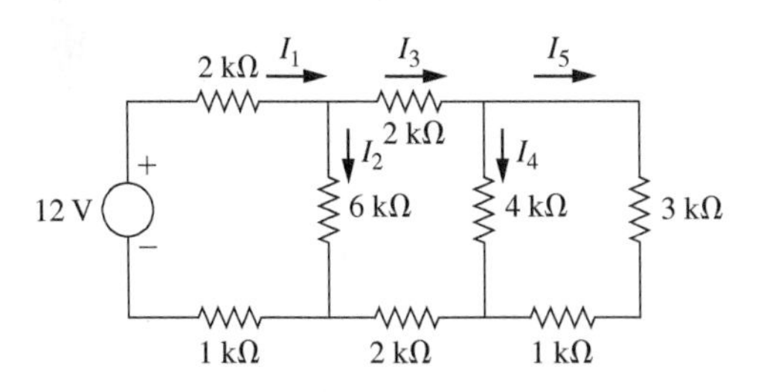

Figure P2.15

Section 2.5 Equivalent Circuits

P2.16 Determine the Thevenin equivalent circuit at the nodes *a* and *b* for each of the two circuits shown in Figure P2.16.

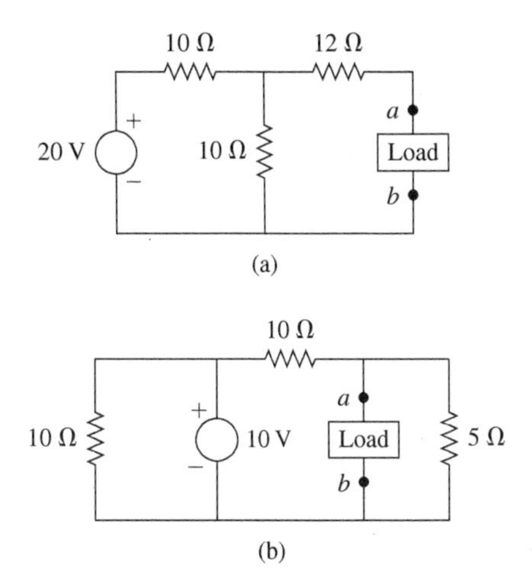

Figure P2.16

P2.17 Determine the Thevenin equivalent circuit at the nodes *a* and *b* for the circuit shown in Figure P2.13. Consider the load to be the 3 Ω resistor.

Section 2.6 Impedance

P2.18 Determine the impedance for the following components at 60 Hz. What is the total impedance of these components if they were connected in series?

a. 1000 Ω resistor

b. 500 mH inductor

c. 1 μF capacitor

P2.19 An inductor with an inductance of 100 mH is connected in parallel with a capacitor of 1 μF. Determine the resultant impedance of these two components at a frequency of 1 kHz.

P2.20 Determine the impedance of the circuit shown in Figure P2.20 at a frequency of 1 rad/s. Let $R_1 = R_2 = 1\ \Omega$, $L = 1$ H, and $C = 1$ F.

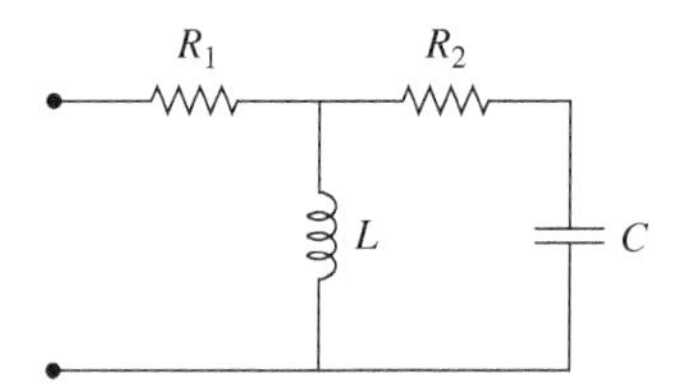

Figure P2.20

P2.21 In electrical circuits, admittance is defined as the reciprocal of the impedance in the circuit. If a circuit has an impedance of $1 + 2j\ \Omega$, what is its admittance?

P2.22 Determine the resultant impedance of the resistor R_S and the capacitor C_S shown in Figure 2.24 at 1, 1000, and 100,000 Hz. Assume the capacitance of C_S is 25 pF. What happens to the impedance of the scope as the frequency increases?

P2.23 The terminal voltage of a power supply is 24 VDC before a load is applied. When a 100 Ω resistor is connected to the power supply, the voltage drops to 23 V. What is the internal resistance of the supply source?

Section 2.7 AC Signals

P2.24 What is the peak voltage for an AC signal with an RMS value of 220 volts?

Section 2.8 Power in Circuits

P2.25 An electric heater is rated at 1500 W for 110 volts RMS operation. Determine the resistance of its heating element.

P2.26 What is the power factor if the three components in Problem P2.18 were connected in series?

P2.27 A load with an impedance of $400 + 300j\ \Omega$ is connected to a 220 V, 50 Hz source. Determine the power factor and the real and reactive powers across the load.

P2.28 A load made up of a resistor and a capacitor connected in series draws 0.2 A from a 60 Hz, 110 V source with a 0.8 power factor. Determine the resistance and reactance of the load.

Section 2.9 Operational Amplifiers

P2.29 Consider the inverting op-amp circuit shown in Figure P2.29. Assume that the saturation voltage for this op-amp is +/−13 volts, and $R_1 = 1\ \text{k}\Omega$ and $R_2 = 5\ \text{k}\Omega$. Compute the voltage output V_O, and the current I_1 for the following input voltages V_{in}.

a. 1 V

b. −2 V

c. 5 V

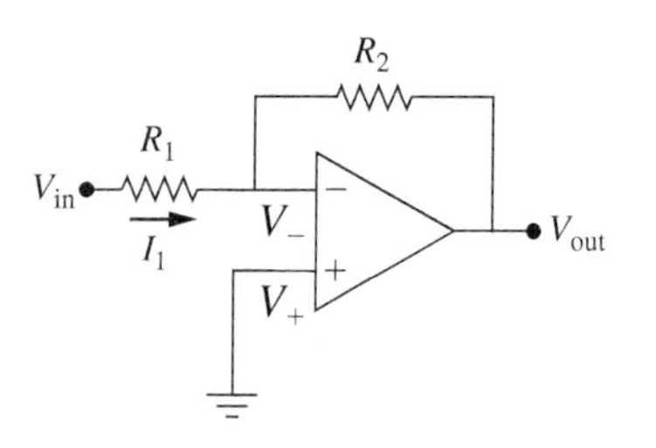

Figure P2.29

P2.30 For the inverting op-amp considered in Problem P2.29, assume that the resistors R_1 and R_2 have a 5% tolerance. Determine the maximum percent change in the gain of this op-amp.

P2.31 Plot the output of the circuits shown in Figure P2.31 for the following input signal. Note that $R_1 = R_2 = 100\ \text{k}\Omega$, and $C = 10\ \mu\text{F}$.

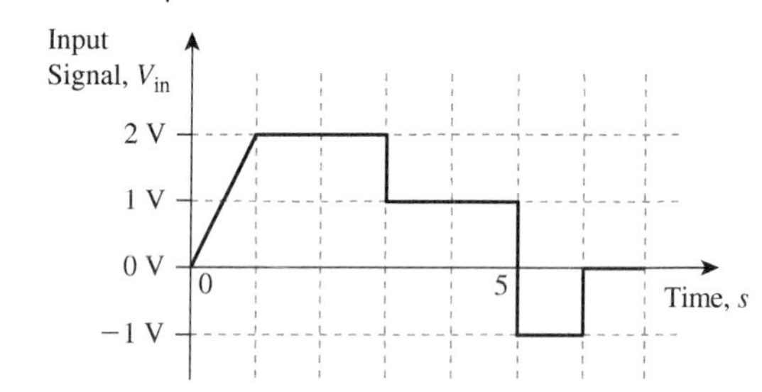

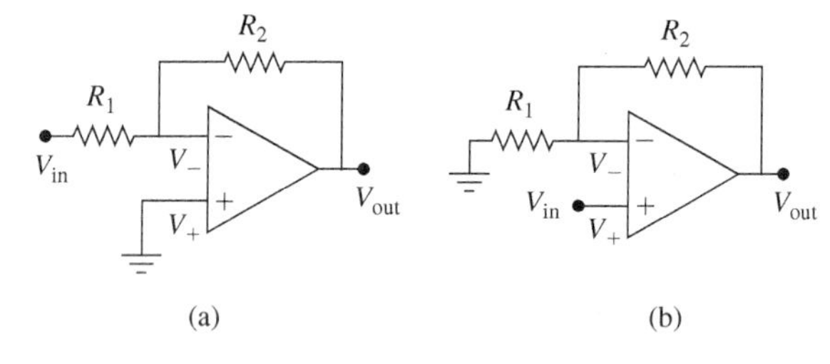

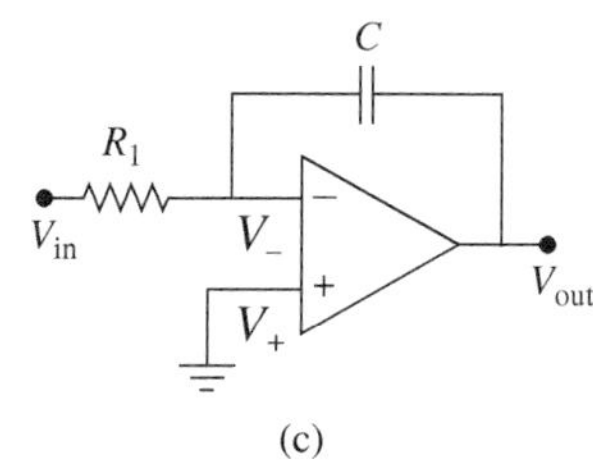

Figure P2.31

P2.32 The input signal shown in Figure P2.32 is applied to a non-inverting Schmitt trigger with $V_{T+} = 3.1$ V and $V_{T-} = 2.9$ V. Plot the output signal if V_{max} is 5 V and V_{min} is 0 V.

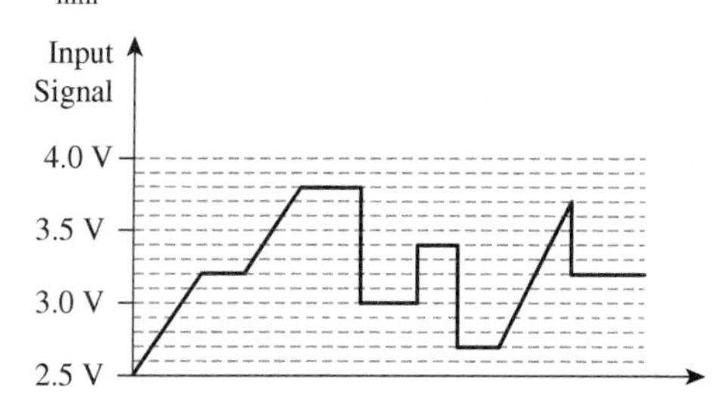

Figure P2.32

P2.33 For the ia op-amp shown in Figure 2.39, if $R = 30\ k\Omega$, what should be the value of R_1 so the amplifier gain is 10?

P2.34 For the INA128 op-amp shown in Figure 2.40, what should be the value of R_G so the amplifier gain is 100?

P2.35 For an ia op-amp with $V_1 = 0.99$ V, $V_2 = 1.01$ V, $G_d = 1000$, and CMRR = 100 dB, compute the output voltage of this op-amp.

P2.36 An ia op-amp with a CMRR value of 80 dB has a common voltage input of 1 volt. Compute the output voltage of this ia op-amp for a differential input gain of:

a. 10

b. 100

c. 1000

P2.37 An engineer has proposed the circuit shown in Figure P2.37 for performing closed-loop proportional control of the speed of a DC motor using a tachometer as the feedback signal. Show if this circuit will operate as proposed.

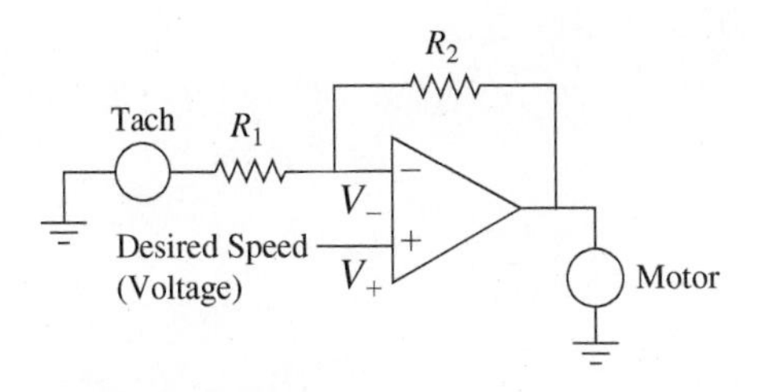

Figure P2.37

P2.38 Design an op-amp circuit to implement an analog PD controller action. Select components to give a K_p gain value of 5 and a K_d value of 0.1.

P2.39 Design an amplifier circuit that uses LM741 op-amps. The circuit should take an input voltage V_{in} and produce an output voltage V_{out} that is equal to $k\ V_{in}$ where $0 \leq k \leq 10$. Specify all of the components that are used in the circuit, the output voltage, and the current limits of the amplifier.

Section 2.11 Power Supplies and Batteries

P2.40 Redo Example 2.10 but assume that the battery voltage decreases linearly as the charge capacity is used and reaches a value of 0.75 volts when the charge is completely depleted.

Laboratory/Programming Exercises

L/P2.1 Build and test the circuit described in Problem P2.8.

L/P2.2 Build the circuit shown in Figure P2.11 and measure the voltages and currents in the circuit. Compare the measured values to the computed ones.

3 Chapter

Semiconductor Electronic Devices and Digital Circuits

Chapter Objectives:

When you have finished this chapter, you should be able to:

- Explain the function of diodes and thyristors
- Predict the output of simple circuits involving regular and Zener diodes
- Explain the use of transistors in switching applications
- Analyze circuits containing bipolar junction and MOSFET transistors
- Analyze digital combinational logic circuits and generate a logic circuit from a truth table specification
- Generate the timing diagrams for various types of flip-flops
- Explain the properties of TTL and CMOS circuit families, their characteristics, and how to interface them
- Draw a wiring circuit for digital devices (such as timers and H-bridge drives)

3.1 Introduction

The previous chapter considered the design and analysis of analog circuits. This chapter discusses the operation of semiconductor electronic devices (such as diodes, thyristors, and transistors) that are used in many circuits and devices for switching or amplification purposes. A semiconductor is a material whose properties are in between a conductor and an insulator. Examples of naturally available semiconductor materials include silicon and germanium. For use in semiconductor electronic circuits, small quantities of other elements (such as boron and phosphorus) are added to silicon or germanium crystals to alter their properties. Semiconductor electronic devices have properties that depend on temperature, lighting conditions, or the amount and direction of voltage applied to them. A basic and important semiconductor device is the transistor, whose invention has led to the development of digital circuits in which transistors form the building blocks. An important feature of a transistor is that it can amplify an input signal. Semiconductor electronic devices (such as transistors) are commonly used as an interface in the operation of real devices (such as motors and heaters). Digital circuits are widely used in devices such as computers, wireless phones, and digital cameras. This chapter considers both combinational and sequential digital logic circuits as well as digital devices. Digital circuits form the foundation for microprocessors and microcontrollers, and the next chapter will discuss microcontrollers that give a flexible but complicated method of implementing control logic. For further reading on the topics covered in this chapter, see [8-10].

3.2 Diodes

Diodes and transistors are examples of solid-state switches. Solid-state switches are devices in which the switching action is caused by non-mechanical motion and is due to the change in the electrical characteristics of the device. A diode (▶ available—V3.1) is a directional element that allows current to flow in one direction. The characteristics of both ideal and real diodes are shown in Figure 3.1. Unlike a resistor or a capacitor, the diode current–voltage relationship for both ideal and real diodes is highly nonlinear and does not follow Ohm's law. Figure 3.1 shows that when the diode is **forward biased,** or the anode voltage is positive with respect to the cathode, current flows in the diode in the direction of the arrow of the diode symbol. An **ideal diode,** which is a simplified model of the real diode, allows current to flow freely in one direction, from its anode to its cathode, while completely blocking any current flow in the opposite direction. In other words, an ideal diode has zero resistance when forward-biased and infinite resistance when reverse-biased. The **forward voltage**, V_F, which is the voltage at which the diode starts conducting current, is 0 V for an ideal diode. On the other hand, a **real diode** is a physical semiconductor device that exhibits similar behavior to an ideal diode but with certain practical limitations. Real diodes have a small but finite forward voltage drop when conducting current in the forward direction. The forward voltage value is dependent on the material from which the diode is made. For silicon diodes (such as the 1N914 diode), the V_F voltage is about 0.6 V (at T = 300 K). The non-zero forward voltage results from the inherent resistance of the diode material. Furthermore, real diodes have a non-zero reverse current (leakage current) when subjected to reverse voltage or reversed biased. The amount of reverse current depends on the diode's construction and material properties. While real diodes provide a substantial barrier to reverse current, they cannot achieve perfect isolation, as they do not have infinite resistance in the reverse-biased state. When a reverse-biased voltage is applied to the diode, very little current (in the nano-ampere range—see the scale in Figure 3.1) flows unless the applied voltage reaches the breakdown voltage (or V_R about 75 V for the 1N914), which causes the diode to break down. Regular diodes are not designed to operate with a voltage lower than V_R unless a Zener diode is used.

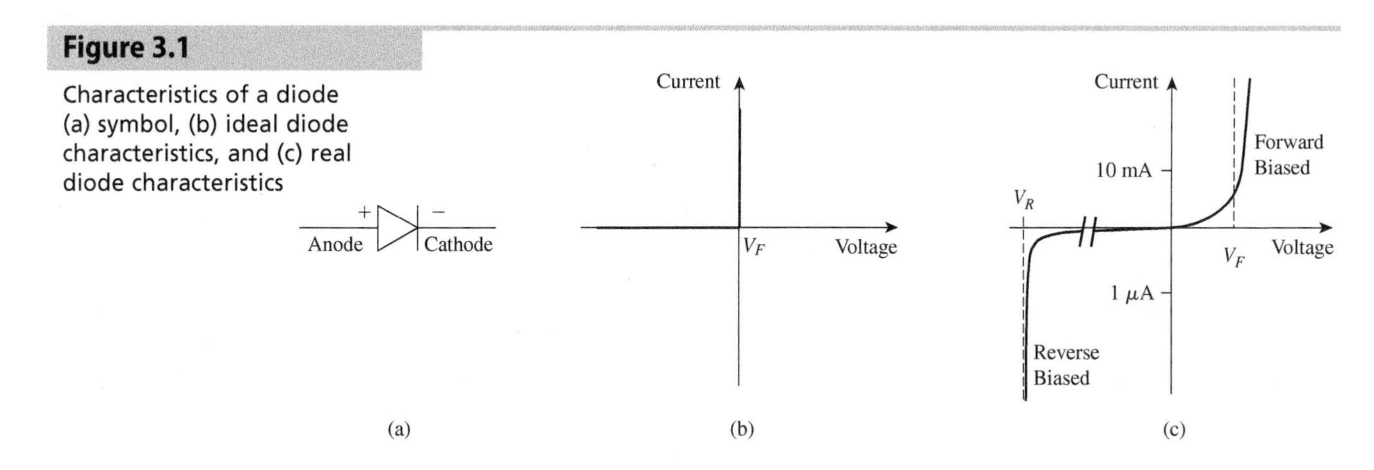

Figure 3.1

Characteristics of a diode (a) symbol, (b) ideal diode characteristics, and (c) real diode characteristics

One common use of diodes is to change AC voltages to DC voltages, which is a process called **rectification**. Figure 3.2(b) shows the output of the circuit shown in Figure 3.2(a), which is called a half-wave rectifier. The rectification occurs only if the amplitude of the sinusoidal signal exceeds the forward voltage (V_F) value for the diode. Notice how the negative portion of the sinusoidal input voltage is eliminated and how the amplitude of the positive portion of the output signal is smaller than the input voltage.

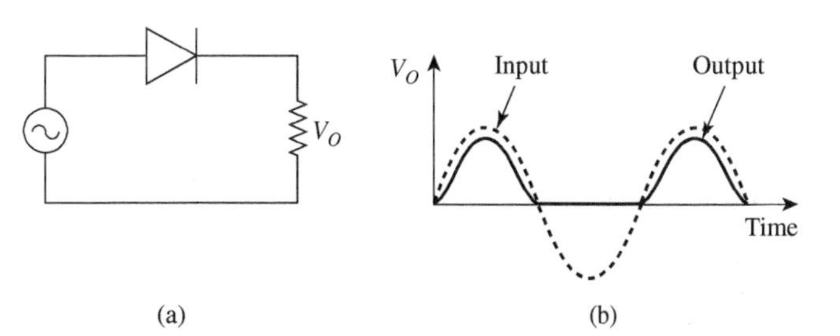

Figure 3.2

Half-wave rectification: (a) circuit and (b) output voltage

Another use of diodes is to limit the range of a signal in a circuit. This is called voltage clamping, and the diode is called a **diode clamp** when used in such a circuit. Figure 3.3 shows such a circuit.

Figure 3.3

Diode clamp circuit

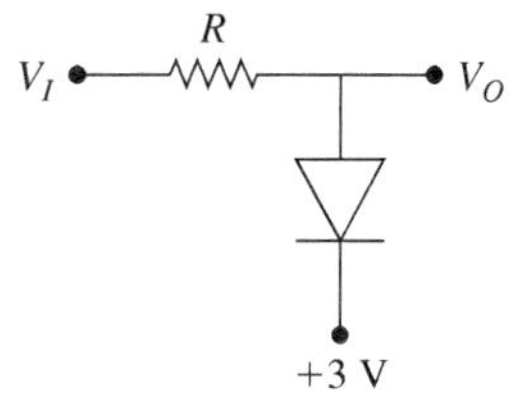

Notice that because a diode conducts a large current only when the forward-bias voltage is greater than V_F (or about 0.6 V for 1N914 diode), the above circuit will limit the output voltage V_O to 3.6 V (0.6 V plus the voltage applied at the cathode). Any input voltage lower than that (including negative voltage above the breakdown voltage) is passed as an output. This is because when the diode is not conducting, no current is flowing across the resistor R in the circuit of Figure 3.3, and thus, $V_O = V_I$.

A further application of diodes is to limit voltage spikes generated when switching off inductive loads (such as DC motors or relay coils). The use of the diode in this case is called a **flyback diode** (available—V3.2). Typical wiring is shown in Figure 3.4.

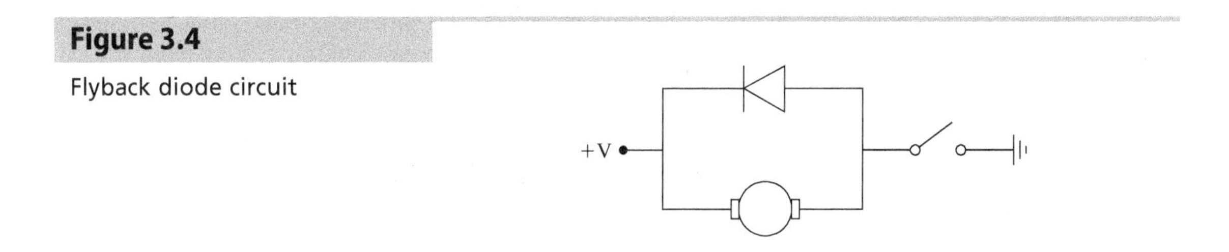

Figure 3.4

Flyback diode circuit

To understand the function of the flyback diode, first assume that the diode is not present. When the switch is opened, the current in the motor coil starts to decrease. Since a coil has inductance, a large voltage of opposite polarity develops across the motor lead according to the relationship $V = L\, di/dt$. Thus, the two sides of the switch will have voltages of opposite polarity, leading to arcing and premature wear of the switch contact. Alternatively, if a diode was added as shown in Figure 3.4, the diode provides a path for the current in the coil leading to a reduced voltage spike at the switch leads. Example 3.1 illustrates the selection of a flyback diode.

Example 3.1 Selection of a Flyback Diode for a Relay

Select an appropriate flyback diode to use with the G5V-2 OMRON relay discussed in Section 2.3.2.

Solution:

As shown in the specifications listed in Table 2.3, the G5V-2 OMRON relay requires 5 VDC for operation. Thus, the reverse voltage rating of the flyback diode should be at least 5 volts. The flyback diode current rating should be at least the coil current which for this relay is 100 mA. The 1N4001G is a commonly used diode that has a reverse voltage rating (V_R) of 50 volts, and an average forward rectified output current (I_O) of 1.0 A will meet these requirements with a comfortable margin.

3.2.1 Zener Diode

A special type of diode is called the **Zener** diode, and its symbol is shown in Figure 3.5. A Zener diode behaves like a normal diode when it is forward-biased, but it can conduct current without destroying itself when the reverse-biased voltage exceeds the breakdown voltage, V_R. The breakdown voltage or Zener voltage, V_Z, can be smaller than that of a normal diode. Common low Zener voltages include 2.7, 3.0, 3.6, 3.9, 4.7, 5.1, 5.6, and 6.2 V, but Zener voltages of 20, 51, 100, and even 200 V are available (for example, the 1N5221 to 1N5281 silicon Zener diodes series from Sematech Electronics LTD).

Figure 3.5

Zener diode symbol

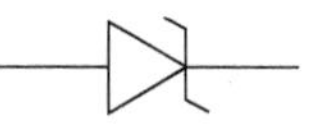

A common use of Zener diodes is to regulate the output voltage in a circuit when the supply voltage is variable or unstable. The circuit for performing this operation is shown in Figure 3.6. Note that the output voltage of this circuit or the voltage drop across the load resistor R_2 will always be V_Z if the voltage drop across R_2 is about to exceed V_Z. Example 3.2 illustrates the use and sizing of such a diode.

Figure 3.6

Voltage regulation using a Zener diode

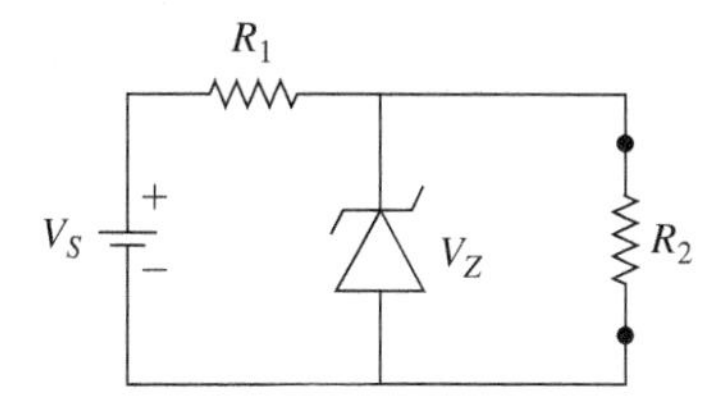

Example 3.2 Voltage Regulation Using a Zener Diode

For the circuit shown in Figure 3.6, assume that the voltage source is an unregulated supply that varies between 10 and 12 V. Select a Zener diode and appropriate resistors to give close to a 5 V drop across the load resistor R_2 using this supply.

Solution:

Select a Zener diode with a breakdown voltage of 5.1 V (which is the closest to 5 V). Select R_2 to be 100 Ω. Then the current across the load resistor R_2 is 51 mA, because the potential drop across the R_2 resistor is the same as that across the Zener diode for cases when the voltage drop across R_2 is about to exceed V_Z. The current through the resistor R_1 has to be greater than the current through the load resistor because the Zener diode will not operate unless some current flows through it, since $I_{R_1} = I_Z + I_{R_2}$. This implies that R_1 has to be less than 96. Ω for $V_S = 10$ V from the requirement that

$$I_{R_1} > I_{R_2}$$

or

$$\frac{V_S - V_Z}{R_1} > \frac{V_Z}{R_2}$$

If we select R_1 to be 91 Ω (a standard resistor value, see Appendix A), then the current through R_1 will be 54 mA for $V_S = 10$ V and 76 mA for $V_S = 12$ V. Notice that the current that is not passing through the load is being dissipated through the Zener diode. For the diode not to heat up, the diode power rating must be greater than 5.1 V × (76 − 51) mA or 0.13 W. A 0.25 W diode will do this. A commercially available diode with such specifications is the 1N4625.

3.2.2 LED

One common form of a diode is the light-emitting diode (**LED**) (▶ available—V3.3). These diodes emit light when forward-biased, and the amount of light they emit is proportional to the current passing through the LED. They are typically encased in a colored plastic casing. An advantage of an LED over other light sources is that it takes only a few milliamps to light the diode. They also can be powered by a digital power supply (5 VDC) since the voltage drop across the LED when it is on is about 2 V. A typical LED and its symbol are shown in Figure 3.7. Note that the anode or the positive terminal is the one that has a longer lead.

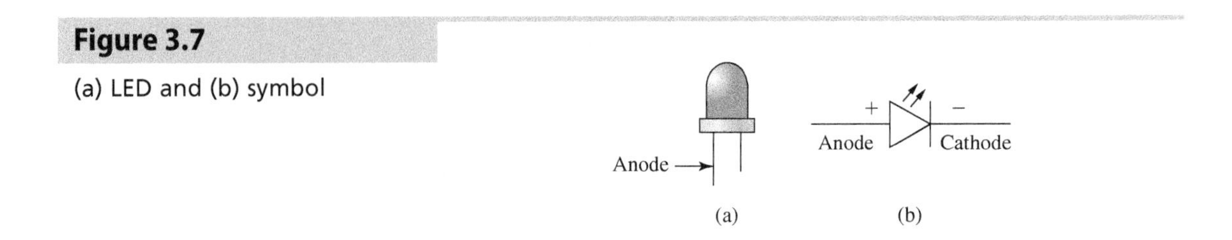

Figure 3.7
(a) LED and (b) symbol

3.2.3 Photodiode

Another form of a diode is the **photodiode**. A photodiode (see Figure 3.8 for a symbol) behaves like an LED but in the opposite fashion. The amount of current that the photodiode passes is proportional to the amount of light it receives, and the current flows from the cathode to the anode (reverse biased). Photodiodes are commonly used as light sensors.

Figure 3.8
Symbol of a photodiode

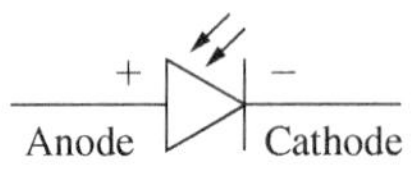

+ −
Anode Cathode

3.3 Thyristors

A **thyristor** (silicon-controlled rectifier or SCR) is a three-terminal semiconductor device that behaves like a diode but with an additional terminal. The additional terminal is called a *gate*, and when a small current flows into the gate, it allows a much larger current to flow from the anode to the cathode (provided that the voltage between the anode and the cathode is forward-biased). The symbol and a typical component form of a thyristor (▶ available—V3.4) are shown in Figure 3.9.

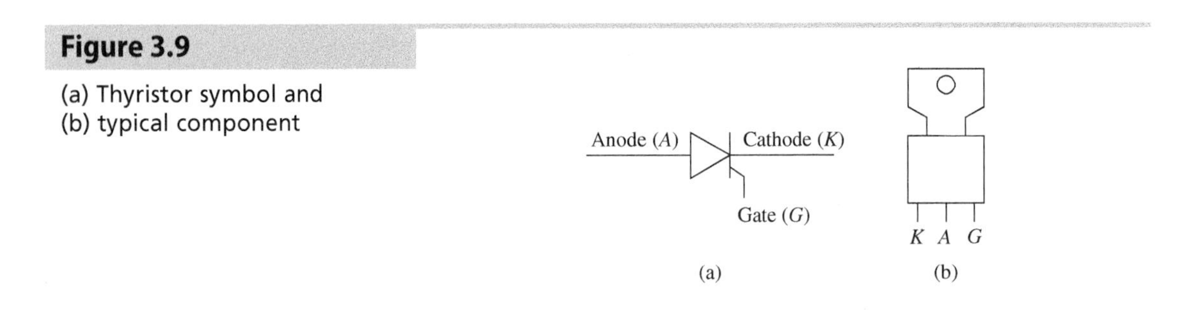

Figure 3.9
(a) Thyristor symbol and
(b) typical component

The current–voltage characteristics of a thyristor are shown in Figure 3.10. When no current is applied to the gate (off state), the current flow between the anode and the cathode in the thyristor is negligible for voltages greater than V_R and less than V_F. Note that the forward voltage (V_F) of a thyristor is quite large (from 50 up to several hundred volts), unlike that of a regular diode. When a small current (mA range) is applied to the gate, the thyristor conducts if the voltage applied to it causes it to be forward-biased. When the thyristor is conducting (on state), the forward voltage across the thyristor is small (1 to 2 V), and the thyristor current can be in several ampere ranges. Note that if the current to the gate is cut off, the thyristor continues to conduct as long as the voltage applied to it causes it to be forward-biased. The thyristor is turned off only when the current between the anode and the cathode drops below a certain value called the **holding current** (I_H). The **gate current** (I_{GT}) that causes the thyristor to conduct is small and is typically a few milliamps or less. For example, for the 2N6401 SCR, V_F is 100 V when not conducting and 1.7 V when conducting, I_{GT} is about 30 mA, and I_H is about 20 mA. The ability of the thyristor to remain on even though the gate current is switched off is called latching.

Figure 3.10

Current–voltage relationship for a thyristor

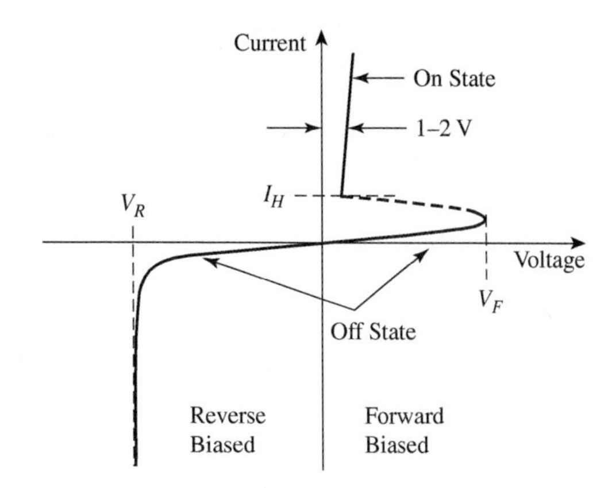

Thyristors are commonly used in power control applications to control heaters, dimming switches, and motors. They are particularly useful in controlling the current from an AC source to control the speed of AC-driven motors, such as the Universal Motor. Figure 3.11 shows such a circuit used for this purpose.

Figure 3.11

Half-wave variable-resistance phase-control circuit

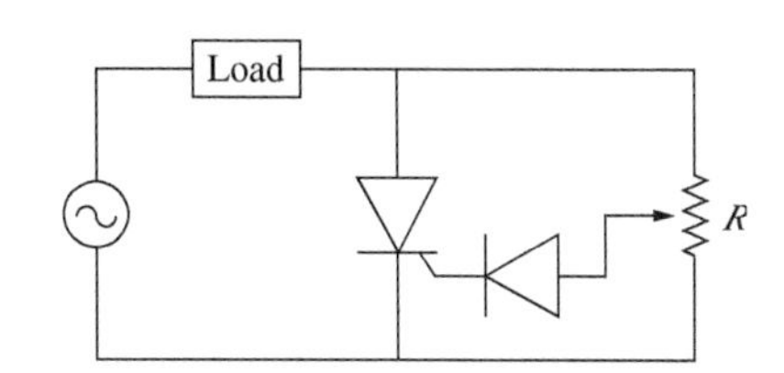

This circuit is called a **half-wave variable-resistance phase-control circuit**. Because a thyristor is unidirectional, the circuit only affects the positive portion of the AC signal or half-wave. When the voltage crosses the zero mark, the thyristor stops conducting. As seen in Figure 3.11, the gate current to the thyristor is controlled by a potentiometer through a diode. The diode is used to prevent the negative half of the AC signal from affecting the gate. Changing the resistance of the potentiometer causes the gate current to change. Since the applied voltage is sinusoidal (not constant), this causes the gate to trigger at different times with respect to the AC signal (hence the name phase control). If the triggering occurs at the beginning of the positive half of the voltage signal, then all of the current is passed, as shown in Figure 3.12(b). If the triggering occurs at a later time, then only a portion of the current is passed, as shown in Figure 3.12(c). Since the duration of the current passed affects the speed or delivered power to the controlled device, the thyristor offers a simple way to control the power.

Figure 3.12

Current output of a half-wave variable-resistance phase-control circuit

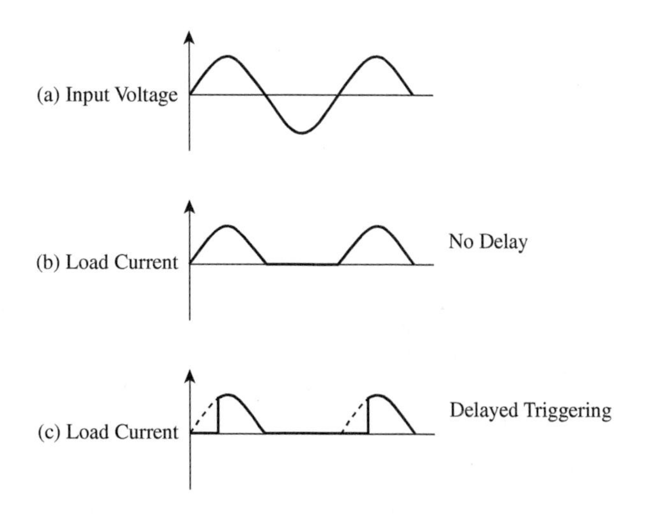

Another application of thyristors is in overvoltage protection circuits. The circuit shown in Figure 3.13 is commonly referred to as a **crowbar circuit**. It is used to protect from power surges or power-supply malfunction problems. It uses a Zener diode in combination with a thyristor. The Zener voltage should be selected to be higher than the nominal supply voltage. When the Zener diode switches on due to overvoltage conditions, this causes a current to flow to the gate of the thyristor. The thyristor will conduct, allowing a large current to flow, thus acting as a short circuit and blowing the circuit fuse. The capacitor is added to prevent the thyristor from triggering when powered up. Because of component tolerances, this circuit will operate reliably only if the overvoltage is 30 to 40% higher than the nominal voltage.

Figure 3.13

Crowbar circuit for overvoltage protection

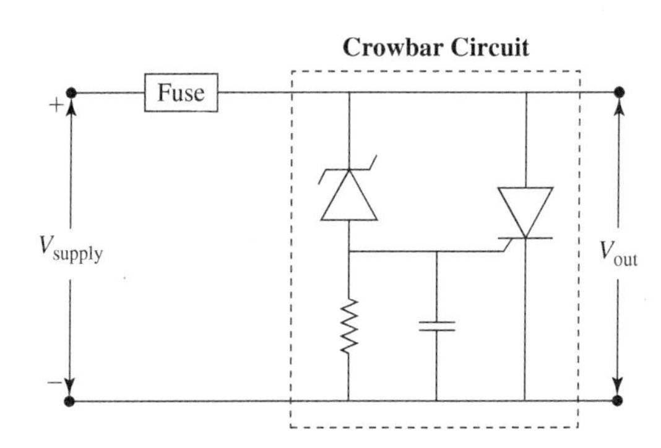

3.4 Bipolar Junction Transistor

A transistor is a solid-state switch that opens or closes a circuit. Unlike an electromechanical relay, the switching action in a transistor is caused by non-mechanical motion and is due to the change in the electrical characteristics of the device. A transistor is a three-terminal device. One terminal is used as the control input, another is connected to the load voltage, and the third is connected to the ground or a constant potential. There are several types of transistors available: the bipolar junction transistor (BJT), the insulated gate bipolar transistor (IGBT), the field effect transistor (FET), and the metal-oxide-semiconductor field effect transistor (MOSFET). We will discuss the BJT and the MOSFET, which are widely used. We will start by discussing the BJT (▶ available—V3.5). The BJT comes in two types: *npn* and *pnp*, which refer to the arrangement of *n*-type (negative) and *p*-type (positive) semiconductors in the transistor's construction. In semiconductors, ***n*-type materials** have extra electrons, and thus the term *n* for negative because electrons have a negative charge, while ***p*-type materials** have electrons removed, creating 'holes' (or absence of electrons), and thus the term *p* for positive. These properties are achieved through a process called '**doping**,' where impurities are introduced into the material. Transistors are built by stacking three different layers of semiconductor material together. Transistors can be created by stacking an *n*-type layer on top of a *p*-type layer, which is then followed by another *n*-type layer, or by stacking a *p*-type layer on top of an *n*-type layer, which is then followed by another *p*-type layer. These configurations determine whether the BJT is *npn* or *pnp*. We will limit the discussion to the *npn* configuration, which is more widely used. A schematic of an *npn* BJT is shown in Figure 3.14(a) while Figure 3.14(b) shows the stacking of the three semiconductor layers. The junctions between *n* and *p* layers are similar to the junctions in diodes and they can be forward-biased or reverse-biased as well. The terminals of the BJT are labeled as the emitter (*E*), base (*B*), and collector (*C*).

Figure 3.14

(a) Schematic of an *npn* bipolar junction transistor (BJT) and (b) *n* and *p* layers in the *npn* configuration

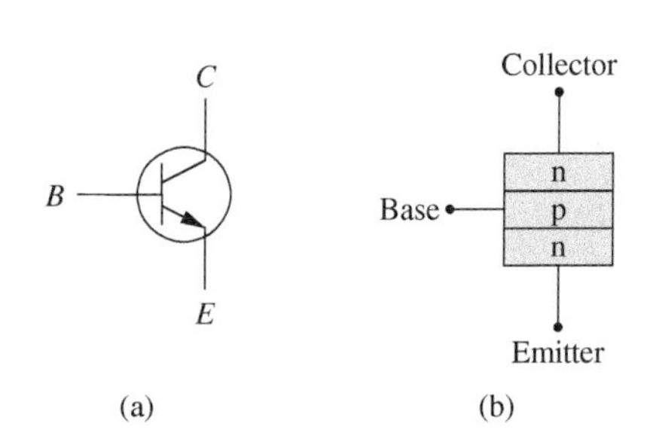

Let us label the voltages applied to the B, C, and E terminals as V_B, V_C, and V_E, respectively. Let us further define the following voltage differences:

$$V_{BE} = V_B - V_E$$

$$V_{CE} = V_C - V_E$$

Some general characteristics of a BJT are as follows.

- A BJT is an active device that requires power to operate.
- The BJT is a current-controlled device whose operation depends on the magnitude of the current supplied to the base.
- A small base current allows a much larger current to flow between the collector and the emitter.
- The BJT has three states of operation. These include the off or non-conducting state, the linear state, and the saturation state. These states of operation are determined by the magnitude of the V_{BE} and V_{CE} voltage. The former is set by the current supplied to the base.
- The voltage at the emitter (V_E) is always lower than the voltage at the base (V_B) by about 0.6 V.
- The collector voltage (V_C) has to be more positive than the emitter voltage (V_E).
- If AC voltages are applied to the base input, then a DC offset voltage (called a bias voltage) needs to be added in series to the AC voltage to enable the transistor to be controlled by both the positive and negative parts of the AC signal.

Two of the most common standard BJT circuits are called the transistor switch (or common emitter circuit) and the emitter follower circuit (or common collector circuit). These circuits are discussed next.

3.4.1 Transistor Switch Circuit

The **transistor switch** or **common emitter circuit** is shown in Figure 3.15. In this circuit, V_{in} is the control voltage, V_{out} is the output voltage, and V_{CC} is the supply voltage. This circuit is also called the *common emitter circuit* because both the emitter and the supply voltage ground are connected to the same common point. In this circuit, a resistor (R_C) is always placed between the supply voltage lead and the collector. In practice, this resistor represents the resistance of a load (such as a light or motor) that needs to be switched on and off, hence, the name of this circuit.

Figure 3.15

Common emitter circuit

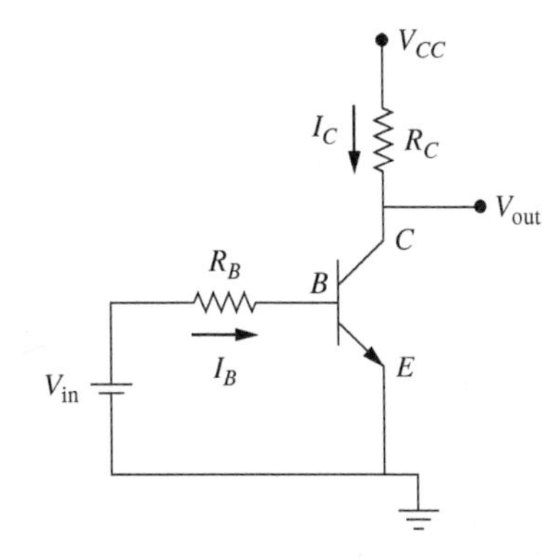

The **transfer** and **output characteristics** of a BJT are shown in Figure 3.16. In Figure 3.16(a), the collector current (I_C) is plotted against V_{BE}. The figure shows that the collector current (I_C) is zero unless V_{BE} exceeds 0.6 V, at which point I_C starts increasing. Figure 3.16(b) shows the relationship between I_C and V_{CE} as a function of the base current (I_B). The figure shows that away from the vertical axis or in the linear region, the collector current is mainly a function of the base current and does not change appreciably with an increase in V_{CE}. This region is called the *active* or *linear* region. Close to the vertical axis or in the saturation region, I_C is a function of both V_{CE} and I_B.

As mentioned before, a BJT has three states of operation. When $V_{BE} < \sim 0.6$ V, the transistor is said to be in the **off state** (non-conducting state). In this state, no current flows between the collector and the emitter, so $I_C = 0$. The V_{out} voltage will be the same as the V_{CC} voltage because no current flows between V_{CC} and V_C.

Figure 3.16

(a) Transfer and (b) output characteristics of a BJT

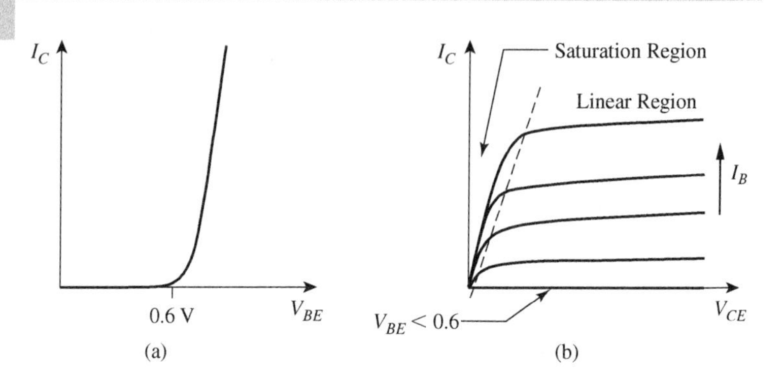

When V_{BE} is $>= \sim 0.6$ V and $< \sim 0.7$ V and $V_{CE} > 0.2$ V, the transistor is in the **linear operation or active state**. In the linear operation state, the collector current (I_C) is linearly related to the base current (I_B) by the following relationship:

(3.1)
$$I_C = \beta\, I_B = h_{FE}\, I_B$$

Where β or h_{FE} is the current gain. The current gain is not constant, and its value is dependent on the current (I_C) as well as V_{CE} and temperature. A typical value is 100, but it could range from 50 to 1000 due to variations in the manufacturing process.

When V_{BE} is $>= \sim 0.7$ V, the transistor is in the **saturation state**. In the saturation state, current flows between the collector and the emitter, and V_{CE} has a value of ~ 0.2 V. V_{out}, in this case, will be the same as V_{CE}, and the output voltage will switch from V_{CC} to ~ 0.2 V when the transistor switches from the off state to the saturation state.

In the transistor switch circuit, the transistor is normally designed to operate in either the off state or the on (saturation) state, but not in the linear state. The question is then what is the minimum V_{in} voltage needed to cause the transistor to saturate? By referencing Figure 3.15 and using KVL, we get

(3.2)
$$V_{in} = I_B\, R_B + V_{BE}$$

and just before saturation, I_B is related to I_C by

(3.3)
$$I_B = I_C/\beta$$

where I_C is determined from

(3.4)
$$I_C = (V_{CC} - V_{CE})/R_C$$

These equations can be solved to find V_{in} to cause saturation. Example 3.3 illustrates these calculations with data from a widely used small transistor, the 2N3904 (see website for complete data sheet). Example 3.4 illustrates the computation of the currents and voltages in this circuit while operating in the linear operation state.

Example 3.3 Voltage Saturation Calculations for the 2N3904 Transistor

Using the data sheet for the 2N3904 transistor and with reference to Figure 3.15, determine the input voltage needed to cause the transistor to saturate. What is the output voltage (V_{out}) of this circuit during the saturation and the off states if $V_{CC} = 10$ V? Let R_C be 1 kΩ, and R_B be 5 kΩ. Also, determine the power output of this transistor.

Solution:

From the data sheet, $V_{CE} = 0.2$ V at saturation for $I_C = 10$ mA. From Equation (3.4), I_C is given by

$$I_C = (10 - 0.2)/1000 = 9.8 \text{ mA}$$

which is close to the assumed value for I_C. Notice that I_C has to be smaller than the 200 mA limit for the collector current, which is satisfied in this case. Also from the data sheet, β or h_{FE} is 100 for $I_C = 10$ mA, and V_{BE} is greater than 0.65 V at saturation. We set $V_{BE} = 0.7$ V. From Equation (3.3), I_B just before saturation is given by

$$I_B = 9.8/100 = 0.098 \text{ mA}$$

Now plugging this value into Equation (3.2) gives

$$V_{in} = 0.098 \times 5 + 0.70 = 1.19 \text{ V}$$

To ensure saturation, V_{in} has to be greater than 1.19 V. This can be achieved easily if we let V_{in} be 2 V for example.

When this transistor is off, V_{in} has to be less than V_{BE} when the transistor just turns on (less than 0.6 V). In this case, V_{out} will be equal to V_{CC} (10 V). When the transistor is in saturation, $V_{out} = V_{CE} = 0.2$ V.

The output power of this transistor is the power that is dissipated by the load, which in this case is the R_C resistor. Power is then computed from $I_C^2 R$:

$$\text{Power} = (9.8 \times 10^{-3})^2\, 1000 = 9.6 \times 10^{-2} \text{ W}$$

Example 3.4 Current and Voltage Calculations in Linear Operation State

With reference to Figure 3.17, determine I_B, I_C, and V_{CE} assuming that the BJT transistor is operating in the linear operation state. Let $V_{in} = 3$ V, $V_{CC} = 8$ V, $R_C = 200\ \Omega$, $R_B = 10$ kΩ, and $\beta = 80$. Assume that V_{BE} is 0.65 V in the linear operation state.

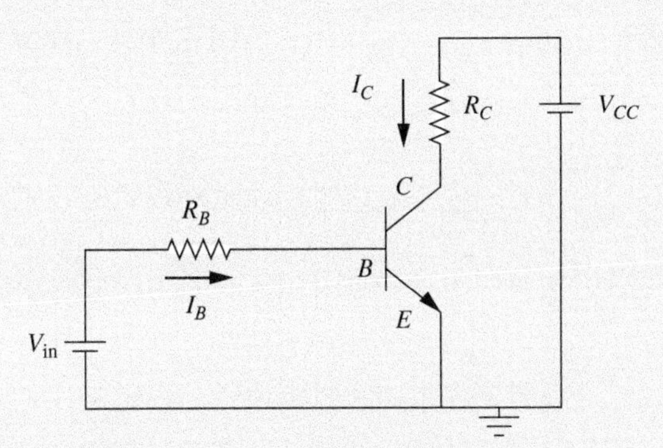

Figure 3.17 Circuit for Example 3.4

Solution:

Applying KVL to the input loop, we get

$$3 - I_B \times 10 \times 10^3 - V_{BE} = 0 \tag{1}$$

Since V_{BE} is 0.65 V, Equation (1) gives I_B as 0.235 mA.

But I_C is linearly related to I_B during the linear operation state, and thus I_C can be computed as

$$I_C = I_B \times \beta = 0.235 \times 10^{-3} \times 80 = 18.8 \text{ mA}$$

Now applying KVL to the output loop gives

$$V_{CC} - I_C \times R_C - V_{CE} = 0 \tag{2}$$

Solving Equation (2) for V_{CE} gives

$$V_{CE} = 8 - 18.8 \times 10^{-3} \times 200 = 4.24 \text{ V}$$

V_{CE} being much larger than 0.2 V means that this transistor is not in saturation as was assumed. Problem P3.13 determines the voltage input V_{in} needed to saturate this transistor.

3.4.2 Emitter Follower Circuit

The **emitter follower circuit**, also known as the **common collector circuit**, is shown in Figure 3.18. In this configuration, the input voltage signal is still applied to the base terminal as before. However, it is important to note that the output is connected to the emitter terminal, and there is no resistor between V_{CC} and the collector terminal. The circuit is termed the 'emitter follower' because the output voltage closely follows the input voltage, typically with a small voltage difference of about 0.6 V due to the base-emitter voltage drop of the transistor. Furthermore, it is referred to as the 'common collector' circuit because the collector terminal is shared or common to both the input and output circuits. The collector terminal is connected to the power supply (V_{CC}) and acts as a reference point, which is often considered 'grounded' or 'earthed' through the power supply connection. The emitter follower circuit offers certain advantages, including a high input impedance, low output impedance, and a voltage gain that is slightly less than unity. It is commonly used in various applications, such as impedance matching, voltage buffering, and signal amplification with minimal signal inversion.

Figure 3.18

Emitter follower circuit

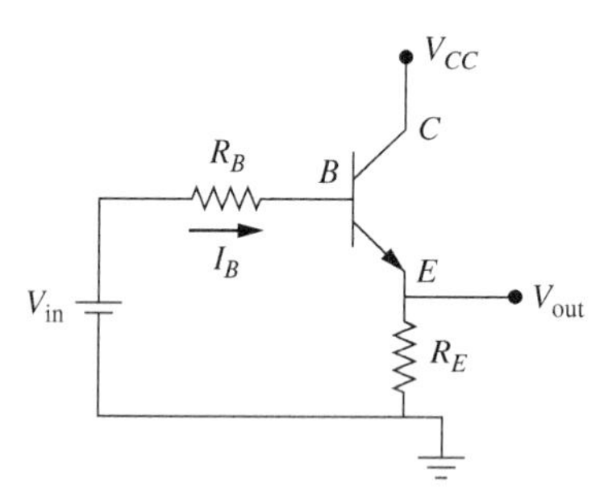

To analyze this circuit, assume first that there is no resistor R_B in this circuit. Then $V_B = V_{in}$,

(3.5)
$$V_{out} = V_E = V_{in} - V_{BE} = V_{in} - 0.6 \text{ for } V_{in} > 0.6\text{ V}$$

and

(3.6)
$$V_{out} = 0 \text{ for } V_{in} < 0.6\text{ V}$$

Now if the resistor R_B was present, we need to account for the voltage drop across this resistor and V_{out} is then equal to

(3.7)
$$V_{out} = V_{in} - I_B R_B - 0.6 \text{ for } V_{in} > 0.6\text{ V}$$

But $I_E = I_C + I_B$, $I_E = V_{out}/R_E$, and $I_C = \beta I_B$ when the transistor is in the linear state. This gives

(3.8)
$$I_B = \frac{V_{out}}{(1 + \beta)R_E}$$

and

(3.9)
$$I_E = I_B(1 + \beta)$$

Equation (3.9) shows the current gain of this circuit is $(\beta + 1)$. Substituting Equation (3.8) into Equation (3.7) and solving for V_{out}, we get

(3.10)
$$V_{out} = (V_{in} - 0.6)\frac{(1 + \beta)R_E}{R_B + (1 + \beta)R_E}$$

Equation (3.10) shows that the output voltage is linearly related to the input voltage and is independent of the supply voltage V_{CC}. The output voltage is also in phase with the input voltage, and the voltage gain is slightly less than 1 as seen from Equation (3.10). Equations (3.8) through (3.10) apply as long as the transistor is not in saturation. When the transistor saturates, V_{out} is equal to $V_{CC} - 0.2$ because V_{CE} is about 0.2 volts at saturation. Example 3.5 illustrates the voltage and current calculations for the emitter-follower circuit, while Example 3.6 further illustrates the voltage calculations for a BJT.

BJT transistors have certain parameters that should not be exceeded. These parameters include maximum collector current and power dissipation capability. There is a list of these parameters in Table 3.1 for three common *npn* transistors. The TIP102 transistor is called a **Darlington transistor** and consists of two cascaded BJT transistors to amplify the collector current (see Figure 3.19). Note that the power dissipation capability of a transistor is dependent on the environment temperature. In Table 3.1, the power is listed for an air temperature of 25°C. The power dissipation decreases with increasing temperature.

Table 3.1

Characteristics of common *npn* BJT transistors

Part #	Max V_{CE}	Max V_{BE}	Max I_C	β	Power Dissipation at $T_A = 25°C$
2N3904	40 V	6 V	200 mA	30–300	0.625 W
TIP29	40 V	5 V	1 A	15–75	30 W
TIP102	100 V	5 V	8 A	200–20000	80 W

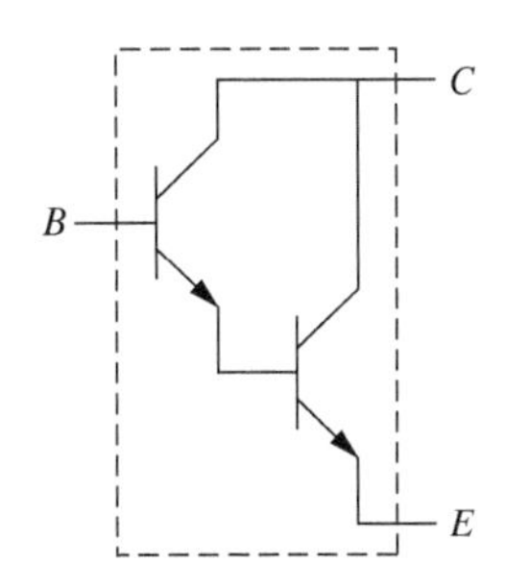

Figure 3.19

Schematic of a Darlington transistor

In a BJT, the power that is seen by the load is computed from the product of the square of the collector current and the load resistance. BJTs are typically used for low-power applications. Note that while a transistor has a very fast switching time in the order of 10's of nanoseconds (ns) or less, there is no electrical isolation between the base circuit and the load circuit unlike that of an electro-mechanical relay. In a relay, the coil circuit and the contact circuits are electrically isolated, so they are better suited to handle noise.

Example 3.5 Voltage and Current Calculations for the Emitter Follower Circuit

For the emitter follower circuit shown in Figure 3.18, compute the output voltage V_{out}, the base current I_B, and the output current I_E. Let $V_{in} = 10$ V, $V_{CC} = 8$ V, $R_E = 1$ kΩ, $R_B = 5$ kΩ, and $\beta = 100$.

Solution:

From Equation (3.10), the output voltage is

$$V_{out} = (10 - 0.6)\frac{(1 + 100)10^3}{5 \times 10^3 + (1 + 100)10^3} = 9.4 \times 0.95 = 8.9 \text{ V}$$

Using Equation (3.8), the base current I_B is

$$I_B = \frac{8.9}{(1 + 100)10^3} = 88\ \mu A$$

and the output current I_E is $88 \times 10^{-6}(1 + 100) = 8.9$ mA using Equation (3.9).

If R_E increases, then the output voltage will increase and will follow the input voltage minus a small voltage difference of about 0.6 V for large values of R_E. For example, if R_E is 10 kΩ instead of 1 kΩ, then $V_{out} = \approx 9.4$ V. For this case I_B will be 9.3 μA and I_E will be 0.94 mA. Increasing the resistance R_E by a factor of 10 results in a decrease in I_B and I_E by almost the same factor.

Example 3.6 Analysis of a BJT Circuit

Determine the voltages at points 1 and 2 in the circuit shown in Figure 3.20 for (a) $V_{in} = 0.1$ V and (b) $V_{in} = 3$ V. Let $R_C = 1$ kΩ, $R_B = R_E = 100$ Ω, and $V_{CC} = 10$ V.

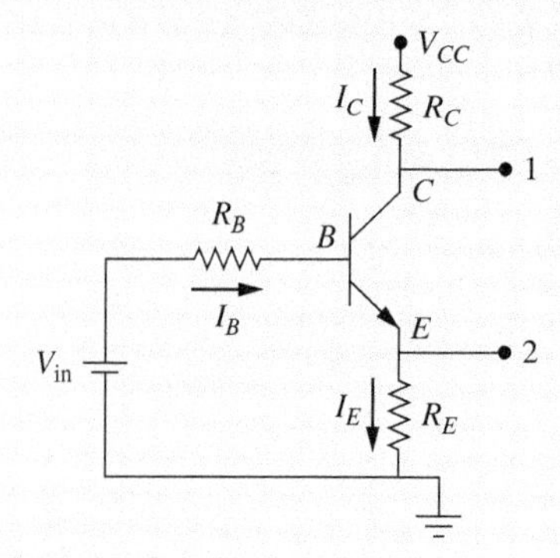

Figure 3.20 Circuit for Example 3.6

Solution:

(a) For $V_{in} = 0.1$ V, the transistor is off because V_{in} has to be larger than 0.6 V to cause the transistor to start conducting. So $V_1 = V_{CC} = 10$ V since the current I_C is zero. V_2 is equal to zero because V_{BE} is less than 0.6 V.

(b) For $V_{in} = 3$ V, the transistor is either operating in the linear range or saturated. We will assume that the transistor is just at the point of being saturated, and we will check this assumption by comparing the currents I_C and I_B.

Applying KVL to the V_{CC} loop gives

$$V_{CC} = I_C R_C + V_{CE} + I_E R_E \quad (1)$$

Similarly, applying KVL to the V_{in} loop gives

$$V_{in} = I_B R_B + V_{BE} + I_E R_E \quad (2)$$

Noting that $I_E = I_C + I_B$, using the given values for R_B, R_C, R_E, V_{in} and V_{CC}, and assuming $V_{CE} = 0.2$ V and $V_{BE} = 0.7$ V at saturation, Equations (1) and (2) give

$$10 = 1100\ I_C + 0.2 + 100\ I_B \quad (3)$$

$$3 = 100\ I_C + 0.7 + 200\ I_B \quad (4)$$

Solving Equations (3) and (4) for I_B and I_C, we get $I_C = 8.24$ mA, and $I_B = 7.36$ mA. For $I_C = 10$ mA, the current gain is about 100 if the transistor is operating in the linear range. Since $I_C \neq \beta\ I_B$, the transistor is in saturation and the assumption is correct. This gives $V_1 = 1.76$ V and $V_2 = 1.56$ V.

3.4.3 Open-Collector Output

Many sensors used in mechatronic applications such as proximity sensors (see Section 8.4) have electronic circuits that use an internal BJT transistor as an interface. The output of the sensor electronic circuit drives the base of the transistor. These circuits are normally known as **open-collector output** voltage circuits. To get an output from these sensors, an appropriate 'pull-up' resistor or load, and the supply voltage needs to be applied to the terminals of the sensor. Figure 3.21 shows typical wiring for such a sensor. In this example, a positive voltage needs to be applied to terminal 1 and a 'pull-up' resistor needs to be connected between terminals 1 and 2. When the proximity sensor is OFF, the transistor is not in saturation, and there will be no voltage drop across the load resistor, since the collector terminal (2) is open with respect to the emitter terminal (3). The output in this case will be 'pulled up' by the load resistor to the value of the external supply voltage V. When an object is detected by the proximity sensor, the transistor conducts and a voltage drop develops across the load resistor, resulting in the output voltage changing from the value of the supply voltage to almost zero. Example 3.7 illustrates an open-collector circuit.

Figure 3.21

Typical circuit for an *npn* type non-contact, capacitive-type proximity sensor

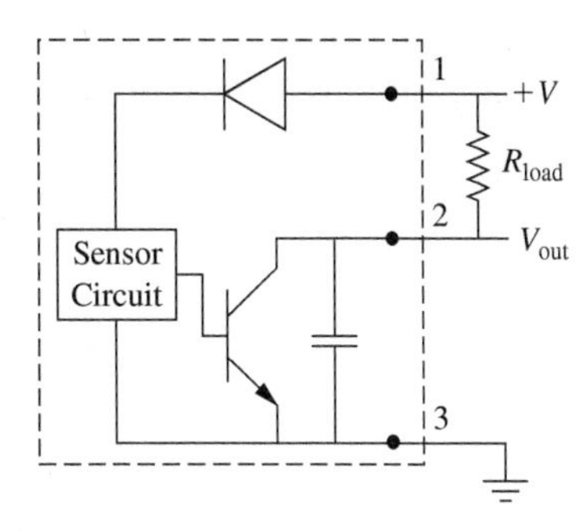

Example 3.7 Open-Collector Calculation

A proximity sensor with an output circuit similar to that shown in Figure 3.21 requires a 24-VDC supply voltage. Select a 'pull-up' resistor and a wiring scheme for this sensor so the voltage output of the sensor can be read by a digital input port with 0 and 5 V logic levels. The maximum current through the pull up load resistor should be limited to 100 mA.

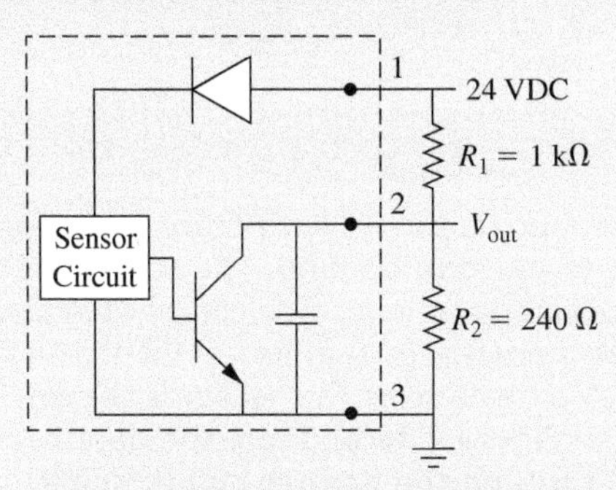

Figure 3.22 Wiring for Example 3.7

Solution:

The wiring scheme to perform this function is shown in Figure 3.22. Since the output voltage should be limited to 5 V when the sensor is off, we need to use a voltage dividing circuit in this setup. The resistors R_1 and R_2 are selected to have a resistance of 1 kΩ and 240 Ω respectively to achieve this. Note that these resistance values are standard (see Appendix A). When the sensor is off, V_{out} will be

$$24 - 1000\frac{24}{1240} = 4.65\ \text{V}$$

which is well within the voltage threshold for the high logic (see Table 3.12). When the sensor is on, V_{out} is about 0.2 V. The maximum current through the load resistor occurs when the sensor is on. The current in this case is

$$\frac{24 - 0.2}{1000} = 23.8\ \text{mA}$$

This is well within the desired specification.

3.4.4 Phototransistor, Photo Interrupter, and Optoisolator

Instead of using a voltage source to saturate the transistor, a **phototransistor** (see Figure 3.23(a)) uses light that is received by a photodiode to do the same thing. Typically, a phototransistor and an LED are packaged together to make optical sensors that can be used to detect the presence of objects. In these sensors, which are commonly referred to as **photo interrupters** (see Figure 3.23(b)), the LED provides a light source that is received by the phototransistor. An interruption of the light received by the phototransistor causes the phototransistor to change state, thus indicating the presence of an object in the path between the LED and the phototransistor.

Figure 3.23

(a) Phototransistor and
(b) photo interrupter

(Courtesy of ROHM Semiconductor, USA, San Diego, CA)

(a) (b)

Case Study I.A discusses the use of a photo interrupter sensor as a homing sensor for the Stepper Motor-Driven Rotary Table Case Study.

Integrated Case Study I.A: Photo Interrupter as a Homing Sensor for the Table

For Integrated Case Study I (Stepper Motor-Driven Rotary Table—see Section 11.2), the table uses a photo interrupter as a sensor to perform homing. The sensor (H21A1 from Fairchild), which consists of an infrared emitting diode coupled with a silicon phototransistor, detects the presence of a notch that is made on the outer diameter of the optical CD that is mounted on the stepper-motor shaft, as shown in Figure 11.1. A schematic of the sensor is shown in Figure 3.24(a) and the circuit for wiring this sensor is shown in Figure 3.24(b). From the datasheet for this sensor, the maximum continuous forward current (I_F) through the emitter LED is 50 mA. Using a 5-volt signal connected to pin 1 requires that at least a 100-ohm resistor should be added. A 220-ohm resistor was used in this circuit, which limits the maximum current to 23 mA. For the output circuit, the maximum collector current I_C is 20 mA. Connecting 5 volts to pin 3 through a 330-ohm resistor will result in a maximum collector current of 15 mA ((5 − 0.2)/330). The output from this sensor is connected to one of the digital pins on the Arduino Uno board (see Section 4.5). When the optical disk obstructs the infrared light sent to the phototransistor, the phototransistor is not conducting, and the sensor output will be the same as the voltage connected to the collector, or 5 VDC in this case. When the notch in the CD is aligned with the opening in the sensor or the table is at the 'home position,' the voltage output should drop to the saturation voltage (about 0.2 V but the actual voltage was measured to be about 1.6 V) which is read as a low logic in the Arduino. Note that a similar sensor is used in Integrated Case Study II as a limit switch to shut off the power to the system in case of accidental overtravel of the slide.

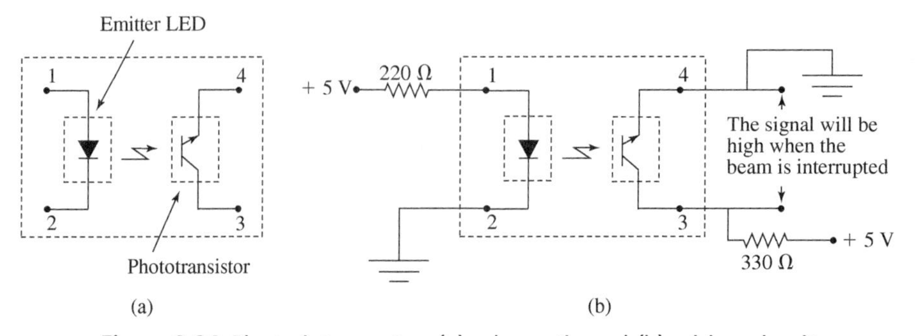

Figure 3.24 Photo interrupter: (a) schematic and (b) wiring circuit

An **optoisolator** or an **optocoupler** combines two elements (a light-emitting device such as a diode and a light-sensitive device) similar to a photo interrupter but in an enclosed package. An optoisolator is also designed for a different purpose, which is to provide an optical coupling between the input and the output sides. The light emitter on the input side takes a voltage signal and converts it into a light signal. On the output side, the light-sensitive device detects the light from the emitter and converts it back to a voltage signal. The light-sensitive device could be a phototransistor, a photodiode, or a thyristor. This optical coupling provides electrical noise isolation between the input and the output sides. To take advantage of this isolation, a separate power supply should be used for the input and output sides. Opto-isolators are used to prevent voltage spikes on one side of the device to damage or affect components on the other side, so they provide a safe electrical coupling between one circuit section and another. Opto-isolators are available with isolation of 5 kV or more between the input and output sides and they are typically used to protect data acquisition systems (see Chapter 5) from damage caused by accidental contact with high external voltages. Figure 3.25 shows a commercial 16-channel optoisolated digital input board (PCLD-782) made by Advantech.

Figure 3.25

Commercial data acquisition card with optoisolated inputs

(Courtesy of Advantech PLC)

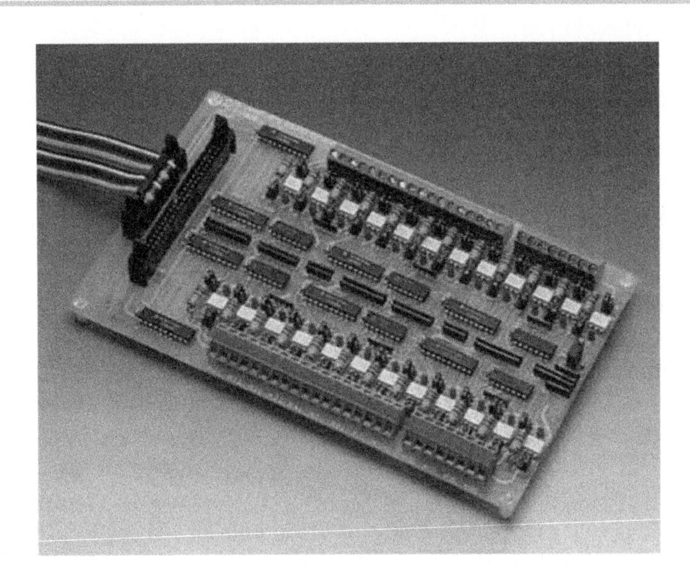

3.5 Metal-Oxide-Semiconductor Field Effect Transistor

Metal-oxide semiconductor field effect transistors (MOSFETs) are the other family of transistors that are common. The MOSFET (▶ available—V3.6) is based on the original field effect transistor (FET) that was introduced in the 1960s. Similar to the BJT family, they are also three-terminal devices, but they have different names for the terminals, and they operate differently. The three terminals are the gate (similar to the base), drain (similar to the collector), and source (similar to the emitter). The naming of the terminals comes from the flow of electrons between the source and drain when the transistor is conducting. The most commonly used MOSFET is the enhanced type and is available as *n*- or *p*-type. We will limit the discussion to the *n*-type enhanced MOSFET here. Figure 3.26 shows the symbol of the *n*-type MOSFET.

Figure 3.26

Symbol of an *n*-type MOSFET

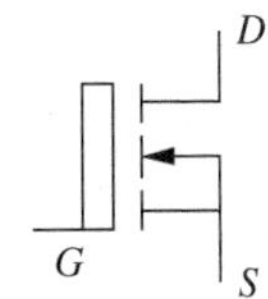

MOSFETs have the following characteristics.

- The voltage applied to the gate (or the electric field) is the signal that controls the operation of the transistor and hence the name field effect transistor. This is in contrast to a BJT where the current applied to the base controls its operation.
- The gate is isolated from the drain-source circuit. This is indicated in the symbol by the separation of the gate terminal from the drain-source connection. The gate has very high internal resistance ($R_{Gate} = \sim 10^{14}\ \Omega$), such that almost no current flows into the gate. This isolation makes it easy to analyze MOSFET circuits because the gate circuit can be analyzed separately from the drain-source circuit. In addition, the high input resistance means that the gate draws no current except for a small leakage current in the nanoampere range.
- The high-input impedance of a MOSFET gives it an advantage in interfacing with other logic circuits.
- MOSFETs have three states: cutoff, active, and saturation, which are similar to BJTs.
- They act as voltage-controlled resistors. When the transistor is OFF, the drain-source resistance is very high, and when the transistor is fully ON, the drain-source resistance is very low (can be less than 1 Ω). When the transistor is ON, current flows from the drain to the source (electrons travel in the opposite direction).
- *n*-type enhancement MOSFETs operate with a positive voltage applied to the gate.
- MOSFETs have a higher power rating and generate less heat than BJTs.

The **transfer** and **output characteristics** of a MOSFET are shown in Figure 3.27. Figure 3.27(a) shows the relationship between the drain current (I_D) and the voltage between the gate and the source (V_{GS}). The figure shows that the drain current is zero until V_{GS} exceeds the threshold voltage (V_T) for the MOSFET. When $V_{GS} < V_T$, the MOSFET is said to be in the **cutoff or non-conducting state**. The threshold voltage for normal MOSFETs (such as 2N4351) is between 2 and 5 V, while for logic-level MOSFETs (ones that are designed to be driven directly from outputs of logic gates, such as microcontrollers) the threshold voltage is about 0.3 to 1.0 V. As V_{GS} increases above V_T, the drain current increases. Figure 3.27(b) shows the relationship between the drain current and the voltage between the source and the drain (V_{DS}) for different values of V_{GS}. For a given $V_{GS} > V_T$, the drain current increases with V_{DS} for a small V_{DS}. This is called the **active** or **ohmic region**, where the MOSFET acts like a variable resistor whose resistance is controlled by V_{GS}. However, as V_{DS} increases, the drain current levels off and stays constant. This is the **saturation region**, where the drain current value is independent of V_{DS}. When V_{GS} is significantly higher than the threshold voltage (approximately 10 V for a normal MOSFET), the transistor is said to be in the fully ON state where the drain current is maximum.

Figure 3.27

(a) Transfer and (b) output characteristics of a MOSFET

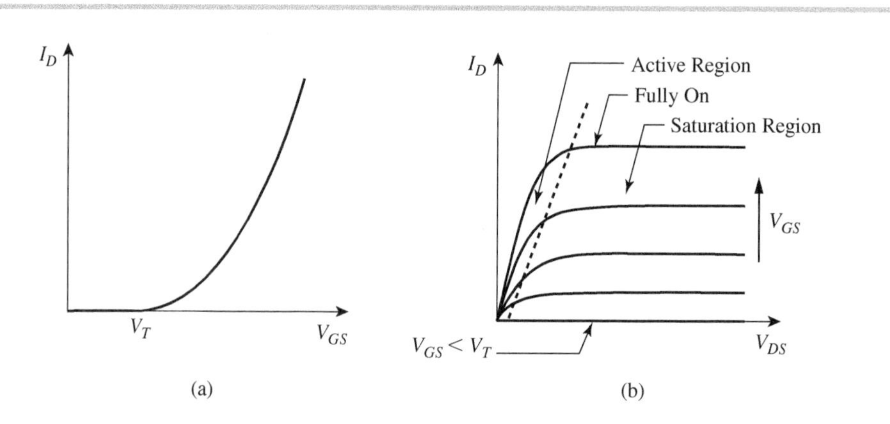

MOSFETs are typically used for switching applications (ON/OFF) to drive motors or LEDs. A typical circuit is shown in Figure 3.28. When the transistor is OFF, no current flows from the drain to the source, and the motor is OFF. When the transistor is fully ON, current flows to the motor, and the motor is ON. Note the use of the flyback diode (see Section 3.2) in the circuit to protect the transistor from the large voltage build-up that occurs when the transistor is switched off. The resistor at the input is used to drive the gate input to the ground and completely turns OFF the transistor when the input voltage is zero.

Figure 3.28

MOSFET circuit for driving a motor

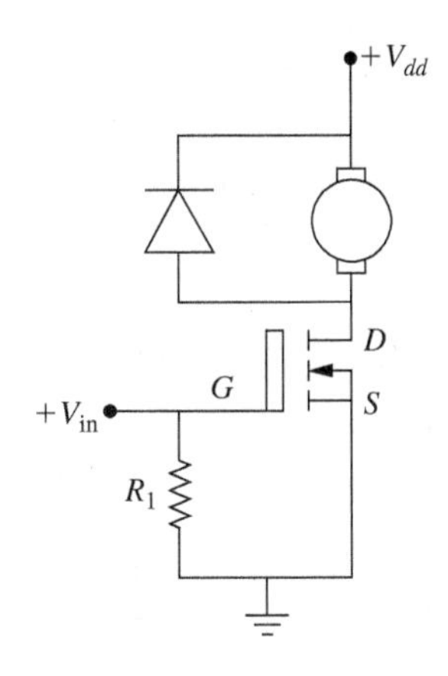

Some of the parameters of a logic-level MOSFET (NTE2980), normal MOSFET (2N4351), and a Power MOSFET (IRFZ14) are shown in Table 3.2.

Table 3.2

Parameters of selected MOSFETs

Part Number	On Resistance $R_{ds(ON)}$	Max. Drain Current $I_{D(max)}$	Power dissipation P_d	Gate Threshold Voltage $V_{GS(th)}$	Drain–to–Source Breakdown Voltage V_{dss}
2N4351	<= 300 Ω	100 mA	375 mW	1–5 V	>= 25 V
NTE2980	0.2–0.28	6.7 A (VGS = 5 V)	25 W	1–2 V	>= 60 V
IRFZ14	0.2 Ω	10 A	43 W	2–4 V	>= 60 V

These parameters are defined here.

$R_{ds(ON)}$ The resistance between the drain and source terminals when the MOSET is fully turned on

$I_{D(max)}$ The maximum current between the drain and source that can be passed by the transistor. It is a function of $R_{ds(ON)}$ and the package type of the transistor

P_d The power dissipation rating for the transistor

$V_{GS(th)}$ The minimum voltage between the gate and the source that causes the transistor to start conducting

V_{dss} The maximum voltage between the drain and source when the transistor is OFF

Similar to BJTs, the power-handling capacity of MOSFET transistors is a very important consideration in the selection of these components. When the power to be dissipated is above 1 W, the MOSFET is mounted on a heat sink. In these MOSFETs, the package has a metal tab that is mounted against the heat sink.

The dissipated power in a MOSFET is the product of the drain current and the voltage across the transistor. Since the voltage across the transistor is equal to the product of the current and the resistance, the power is then given by

$$P_d = I_D^2 R_{ds(ON)} \tag{3.11}$$

when the transistor is fully conducting. Case Study III.A discusses the use of a MOSFET to switch the power to a heater.

Before moving on to other types of circuits, it is important to clarify the terms V_{DD} and V_{CC}, which are common voltage supply designations used in electronic circuits, but they are typically associated with different types of circuits. $\boldsymbol{V_{DD}}$ or $\boldsymbol{V_{dd}}$ stands for 'drain voltage' or 'drain-to-source voltage' and is commonly used in the context of MOSFETs. It represents the supply voltage applied to the drain terminal of a MOSFET. V_{DD} is often used to designate the power applied to digital circuitry in ICs and microcontrollers. V_{CC} or V_{cc} stands for 'collector voltage' or 'collector-to-emitter voltage' and is primarily used in the context of BJTs. V_{CC} represents the voltage level applied at the collector terminal of the transistor and V_{CC} is typically used to designate the power applied in analog circuitry in ICs and other electronic devices. In some cases, the terms V_{DD} and V_{CC} are used interchangeably to refer to the positive power supply voltage in a circuit, regardless of the type of circuit.

Integrated Case Study III.A: Transistor for Controlling the Voltage

For Integrated Case Study III (Temperature-Controlled Heating System—see Section 11.4), the microcontroller controls the voltage applied to the heating element which in turn controls the heat energy applied to the system. The heating element is a resistance heater that supplies a power of 10 W at an input voltage of 12 VDC and a current of 0.83 A. Both the current and voltage input levels are beyond the output limits of the PWM pins on the Arduino Uno board (see Section 4.5), so an interface circuit between the microcontroller and the heater using an IRFZ14 power transistor as the switching element is used. The interface circuit is like Figure 3.28 but without the use of the flyback diode across the leads of the heating element since the heating element is a resistive type. The heating element is connected to the drain line of the transistor. The PWM output of the MCU is connected to the gate input of the IRFZ14 transistor to modulate the 12-volt external supply to the heater. For a resistive-type element, the direction of the voltage does not affect its operation, so there is no need to use an H-bridge circuit in this application. The IRFZ14 has a 10 A maximum drain-current rating and a power rating of 43 W, which is more than sufficient to drive the 10 W heater. The R_1 resistor is a 1 kΩ resistor that is used to drive the gate input to the ground and completely turns OFF the transistor when the input voltage is zero. Since this transistor is a CMOS type, inputs should not be left floating.

3.6 Combinational Logic Circuits

The invention of the transistor has led to the development of digital circuits (▶ available—V3.7) in which transistors form the building blocks. Digital logic circuits can be classified into two categories. These are combinational logic circuits and sequential logic circuits. In a **combinational logic circuit** (▶ available—V3.8), the output is not dependent on the history of the input, and the circuit uses rules of mathematical logic to generate the output. On the other hand, in a **sequential logic circuit**, signal history is important and determines the output of the system. We start with combinational logic circuits. Table 3.3 lists the basic combinational logic devices that are used along with their symbols, logic function expressions, and truth tables. A **truth table** gives the output logic for all combinations of the input logic. Note that a bar above a logic variable means the inverse of that variable.

Table 3.3

Basic combinational logic devices

Device	Symbol	Logic Function Expression	Truth Table
AND gate	A, B → C	$C = A \cdot B$	A B C 0 0 0 1 0 0 0 1 0 1 1 1
NAND gate	A, B → C	$C = \overline{A \cdot B}$	A B C 0 0 1 1 0 1 0 1 1 1 1 0

Table 3.3

(*Continued*)

Device	Symbol	Logic Function Expression	Truth Table
OR gate	A, B → C	$C = A + B$	A B C 0 0 0 1 0 1 0 1 1 1 1 1
NOR gate	A, B → C	$C = \overline{A + B}$	A B C 0 0 1 1 0 0 0 1 0 1 1 0
XOR gate	A, B → C	$C = A \oplus B$	A B C 0 0 0 1 0 1 0 1 1 1 1 0
Buffer	A → C	$C = A$	A C 0 0 1 1
Inverter	A → C	$C = \overline{A}$	A C 0 1 1 0

An example of a logic circuit that uses these devices is shown in Figure 3.29. The associated truth table and logic function are also shown.

Figure 3.29

An example of a combinational logic circuit

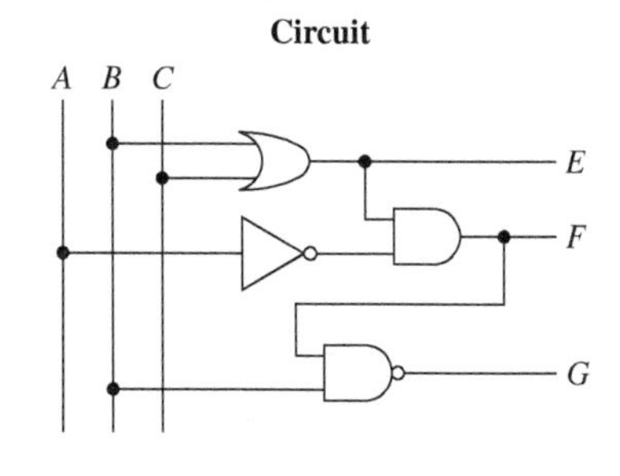

Truth Table

A	B	C	E	F	G
0	0	0	0	0	1
0	0	1	1	1	1
0	1	0	1	1	0
0	1	1	1	1	0
1	0	0	0	0	1
1	0	1	1	0	1
1	1	0	1	0	1
1	1	1	1	0	1

Logic Functions

$G = B + C$

$F = \overline{A} \cdot E = \overline{A} \cdot (B + C)$

$G = \overline{B \cdot F}$

The gates shown in Table 3.3 are available in packages. For example, Figure 3.30 shows the SN7402 package from Texas Instruments, which contains four independent two-input NOR gates. The A's and the B's are the gate's input, and the Y's are the gate's output. V_{CC} is the supply-voltage connection pin. Note that the semicircular notch at the top of the IC is used as a mark for the orientation of the device. Logic AND, NAND, OR, and NOR gates are also available with three and four inputs.

Figure 3.30

SN7402 package

(Courtesy of Texas Instruments, Dallas, TX)

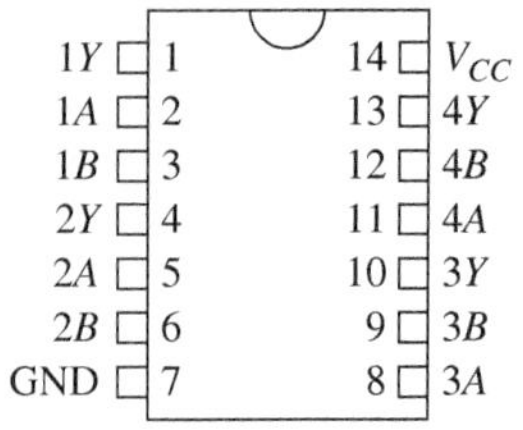

	Pin	Pin	
1Y	1	14	V_{CC}
1A	2	13	4Y
1B	3	12	4B
2Y	4	11	4A
2A	5	10	3Y
2B	6	9	3B
GND	7	8	3A

3.6.1 Boolean Algebra

Since digital circuits perform logic operations, we need to understand the logic rules that govern these operations. Here is a list of rules that can be used to simplify Boolean expressions.

1. $A + A = A$
2. $A + 1 = 1$
3. $A + 0 = A$
4. $A \cdot A = A$
5. $A + B = B + A$
6. $A \cdot (B + C) = A \cdot B + A \cdot C$
7. $A + (B \cdot C) = (A + B) \cdot (A + C)$
8. $A + \overline{A} = 1$
9. $A \cdot \overline{A} = 0$
10. $A \cdot 0 = 0$
11. $A \cdot 1 = A$

The following two rules are called De Morgan rules and are useful in converting between AND and OR gates:

12. $\overline{A + B + C + \ldots} = \overline{A} \cdot \overline{B} \cdot \overline{C} \ldots$
13. $\overline{A \cdot B \cdot C \ldots} = \overline{A} + \overline{B} + \overline{C}$

To illustrate these rules, consider the following example. Assume we are given a circuit, as shown in Figure 3.31.

Figure 3.31

Two-gate circuit

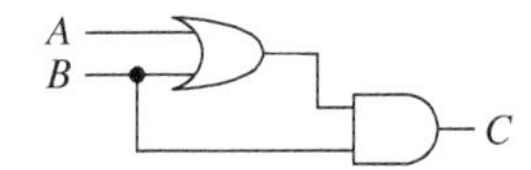

The output of this circuit in terms of the inputs is given by

$$C = (A + B) \cdot B$$

Using rule 6 and then rule 4, we can write the above expression as

$$C = A \cdot B + B \cdot B = A \cdot B + B$$

Also applying rule 11 and then rule 6, we get

$$C = A \cdot B + B = A \cdot B + B \cdot 1 = (A + 1) \cdot B$$

Since $A + 1 = 1$ by rule 2 and $(1 \cdot B) = B$ by rule 11, the output of this circuit is $C = B$. Thus, by using the above rules, we were able to convert a circuit that has two gates into one that requires no gate—just a wire between B and C.

As a further illustration of the above rules, consider Example 3.8.

Example 3.8 Boolean Logic Simplification

Simplify the following output function:

$$Q = (A \cdot C) + (B \cdot C) + (A \cdot \bar{B} \cdot C)$$

Solution:

Q can be written as

$$Q = (A + B) \cdot C + (A \cdot \bar{B}) \cdot C$$

by applying rule 6 to the first two expressions and factoring C from the third expression. Application of rule 6 again gives

$$Q = [(A + B) + (A \cdot \bar{B})] \cdot C$$

Factoring out A and then applying rules 2 and 11 gives

$$Q = [A \cdot (1 + \bar{B}) + B] \cdot C = (A \cdot 1 + B) \cdot C = (A + B) \cdot C$$

Thus, the original expression which would need one inverter gate, four two-input AND gates, and two two-input OR gates now can be realized using just one OR gate and one AND gate.

3.6.2 Boolean Function Generation from Truth Tables

In this section, we look at the problem of finding a logic gate system with the minimum number of gates that can be used to realize a logic circuit operation that is specified in terms of a truth table. An application of this would be the design of a circuit to process the output from several ON/OFF sensors. The basic approach is to manipulate the logic functions into one of two equivalent forms.

The sum of products form: $A \cdot B + A \cdot C$

Or

The product of sums form: $(A + B) \cdot (A + C)$

In the **sum of products form**, different combinations of the inputs are AND-ed together to form products, and these products then are OR-ed together to generate the output. In the **product of sums** form, different combinations of the inputs are OR-ed together to form sums, and sums then are AND-ed together to generate the output. The sum of products form is more commonly used. As an example, consider the following truth table (Table 3.4) that defines the output Q in terms of the inputs A and B. Using the sum of products form, we included in each row of the table the product form that generates the output in that row. If the input is low (or zero), then that input is shown with a bar above it in the product expression. Based on the table, this logic system is given by

$$Q = \bar{A} \cdot B$$

Note that we only considered rows that have an output of 1. Rows with zero output do not contribute to the final expression.

Table 3.4

Logic truth table

A	B	Output - Q	Products
0	0	0	$\bar{A} \cdot \bar{B}$
0	1	1	$\bar{A} \cdot B$
1	0	0	$A \cdot \bar{B}$
1	1	0	$A \cdot B$

In Table 3.5, this example is solved graphically where a table is constructed that has all of the input combinations. The output of the system is decided by the cells that have a non-zero output. In this case, there is only one non-zero cell, so the output is read as determined before as $Q = \overline{A} \cdot B$.

Table 3.5

Graphical representation of Table 3.4

	$\overline{B}$	B
$\overline{A}$	0	1
A	0	0

This graphical technique works very well for logic functions with many inputs and is part of a method using **Karnaugh maps** (K-maps). A Karnaugh map (▶ available—V3.9) is a graphical method that can be used to produce simplified Boolean expressions from sums of products obtained from truth tables. There will be a brief overview of Karnaugh maps in this text. For more detail, the reader should consult textbooks on digital logic design (see, for example, [9]). To illustrate the Karnaugh map approach, consider the truth table for a three-variable input problem that is shown in Table 3.6. We added a column to that table to show the product form for rows that have non-zero output. Unlike the truth table shown in Table 3.4, which has only one non-zero output, this truth table has five non-zero outputs. Trying to simplify these five products using the Boolean rules discussed earlier is not always straightforward. Instead, we will use the Karnaugh map approach for this example since it offers a simpler method to obtain the output.

Table 3.6

Three input variables truth table

ABC	Output	Products
000	0	
001	1	$\overline{A} \cdot \overline{B} \cdot C$
010	0	
011	1	$\overline{A} \cdot B \cdot C$
100	0	
101	1	$A \cdot \overline{B} \cdot C$
110	1	$A \cdot B \cdot \overline{C}$
111	1	$A \cdot B \cdot C$

The Karnaugh map for the data in Table 3.6 is shown in Table 3.7. Because we have three input variables in this problem, the Karnaugh map has eight elements. Note how the rows in the map are labeled such that only one variable changes in adjacent rows (i.e., $\overline{A} \cdot \overline{B} \rightarrow \overline{A} \cdot B \rightarrow A \cdot B \rightarrow A \cdot \overline{B}$). In the first step in this approach, cells corresponding to non-zero output are assigned a value of 1. The next step is to group adjacent horizontal or vertical map cells that have a 1 in them. For this example, we have two groupings: a vertical group that has four cells and a horizontal group that has two cells. The last step is to derive a logical expression from the cell groupings. As seen in Table 3.7, the 1-output in the vertical group is independent of the values of A and B, so Q corresponding to this group is C. Also from the horizontal group, the 1-output is independent of the value of C, so Q is $A \cdot B$.

Table 3.7

Karnaugh map for data in Table 3.6

	$\overline{C}$	C
$\overline{A} \cdot \overline{B}$		1
$\overline{A} \cdot B$		1
$A \cdot B$	1	1
$A \cdot \overline{B}$		1

Combining these expressions, the output of the truth table is then $Q = C + A \cdot B$. This output can be realized by the circuit shown in Figure 3.32.

Figure 3.32

Circuit corresponding to Table 3.6

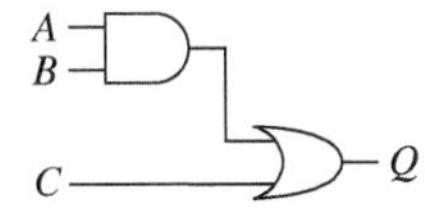

Example 3.9 illustrates the use of combinational logic circuits.

Example 3.9 Application of Combinational Logic Circuits

Design a combinational logic circuit to process the output from three sensors (A, B, and C). The circuit output should be on if the following is true:

- *A* is ON, but both *B* and *C* are OFF
- *A* is OFF, but either *B* or *C* is ON

Solution:

First construct a truth table (Table 3.8) that has all of the input combinations and follows the above rules. Then construct a Karnaugh map (Table 3.9) to map all the non-zero outputs.

Table 3.8

Truth table

ABC	Output	Products
000	0	
001	1	$\bar{A} \cdot \bar{B} \cdot C$
010	1	$\bar{A} \cdot B \cdot \bar{C}$
011	1	$\bar{A} \cdot B \cdot C$
100	1	$A \cdot \bar{B} \cdot \bar{C}$
101	0	
110	0	
111	0	

Table 3.9

Karnaugh map

	$\bar{C}$	C
$\bar{A} \cdot \bar{B}$		1
$\bar{A} \cdot B$	1	1
$A \cdot B$		
$A \cdot \bar{B}$	1	

Then group adjacent cells that have a 1 in them. From the Karnaugh map, it is clear that the output corresponding to the vertical group is $\bar{A} \cdot C$ because it is independent of the value of *B*. Similarly, the output corresponding to the horizontal group is $\bar{A}. B$ because it is independent of the value of *C*. From the last row, the output is $A \cdot \bar{B} \cdot \bar{C}$. Combining these expressions, the output can be written as:

$$Q = \bar{A} \cdot C + \bar{A} \cdot B + A \cdot \bar{B} \cdot \bar{C} = \bar{A} \cdot (B + C) + A \cdot \bar{B} \cdot \bar{C}$$

This output can be realized using the appropriate gates (see Problem P3.26). Note the correspondence between the circuit output-logic expression and the problem statement.

Combinational logic circuits are used in a variety of useful applications, including multiplexers, decoders, and converters. A discussion of multiplexers and decoders follows.

3.6.3 Multiplexers and Decoders

A **multiplexer** (▶ available—V3.10) is a circuit that selects one input out of the several available to be connected to the output (see Figure 3.33). It is commonly used in the design of analog-to-digital converters and in microcontroller circuits to select the timing source.

Figure 3.33

A multiplexer

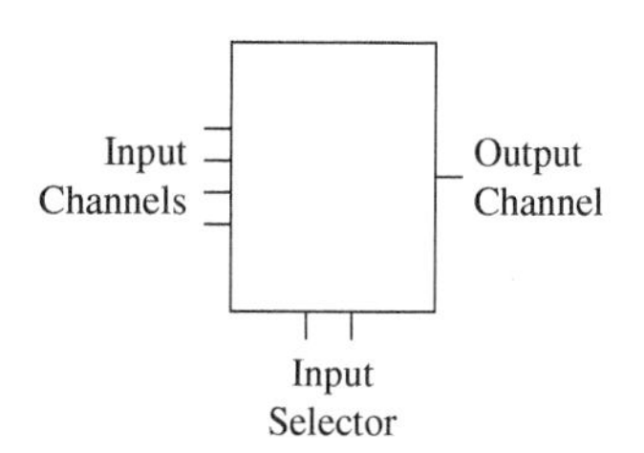

Multiplexer circuits are built from a combination of basic logic gates. Figure 3.34 shows the circuit for a two-input channel multiplexer. The desired channel number is selected by setting the value of the *Channel Selector* input. If the *Channel Selector* input value is 0, then channel 0 is selected. The input connected to channel 0 will then be transmitted to the output channel in this case. Similarly, if the *Channel Selector* input value is 1, then channel 1 is selected. If the multiplexer has four input channels instead of two, then we can see that we need to have two channel selector inputs. We can generalize this to a multiplexer with 2^n input channels, which will need *n-channel* selector inputs.

Figure 3.34

Two-input channel multiplexer circuit

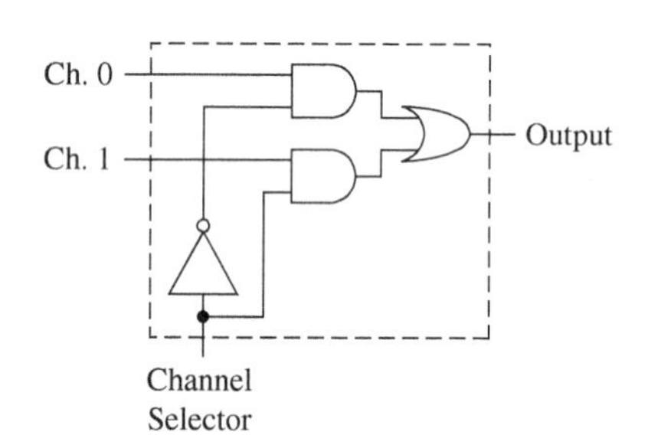

A **decoder** or a **demultiplexer** operates oppositely to a multiplexer. A decoder with n inputs and m outputs will activate only one of the m outputs for a specified pattern of the n inputs. For example, a decoder with four inputs and one output will have a high output for only one combination of the four inputs. For all other combinations, the output will be low. Decoder circuits are used to select devices that are connected to a common line or a bus system. To select a particular device, the address of the device is placed on the bus that connects all of the devices. If the binary pattern of the address placed on the bus matches the individual address for that device, then that device is selected.

A particular use of decoder circuits is in binary-coded decimal (BCD) (see Section 4.2) to decimal format conversion applications. A popular commercial chip is the BCD-to-7 chip, such as the CD74HC4511 IC (see Figure 3.35) which is used to drive seven-segment digital displays (see Figure 3.36). Here the input to this IC is the BCD corresponding to the digit to be displayed on the display, and the output is a combination of the segments a through g.

Figure 3.35

Pin layout for the BCD-to-7 decoder CD74HC4511 IC

(Courtesy of Texas Instruments, Dallas, TX)

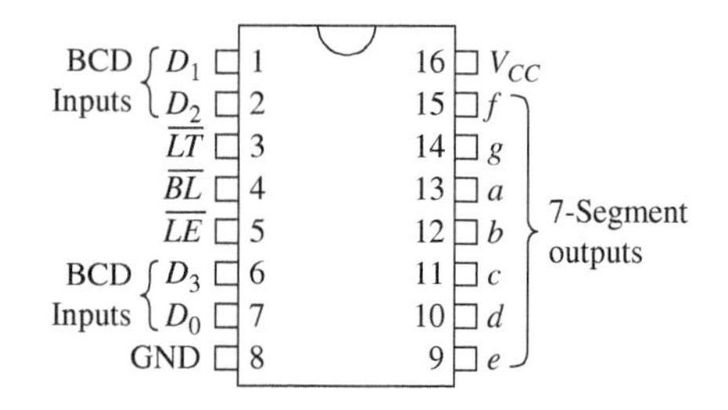

Figure 3.36

Seven-segment digital display

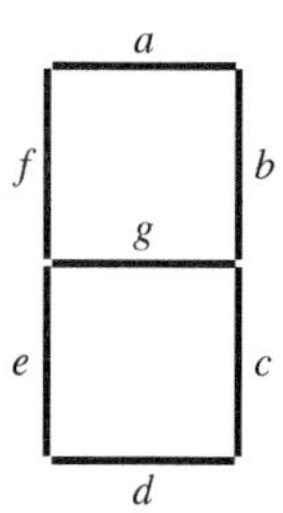

3.7 Sequential Logic Circuits

Unlike a combinational logic circuit, a sequential-logic circuit output is dependent on the history of the input. A sequential logic circuit (▶ available—V3.11) can be thought of as consisting of a combinational logic circuit and memory. A basic sequential logic circuit is the **flip-flop**, which is a sequential logic device that can store and switch between two binary states. Examples of other sequential circuits include counters, shift registers, and microprocessors. There are several types of flip-flops including the SR, the clocked SR, the JK, the D, and the T. We will start by talking about the SR flip-flop.

SR Flip-Flop The **set-reset (SR)** flip-flop has the symbol shown in Figure 3.37. It has two inputs, called S and R, and two outputs, called Q and complementary $\overline{Q}$. The operation of the SR flip-flop is set by the following rules.

1. When $S = 0$ and $R = 0$, the output of the flip-flop does not change.
2. When $S = 1$ and $R = 0$, the flip-flop is set to $Q = 1$ and $\overline{Q} = 0$.
3. When $S = 0$ and $R = 1$, the flip-flop is reset to $Q = 0$ and $\overline{Q} = 1$.
4. S and R are not allowed to be set to 1 simultaneously, since the output will not be predictable.

Figure 3.37

SR flip-flop

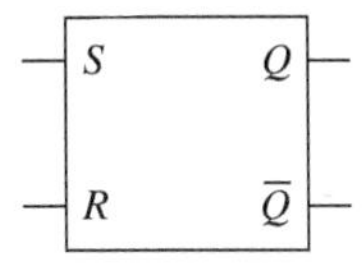

To understand the operation of the SR flip-flip, consider the equivalent circuit made up of two NOR gates with feedback shown in Figure 3.38. Assume that we started with $Q = 0$ and $\overline{Q} = 1$. Now if S is set to 1 and R is 0, $\overline{Q}$ will reset to zero. Due to feedback, Q will also change from 0 to 1. By tracing the output of this circuit for different combinations of S and R, we can verify all of the above rules.

Figure 3.38

Equivalent circuit for an SR flip-flop

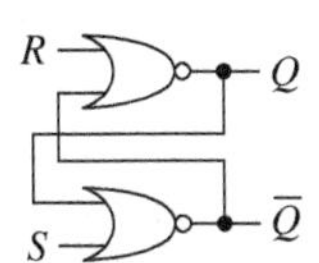

Several types of flip-flops use a clock signal to control their operation. A **clock signal** is a two-state signal. It is commonly a periodic square-wave signal, but it also can be a non-periodic signal made up of a collection of pulses. A periodic square-wave clock signal is shown in Figure 3.39. When a device that uses a clock input responds to the low-to-high change in the clock signal, it is called a **positive edge-triggered** device. Similarly, a device that responds to the high-to-low clock transition is called a **negative edge-triggered** device.

Figure 3.39

Clock transitions

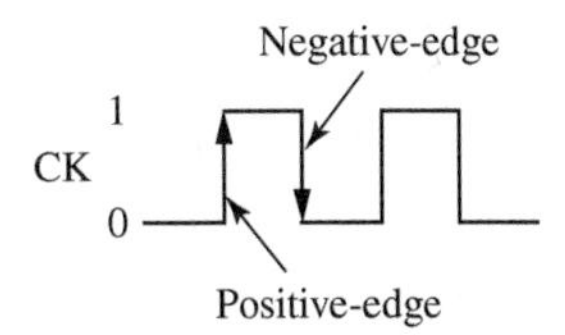

Clocked SR Flip-Flop A clocked SR flip-flop is an SR flip-flop with added clock input. In a clocked SR flip-flop, the output changes state at clock transitions. This is done to provide synchronization of the output change in complex circuits. Note that a flip-flop with no clock input is called a simple or transparent flip-flop. Two variations of a clocked SR flip-flop are shown in Figure 3.40.

Figure 3.40

Clocked SR flip-flop: (a) positive edge-triggered and (b) negative edge-triggered

The flip-flop shown in Figure 3.40(a) is said to be positive edge-triggered. In a circuit diagram, this is shown with a small triangle or wedge at the clock (CK) input. A negative edge-triggered flip-flop is shown in Figure 3.40(b), where the negative edge transition is indicated with a small circle and a triangle at the clock input. Figure 3.41 shows a timing diagram for a positive edge-triggered SR gate. Notice how the output Q changes state at the instant of the positive edge transitions of the clock signal and not when the input S and R change states. In reality, the output does not change instantaneously. There is a small propagation delay on the order of a few nanoseconds or less.

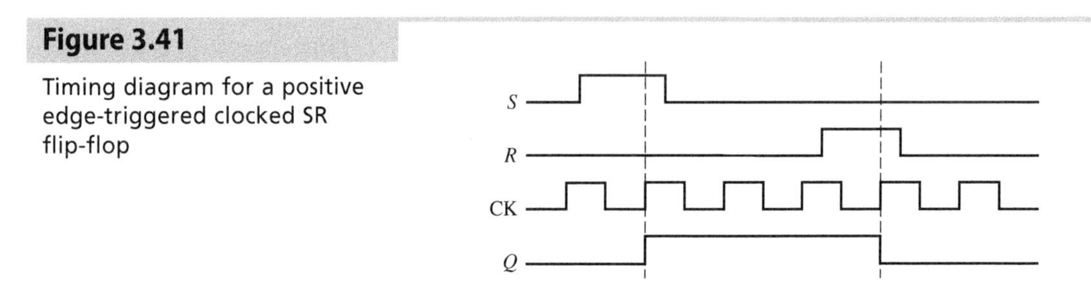

Figure 3.41

Timing diagram for a positive edge-triggered clocked SR flip-flop

The truth table for a positive edge-triggered SR flip-flop is shown in Table 3.10. The up arrow (↑) in the clock column refers to the positive edge transition of the clock, while 0 and 1 in that column refer to the clock state. In Table 3.10, an X entry in any of the cells means that the output is not affected by that entry.

Table 3.10

Truth table for a positive edge-triggered SR flip-flop

Clock	S	R	$Q_t \rightarrow Q_{t+1}$
↑	0	0	$0 \rightarrow 0$
↑	0	0	$1 \rightarrow 1$
↑	0	1	$0 \rightarrow 0$
↑	0	1	$1 \rightarrow 0$
↑	1	0	$0 \rightarrow 1$
↑	1	0	$1 \rightarrow 1$
↑	1	1	Not Allowed
0,1	X	X	$0 \rightarrow 0$
0,1	X	X	$1 \rightarrow 1$

A common use for the SR flip-flop is to debounce switch input. Example 3.10 illustrates this case. Example 3.11 shows a further application of flip-flops.

Example 3.10 SPDT Switch Debouncing Circuit

Illustrate how an SR flip-flop can be used as a switch debouncer for an SPDT switch.

Solution:

The circuit that does the job is shown in Figure 3.42. The switch leads are connected to the flip-flop inputs through an inverter. When the switch pole is in the upper position, *D* is low and *E* is high. *S* is 1 in this case, and *R* is 0. The flip-flop output *Q* will be 1. As the pole leaves the upper switch position, the switch bouncing at that position will not affect the output of the flip-flop, since *S* becoming 0 does not reset the output. When the switch pole reaches the lower position, the first contact at the position will cause the flip-flop output to be turned off, since *R* will be 1, and *S* will be zero. Any subsequent bouncing at the lower position will not change the output since *S* is 0. Note that for this circuit to work, the bouncing at the upper switch position should be completed before the pole reaches the lower position, which is the case in reality. Thus, using this circuit, the output of the flip-flop changes with no bouncing as the switch is rotated from the *D* to the *E* position.

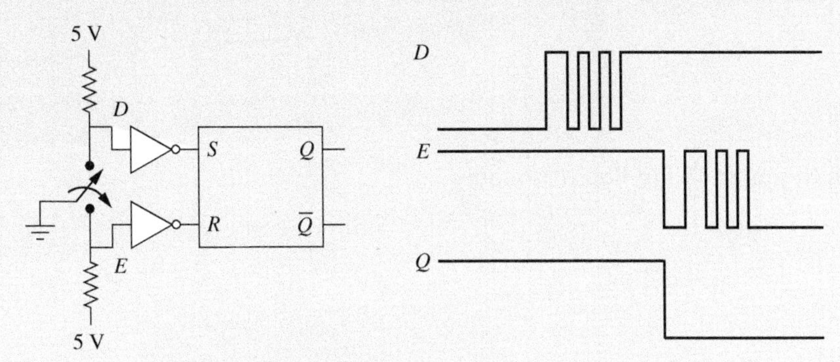

Figure 3.42 SPDT switch debouncing circuit

Example 3.11 Wire Game

An interesting skill game is to guide a closed ring around a wire path without the wire and the ring touching. Design a circuit that uses an SR flip-flop so that, when the ring and the wire touch, a buzzer is turned on and remains on even when the two no longer touch. The buzzer can be turned off by pressing a reset switch.

Solution:

The circuit that performs this function is shown in Figure 3.43. The ring and wire contact are represented by a NO push-button switch. When the ring and the wire are not in contact, the voltage at the *S* and *R* leads is zero, and the buzzer will be off, since output *Q* will be zero (the flip-flop can be initialized to this state through the reset switch). The instant the ring and the wire contact, the input *S* will be set, causing the buzzer to be turned on. The buzzer will remain on even if the ring and the wire are no longer in contact. When the NC push-button reset switch is activated and assuming that the ring and wire are not in contact, *R* will be set and *S* will be reset, causing the buzzer to turn off.

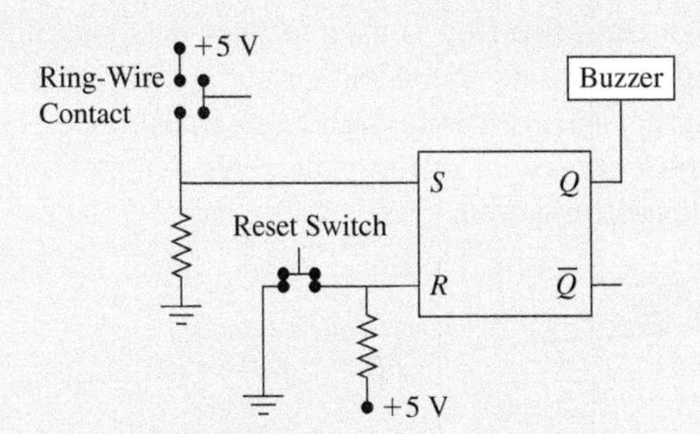

Figure 3.43 Wire game circuit

JK Flip-Flop A JK flip-flop is similar to an SR flip-flop but allows a simultaneous input of $J = 1$ and $K = 1$ similar to $S = 1$ and $R = 1$ in an SR flip-flop. The symbol for a JK flip-flop is shown in Figure 3.44, and a truth table for its operation is listed in Table 3.11. The figure also shows two other input lines for this flip-flop, called *Preset* and *Clear*. Many flip-flops have these additional inputs which can be used to force the output of the flip-flop irrespective of the clock signal. These types of inputs are called *asynchronous inputs* since they are not synchronized with the clock signal. The *Preset* input sets the output Q to 1 or high when it is activated, and the *Clear* input sets the output Q to 0 or low. In Figure 3.44, both of these inputs are active low-type (indicated by the small circle at the input lead), which means that the desired action occurs when the input is low. These inputs can be used to initialize the flip-flop output at power-up.

Figure 3.44

JK flip-flop

Table 3.11

Positive edge-triggered JK flip-flop truth table

Preset	*Clear*	*Clock*	*J*	*K*	$Q_t \rightarrow Q_{t+1}$
1	0	X	X	X	0→0
1	0	X	X	X	1→0
0	1	X	X	X	0→1
0	1	X	X	X	1→1
0	0	Not Allowed			
1	1	↑	0	0	0→0
1	1	↑	0	0	1→1
1	1	↑	0	1	0→0
1	1	↑	0	1	1→0
1	1	↑	1	0	0→1
1	1	↑	1	0	1→1
1	1	↑	1	1	0→1
1	1	↑	1	1	1→0
1	1	0,1	X	X	0→0
1	1	0,1	X	X	1→1

In Table 3.11, an X entry in any of the cells means that the output is not affected by that entry. Notice that when $J = 1$ and $K = 1$, the output will toggle between 0 and 1.

D Flip-Flop A D flip-flop or **data flip-flop** is used to store data and make it available at clock transitions. The symbol of the D flip-flop and its equivalent circuit are shown in Figure 3.45. Due to the use of the clock input, the output only changes at low-to-high clock transitions. The D flip-flop is typically used to implement data registers, which are sets of memory elements that are used to hold information until it is needed. Example 3.12 illustrates this function.

Figure 3.45

(a) D flip-flop and (b) its equivalent circuit

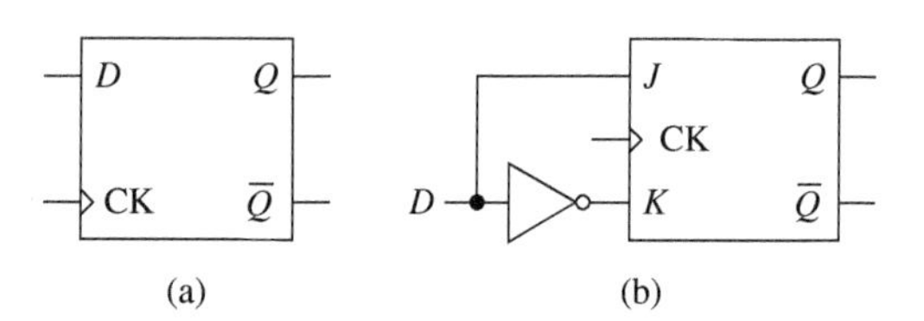

A flip-flop that looks similar to the D flip-flop (but its clock input is not edge-triggered) is called a **latch**. The symbol for the latch is shown in Figure 3.46. Notice that there is no triangle shown at the clock lead. A latch is a level-sensitive device which means it responds to the level of its clock or control signal (also called the enable signal). In a D-latch, when the clock signal is high, the output of the latch changes as the input changes. When the clock signal is low, the latch retains its previous state, i.e., it 'latches' onto the previous input and holds it. With reference to the timing diagram shown in Figure 3.47, the latch flip-flop output changes when the clock signal is high and not at the clock transition. Latches are commonly used to hold and stabilize the digital input data in a digital-to-analog converter (see Chapter 5).

Figure 3.46
Latch

Figure 3.47
Latch timing diagram

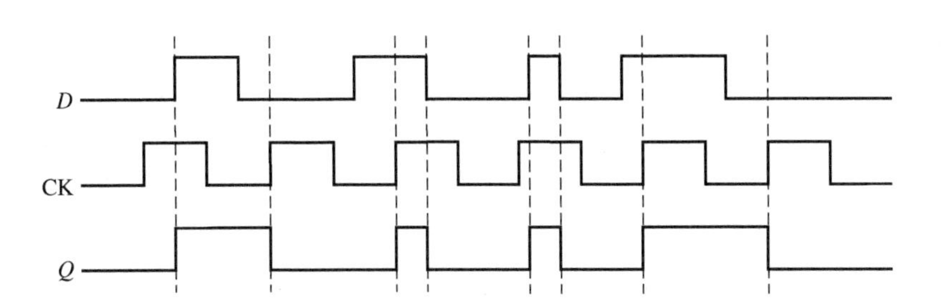

Example 3.12 Three-bit Data Register

Show how a D flip-flop can be used as a 3-bit data register.

Solution:
The circuit to perform this operation is shown in Figure 3.48. Each bit input is connected to the D lead on the D flip-flop. The clock input is combined with a *load* input using an *AND* gate. In this way, the outputs are updated at the low-to-high clock transitions but only when the *load* line is high.

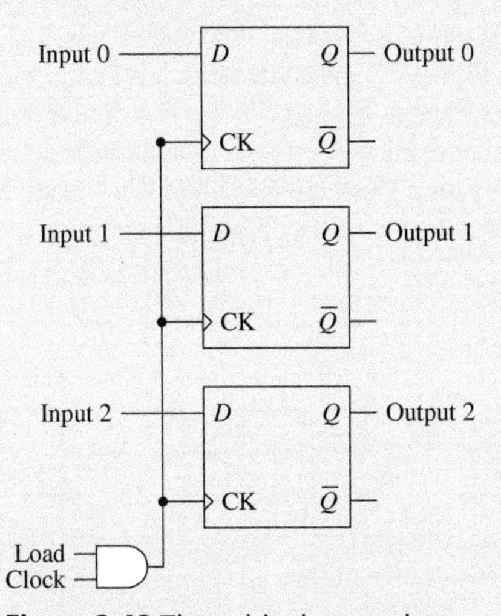

Figure 3.48 Three-bit data register using D flip-flops

T Flip-Flop If the J and K inputs of the flip-flop are permanently set to 1, and the input is applied at the clock input, we get what is called a 'T' or **toggle flip-flop**. The symbol for a positive edge-triggered T flip-flop and its equivalent circuit are shown in Figure 3.49. The T flip-flop has a single input, and its output (Q) changes state (or toggles) at each low-to-high clock transition. The timing diagram for a T flip-flop is shown in Figure 3.50. A characteristic of the T flip-flop is that the output changes its state at a frequency that is half of the input clock frequency. This feature is utilized in the construction of binary counters and frequency dividers.

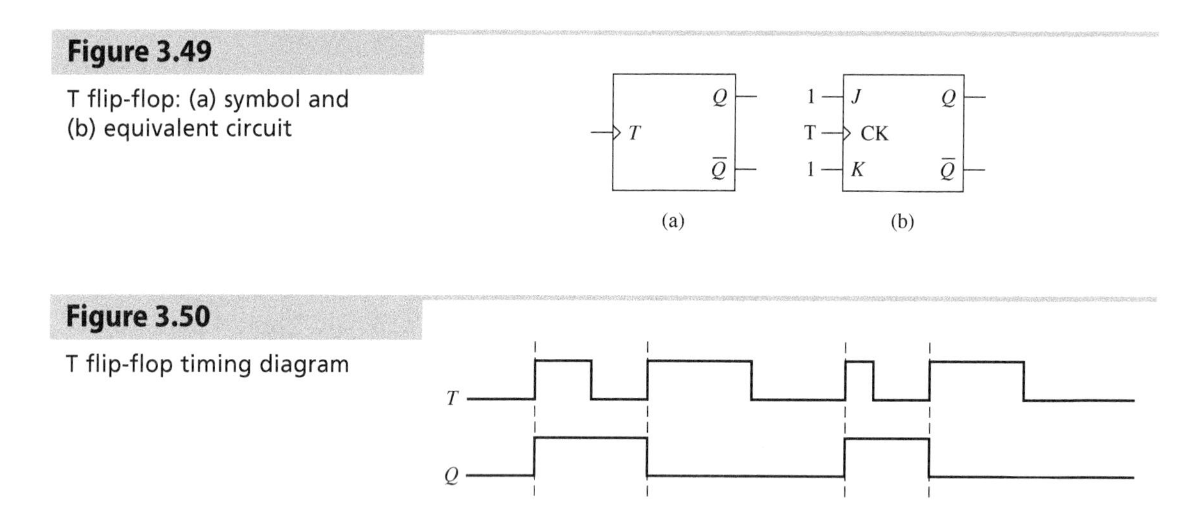

Figure 3.49

T flip-flop: (a) symbol and (b) equivalent circuit

Figure 3.50

T flip-flop timing diagram

As an example of using T flip-flops for counting, consider a 3-bit counter that uses T flip-flops. The circuit for this **binary counter** is shown in Figure 3.51. A clock signal is applied only to the first flip-flop T input, but the output of each flip-flop is fed to the T input of the next flip-flop in the circuit.

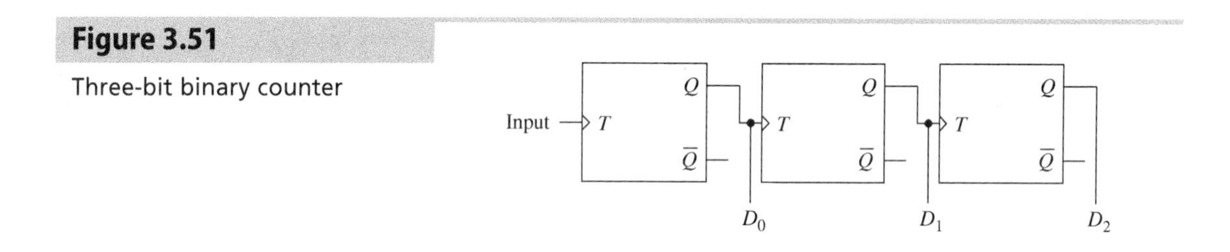

Figure 3.51

Three-bit binary counter

Figure 3.52 shows the timing diagram for the output of this circuit when square pulses are applied to the leftmost T flip-flop. The outputs D_0 through D_2 count the applied pulses in a count-down fashion with D_0 being the least significant bit. As seen in Figure 3.52, the counter reads 7 after the first pulse, 6 after the second pulse, and so forth. If we had used a negative edge-triggered T flip-flop (see Problem P3.37), this counter would count in a count-up fashion with the counter reading 1 after the first pulse. Note how the frequency of each output line is half the frequency of the previous one. Thus, the D_0 output is a 'divide by 2' line, and the D_1 is a 'divide by 4' output, and so forth. Thus, this circuit can be used for either counting or frequency division.

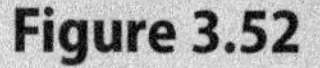

Figure 3.52

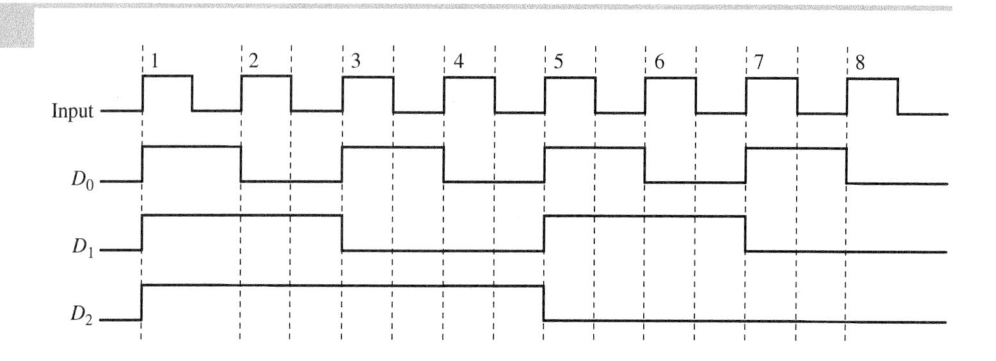

Timing diagram for the 3-bit counter

There are many commercially available counter ICs. These include 4-bit binary counters that count from 0 to 15 and decimal (or decade) counters that count from 0 to 9. The 7490 IC is an example of a 0 to 9 decimal counter (see Figure 3.53). The signal to be counted is applied to the clock input of the 7490, and the count is available from the output lines labeled Q_A through Q_D as a 4-bit BCD. The 7490 counts from 0000 (0) to 1001(9), and then resets back to 0. At the reset from 1001 to 0000, the Q_D line goes from high to low. Since the clock input on the 7490 is negative edge-triggered, the Q_D line from one 7490 can be fed to the clock input of another 7490 to create a counter that can also count the tens digit. In Figure 3.53, the counter counts from 0 to 999 using three cascaded 7490 ICs. Note that this IC has other inputs and other modes of operation, and the reader should consult the data handbook for detailed information on this IC. The 7490 IC is an example of a **modulo-*n* counter**, which is a counter that wraps around back to zero when it has counted *n* counting states. The I7490 is a modulo-10 counter since it has ten counting states (0 to 9).

The counter circuit shown in Figure 3.53 is called a **ripple counter** because the counter output lines for each decade or digit in one IC do not change at the same time as the lines in the next IC. This is because each IC triggers the next IC in series in the order of the connection of the ICs. This creates no problem if the output from each IC is connected to a digital display and is read by the human eye since the 'ripple' effect is too fast to be seen by the human eye. However, if the output lines were connected to other digital circuits, this creates glitches. To circumvent this problem, a synchronous counter should be used. In a **synchronous counter**, all of the digits of the counter change at the same time. This is done by feeding the same clock signal (or the signal to be counted) to the clock input line on each of the cascaded ICs. This causes all of the output lines on all ICs to change at the same clock transition. An example of a synchronous counter IC is the 74160 IC.

Figure 3.53

A 0-999 counter using three 7490 ICs

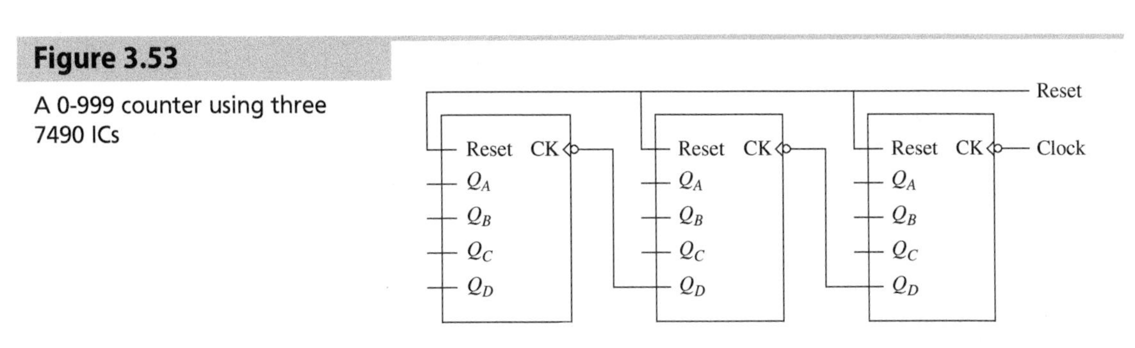

There are many other applications for flip-flops. For example, JK flip-flops can be used in **serial-to-parallel conversion** or **parallel-to-serial conversion**. In serial-to-parallel conversion, the goal is to convert a serial data stream into parallel data. This is commonly encountered in communication systems such as the SPI interface (see Section 5.7) where data is transmitted serially but needs to be processed or stored in parallel form. JK flip-flops can be used in a **shift register** configuration to achieve this conversion. Each flip-flop in the shift register represents a parallel bit position, and as the serial data is shifted in, it gets distributed to the respective flip-flops, resulting in parallel data at the output of the register. In parallel-to-serial conversion, the objective is to convert parallel data into a serial data stream. This is often required when transmitting data over serial communication channels or when storing data in a serial memory device. JK flip-flops, again in a shift register arrangement, can be utilized for this purpose. The parallel data is applied to the inputs of the flip-flops, and as the clock pulses, the flip-flops shift the parallel data out in a serial fashion.

It is very important in digital circuits to reduce the noise in the power supplied to the circuit. This is done with the help of a **bypass** capacitor. Typically, a 0.1 μF capacitor is placed between the voltage source and ground on the power line to the circuit, as seen in Figure 3.54. The purpose of the bypass capacitor is to dampen any AC component in the DC power signal.

Figure 3.54

Bypass capacitor

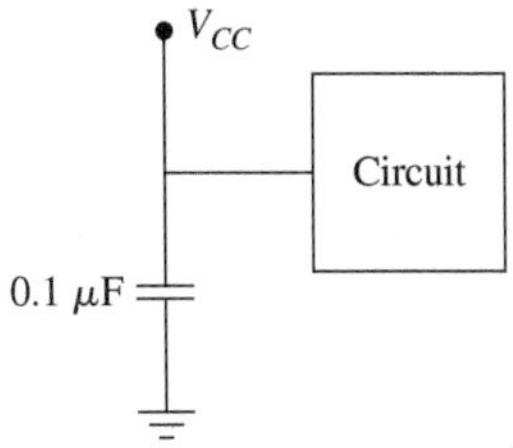

3.8 Circuit Families

Digital circuits are commonly available in two families (▶ available—V3.12): **transistor-transistor logic (TTL)** and **complementary metal-oxide semiconductor (CMOS)**. There are other families, such as emitter-coupled logic (ECL), but they are not as widely used. TTL devices are based on the bipolar junction transistor technology, while CMOS devices are based on the FET transistor technology. Table 3.12 shows the voltage levels for the two families for a 5 V supply. The values define the allowable voltage ranges for the low and high logic states.

Table 3.12

TTL and CMOS* voltage levels

	Low State Voltage Range		High State Voltage Range	
Operation	**TTL**	**CMOS**	**TTL**	**CMOS**
Input	0–0.8 V	0–1.5 V	2.0–5.0 V	3.5–5.0 V
Output	0–0.5 V	0–0.05 V	2.7–5.0 V	4.95–5.0 V

*CMOS data is for operation using a 5 VDC supply

Note how the output level range for either logic state is smaller than the corresponding input level range for that state. This is to allow for noise and signal variation in the output voltage values. Table 3.13 compares the two families in several categories. In general, CMOS ICs consume less power than TTL ICs and can operate over a wider voltage supply range, but they can be easily damaged by static electricity and proper grounding is needed when handling CMOS ICs. In Table 3.13, **gate propagation delay** refers to the time it takes for a gate to switch logic levels from either high to low or from low to high. The propagation delay time from low to high (t_{PLH}) and from high to low (t_{PHL}) are generally not the same, and the largest of the two is used. **Fan out** refers to the number of inputs that can be driven by one output. CMOS devices have a higher fan-out than TTL devices. The terms current **sinking** and **sourcing** are commonly used when listing specifications about IC. A device is said to be sinking current if the current flows *into* the output gate of the IC device when the output is low, while a device is said to be sourcing current if the current flows *from* the output gate of the IC device when the output is high. As listed in Table 3.13, TTL devices can sink much more current than CMOS devices.

Table 3.13

Comparison between TTL and CMOS families

	TTL	**CMOS**
Supply Voltage	Tight supply voltage (about 5 V)	Can operate over a wide supply range from 3 to 18 volts
Power Consumption	High but power consumption does not increase with signal frequency	Much lower power consumption than TTL, but power consumption increases with frequency
Static Sensitivity	Not sensitive	Very sensitive
Unused Inputs	Can be left floating	Should be tied to ground or +Vs
Gate Propagation Time	~ 10 ns	Slower than TTL*
Input Current	High current draw	Very low current draw at the gate input
Output Current	Source about 2 mA but can sink about 16 mA	Can sink or source about 4 mA
Fan-out	One output can drive about 10 inputs	One output can drive about 50 inputs

*Advanced CMOS logic features gate delays of less than 0.1 ns

Within each family, there are several sub-families or series of ICs. The different sub-families are listed in Table 3.14 along with some information about each sub-family. This table is not comprehensive, as there exist over 30 sub-families. Most TTL and CMOS devices are designated using the notation:

mmNNssddp

where

- *mm* is a 2- or 3-letter code for the manufacturer (such as *SN* for Texas Instruments)
- *NN* is either 74 or 54 and refers to the operating temperature range, where 74 is for industrial applications (0 to 70°C), and 54 is for military applications (−55 to 125°C).
- *ss* refers to the sub-family (such as *LS* for low-power Schottky)
- *dd* is a 2- to 4-digit device number (such as 08 for an AND gate)
- *p* is the designation for the type of package in which the device is available (such as *N* for plastic DIP)

For example, the *SN74LS08N* is a TTL AND gate of the low-power Schottky series designed for commercial applications and manufactured by Texas Instruments in the DIP plastic package type. Note that TTL and CMOS devices have been in existence for many years, and several of the sub-families listed in Table 3.14 are now obsolete but are mentioned for reference.

Table 3.14

Listing of TTL and CMOS circuit families

		TTL		
Series	**Designation**	**Example Device**	**Feature**	**Notes**
Regular TTL	—	7408	Original TTL series; has high power consumption	Obsolete
Low-Power TTL	L	74L08	Lower power than Regular TTL but also has a lower speed	Obsolete
High-Speed TTL	H	74H08	Double the speed and power of the regular TTL series	Obsolete
Schottky TTL	S	74S08	Uses more power than Regular TTL but is faster	
Low-Power Schottky TTL	LS	74LS08	Lower power version of the S series	Very commonly used
Advanced Schottky TTL	AS	74AS08	Faster than the S series with lower input current requirements	
Advanced Low-Power Schottky TTL	ALS	74ALS08	Very low power dissipation	
Fast TTL	F	74F08	Lower power than S and LS series	
		CMOS		
Metal Gate	C	74C08	Pin-compatible with TTL	Can use 3 to 15 V power supply
High-Speed Silicon-Gate	HC	74HC08	Pin-compatible with TTL and has the same speed as the 74LS	Requires 2–6 V power supply. Can drive 74LS devices but not driven by them.
High-Speed Silicon-Gate TTL Compatible	HCT	74HCT08	Pin compatible with TTL	Requires 5 ± 0.5 V supply. Can be interfaced with 74LS devices for both input and output
Advanced CMOS	ACT	74ACT08	Inputs are TTL-voltage compatible	Requires 5 ± 0.5 V supply

In data sheets for TTL and CMOS devices, several parameters are defined:

- V_{CC} Supply voltage
- V_{OL} Output voltage when the output is LOW
- V_{OH} Output voltage when the output is HIGH

V_{IL} Input voltage when the input is LOW
V_{IH} Input voltage when the input is HIGH
I_{OL} Output current when the output is LOW
I_{OH} Output current when the output is HIGH
I_{IL} Input current when the input is LOW
I_{IH} Input current when the input is HIGH

As an example, Table 3.15 lists the values of these parameters for the SN74LS08 and the SN74HCT08 AND gates.

Table 3.15

Voltage and current parameters for an AND gate

Parameter	SN74LS08	SN74HCT08
V_{CC}	4.75–5.25 V	4.5–5.5 V
V_{OL} (max)	0.5 V	0.1 V
V_{OH} (min)	2.7 V	4.4 V
V_{IL} (max)	0.8 V	0.8 V
V_{IH} (min)	2 V	2 V
I_{OL} (max)	8 mA	4 mA
I_{OH} (max)	−0.4 mA	−4 mA
I_{IL} (max)	−0.4 mA	±1 μA
I_{IH} (max)	20 μA	±1 μA

For current, the convention is that the current entering a device (sinking) is positive, and the current leaving a device (sourcing) is negative. By examining the values of the input and output currents at low and high logic states, one can determine how many inputs can be connected to the output of one gate or fan out. For example, for the TTL AND gate in Table 3.15, one output can drive up to 20 TTL AND inputs (I_{OL}/I_{IL} = 8 mA/0.4 mA = 20). Table 3.15 shows that for the CMOS AND gate, the input current is significantly smaller than that for the TTL gate.

TTL devices are available with different types of outputs. These include totem-pole, open-collector, and tristate. **Totem-pole** is the most commonly used construction. The output gate has two transistors stacked on top of each other, as seen in Figure 3.55(a) and, hence, the name totem pole. When the output is high, transistor Q_4 is ON, and transistor Q_3 is OFF. In this case, current flows through Q_4 and out of the device. Thus, when the output is high, the gate is sourcing current, and I_{OH} is negative. When the output is low, transistor Q_4 is off, and transistor Q_3 is on. Current flows into the output gate through transistor Q_3. Therefore, the output gate is sinking current, and in this case, I_{OL} is positive.

Figure 3.55

(a) Totem-pole output and
(b) open-collector output

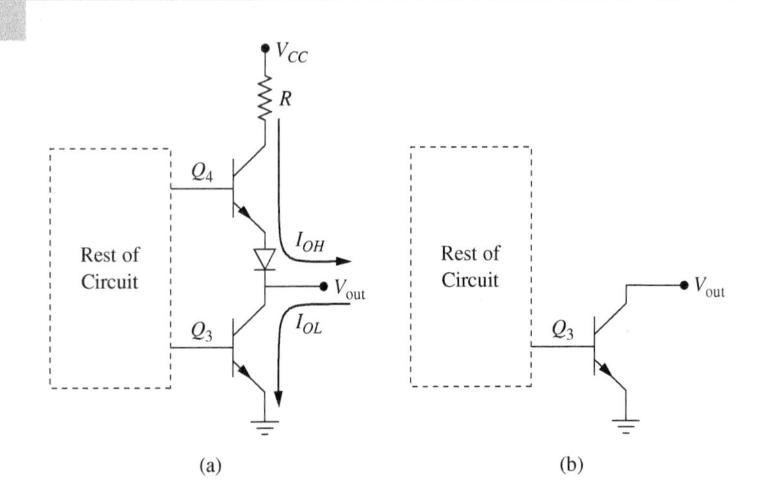

In the **open-collector** configuration, an external 'pull-up' resistor needs to be connected to the output gate. This situation is similar to the totem-pole configuration of Figure 3.55(a) with transistor Q_4 not present

(Figure 3.55(b)). Figure 3.56 shows a typical wiring of an open-collector AND gate. When the transistor Q_3 is ON, the output voltage of the gate will be at a low logic level or close to 0 V (about 0.2 V). When transistor Q_3 is OFF, the output voltage of the gate will be pulled to the supplied voltage (V_S). With open-collector output, the high logic-state voltage output is not limited to the IC V_{CC} voltage but could be any voltage higher or lower than V_{CC}. An example of an open-collector AND gate is the DM74LS09 IC. ICs with open-collector output are typically used to interface ICs from different logic families (such as TTL and CMOS).

Figure 3.56

Wiring of open-collector AND gate

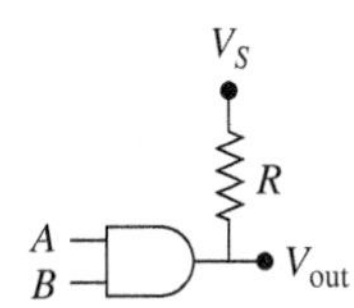

In the **tristate** output, the gate has an additional input called *enable*. When the enable input is low, the output can be either low or high, depending on the input applied to the gate. When the enable input is high, the output is disconnected from the rest of the circuit. The gate will have a high output impedance in this state. Three-state output is used in cases where data from several digital devices is transferred on a common line or bus. The enable signal is then used to connect/disconnect these devices from the bus. Examples of such devices include buffers, flip-flops, and memory chips.

The question that arises is whether devices from different families can be interfaced together. The answer depends on the sub-family of the device. In general, most TTL and CMOS sub-families cannot be directly interfaced due to voltage-level incompatibility, but the CMOS 74HCT and the TTL 74LS sub-families can be mixed without using any additional components. For interfacing a TTL output to a CMOS input where there is a voltage incompatibility, a pull-up resistor is added to the output of a TTL device before it is interfaced with the CMOS device. This ensures that the high-output voltage of the TTL device (V_{OH}) is higher than the high-input voltage (V_{IH}) of the CMOS device. The pull-up resistor resistance value is selected such that the I_{OL} for the TTL device is not exceeded. A CMOS device from the HC or HCT series can drive a single LS device. For driving multiple LS devices, a buffer is inserted between the CMOS output and the TTL inputs to meet the current requirements.

3.9 Digital Devices

In addition to logic gates such as AND or NOR ICs, many specialized ICs are commercially produced. We previously discussed a few of them (such as digital counters and multiplexers). This section discusses another commonly used digital device; the 555-timer chip (▶ available—V3.13). The **555-timer chip** (such as the NE555 8-pin chip from Texas Instruments) is an IC that uses a transistor, resistors, flip-flops, comparators, and capacitors to produce a variety of clock signals, including a fixed pulse, a periodic signal, and a frequency dividing signal. The NE555 can operate over a wide voltage supply range (5 to 15 VDC). With a 5 V supply, it has a TTL-compatible output that can sink or source up to 200 mA. The pin layout and a functional block diagram of the NE555 timer are shown in Figure 3.57.

Figure 3.57

(a) Pin layout and (b) a functional diagram of the NE555 timer chip

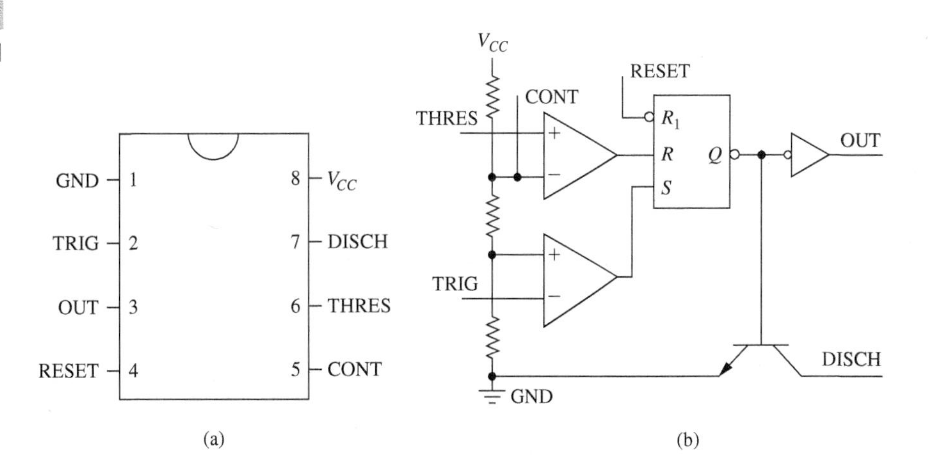

The chip operation is controlled by the inputs applied to the trigger (TRIG) and threshold (THRES) pins. Each of these inputs is connected to a two-input comparator. The comparator outputs are attached to the set (S) and reset (R) inputs of the SR flip-flop. The trigger and threshold inputs are irrelevant if the RESET input is low. When the RESET input is high, the timer output changes according to the trigger and threshold levels. The functional operation is shown in Table 3.16. The trigger input sets the timer output to high if the trigger voltage level is below one-third of the supply voltage (V_{CC}) regardless of the voltage level applied to the threshold input. When the trigger voltage is larger than one-third of the supply voltage, the timer switches from high to low if the threshold voltage exceeds two-thirds of the supply voltage and maintains its output if the threshold voltage is below two-thirds of the supply voltage.

Table 3.16

Functional operation of the 555 timer chip

RESET	Trigger Voltage	Threshold Voltage	Timer Output	Discharge Switch
Low	X	X	Low	On
High	$< 1/3\ V_{CC}$	X	High	Off
High	$> 1/3\ V_{CC}$	$> 2/3\ V_{CC}$	Low	On
High	$> 1/3\ V_{CC}$	$< 2/3\ V_{CC}$	No change	No change

While the NE555 timer chip has several modes of operation, we will focus on two of them here. These are the **monostable** (or fixed-pulse generation) mode and the **astable** (or self-generating periodic signal) mode. In the monostable mode, the pulse properties are controlled by one external resistor and one capacitor. In the astable mode, two external resistors and one capacitor control the duty cycle and the frequency of the timing signal. The wiring diagram for monostable operation is shown in Figure 3.58(a), and the timing diagram is shown in Figure 3.58(b). Here the output of the timer is controlled by the input signal applied to the trigger input. Initially, the internal SR flip-flop output is OFF, and the external capacitor C is held in an uncharged state by the internal transistor inside the timer. When a falling-edge pulse signal is applied to the trigger input with a voltage level less than one-third of the supply voltage, it causes the internal SR flip-flip to turn ON, and the timer output will be high. This happens because the internal, lower-comparator output will be high, which will set the S input of the

Figure 3.58

(a) Wiring diagram for monostable operation and (b) timing diagram

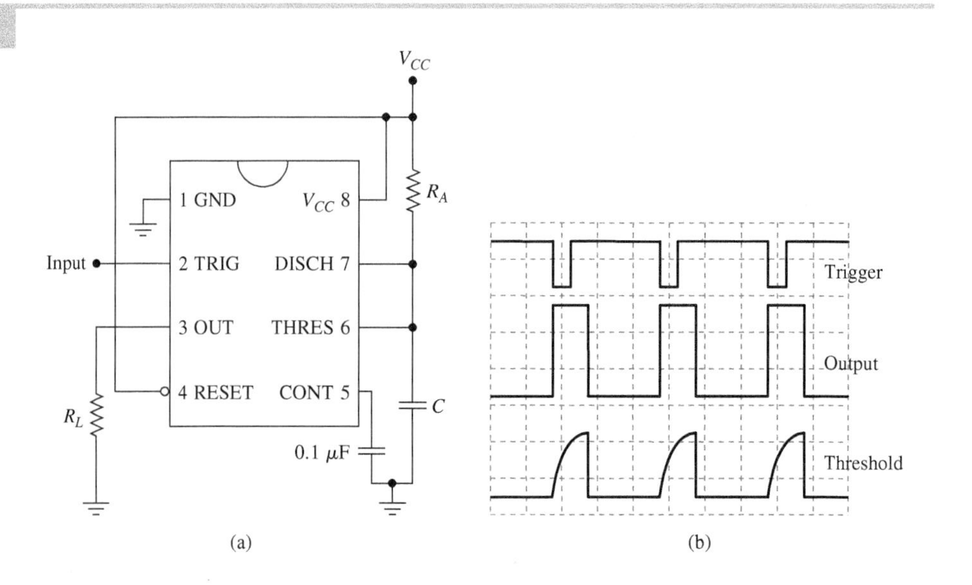

internal RS flip-flop to high. When the timer output turns high, the internal transistor is not conducting. This causes the capacitor C to charge through the resistor R_A since the DISCH pin is not connected to ground voltage in this case. When the voltage level at capacitor C has reached two-thirds of the supply voltage, the internal flip-flop will reset, because the internal upper-comparator output will be high and the internal lower comparator output will be low (provided that the trigger input has returned to high at this point). This causes the output of the timer to go low, and the voltage across the capacitor C will discharge through the internal transistor. This cycle is repeated for every application of a falling-edge trigger pulse. The **output pulse duration** is approximately given by

$$T_H = 1.1\ R_A C \tag{3.12}$$

Note that the trigger signal duration has to be smaller than the output pulse duration. Otherwise, the timer output will remain high.

The wiring diagram for astable operation is given in Figure 3.59(a), and the timing diagram is shown in Figure 3.59(b). A second resistor (R_B) is added to the monostable circuit of Figure 3.58(a), and the threshold and trigger inputs are connected causing the timer to self-trigger. In this configuration, the capacitor C charges through the resistors R_A and R_B and discharges through R_B only. When V_{CC} is first turned on, the capacitor C is discharged, and the trigger input voltage level is zero. The timer output will be high. When the voltage across the capacitor reaches two-thirds of the supply voltage, the internal SR flip flop resets, the timer output switches to low, and the voltage across the capacitor C discharges through the internal transistor. In the astable mode, the capacitor C alternates between charging and discharging states with the charging time being a function of the values of the resistors R_A and R_B and the capacitor C, and the discharging time is a function of the resistor R_B and the capacitor C.

Figure 3.59

(a) Wiring diagram for astable operation and (b) timing diagram

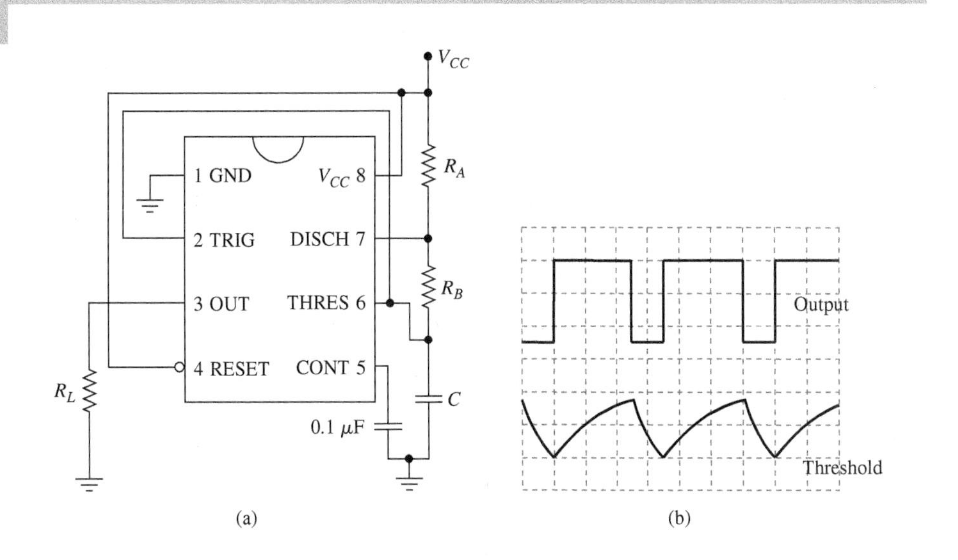

The on-time period (T_H) and the off-time period (T_L) are given by

$$T_H = 0.693(R_A + R_B)C \tag{3.13}$$

and

$$T_L = 0.693\ R_B C \tag{3.14}$$

Example 3.13 illustrates the selection of components for a 555 timer.

Example 3.13 555-Timer

Design a 555-timer circuit to produce a timing signal at a frequency of 1 kHz, and a duty cycle of:

(a) 50%
(b) 75%

Solution:

(a) At 50% duty cycle, the on time (T_H) and the off time (T_L) are equal, or from Equations (3.13) and (3.14), we obtain

$$R_A + R_B = R_B$$

The above equation implies that R_A is zero or not used. Setting the capacitance of the capacitor C to 0.47 μF (a common capacitor value), R_B is obtained from Equation (3.14) as 1.54 kΩ for 1 kHz timer frequency. This is one of the standard resistor values for resistors with 1% tolerance (see Appendix A). Note for the astable wiring circuit, a duty cycle less than 50% cannot be obtained. To obtain less than 50% duty cycle, the control voltage applied to the CONT pin (pin #5) need to be modulated (see manufacturer data sheet).

(b) For 75% duty cycle, we obtain from Equations (3.13) and (3.14)

$$0.75 \times 10^{-3} = 0.693(R_A + R_B)C$$

$$0.25 \times 10^{-3} = 0.693\, R_B C$$

Dividing these equations, we obtain

$$3\, R_B = R_A + R_B$$

Selecting C as 0.47μF and solving, we obtain R_B as 768 Ω and R_A as 1536 Ω. 768 Ω and 1.54 kΩ are standard 1% resistor values which can be used.

3.10 H-Bridge Drives

A very common application of transistors is to construct drivers to drive motors. One such circuit is the H-bridge driver circuit, which is commonly used to drive motors in both directions. The circuit looks like the letter 'H' in circuit schematics, so it is called an H-bridge. An H-bridge circuit (▶ available—V3.14) is constructed using four switching elements that are situated at the corners of the 'H' as the main components. The switching elements used are transistors. The transistors can be of the BJT type or MOSFET type, but depending on the transistors used, the circuit will have different power ratings. To understand the operation of an H-bridge circuit, consider first the simple schematic shown in Figure 3.60, which shows four switches connected to a DC motor. Let us assume that for this motor to turn clockwise, a positive voltage needs to be applied to the left lead, and the right lead should be grounded. This can be achieved by closing switches 1 and 4 and keeping switches 2 and 3 open. This causes current to flow from the power source to the ground through switch 1, the motor leads, and switch 4. If we want to rotate the motor in a counterclockwise fashion, then switches 2 and 3 should be closed, and switches 1 and 4 should be open. This reverses the voltage polarity across the motor and causes the motor to rotate in the opposite direction. Closing switches 1 and 3 or 2 and 4 at the same time should not be done, as this will lead to a short circuit.

Figure 3.60

H-bridge circuit using switches

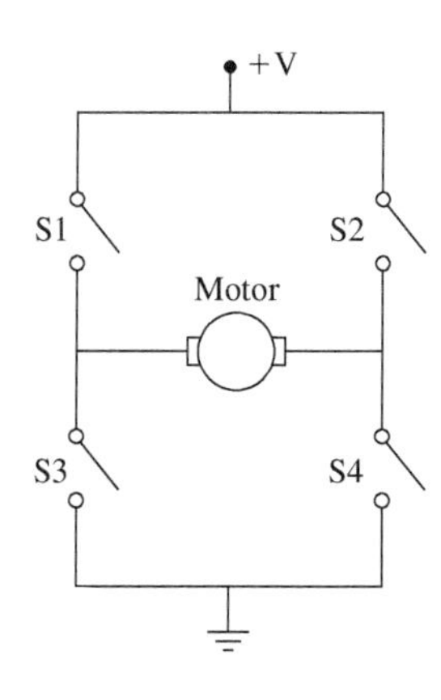

Figure 3.61 shows an implementation of the H-bridge circuit using the G5V-2 Omron® relay as the switching element. The wiring for the relay coil is not shown in the figure. Notice that using two relays, there are four states of operation for this circuit. When both relays are OFF, the switch configuration is as shown in the figure, and the motor will not operate. Similarly, if both relays' coils are ON, the motor still will not operate. In the third state when the left relay is ON and the right relay is OFF, the motor will rotate in one direction. In this case, a positive voltage is applied to the left lead of the motor through pin #9 on the left relay, while the right lead of the motor is grounded through pin #6 on the right relay. In the fourth state, the left relay is OFF, and the right relay is ON. The polarity of the voltage applied to the motor will be opposite to that of state three, and the motor will rotate in the opposite direction.

Figure 3.61

H-bridge implementation using DPDT relays

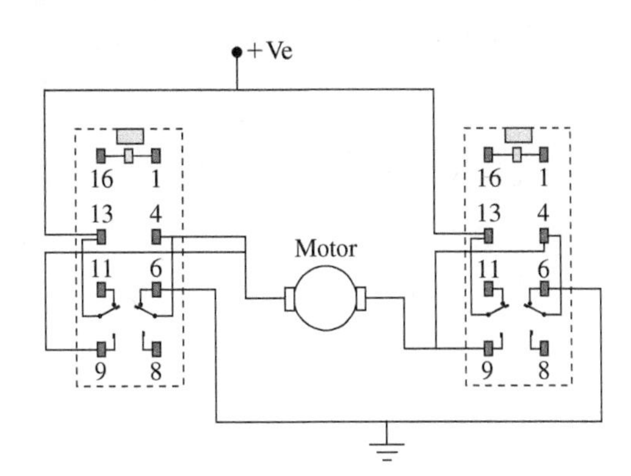

While Figure 3.61 shows how an H-bridge can be implemented using electromechanical relays, typical H-bridge circuits are implemented using transistors due to the fast-switching time of transistors compared to electromechanical relays. Figure 3.62 shows such an implementation using transistors. This **L298 chip** from STMicroelectronics has two full H-bridges that can be independently controlled. Each full bridge can supply up to 2 A current, which is enough to drive small inductive loads such as relays, solenoids, DC, and stepping motors. The supply voltage can be up to 46 V. For illustrating how this chip works, let us consider the left H-bridge. There are two input signals *In1* and *In2* and one enable input signal (*EnA*) that control the operation of the left H-bridge. Similar signals are used to control the right H-bridge. The *In1* and *In2* signals are used to select the particular leg of the H-bridge. The enable input has to be high for the H-bridge to operate. The inductive load or motor leads are connected to *OUT1* and *OUT2* leads. If *In1* is high and *In2* is low, *OUT1* will be at the supply voltage (*Vs*), and *OUT2* will be grounded. In this case, the upper-left and

Figure 3.62

Block diagram of the L298 dual H-bridge

(Courtesy of STMicroelectronics, Geneva, Switzerland)

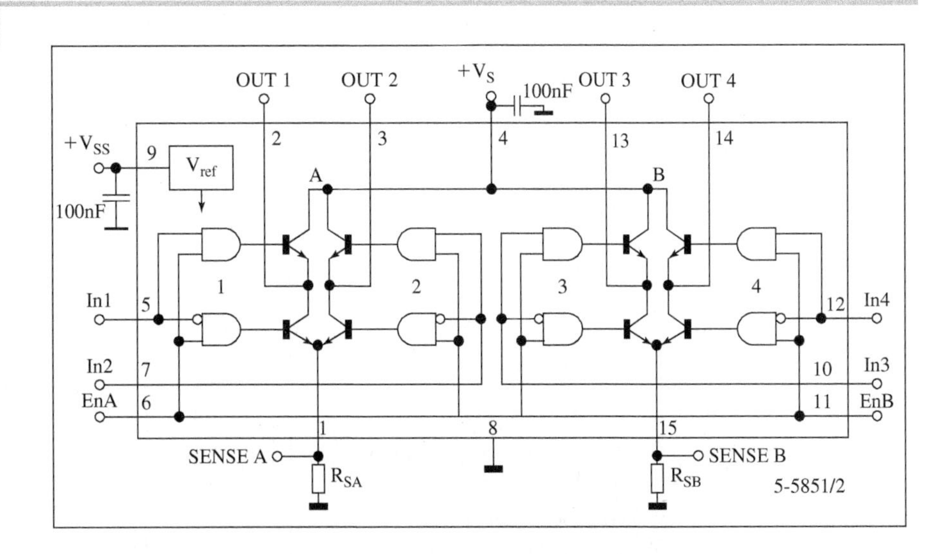

lower-right transistors of the left bridge will be conducting, and the upper-right and lower-left transistors will not be conducting. Similarly, if *In1* is low and *In2* is high, *OUT1* will be grounded, and *OUT2* will be at the supply voltage (*Vs*), thus reversing the polarity of the voltage supplied to the load. If both *In1* and *In2* are high or low at the same time while the enable signal is high, it causes a fast stop or braking of the inductive load such as a motor that is connected to *OUT1* and *OUT2*. This is because setting both inputs to the same value effectively shorts the motor terminals, creating a braking effect. Turning off the enable signal will cause a motor to coast to a stop regardless of the values of the *In1* and *In2* signals. This chip does not come with built-in or intrinsic flyback diodes, but these should be used when connecting this chip to drive inductive loads (see data sheet for more details).

H-bridge drivers are used to control various types of actuators, including brush DC, brushless DC, and stepper motors. H-bridge drivers are typically packaged with additional components to produce what is called a **servo drive**, which is used in controlling actuators. Actuators and servo drives are part of Chapter 9. Section 9.2.4 has additional information on using H-bridge drives in DC motor control, and on a commercial board that uses this chip for motor control. Case Study IV.A discusses another H-bridge IC that is available on the control board for the mobile robot discussed in that case study.

Integrated Case Study IV.A: H-bridge Driver for the Motor

For Integrated Case Study IV (Mobile Robot—see Section 11.5), the control board on the robot uses two Texas Instruments DRV8838 motor driver's chips to drive the two geared motors. One feature of the DRV8838 driver is that it can operate with low motor power supply voltage in the range of 0 to 11 volts which works well for battery-powered devices such as the ROMI robot which uses six 1.5 V batteries (total 9 V) to provide the power. A functional block diagram of the DRV8838 is shown in Figure 3.63(a), and a simplified interface diagram is shown in Figure 3.63(b). The chip has three control lines. The Enable signal (*En*) is used to control the speed of the motor by providing a PWM signal to it

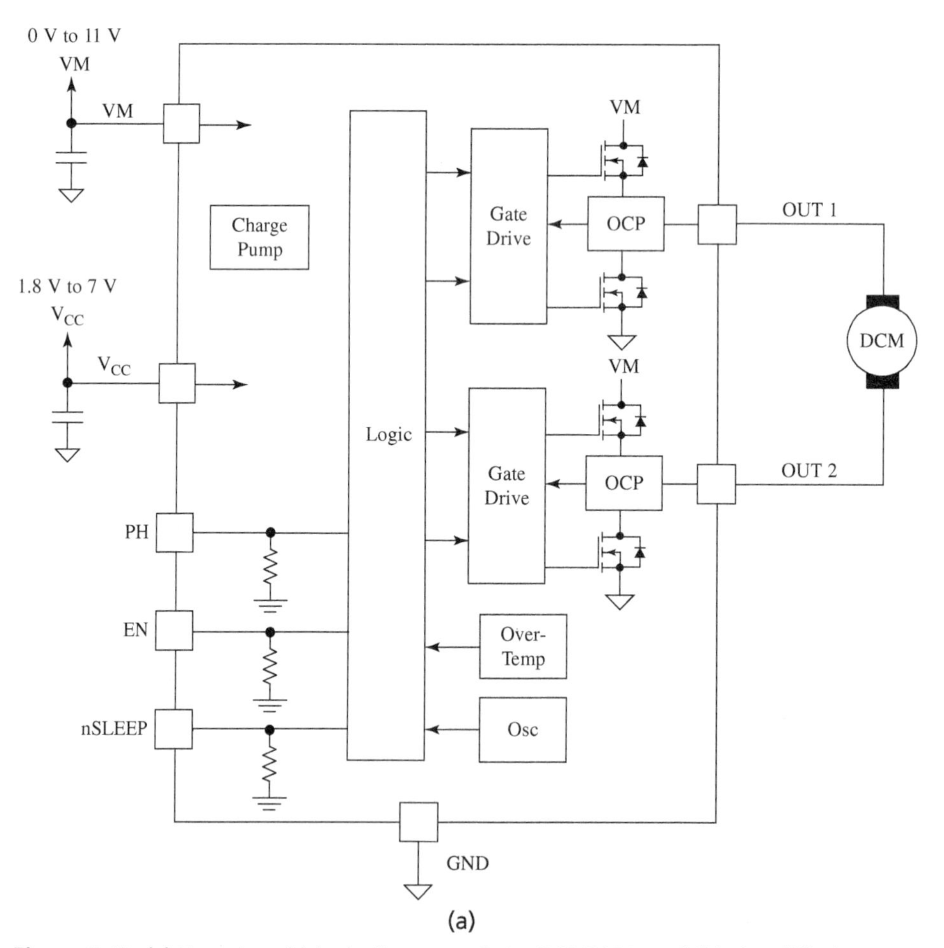

(a)

Figure 3.63 (a) Functional block diagram of the DRV8838 and (b) simplified interface diagram

(Courtesy of Texas Instruments, Dallas, TX)

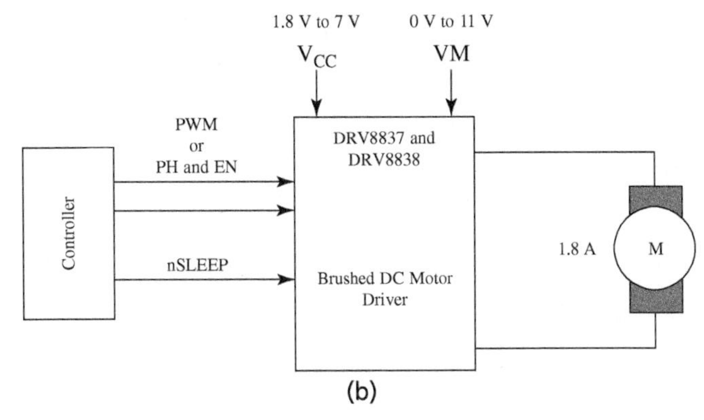

Figure 3.63 *(Continued)*

from the controller, while the phase line (*PH*) is a digital signal to control the direction of rotation of the motor. The *nSLEEP* signal which is active low sets the driver in a low-power sleep mode. For normal operation of the driver, the *nSLEEP* should be set high. The driver can supply a maximum current of 1.8 A but only for a short interval. To prevent overheating of the driver and its shutdown, the output current should be limited to about 1.7 A. Unlike the L298 driver, the DRV8838 driver chip includes built-in flyback diodes (called intrinsic didoes) that are placed between the source and the drain leads of the MOSFET transistors on the chip. The flyback diode purpose is to protect the transistor when the transistor switches its state with inductive loads attached to the chip.

The DRV8838 is not available in the PDIP package format, so it cannot be plugged directly into a breadboard, but Pololu makes a tiny breakout carrier board (Item # 2990) for this motor driver that can be used to control a single brushed DC motor. The carrier board can be mounted on a breadboard since it has the form factor of a 10-pin DIP package.

3.11 Chapter Summary

This chapter dives into the inner workings of semiconductor electronic devices like diodes, thyristors, and transistors, which are integral components of numerous circuits and devices. They function as solid-state switches, initiating the switching action through non-mechanical means, triggered by alterations in their electrical characteristics. Solid-state devices do not obey Ohm's law.

Focusing on diodes, the chapter describes them as two-terminal elements which direct current flow in a single direction. The classification and distinctive characteristics of various diodes, encompassing regular diodes, Zener diodes, LEDs, and photodiodes, are detailed. In addition, the chapter investigates thyristors; these three-terminal semiconductor devices function similar to diodes, but with a crucial differentiator—an additional terminal known as a gate that dictates their operation. Transistors, another category of three-terminal devices, are discussed at length in the chapter. Outlined are the three operational states—off, linear, and saturation—of transistors. In particular, the chapter introduces bipolar junction transistors (BJTs) and metal-oxide-semiconductor field-effect transistors (MOSFETs). The crucial difference between them is that while BJTs are governed by current, MOSFETs are controlled by voltage. They are fundamental to the creation of digital circuits.

The chapter differentiates between two types of logic circuits. In combinational logic circuits, the output is a direct result of the current input and mathematical logic rules determine the output. In contrast, sequential logic circuits account for the history of inputs to determine the system's output. The chapter also explores the practical applications of digital circuits in the design of devices such as digital counters, multiplexers, and timers. It sheds light on the two most common families used for digital circuits: transistor-transistor logic (TTL) and complementary metal-oxide-semiconductor (CMOS), guiding how to interface between them. Furthermore, the chapter discusses the 555-timer chip, an integral component in many electronic devices. It also explores the H-bridge driver circuit, often employed to drive motors bidirectionally. It leverages transistors as the switching elements, demonstrating the versatile applications of transistors in real-world scenarios.

Questions

3.1 What is the difference between an ideal and a real diode?

3.2 What is the functional difference between a normal diode and a Zener diode?

3.3 What is a thyristor?

3.4 What is the difference between a relay and a transistor?

3.5 What are the three states of operation of a BJT?

3.6 In what state is a BJT operating if the collector current is zero?

3.7 For a BJT, is I_C controlled by I_B or V_{CC} when the transistor is operating in the active region?

3.8 How does a phototransistor operate?

3.9 Name one major difference between a BJT and a MOSFET.

3.10 What is meant by an open-collector output circuit? Why do manufacturers make devices with such an output type?

3.11 What information does a timing diagram give?

3.12 What is the difference between combinational and sequential logic circuits?

3.13 What is a Karnaugh map?

3.14 Why do many digital circuits have clocked input?

3.15 Explain the function of a multiplexer.

3.16 What is the difference between an SR and a JK flip-flop?

3.17 How is a latch different from a D flip-flop?

3.18 What is a T flip-flop?

3.19 List the names of digital circuit families.

3.20 Define what is meant by 'fan out'.

3.21 List the different output methods of digital circuits.

3.22 Define current sinking and current sourcing.

3.23 Name two modes of operation of the 555-timer chip.

3.24 For what purpose is an H-bridge driver circuit used?

Problems

Section 3.2 Diodes

P3.1 Determine whether the ideal diode in Figure P3.1 is conducting.

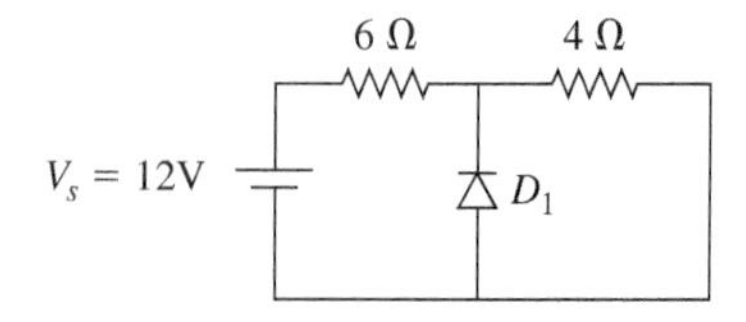

Figure P3.1

P3.2 For the circuits shown in Figure P3.2, plot the output voltage of the circuit for an AC sinusoidal input signal with an amplitude of 5 V and a frequency of 1 Hz. Assume an ideal diode is used.

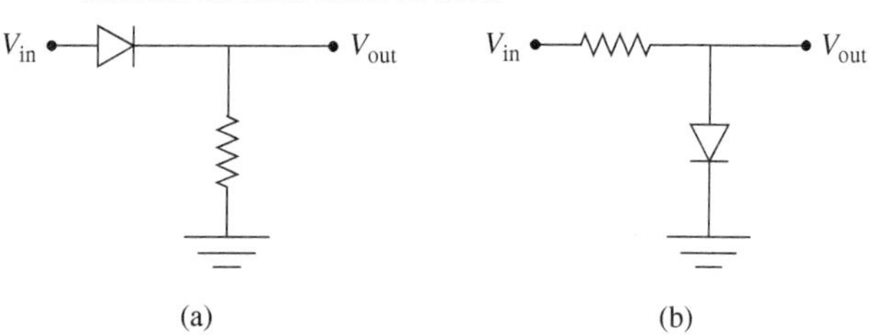

Figure P3.2

P3.3 Consider the circuit in Figure P3.3. Assume the diode D_1 has a V_F value of 0.6 V, and its forward resistance is zero. Determine the current in the circuit.

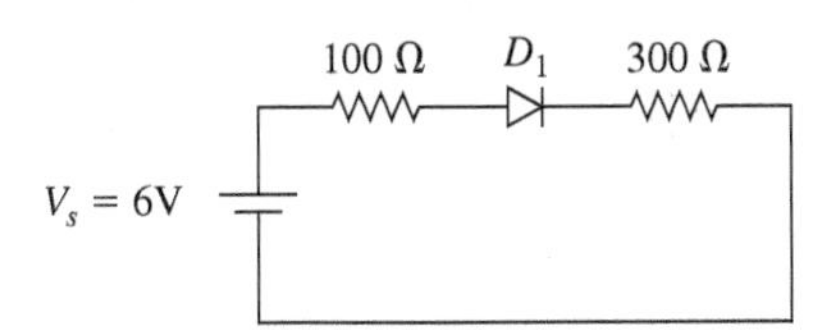

Figure P3.3

P3.4 Consider the half-wave diode rectifier circuit shown in Figure 3.2. Assume a V_F value of 0.6 V. Plot the output voltage of the circuit for an AC sinusoidal signal with the following amplitudes.

a. 0.5 V

b. 2 V

c. 5 V

P3.5 Consider the signal conditioning op-amp circuit shown in Figure P3.5 with an ideal diode in the feedback loop. Show that, for the case $V_o < 0$ and $V_i > 0$, the input-output relationship is given by

$$\frac{V_o}{V_i} = -\frac{R_2R_3}{R_1(R_2 + R_3)}$$

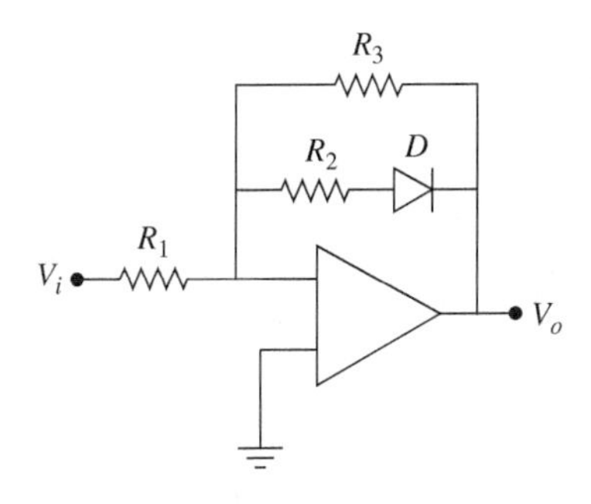

Figure P3.5

P3.6 For the diode clamp circuit shown in Figure P3.6, determine the output voltage V_o as V_i is increased from 0 to 10 V in increments of 1 V. Assume that the diode has a V_F value of 0.6 V.

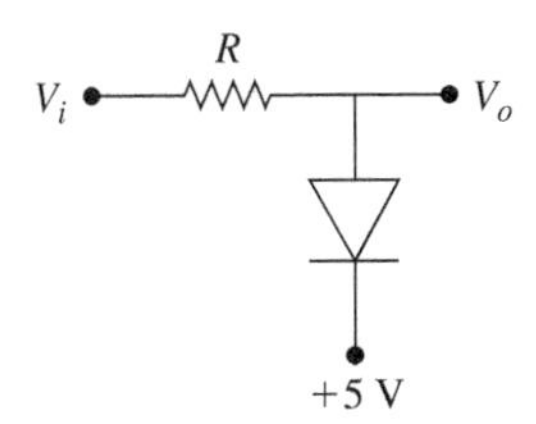

Figure P3.6

P3.7 Consider a simple circuit that powers a red LED using a 9 V battery. The LED requires a forward voltage of 2 V and has a forward current rating of 20 mA. Assume that resistors of various values are available. Determine the following.

a. The value of the resistor that needs to be connected in series with the LED to ensure that the current does not exceed the forward current rating.

b. Repeat part (a) but assume that two such LEDs need to be connected in series.

c. Repeat part (a) but assume that two such LEDs need to be connected in parallel.

P3.8 For the Zener diode circuit shown in Figure 3.6, determine the voltage across the R_2 resistor for the following two cases: (a) $V_S = 8$ V, and (b) $V_S = 12$ V. Assume $R_1 = 82\ \Omega$, and $R_2 = 100\ \Omega$ and a Zener diode with a breakdown voltage of 5.1 V is used.

Section 3.3 Thyristors

P3.9 A Silicon Controlled Rectifier (SCR) is connected in series with a 100 Ω resistor. The SCR has the following specifications: a forward voltage drop (V_F) of 1.6 V when conducting and a gate trigger current (I_{GT}) of 30 mA. A power supply is connected to this series circuit, and a triggering device is connected to the gate terminal of the SCR, which can provide up to 40 mA of current. For this problem,

a. What is the minimum voltage the power supply should provide to ensure the SCR is fully conducting, considering the SCR and the resistor in the circuit?

b. Calculate the minimum gate current required to trigger the SCR. Would the triggering device connected be able to do this? Explain your answer.

c. If the power supply provides a voltage of 6 V, calculate the current flowing through the circuit when the SCR is conducting.

Section 3.4 Bipolar Junction Transistor

P3.10 Draw a circuit that uses a solar cell (which can be represented as a voltage source) with a small output current to turn on a lamp (that uses a much larger current). We want the brightness of the lamp to be controllable by the amount of light received by the solar cell.

P3.11 Consider the BJT circuit shown in Figure P3.11 where $V_{CC} = 15$ V, $R_C = 5$ kΩ, $I_B = 40\ \mu$A, and $\beta = 70$. Determine I_C and V_{CE}.

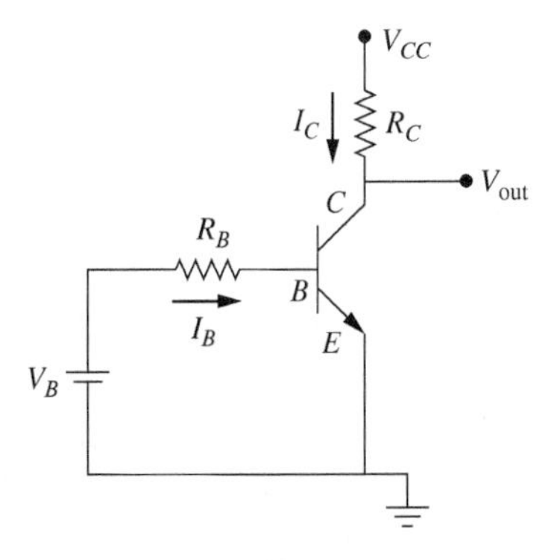

Figure P3.11

P3.12 For the BJT circuit shown in Figure P3.11, determine the minimum input voltage V_B required to saturate the transistor. Assume $V_{CC} = 5$ V, $R_C = 2\ k\Omega$, $R_B = 50\ k\Omega$, V_{CE} sat $= 0.2$ V, V_{BE} sat $= 0.7$ V, and $\beta = 50$.

P3.13 For the BJT transistor considered in Example 3.4, determine the minimum input voltage V_{in} required to saturate the transistor. Assume V_{CE} sat $= 0.2$ V, V_{BE} sat $= 0.7$ V, and use the other parameters given in the example.

P3.14 For the BJT circuit shown in Figure P3.14, determine the current across the resistor R_2 if (a) the transistor is off, and (b) the transistor is in saturation. Let $R_C = 1$ kΩ, $R_2 = 1$ kΩ, $V_{CC} = 10$ V, and V_{CE} sat $= 0.2$ V.

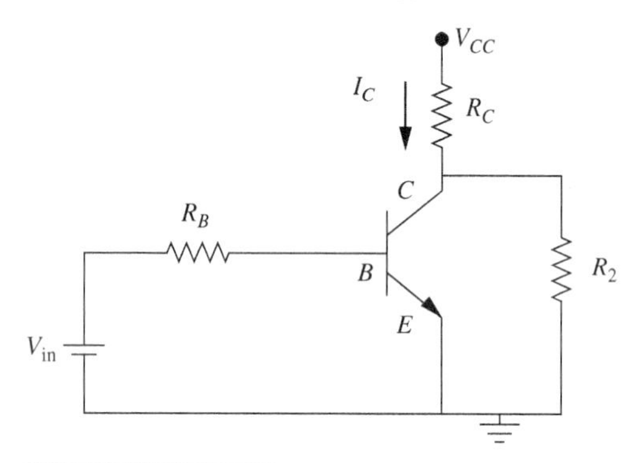

Figure P3.14

P3.15 Figure P3.15 shows a BJT amplifier with a base bias. For this circuit, compute I_B, I_C, and V_{CE}. Assume V_{BE} is 0.65 volts, and let $V_{CC} = 16$ V, $R_C = 1$ kΩ, $R_B = 200$ kΩ, and $\beta = 100$.

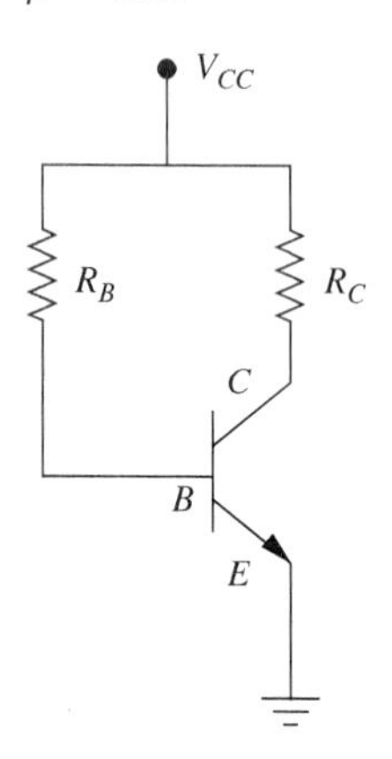

Figure P3.15

P3.16 For the emitter follower circuit shown in Figure 3.18, compute the output voltage V_{out}, the base current I_B, and the output current I_E. Let $V_{in} = 8$ V, $V_{CC} = 5$ V, $R_E = 2$ kΩ, $R_B = 10$ kΩ, and $\beta = 100$.

P3.17 Figure P3.17 shows a BJT transistor circuit with a bias current that uses two resistors (R_1 and R_2) connected in series between the supply voltage (V_{CC}) and ground. Determine the voltages at points 1 and 2 in the circuit. Let $\beta = 50$, $V_{BE} = 0.6$ V, $R_C = 2$ kΩ, $R_E = 100$ Ω, $R_1 = 2$ kΩ, $R_2 = 200$ Ω, and $V_{CC} = 10$ V.

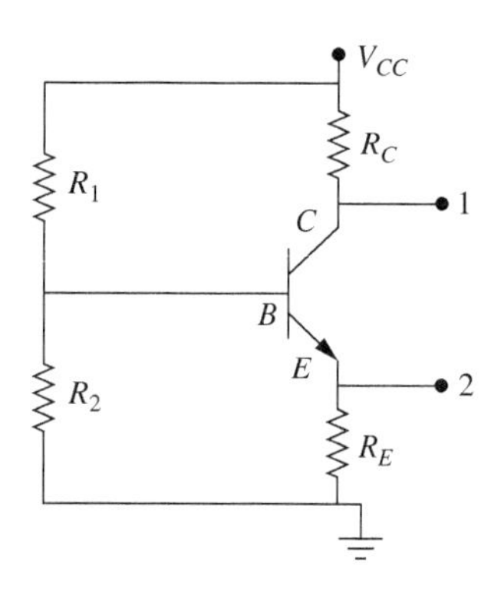

Figure P3.17

P3.18 For the proximity sensor in Example 3.7, rework that example, but assume the supply voltage is 12 VDC.

P3.19 Draw a circuit to show how to use the proximity sensor shown in Figure 3.21 to turn on the coil of the relay shown in Figure 2.11. Assume that the proximity sensor uses a 24 VDC supply. Specify the value of all the resistors that are needed. Make the coil current compatible with the relay characteristics in Table 2.3.

P3.20 Design a circuit that uses the 2N3904 transistor to activate the coil of the relay shown in Figure 2.11. Design your circuit so that it is compatible with the 2N3904 transistor and the relay characteristics.

Section 3.5 MOSFET

P3.21 Figure P3.21 shows a simple inverter circuit that uses a MOSFET transistor. Assume that this transistor has a gate threshold voltage V_T of 1 volt and a very low $R_{ds(ON)}$. For $V_{DD} = 5$ V,

a. What will be the state of the transistor (ON or OFF) and what will be the output voltage at 1 if the input to the inverter circuit (V_{in}) is 0 V?

b. What will be the state of the transistor (ON or OFF) and what will be the output voltage at 1 if the input to the inverter circuit (V_{in}) is 5 V?

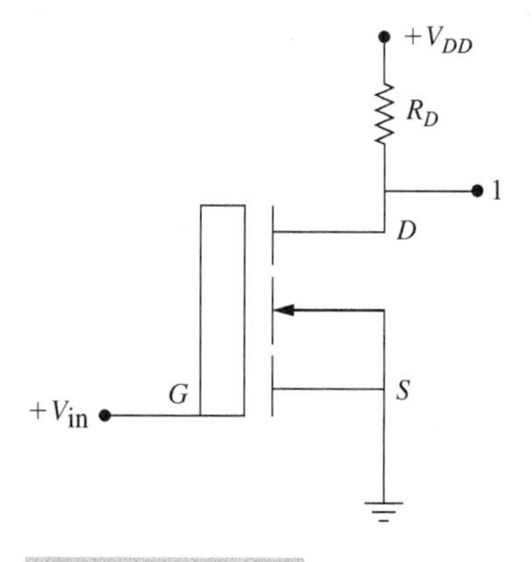

Figure P3.21

P3.22 Figure P3.22(a) shows the output characteristics of a power MOSFET transistor with a $V_T = 2$. Assume $V_{DD} = 5$ V. Determine the maximum I_D current in the circuit in Figure P3.22(b) for the following cases.

a. $V_{in} = 1$ V

b. $V_{in} = 4$ V

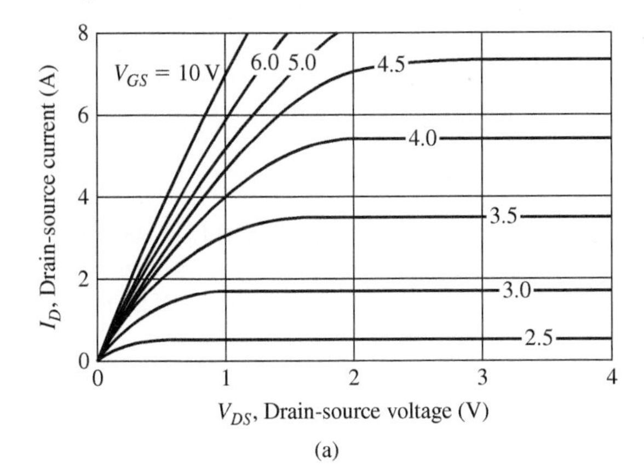

(a)

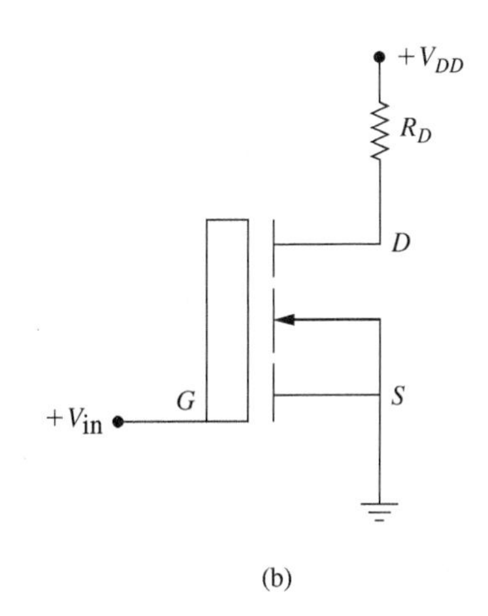

(b)

Figure P3.22

(Courtesy of Fairchild Semiconductor, South Portland, ME)

Section 3.6 Combinational Logic Circuits

P3.23 Provide a truth table for the following logic functions.

a. $Q = A + B \cdot C$

b. $Q = (A + B) \cdot C$

P3.24 Using the Boolean rules, simplify the following expressions.

a. $Q = A \cdot B + \overline{A} \cdot B + \overline{A} \cdot C$

b. $Q = A \cdot B + A \cdot C + A \cdot \overline{B} \cdot C$

P3.25 Draw a circuit realization of the output Q in Problems P3.23(a) and P3.23(b).

P3.26 Draw a circuit realization of the output Q in Example 3.9.

P3.27 Using only NAND gates, draw a circuit that operates as follows.

a. AND gate

b. OR gate

P3.28 An overheating monitoring system uses three digital temperature sensors operating in ON/OFF mode. The output of each sensor is turned ON when the temperature exceeds a specified value and is OFF otherwise. Design a combinational logic circuit to process the output from these sensors such that the circuit output should go high when the output of *any two* of the three temperature sensors goes high.

P3.29 Redo Problem P3.28, but assume the system uses four temperature sensors and the circuit output should go high when the output of *any three* of the four temperature sensors goes high.

P3.30 Provide the logic expression for the output of a simplified 2-bit comparator. The simplified comparator compares two 2-bit numbers and produces an output of 1 only if the two numbers are identical.

P3.31 Draw a combinational logic circuit that implements a four-channel multiplexer that uses two input lines to select the input channel to be connected to the output.

Section 3.7 Sequential Logic Circuits

P3.32 Complete the timing diagram in Figure P3.32 for an SR flip-flop.

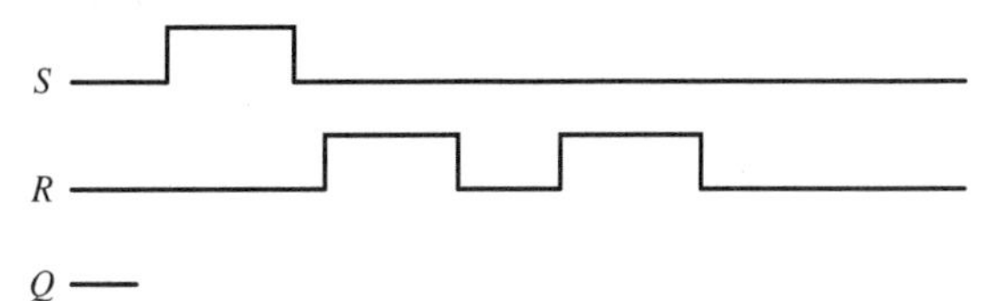

Figure P3.32

P3.33 Complete the timing diagram in Figure P3.33 for a positive edge-triggered clocked SR flip-flop.

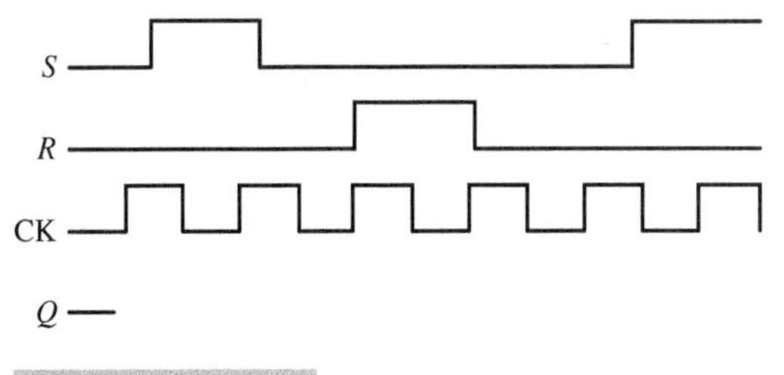

Figure P3.33

P3.34 Complete the timing diagram in Figure P3.34 for a positive edge-triggered clocked JK flip-flop.

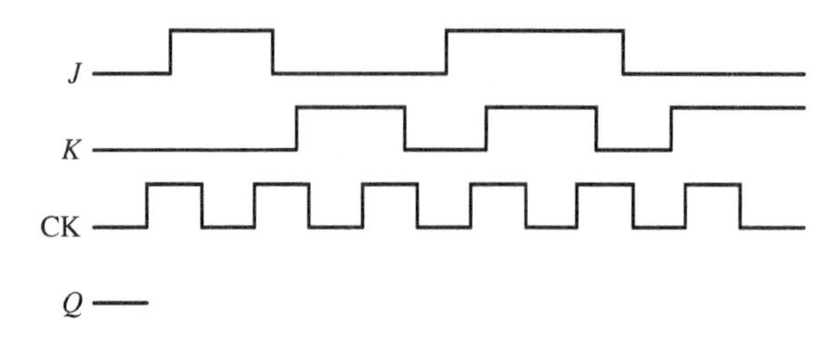

Figure P3.34

P3.35 Redo Problem P3.34 but assume a negative edge-trigged clocked JK flip-flop.

P3.36 Referring to the datasheet for the 7490-decade counter IC, draw a circuit that shows how a single 7490 IC can be used as a divide-by-5 counter.

P3.37 Draw the timing diagram for the counter circuit shown in Figure 3.51 but assume that a negative edge-triggered T-flip flop was used.

P3.38 Research and explain using examples what is meant by a free-running counter.

Section 3.8 Circuit Families

P3.39 Map the following input voltages to low or high logic levels assuming TTL logic (use Table 3.12): 0.1, 0.7, 1.0, 3.0, and 4.0 volts.

P3.40 Map the following output voltages to low or high logic levels assuming TTL logic (use Table 3.12): 0.3, 0.7, 2.0, 3.0, and 4.0 volts.

P3.41 A digital system requires a 4-input AND gate for a particular operation. Unfortunately, you only have SN74LS08 ICs available which are Quad 2-input AND gates. Design a circuit that effectively creates a 4-input AND gate using only 2-input AND gates from the SN74LS08 IC. Do you expect your circuit to be faster or slower than using the single 4-input AND gate?

Section 3.9 Digital Devices

P3.42 Determine the values of standard 1% tolerance resistors (see Appendix A) to use with a 555-timer circuit to produce a timing signal at a frequency of 2 kHz and a duty cycle of 50%.

P3.43 Determine the values of standard 1% tolerance resistors (see Appendix A) to use with a 555-timer circuit to produce a timing signal at a frequency of 500 Hz and a duty cycle of 75%.

Section 3.10 H-Bridge Drives

P3.44 For the dual H-bridge shown in Figure 3.62,

a. What happens if the enable signal is low?

b. What happens if both *In1* and *In2* are high at the same time?

Laboratory/Programming Exercises

L/P3.1 Build the common-emitter transistor circuit shown in Figure 3.15. Use $R_C = 1\ \text{k}\Omega$ and $R_B = 1\ \text{k}\Omega$. Starting from zero, increase the voltage V_{in} in increments of 0.1 V over the range of 0 to 2 V and measure V_{CE}. Change R_B to 10 kΩ, and repeat the procedure. At what input voltage did the transistor turn on in each case? How do the measured results agree with the theory? Use a V_{CC} of 5 V.

L/P3.2 Build the wire-game circuit discussed in Example 3.8 that uses an SR flip-flop.

L/P3.3 Build the circuit shown in Figure L/P3.3 that uses a NO push-button switch and a 555-timer chip that act as a bounceless switch. Pressing the switch should cause the circuit to produce a clean, single pulse. The pulse duration is a function of the resistor and capacitor values used in the circuit. Use 1 μF for C_1 and 0.1 μF for C_2. Try 100 kΩ, and 1 MΩ values for R_1, and measure the duration of the output pulse.

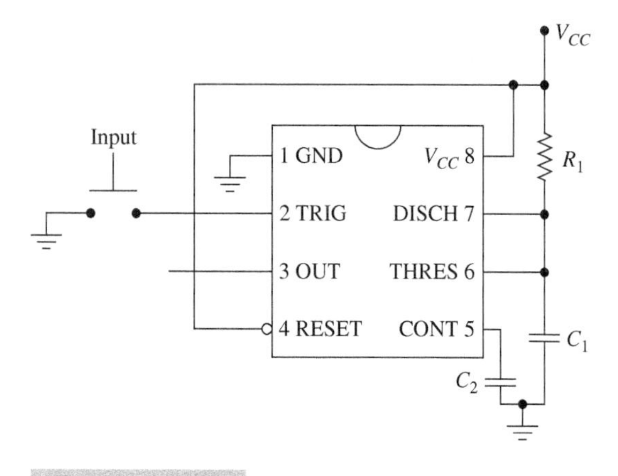

Figure L/P3.3

Chapter 4

Microcontrollers

Chapter Objectives:

When you have finished this chapter, you should be able to:

- Write and interpret data in different numbering systems
- Explain the difference between microprocessors and microcontrollers
- Explain details of the processor, different types of memory and buses, and design architectures in microcontrollers
- Explain the basic components of a microcontroller
- Explain the process of programming a microcontroller
- Explain the features and details of the ATmega328P MCU and the Arduino Uno board
- Explain the different clock sources for an MCU
- Explain the details of digital I/O, A/D, and PWM operations performed by the ATmega328P MCU
- Explain the basic functions provided by the Arduino IDE
- Explain EPROM data, timers, watchdog timer, and reset operations in AVR microcontrollers

4.1 Introduction

This chapter focuses on microcontrollers. Unlike combinational and sequential circuits which were covered in the previous chapter, microcontrollers offer a flexible but complex method to implement control logic. The flexibility comes from the fact that the control logic is software-based and can be changed by changing the software or control program. The complexity comes from the fact that the processor in a microcontroller is made up of a very large number of digital circuits. Microcontrollers are widely used for control applications in vehicles, toys, appliances, and telecommunications devices. Microcontrollers are also known as embedded controllers since they are not normally seen in the control application. Due to their small size, microcontrollers do not have many of the components that a typical PC has such as a display device, a keyboard, or a mass storage device.

This chapter discusses the operation of microcontrollers in detail. The objective of this chapter is to give the reader a complete coverage of the features and capabilities of a typical microcontroller. To understand the operation details and the performance characteristics of microcontrollers, the chapter starts by discussing different numbering systems. This is followed by some background information about microprocessors and microcontrollers including the processor, different types of memory and buses, and design architectures before it goes to the details of a microcontroller. This textbook will discuss the AVR microcontroller units (AVR™ MCUs) manufactured by Microchip Technology, Inc. It focuses on the ATmega328P microcontroller, which is the microcontroller used in the popular Arduino Uno board. The chapter covers the details of the Uno board, as well as the basic operations performed by the ATmega328P microcontroller including digital input and output, analog-to-digital conversion, and pulse width modulation output. It also has coverage of clock sources for an MCU as well as timers and reset operations. For further reading on microcontrollers, see [11-12].

4.2 Numbering Systems

Different numbering systems represent different types of data. For everyday operations the decimal system is appropriate, which is based on ten digits (0, 1, 2,... 8, 9). For computer operations, other numbering systems are used such as the binary and hexadecimal systems. This section will discuss the representation of numbers using different numbering systems (▶ available—V4.1).

4.2.1 Decimal System

In the decimal system or base-10 system, a number is represented as a combination of any of the base ten digits that are in that system. The value of the number is found by summing the product of each digit in the number multiplied by ten raised to the power of the location of that digit in the number where the rightmost digit has a location of zero and increases by 1 for every digit when moving to the left. For example, the value of the number 763 is found from

$$763 \Rightarrow 7 \times 10^2 + 6 \times 10^1 + 3 \times 10^0$$

While the decimal system is convenient for financial, scientific, and everyday mathematical operations, it is not convenient to use when representing numbers that are used in computer systems.

4.2.2 Binary System

The binary system, or base-2 system, uses two digits (0 and 1) to represent numbers. This is similar to how data is stored inside a computer's memory. Similar to the decimal system, the value of the number represented by this system is equal to the sum of the product of each digit multiplied by two raised to the power of the location of that digit. For example, the binary number 10110 is equivalent to decimal 22. This can be seen by writing

$$10110 \Rightarrow 1 \times 2^4 + 0 \times 2^3 + 1 \times 2^2 + 1 \times 2^1 + 0 \times 2^0 \Rightarrow 22$$

To convert from decimal to binary representation, we start by dividing the decimal number by the largest power of 2 that it can be divided by. We then take the remainder and divide by the next highest power of 2 possible. We repeat this process until we divide by 1. Example 4.1 shows the conversion of decimal 59 into a binary representation. In the Arduino programming language, a number with a leading 0b before it is interpreted as a binary number. For example, decimal 6 is written as 0b110.

Example 4.1 Decimal to Binary Conversion

Represent decimal 59 in binary form.

Solution:

Start by dividing 32 into 59, the largest power of 2 that we can divide 59 into. Take the remainder, which is 27, and then divide by the next largest power of 2, which in this case is 16. The remainder after the second division operation is 11. Continue this process until dividing by 1. The details are shown below, where r is the remainder after each division.

59/32 = 1 r 27
27/16 = 1 r 11
11/8 = 1 r 3
3/4 = 0 r 3
3/2 = 1 r 1
1/1 = 1 r 0

The binary equivalent is then formed by arranging all the quotients from the division operations in a left to right fashion starting from the top division operation. So 59 ⇒ 0b111011.

The smallest unit of storage in a computer system is the **bit** (which comes from combining letters from the words **bi**nary and digi**t**). A single bit can store either 0 or 1. A group of eight bits is called a **byte**, and two bytes are called a **word**. The different number of combinations of binary numbers that can be stored in an *n-bit* wide memory location is 2^n. Thus, a byte can store 256 different combinations of 0 and 1. If these values are limited to unsigned integers, then all decimal numbers from 0 to 255 can be stored in a byte. Similarly, a 16-bit memory location can store 65536 different combinations.

When dealing with binary numbers, the rightmost bit of a binary number is called the Least Significant Bit (or **LSB**) since it represents the smallest power of 2, and the leftmost bit is called the Most Significant Bit (or **MSB**) since it represents the largest power of 2. The LSB is referred to as bit zero, the bit adjacent to it as bit 1, and so forth. Binary digits can be added or subtracted in the same manner as decimal digits. When adding, a carry of 1 occurs to the next higher-order bit if the sum of the bits is 2 or higher. When subtracting, a borrow of 2 occurs from the next higher-order bit if the bit being subtracted is larger than the one it's being subtracted from. As an example, the binary sum of 0011 and 0010 is 0101. Here, when 1 was added to 1 in the first-bit location, we get a zero with a carry of 1 to the second-bit location. Example 4.2 further illustrates binary addition and subtraction. When referring to computer memory, the term 1K is normally used, and 1K of memory is 1024 bytes. Similarly, 64K of memory is 65536 or 64×1024.

Example 4.2 Binary Addition and Subtraction

Perform (a) the binary addition of 10110 (decimal 22) and 00111 (decimal 7) and (b) the binary subtraction of 10110 (decimal 22) from 11001 (decimal 25).

Solution:

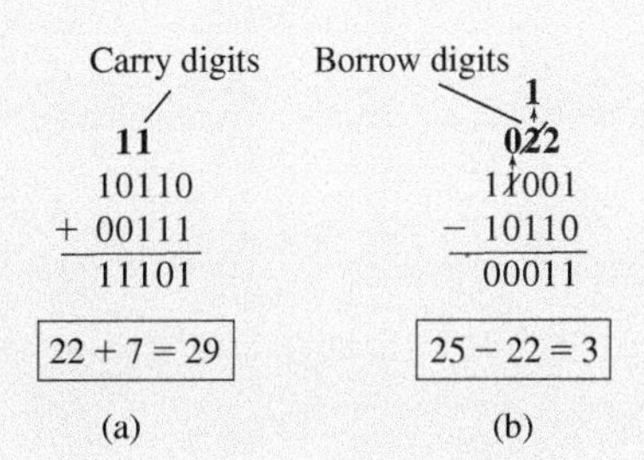

Figure 4.1 (a) Binary addition and (b) binary subtraction

Binary digits can be added in the same fashion as decimal digits with a carry of 1 to the next higher-order bit if the sum of the two bits is 2 or higher. As shown in Figure 4.1(a), when 0 is added to 1 in the zero-bit location, we get 1. When we add the two 1 digits in the first-bit location, we get 2 which is written as zero with a carry of 1 to the second-bit location. When that carry of 1 is added to the two 1 digits in the second-bit location, we get 3. Write 1 down and carry another 1 to the third-bit location.

Like decimal subtraction, if we cannot perform the subtraction for a particular digit, we need to borrow 2 from the next left digit location. As shown in Figure 4.1(b), we cannot subtract 1 from zero in the first-bit location. So, borrow 2 from the next left bit location. However, since that bit is also zero, first borrow 2 from the third-bit location to the second-bit location, which then becomes 1 when we borrow 2 from that bit location to the first-bit location. The relationship between the carry and borrow operations is that in the carry operation, 2 becomes 1 when moved to the next left bit location, while in the borrow operation, 1 becomes 2 when moved to the next right bit location.

4.2.3 Hexadecimal System

When evaluating the contents of a large memory location, such as a 32-bit field, it is more convenient if we can write the values of each 4-bit into one digit. This can be done using the hexadecimal system, or base-16 system, which uses sixteen digits to represent numbers. The first ten digits are the same as the decimal digits 0 through 9, while the last six digits (10, 11, 12, 13, 14, and 15) are represented using the letters A, B, C, D, E, and F, respectively. The value of the number represented by the hexadecimal system is found in a similar way to the decimal and binary systems with the exception that the base used in this case is 16. Table 4.1 below shows the numbers 0 through 15 in decimal, hexadecimal, binary, and binary-coded decimal (BCD) systems. A hexadecimal number is indicated by a suffix *h* or *H* or a prefix *0x* added to the number. For example, 12h (or 12H) is hexadecimal 12 or decimal 18. Similarly, 0x10 indicates hexadecimal 10 or decimal 16.

In the **BCD system**, each decimal digit is coded separately in binary. Thus, the BCD representation of decimal 12 is the binary representation of decimal 1 and decimal 2 digits grouped. The BCD system is normally used to drive decimal display systems.

Table 4.1

Different numbering systems

Decimal	Hexadecimal	Binary	Binary Coded Decimal (BCD)
0	0	0000	0000 0000
1	1	0001	0000 0001
2	2	0010	0000 0010
3	3	0011	0000 0011
4	4	0100	0000 0100
5	5	0101	0000 0101
6	6	0110	0000 0110
7	7	0111	0000 0111
8	8	1000	0000 1000
9	9	1001	0000 1001
10	A	1010	0001 0000
11	B	1011	0001 0001
12	C	1100	0001 0010
13	D	1101	0001 0011
14	E	1110	0001 0100
15	F	1111	0001 0101

Example 4.3 addresses conversions between binary, hexadecimal, and decimal systems.

Example 4.3 Conversions Between Binary, Hexadecimal, and Decimal

A 16-bit port has the following binary pattern. Determine the value of this data in hexadecimal and decimal.

Binary Pattern: 0b0110100101010011

Solution:

Make use of Table 4.1 to solve this problem. Using the table, the above 16-bit binary number will be first converted to a four-digit hexadecimal number. This will be done by finding the hexadecimal digit that corresponds to each 4-bit grouping of the above number starting from the right end. This gives us 0x6953. The next step is to determine the value of the hexadecimal number in decimal. This is obtained by evaluating $6 \times 16^3 + 9 \times 16^2 + 5 \times 16^1 + 3 \times 16^0 = 24576 + 2304 + 80 + 3 = 26963$.

4.2.4 Negative Number Representation

Negative numbers are represented using a method called the **2's complement**. It is calculated for a given number by taking the complement of its bit pattern and adding 1 to it. For example, consider the representation of −1 in binary format. We will assume an 8-bit field. We start by writing the number 1 in binary format using the 8-bit field. This gives us 00000001. Next, find the complement of this number which is the bit pattern that contains the exact opposite values of the given bit pattern (i.e., 1 changes to 0 and 0 changes to 1). So, the complement of 1 is then 11111110. The last step is we add 1 to the complement, which gives us 11111111 as the 2's complement representation of −1. The representation of negative numbers in binary or hexadecimal format is very dependent on the number of bits that are used to represent the number. If we had used 4 instead of 8 bits to represent −1, the answer would be 1111. An alternative method to determine the 2's complement is to find the binary pattern representation for the result of the following operation

$$2^n - number \tag{4.1}$$

where n is the bit field width. So, applying Equation (4.1) to the representation of −1, and using a 4-bit field gives 15 ($2^4 - 1 = 15$), which has a binary representation of 1111.

For an n-bit field, the range of signed numbers that can be represented by that field is from -2^{n-1} to $2^{n-1} - 1$. For example, for $n = 4$, the range is −8 to 7, or a total of 16 numbers including 0. For $n = 8$, the range is −128 to 127.

4.2.5 Representation of Real Numbers

The representation of real or floating-point numbers in binary format is more complicated than the representation of integer numbers. There are several methods available to represent real numbers; the most common is the IEEE-754 floating point method, which is used by all modern CPUs. The representation is dependent on the number of bits that are used to represent the number. We will illustrate this method using a 32-bit field, or single-precision representation. In this method bits 0–22 are used to represent the mantissa, or fraction, bits 23–30 are used to represent the exponent, and the MSB, or bit 31, is used to represent the sign. For positive numbers, the sign bit is 0 and for negative numbers, the sign bit is 1. The value of the exponent is computed from bits 23–30 by subtracting 127. This allows both positive and negative exponents to be represented. A value of 1 in bit 22 represents a ½ fraction; a value of 1 in bit 21 represents a ¼ fraction, etc. In this representation, an invisible leading bit with a value of 1 is assumed to be placed in front of bit 22. Thus, the values of these fractions are added to the invisible one to give a mantissa value between 1 and 2. The value of a floating-point number in this representation is then computed from

$$\text{sign} \times 2^{\text{exponent}} \times \text{mantissa} \tag{4.2}$$

As an example, the binary pattern 0100 0000 0111 1000 0000 0000 0000 0000 is a representation of the number 3.875. Here the exponent value is 1 (128 − 127), the mantissa value is 1.9375 (1 + 1/2 + 1/4 + 1/8 + 1/16), and the value is $2^1 \times 1.9375 = 3.875$.

4.3 Microprocessors and Microcontrollers

The **microprocessor** (▶ available—V4.2), the brain of modern computers, is an integrated circuit (or a chip) that has a processor which consists of many digital circuits. For example, microprocessors such as the Intel Core i7 contain hundreds of millions of transistor elements. For personal computers, the microprocessor is housed on the motherboard of the PC and uses an external bus to interface with memory and other components on the PC such as mass memory and system I/O. A **bus** is a set of shared communication lines (physically it could be tracks on a printed circuit board or wires in a ribbon cable). The combination of the microprocessor and other elements on the motherboard is called a **microcomputer**. The **microcontroller**, on the other hand, is a **single-chip device** that contains a processor along with memory and interface devices on the same integrated circuit chip. The microcontroller uses an internal bus to communicate with memory and other devices on the chip. Note that microprocessors require peripheral chips to interface with I/O devices. Figure 4.2 shows a picture of a motherboard and a microcontroller.

Figure 4.2

A picture of (a) a motherboard and (b) a microcontroller

(Courtesy of eBay); (Jouaneh, University of Rhode Island)

(a)

(b)

While microprocessors and microcontrollers share many features, they had different evolution paths. Microprocessors were developed for use in personal computers and workstations, while microcontrollers were developed for use in control applications in appliances, automotive, entertainment, and telecommunications industries. In microprocessors, the emphasis is on high speed and large word sizes such as 32 or 64-bit, while in microcontrollers, the emphasis is on compactness and low cost. In microprocessors, the RAM size is typically in Gigabytes, while in microcontrollers it is in 1 to 100K of bytes. In microprocessors, clock speed is in the range of several GHz, while in microcontrollers, it is in the tens of MHz.

It should be noted that in addition to MCUs, there exists another type of embedded controller called a **field programmable gate array** or **FPGA**. While similar to an MCU in that it is a single-chip device, compact, and programable by the user, there are also major differences between them. An FPGA allows parallel processing of commands (and thus faster execution) unlike an MCU which only allows serial processing of commands, and its hardware structure is not fixed but can be reconfigured by the user to meet the needs of the application unlike that of an MCU. FPGAs also tend to be more expensive than MCUs and are more complicated to program than MCUs. FPGAs are used in several industries including telecommunications, networking, and automotive, and are typically suited for computationally intensive applications such as image and video processing. This book does not cover FPGAs. For more information about FPGAs, the reader can refer to [13].

4.3.1 Processor, Memory, and Buses

The basic job of a processor is to execute program instructions, which are the low-level code generated by the compiler in translating a high-level computer program, such as C code, into machine instructions that the processor uses. The processor is also known as the **Central Processing Unit** or CPU. The CPU (▶ available—V4.3) has three basic elements: the Control Unit, the Arithmetic and Logic Unit (ALU), and the Registers. The function of each of these elements is described below:

Control Unit: Determines timing and sequence operations. This unit generates timing signals used to fetch a program instruction from memory and execute it.

Arithmetic and Logic Unit (ALU): This unit performs logical evaluations and actual data manipulations, such as adding two numbers.

Registers: These are memory locations inside the CPU that hold internal data while instructions are being executed. A register is also a general term for a memory location that is set for a particular purpose.

As an example, to add two numbers, the following operations typically occur: The first number is brought from memory and held in one of the registers. The second number is brought from memory, and the ALU operates on the two numbers. The result of the operation is first stored in one of the registers before it is transferred back into memory.

Table 4.2 lists the different types of memory devices (▶ available—V4.4) that are used. In Read Only Memory (ROM) or any variation of it, such as Erasable Programmable ROM (EPROM), the data remains in memory even after the power is turned off. In Random Access Memory (RAM) (▶ available—V4.5), the data is lost if the power is turned off. Recent microcontrollers use Flash memory to store program instructions, which are downloaded to the microcontroller through a serial or USB connection.

In addition to memory, means must be provided to transfer data between the different components in a microcontroller or a microprocessor. The data transfer occurs over a bus (▶ available—V4.6). There are different kinds of buses, including:

Data Bus: Used to transport data to and from the CPU and the memory or the input/output devices. Data width can be 4, 8, 16, 32, or 64 bits.

Address Bus: Used to select devices on the bus or specific data locations within memory. Each memory location has an address that must be specified before the contents of that location can be accessed. The size of the address bus determines the number of locations to be addressed. A 16-bit bus can access 2^{16} addresses or 64K locations, while a 32-bit bus can access 4G locations.

Control Bus: Used to synchronize the operation of the different elements. It transmits read and write signals, system clock signals, and other control signals.

Table 4.2

Different types of memory

Memory Type		Description
ROM	Read Only Memory	Nonvolatile memory that is programmed with required content during the manufacture of the IC chip. Data can only be read but cannot be written during use, and it does not lose its data when power is turned off. It is used for fixed programs such as computer operating systems.
PROM	Programmable ROM	Same as ROM but can be programmed by the user once, after which no further changes are allowed.
EPROM	Erasable PROM	Can be programmed more than once during use. Contents can be erased by shining ultraviolet (UV) light through a quartz window on top of the device.
EEPROM	Electrically Erasable PROM	Similar to EPROM but contents can be erased by applying a voltage signal rather than a UV light.
FLASH	Flash memory	Flash memory is a special form of EEPROM. Unlike an EEPROM, where erasure is done byte-wise, an entire block of bytes (such as 4K) is erased in Flash which makes them faster. Flash memory is less expensive than EEPROM, but its lifespan has fewer erase/write cycles than EEPROM.
RAM	Random Access Memory	It is the memory that is used to store data while a program is running. It is a volatile memory that requires power to operate, and data is lost when power is removed. The access time for the data is constant and is not dependent on the physical location of the data. RAM is available in two types: SRAM and DRAM.
SRAM	Static RAM	A RAM in which data does not need to be refreshed as long as the power is applied. The data can be accessed faster than DRAM, but it is more expensive.
DRAM	Dynamic RAM	A RAM that uses capacitors to store data. Data must be refreshed (rewritten) periodically because of charge leakage.

4.3.2 Components of a Typical Microcontroller

This textbook will discuss the 8-bit AVR microcontroller units (AVR MCUs) manufactured by Microchip Technology, Inc. There are many other microcontrollers on the market today such as those made by NXP (LPC and Kinetis), Renesas (RL78), and STMicroelectronics (STM8S). We selected the AVR microcontrollers due to their widespread use, low cost, and ease of use. AVR microcontrollers are also used in the popular **Arduino** microcontroller systems. As mentioned before, a microcontroller is a *single-chip* device that includes a processor, memory, and interface devices. The components of a typical microcontroller are shown in Figure 4.3. These components include the CPU; the nonvolatile **Flash program memory** to store the code or program instructions; the volatile **RAM** to store data (variables and register values) while a program is executing; interface devices such as digital input/output (I/O) ports, analog-to-digital (A/D) converter, serial port, or USB; and clock and timers. Note also that in volatile memory such as RAM, stored information is lost when the power supply is cut off. Thus, many microcontrollers have a **data EEPROM** section, which can be used to store data values during program execution (at a longer access time than RAM storage) but has the advantage that the data will not be lost if the power was lost.

Figure 4.3

Typical components of a microcontroller

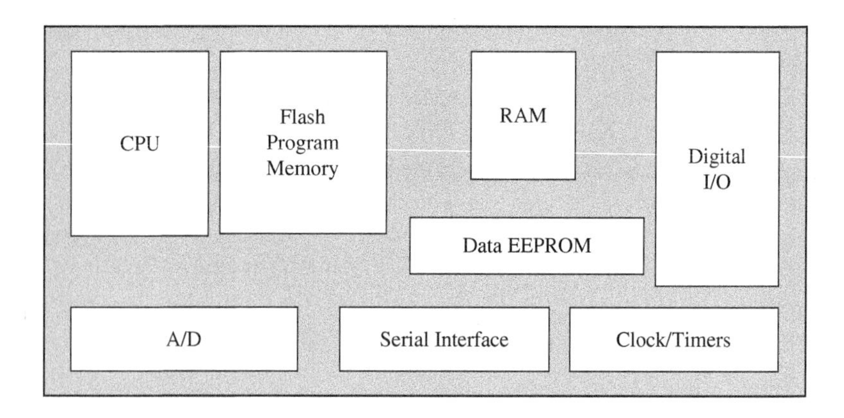

In addition to the 8-bit AVR MCUs, Microchip manufactures several families of 8, 16, and 32-bit microcontrollers. The number of bits refers to the size or width of the data bus which is used to transport data from/to CPU and the memory or the input/output devices. The wider the bus size, the faster the data is transferred between the components of the microcontroller. Also, it takes longer to complete a larger computation with a narrow data bus than with a wider data bus. So, if the goal is to have higher throughput and faster computational speed, then an MCU with a wide data bus should be used such as a 32-bit MCU. Within each family, several microcontrollers are available that differ in their physical size, number of pins, memory size (Program memory, RAM, and EEPROM), and type of interface devices provided.

4.3.3 Design Architectures and MCU Operation

Microcontrollers are designed using either the Harvard or Von Neumann architectures (▶ available—V4.7). The architecture refers to the internal design of the hardware in the microcontrollers which affect the method of exchanging data between the CPU and memory. Figure 4.4 shows the two architectures.

Figure 4.4

Illustration of the Harvard and Von Neumann architectures

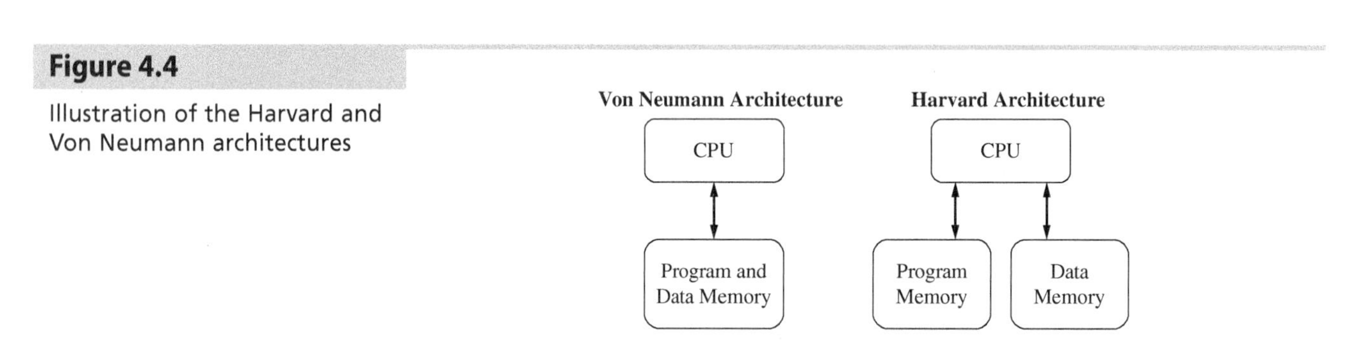

The **Von Neumann architecture** uses a single memory for the program (or instructions) and the data, and a single bus to access them, while the **Harvard architecture** uses separate memories and buses for the program and data. Due to the use of separate memories and buses, the Harvard architecture has the advantage that instructions and data can be accessed at the same time. Since memory can only be accessed once per clock cycle of the microcontroller, this leads to a faster execution than in the Von Neumann architecture where the instructions and data are accessed in separate clock cycles. In Harvard architecture, instructions are executed in every clock cycle. and while one instruction is being executed, another is brought from program memory. The AVR microcontrollers are designed using the Harvard architecture. Within each architecture, instructions (such as adding two numbers or multiplying two numbers) can be designed using two design approaches. These are the **Complex Instruction Set Computer** (**CISC**) approach and the **Reduced Instruction Set Computer** (**RISC**) approach. The CISC approach is characterized by many complicated instructions that can perform more functions, often leading to a more compact program size but requiring more clock cycles per instruction. On the other hand, the RISC approach emphasizes simplicity and speed, using fewer, simpler instructions that allow for faster execution. Even though the compiled program for a RISC processor, like the AVR microcontroller units (MCUs), may be larger in terms of the number of instructions, it can run faster due to the optimization of each instruction.

The main job of the microcontroller unit (MCU) is to execute program instructions. To do this, the instructions must be brought from program memory to the MCU's CPU, under the control of the **program counter** (PrC). The PrC is a special register that holds the address of the next instruction to be executed. When the MCU is powered on or reset, the PrC is cleared and points to the reset vector address (typically 0x0000, though this can vary if a bootloader is used—see next section), which contains the first instruction to execute. While the processor executes the instruction at the current address, the address of the next instruction is automatically loaded into the PrC. The size of the PrC is designed to access all the instructions in program memory. For example, on the ATmega328P (discussed in the next section), the PrC is 14-bits wide, enabling access to up to 16K instructions. Associated with the PrC is the **stack**, a special area of memory distinct from data or program memory, operating in a first in, last out (FILO) fashion. The stack serves as a storage area for the contents of the PrC when executing a function/subroutine call or an interrupt (a hardware-initiated function call). When the program needs to perform a branching operation, such as executing a subroutine call, the contents of the PrC are placed on the stack (pushed), and the address of the subroutine to be executed is loaded into the PrC. Once the routine is complete, the address of the next instruction is retrieved from the stack (popped) back into the PrC. The size of the stack determines the number of subroutine calls or interrupts that can occur. In the ATmega328P, the stack size is variable and is limited only by the total RAM size and its usage.

The speed of execution of the instructions is a function of the clock speed of the MCU. On AVR MCUs, most of the instructions execute in one clock cycle, while some instructions such as those that perform certain branching operations (e.g., a Direct Subroutine Call instruction) take more than one clock cycle to execute. Examples of single clock execution instructions are *ADD* (add two registers), *OR* (logical OR registers), and *INC* (increment).

4.4 AVR Microcontrollers

Microchip manufactures several families of microcontrollers. These include the 8-bit megaAVR family, the 8/16-bit AVR XMEGA family, and the 32-bit AVR UC3 family. This textbook will focus on the ATmega328P, an 8-bit megaAVR microcontroller, which is the microcontroller used in the popular Arduino Uno microcontroller. Table 4.3 shows some of the many microcontrollers that are currently available in the megaAVR family along with some pertinent data on them. The table also lists in parentheses the applicable Arduino board which uses these MCUs. Note that due to market forces, these microcontrollers are continually replaced by ones with improved features. The letter 'P' (stands for pico power) in the microcontroller part number (such as ATmega48P) specifies an MCU with lower power consumption than the same MCU without the letter 'P'.

Table 4.3

A sampling of different microchip megaAVR MCUs

AVR MCU (Arduino Board)	Program Memory (Kbytes)	Data EEPROM	RAM	Max I/O	ADC Chan	ADC Bit	PWM Chan	Pin Count	Max Oper. Freq (MHz)	Interface
ATmega48P	4	256	512	23	8	10	6	32	20	UART, SPI, I2C
ATmega328P (Uno)	32	1024	2048	23	8	10	6	32	20	UART, SPI, I2C
ATmega32U4 (Leonardo or Micro)	32	1024	2.5 K	26	12	10	7	44	16	UART, SPI, I2C
ATmega2560 (Mega 2560)	256	4096	8 K	86	16	10	15	100	16	UART, SPI, I2C

The maximum number of I/O lines refers to the number of digital input/digital output lines available on the chip. These I/O lines are bi-directional and are configured by program code to be either an input or output type. All the above chips come with analog-to-digital converter capability with a 10-bit resolution.

The A/D channels use the same pins as those used for digital I/O lines, but which can be configured by program code to operate as A/D channels. All the above chips have several PWM lines which can be conveniently used to drive motors through an H-bridge driver or a transistor. The **pin count** refers to the number of physical pins on the chip. Some chips are available in different pin counts depending on the chip packaging configuration (see Section 4.4.1). For example, the ATmega328P MCU is available as 28 pins in the PDIP configuration and as 32 pins in the TQFP configuration. The maximum operating frequency refers to the maximum speed of the clock which can drive the chip. The chips listed in Table 4.3 have a maximum clock frequency of 20 MHz or lower. All the listed chips have interfaces for ease of communication with other devices. An explanation of these interfaces is given in Table 4.4.

Table 4.4

A list of interfaces available on some MCUs

Interface Feature	Explanation
USART	Universal Synchronous Asynchronous Receiver Transmitter (USART) module—used for synchronous (data line and clock signal) and asynchronous (data line but no clock signal) serial communication
AUSART	Addressable Universal Synchronous Asynchronous Receiver Transmitter (AUSART) module—Can be configured as Asynchronous (full duplex), Synchronous—Leader (half duplex), or Synchronous—Follower (half duplex) serial communication line
EUSART	Enhanced Universal Synchronous Asynchronous Receiver Transmitter (EUSART) module—Supports RS-485, RS-232, and LIN compatibility—Auto-Baud Detect—Auto-wake-up on Start bit
MSSP	Leader Synchronous Serial Port (MSSP) module which includes an SPI™ and I²C™
SPI™	Synchronous serial port configured as 3-wire Serial Peripheral Interface (SPI™)
I²C™	Synchronous serial port configured as a 2-wire Inter-Integrated Circuit (I²C™) bus
USB	Universal serial bus
LIN	Local Interconnect Network which is a serial communication system

4.4.1 Pin Layout

MCUs are available in different pin layouts (see Figure 4.5). A common layout is the PDIP (**P**lastic **D**ual **I**nline **P**ackage). In this layout, the pins are arranged in two parallel opposite rows, and this layout is normally used with breadboards. The SOIC (**S**mall-**O**utline **I**ntegrated **C**ircuit) layout uses gull-wing pins extending outward that are more tightly spaced than the PDIP. The QFP layout (**Q**uad-**F**lat **P**ackage) has pins on all four sides of the package, while the QFN (**Q**uad-**F**lat **N**o-Leads) layout also has pins on all four sides, but the pins are exposed on the bottom and sometimes on both sides and the bottom. QFP and QFN layouts are also available in other variations including thin QFP (TQFP), low profile QFP (LQFP), or very-thin QFN (VQFN). Integrated circuits in SOIC, QFP, and QFN layouts are classified as surface mount-type devices. These layouts are not designed to be used on breadboards, and special automated tools are used to solder these chips to circuit boards.

Figure 4.5

(a) PDIP (b) SOIC, (c) QFP, and (d) QFN packaging

(Courtesy of Microchip, Chandler, AZ)

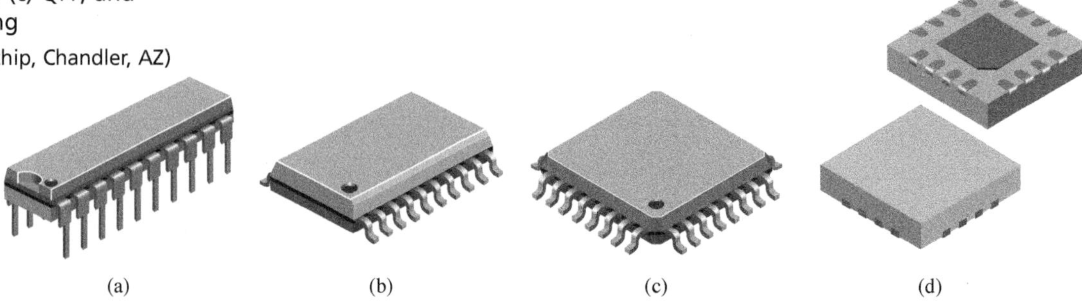

Figure 4.6 show the pin layout for the ATmega328P MCU in the PDIP configuration. The ATmega328P has three I/O ports labeled PB, PC, and PD. Because this MCU has only 28 pins but supports many interface functions, many of the pins are designed for more than one function. For example, pin 23 can be configured as channel 0 of digital input/output port C (PC0), analog-to-digital converter channel 0 (ADC0), or pin change interrupt 8 (PCINT8). The limited number of external pins also means that you cannot have 8 A/D channels and 23 I/O lines operating at the same time. Table 4.5 explains the labels of these pins.

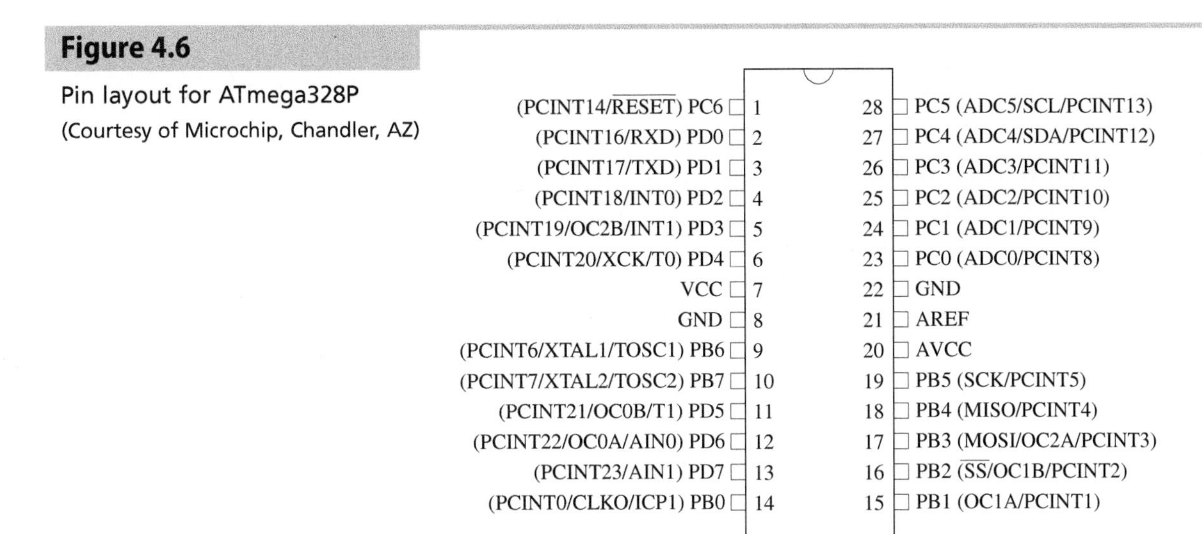

Figure 4.6

Pin layout for ATmega328P

(Courtesy of Microchip, Chandler, AZ)

Table 4.5

Pin functions on ATmega328P MCU

ADC	Analog-to-digital converter input channel
AIN0	Analog comparator positive input
AIN1	Analog comparator negative input
AREF	External voltage reference for A/D
AVCC	Supply voltage pin for A/D
CLKO	Divided system clock output
ICP1	Timer/counter 1 input capture pin
INT	External interrupt input pin
GND	Ground reference line
MISO	SPI bus leader input/follower output
MOSI	SPI bus leader output/follower input
OC0A, OC1A, OC2A	Timer/counter (0, 1, and 2) output compare match A output
OC0B, OC1B, OC2B	Timer/counter (0, 1, and 2) output compare match B output
PB, PC, PD	Digital input/output ports B, C, and D
PCINT	Pin change interrupt
RESET	Reset pin
RXD	USART input pin
SCK, SCL	SPI clock, and I2C clock
SS	Follower select line for SPI

(*Continued*)

Table 4.5

(Continued)

T0, T1	Timer/counter0 (or 1) external counter input
TOSC	Timer oscillator pin
TXD	USART output pin
VCC	Positive lead of supply voltage. This chip operates at an input voltage range of 1.8-5.5 V
XCK	USART external clock input/output
XTAL	Chip clock oscillator pin

4.4.2 AVR MCU Block Diagram

Figure 4.7 shows a block diagram of the main components of the ATmega328P MCU. Some of these components include the three digital I/O ports (Ports B, C, and D), the three timers/counters (T/C 0, 1, and 2), the USART module for RS232 communication, the A/D converter module, the SPI and I2C module, the flash program memory, the SRAM, and the CPU. Most of these components will be explained in detail in later sections of this chapter. Note how the 8-bit data bus connects the CPU to the other components on the MCU.

Figure 4.7

A block diagram of the ATmega328P MCU

(Courtesy of Microchip, Chandler, AZ)

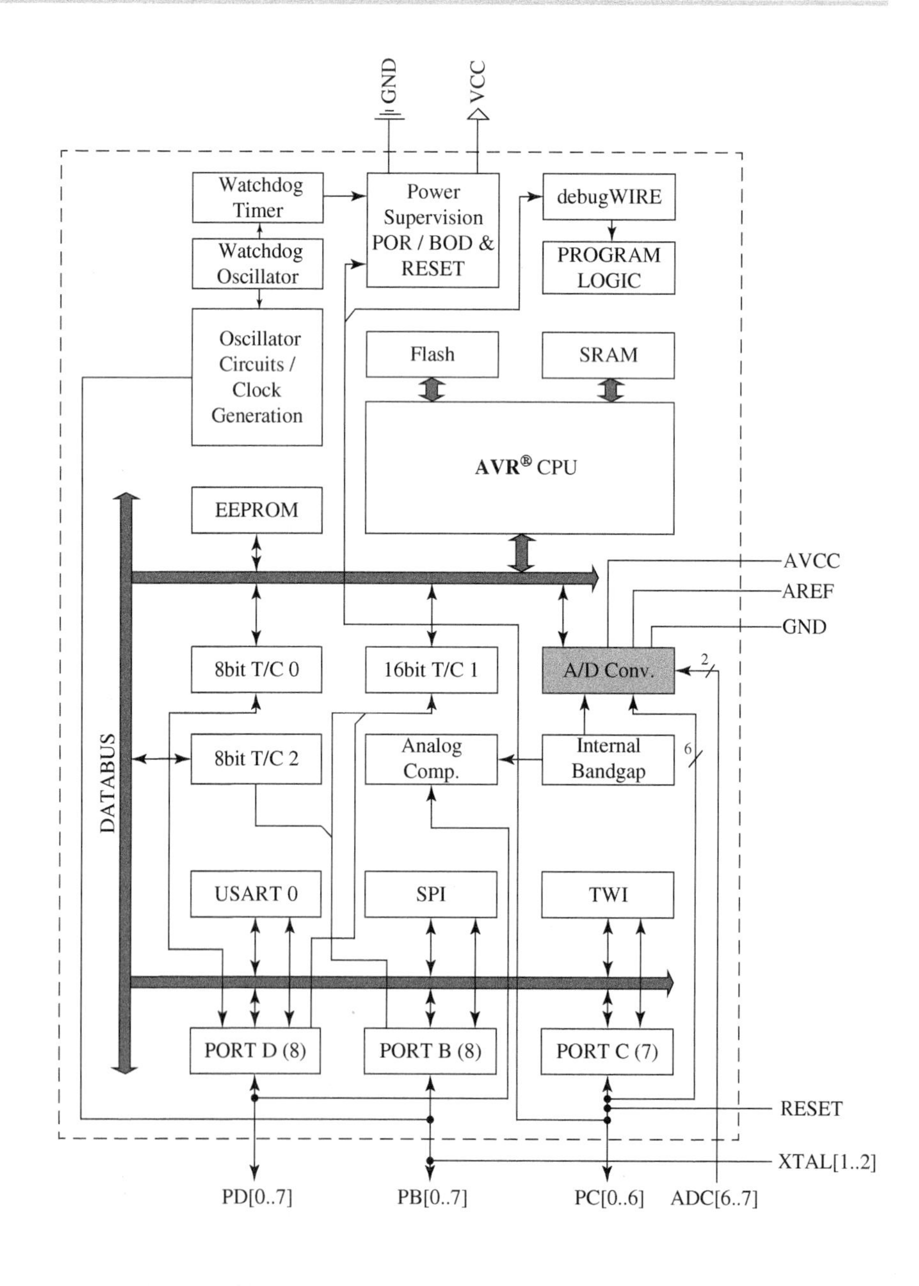

4.4.3 Arduino Uno Board

The **Arduino Uno** board is an open-source MCU-based development platform that includes an AVR ATmega328P MCU built into the board. The Arduino boards (▶ available—V4.8) were first introduced in 2005. Since then, there have been several boards available including the famous Arduino Uno, Arduino Leonardo, and Arduino Nano. While the different Arduino boards have different sizes, and memories, they all share nearly similar board structure and labeling. The Arduino boards have attracted the attention of many users due to the freely available software to program the boards, the ease of connecting wires to the boards, and the ease of connecting the boards to a computer. In addition, the boards take advantage of the large availability of shields, which are interchangeable custom boards that are mounted on the top of the Arduino boards to provide additional functionality such as prototyping, Ethernet communication, or motor drive.

A picture of the original Arduino Uno R3 board is shown in Figure 4.8. There have been several revisions of the Uno board, and the one shown in the figure is revision three or R3. The original boards refer to the ones made by the Arduino consortium in Italy. Note that since the Arduino boards are open source, many companies make similar boards that are usually less expensive than the original ones. The Arduino Uno board has two AVR MCUs. It has an ATmega328P MCU as the main MCU which is available for the user to program, and an ATmega16U2 which is solely used for communication between the board and a PC. In addition to these two MCUs, the board has additional components. These include a 16 MHz crystal oscillator, a 16 MHz ceramic resonator, various connectors, LEDs, and a reset switch. The board connects to a computer via a Type A/B USB cable. The board operates at a 5 VDC voltage level and could be powered by an adapter that supplies 7-12 VDC. The board can also operate on the power provided to it by the USB cable, but the cable can only provide a maximum current of 500 mA to the board.

Figure 4.8

Top view of the Arduino Uno R3 board

(Jouaneh, University of Rhode Island)

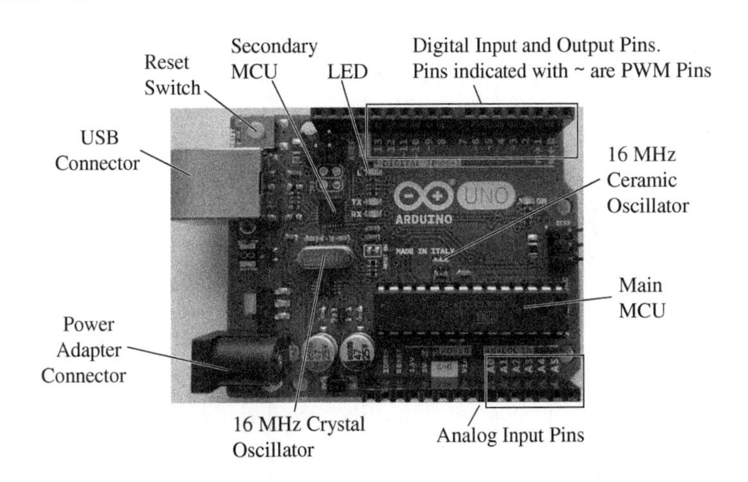

Note the pin numbers shown on the boards are different from the pin numbers on the ATmega328P chip. The correspondence between the pin numbers is shown below in Table 4.6.

Table 4.6

Correspondence between pin numbers on the Arduino Uno board and the ATmega328P chip

Pin # on Uno	ATmega328P Port	Function
Analog A0	PC0	Analog Input
Analog A1	PC1	Analog Input
Analog A2	PC2	Analog Input
Analog A3	PC3	Analog Input
Analog A4	PC4	Analog Input, I2C SDA

(Continued)

Table 4.6

(Continued)

Analog A5	PC5	Analog Input, I2C SCL
Digital 0	PD0	Digital I/O, USART RXD
Digital 1	PD1	Digital I/O, USART TXD
Digital 2	PD2	Digital I/O
Digital 3	PD3	Digital I/O, PWM T2
Digital 4	PD4	Digital I/O
Digital 5	PD5	Digital I/O, PWM T0
Digital 6	PD6	Digital I/O, PWM T0, Analog Comparator positive input
Digital 7	PD7	Digital I/O, Analog Comparator negative input
Digital 8	PB0	Digital I/O, Input Capture
Digital 9	PB1	Digital I/O, PWM T1
Digital 10	PB2	Digital I/O, PWM T1, SPI SS
Digital 11	PB3	Digital I/O, PWM T2, SPI MOSI
Digital 12	PB4	Digital I/O, SPI MISO
Digital 13	PB5	Digital I/O, SPI SCK

An advantage of Arduino boards is that they can be programmed without the need for an external programmer. The Arduino environment software has this feature built into it and it allows the user to upload the code to the MCU by simply clicking on a button.

4.4.4 Clock/Oscillator Source

A microcontroller needs a clock source to function since all CPU operations are synchronized with the clock. A **clock** is any device that can produce a train of pulses at a fixed frequency. Most of these devices are called **oscillators** (▶ available—V4.9) which is the term given to an electronic circuit that uses a DC power source to generate a periodic AC signal. Some MCUs have a built-in clock source while others require or allow an external device to produce the clock pulses. The ATmega328P also has two internal oscillators: an 8 MHz high-frequency oscillator, and a 128 kHz low-frequency oscillator, but it also allows the use of an external clock source. The output frequency of the internal 8 MHz oscillator can be further divided by software. The effective clock frequency of an internal oscillator varies with the supply voltage and the temperature of the MCU.

External devices to produce a clock signal include a quartz crystal resonator, a ceramic resonator, a resistor-capacitor (RC) oscillator circuit, or an external clock source such as the 555-timer chip. The microcontroller allows through software the selection of the clock source.

A **quartz crystal resonator** (▶ available—V4.10) uses the mechanical resonance of a piezoelectric crystal to generate a very accurate timing signal while a **ceramic resonator** uses the mechanical resonance of piezoelectric ceramics (commonly Lead Zirconate Titanate or PZT). Ceramic resonators are small, rugged, and inexpensive while quartz crystal resonators are more expensive but more precise. Ceramic and quartz crystal resonators are available with different clock frequencies. A photo of a quartz crystal resonator is shown in Figure 4.9(a), while that of a ceramic resonator is shown in Figure 4.9(b).

Figure 4.9

(a) A quartz crystal resonator (Courtesy of TXC Technology Inc, Brea, CA) and (b) a ceramic resonator (Courtesy of Murata Manufacturing Co., Ltd., Nagaokakyo, Kyoto, Japan)

(a)

(b)

An **RC oscillator circuit** uses a feedback network of three RC circuits arranged in a ladder fashion and connected to a BJT transistor or an inverting Op-amp to produce a clock signal. It is normally used in applications where clock accuracy (1-10% error) is not very important. An external clock source such as the **555 timer** can also be used as the clock. A 555 timer chip (such as the NE555 8-pin chip from Texas Instruments—see Section 3.9) is an integrated circuit that uses transistors, resistors, and diodes to produce a variety of clock signals including periodic signal output, and time delays.

The Arduino Uno board uses a 16 MHz ceramic resonator as the clock source for the primary MCU (ATmega328P). Since a ceramic resonator is not very precise (accuracy is about 0.5%), the Arduino Uno board should not be used for precision timing applications since it could produce a timing error of about 18 seconds over an hour of operation. The board also uses a 16 MHz external crystal oscillator as the clock source for the secondary MCU (ATmega16U2) which is solely used for USB/serial communication between the board and a PC. USB communication is very sensitive to clock signal accuracy, so that is why the high-precision crystal oscillator (accuracy of 0.003%) is used for timing the operation of the secondary MCU. Figure 4.8 shows these devices on the Uno board.

To **get an MCU to run**, at least three pins need to be wired. These include the power supply pin (VCC pin on AVR MCUs) which should be connected to the positive lead of the supply voltage (1.8 to 5.5 volts for most AVR chips); the ground pin (GND pin on AVR MCUs) which should be connected to ground voltage; and the reset pin (RESET pin on AVR MCUs) which needs to be connected to the supply voltage (through a resistor) to prevent the MCU from resetting itself. For microcontrollers without internal clock sources, you also need to connect the external clock source to the microcontroller.

4.4.5 Programming a Microcontroller

The development of a control program running on a microcontroller has similarities to developing a program to run on a PC. In both cases, high-level programming languages such as C or Python are often used to write the code. Within a development environment, which may be tailored to a specific language or microcontroller, the high-level code is compiled into binary code using the built-in compiler. This binary code is then linked with other necessary files to form the executable program. On a PC, this program can be run simply by calling its name, while on a microcontroller, the compiled code must typically be loaded onto the device. In both cases, the development environment often includes debugging tools that allow developers to identify and correct errors in the code. This approach helps streamline the development process, whether targeting a traditional computing platform or an embedded microcontroller system.

In microcontroller development, programmers have the option to use either high-level programming languages such as C or **low-level assembly language**. Assembly language is tailored specifically to the architecture of the microcontroller being used. Though programming in assembly can be more challenging, it offers greater control over the execution timing of the code, as the number of clock cycles needed to execute each instruction is known. This allows for precise control and optimization of performance. Another benefit of using assembly language is the typically cost-free availability of assembly compilers, as they are often provided without charge by the microcontroller manufacturer or other sources. Whether using high-level or assembly language, the code must be compiled and translated into binary code (or a hex code file). This binary or hex code is then downloaded to the microcontroller, where it is stored in the non-volatile program memory, ready for execution. This dual approach of using either high-level languages or assembly offers flexibility to developers, balancing ease of use and powerful, fine-grained control.

One appealing feature of Arduino boards is the free Integrated Development Environment (IDE) software used to program them. The IDE is a combination of a text editor for writing and editing code and a compiler to validate the code's syntax and convert it into binary code that the Microcontroller Unit (MCU) can read. With Arduino boards, the user simply connects a USB cable between the board and a PC or laptop, then uses commands within the IDE to compile and upload the program (referred to as a 'sketch' in Arduino terminology) to the microcontroller. The Arduino IDE employs a built-in boot-loading application, eliminating the need for an external programmer. A **bootloader application** (▶ available—V4.11) is a special piece of software that facilitates the process of loading the operating system or other necessary programs (known as the 'boot image') into a device's memory. In the context of microcontrollers, a bootloader is a small program that resides in a dedicated section of the device's memory, and its primary function is to load the main application code. AVR MCUs support boot loading because their program memory is divided into two sections: one for program instructions and the other (located at the upper end of program memory) for the boot loading section where the bootloader code resides. When the microcontroller is powered on or reset, the bootloader is the first program that runs. It checks whether there is a new version of the application code to be loaded, usually communicating with an external source like a PC via a specific protocol (e.g., serial communication). If new code is present, the bootloader loads it into the appropriate area of the microcontroller's memory, overwriting the existing application code if necessary. The bootloader then passes control to the main application, which begins executing. Bootloaders are especially useful in embedded systems, where updating the software might be more challenging than in a typical computer environment. By providing a controlled way to load new code onto the device, bootloaders enable updates, bug fixes, and feature enhancements to be applied without requiring specialized hardware or physical access to the device. The complete process of programming a microcontroller, including these steps, is illustrated in Figure 4.10.

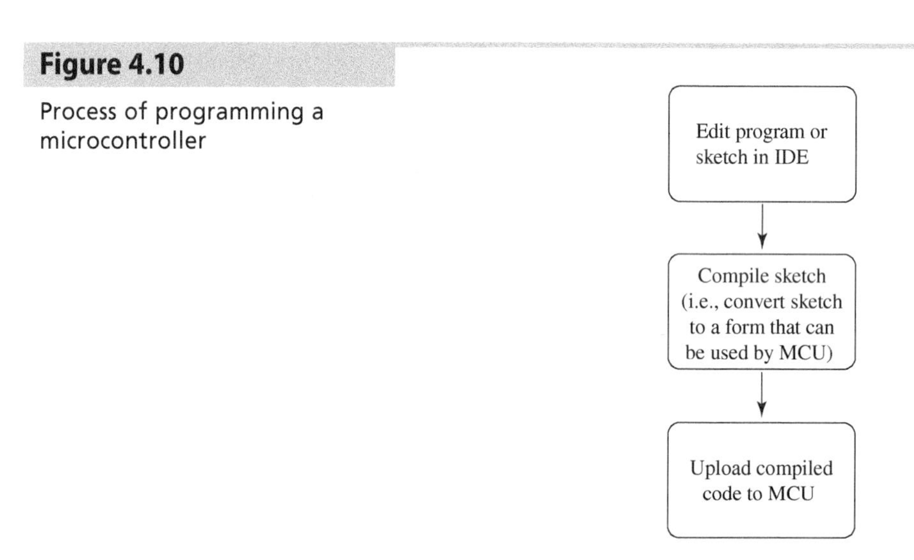

Figure 4.10

Process of programming a microcontroller

Before discussing the details of performing various operations in an MCU, Integrated Case Study IV.B discusses another AVR MCU that is used in that case study.

Integrated Case Study IV.B: Microcontroller to Control the Robot

For Integrated Case Study IV (**Mobile Robot**—see Section 11.5), the mobile robot uses the ATmega32U4 microcontroller to control its operation. This microcontroller belongs to the same AVR family and has the same program or flash memory size (hence the 32 in the name) as the ATmega328P microcontroller used in the Arduino Uno board. Both microcontrollers use the same RISC architecture, but the 32U4 has a slightly bigger SRAM capacity (2.5 KB vs 2 KB for the 328P). The main difference between the 32U4 and the 328P is that the 32U4 has a built-in capability for USB interfacing, while the 328P does not have that capability. That is why the Uno board has another microcontroller (ATmega16U2) that is solely dedicated to USB interfacing, while the 32U4 does not need that additional microcontroller. The 32U4 has onboard support for USB communication. This enables the 32U4 to emulate USB devices like mice and keyboards. Another difference is that the 32U4 has more channels that can be used for A/D conversion (a maximum of 12 channels in TQFP and 8 channels in the VQFN package configuration versus a maximum of 6 channels in the SPDIP Package for the 328P). Also, the 32U4 has more PWM channels than the 328P (7 vs 6).

Arduino manufactures the Leonardo board which uses the same ATmega32U4 microcontroller. The board has a similar size and layout to the Uno board, but it uses a micro-USB connector for connecting the USB cable. Unlike the Uno board, the Leonardo makes available to the user 7 PWM output channels and 12 A/D channels. Because Leonardo does not use a dedicated microcontroller for USB/serial communication, there are some differences in the operation of this board compared to the Uno board. The user should refer to the Arduino site for more details.

4.5 Digital Input/Output and Analog-to-Digital Conversion Operations

One of the basic operations that an MCU does is digital input and digital output. This is used when a microcontroller needs to interface with external devices such as switches, relays, and lights. A digital input operation reads an input voltage signal and converts it into one of the two digital logic levels, while a digital output operation takes a logic signal (either high or low) and converts it into an output voltage level. The low and high logic values correspond to different voltages depending on the construction family of the microcontroller. The Arduino MCU is built using the CMOS family. However, it has slightly different voltage level ranges than a standard CMOS (see Table 3.12). Figure 4.11 shows the voltage levels for the Arduino Uno MCU using a 5 V supply voltage. The values define the allowable voltage ranges for the low and high logic states.

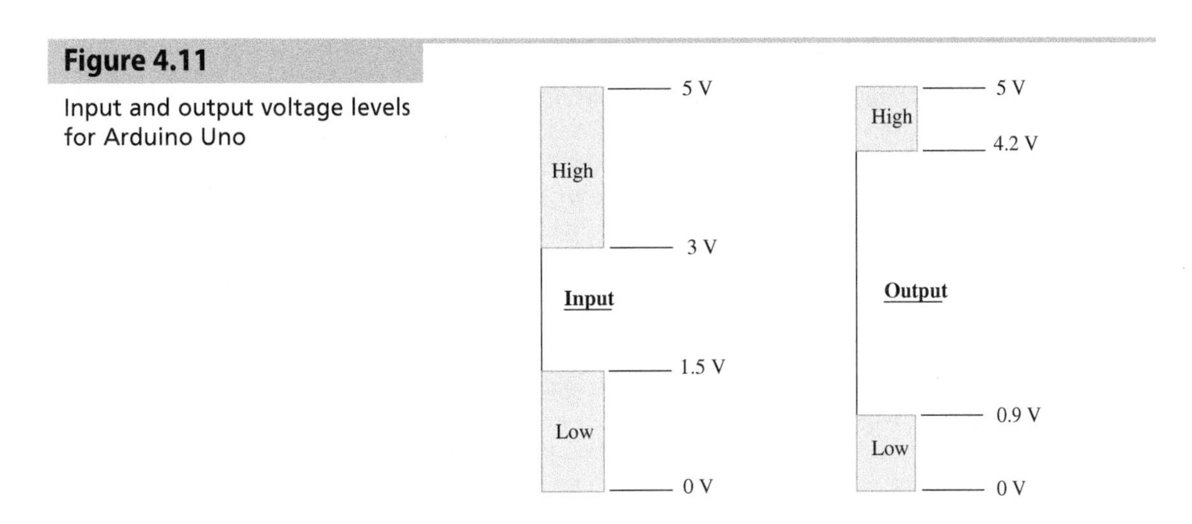

Figure 4.11

Input and output voltage levels for Arduino Uno

While the ATmega328P MCU has a maximum of 23 pins that can be used for digital I/O (see Section 4.4), the Arduino Uno board makes available to the user a maximum of 14 pins (pins 0 through 13) that can be used for digital input and output operations (see Figure 4.8). Pins 0 and 1 will not operate properly if serial communication (see next chapter) is active during program execution as these two pins are used for serial communication between the MCU and the PC. Each of the digital I/O pins can be used for either input or output. If the direction is not explicitly set, the input/output pins on the Arduino default to input. Arduino provides a function that can be used to configure the pins. The function is called *pinMode()* and is listed below, where pin is the pin number such as 3 or 8 and mode is either OUTPUT, INPUT, or INPUT_PULLUP.

pinMode(pin, mode)—Configures the specified pin to behave either as an input, an output, or an input with pull-up mode activated.

Without using the Arduino-provided pin configuration function, the user must write code to configure the **data direction register** associated with the port to which the pin belongs. For example, the ATmega328P has three I/O ports referred to as ports B, C, and D, and three data direction registers associated with these ports and referred to as DDRB, DDRC, and DDRD, respectively. On the AVR MCU, a value of 1 written to the data direction register bit corresponding to a particular I/O pin causes that pin to function as digital output while a value of zero causes that pin to function as digital input.

The Arduino pin configuration function can also be used to configure a pin as an input with a special mode of operation called **pull-up mode**. The special mode makes use of a built-in 20 kΩ pull-up resistor that is attached to each I/O pin. The primary purpose of the pull-up resistor is to steer the input pin to a known state when no input is present. As seen in Figure 4.12, when the pull-up resistor is activated, and the switch is open, the input pin voltage will be close to V_{CC}. When the switch is closed, current flows through the pull-up resistor to the ground, and the voltage at the input pin will be zero. There is a chance for the input to float, which could cause fluctuation in the reading if this mode is not used and no input signal was connected to the pin.

Figure 4.12

Pull-up resistors on AVR I/O pins

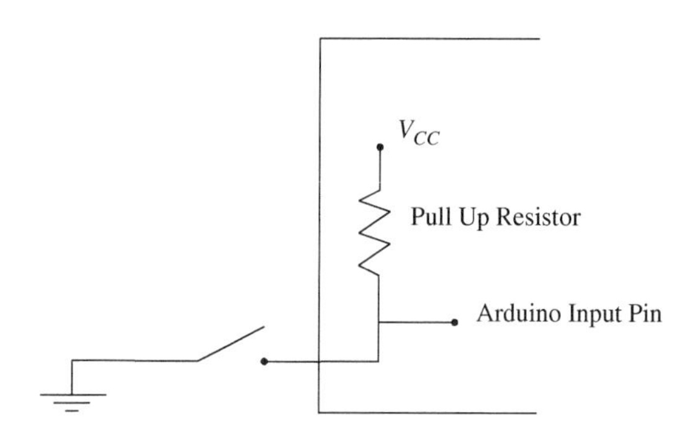

To read or send digital values, Arduino provides the following two functions:

digitalWrite(pin, value)—Write a HIGH or a LOW value to a digital pin

digitalRead(pin)—Reads the value from a specified digital pin

Many MCUs including the ATmega328P can convert an analog voltage to a digital value through a device called an **analog-to-digital (A/D) converter**. An A/D converter is used to read the output from analog sensors such as potentiometers and tachometers. A/D converters are covered in detail in the next chapter (see Section 5.3), but in short, an A/D converter performs a mapping operation between analog voltages (such as between 0 and 5 volts) and a set of digital values (such as between 0 and 1023) similar to how an instructor maps numerical grades between 0 and 100 to a set of specific letter grades (such as 94 to 100 is mapped to A, 90 to 93 is mapped to A-, ... etc.). The analog voltage range that is mapped is typically between 0 volts and the supply voltage to the MCU (called either V_{CC} or V_{DD}), but a different range can be used if an external voltage was used as the reference signal (connected to the AREF pin on AVR MCUs). The set of digital values is a function of the resolution of the A/D converter. On the ATmega328P, the resolution is 10 bits, which means that the analog voltage range is mapped into 1024 (or 2^{10}) values.

The Arduino Uno board has six channels dedicated to A/D conversion that are labeled channels A0 through A5 on the board (see Figure 4.8). Arduino provides the *analogRead(pin)* function to read the analog voltage applied to these pins, where pin is the analog pin number (0 to 5 on the UNO board). It takes approximately 0.1 ms to do a single A/D conversion.

4.6 PWM Operation

None of the MCUs in the AVR ATmega family has a digital-to-analog (D/A) conversion capability that converts digital values to analog output voltages (see Section 5.4 for more detail on D/A conversion), but some microcontrollers made by Microchip (such as those in the AVR XMEGA series and the AVR ARM series) have a built-in digital to analog conversion capability. However, many AVR MCUs have a built-in module to generate a pulse width modulated (or PWM) output that can approximate the operation of a D/A converter. A **PWM signal** (▶ available—V4.12) is a square-wave signal of fixed amplitude and frequency, but the width of the on and off parts of the signal (or duty cycle) can be varied (see Figure 4.13). Varying the PWM duty cycle (see Figure 4.14) will result in varying the effective voltage output from the PWM port. For example, at 50% duty cycle and 0 to 5 volts PWM signal, the effective output is 2.5 volts (0.5 × 5 volts). The PWM output is used to conveniently drive H-bridge circuits that are used in bidirectional DC motor speed control (see Section 9.2.4) as well as to modulate sound and light intensity.

Figure 4.13

Illustration of a PWM signal

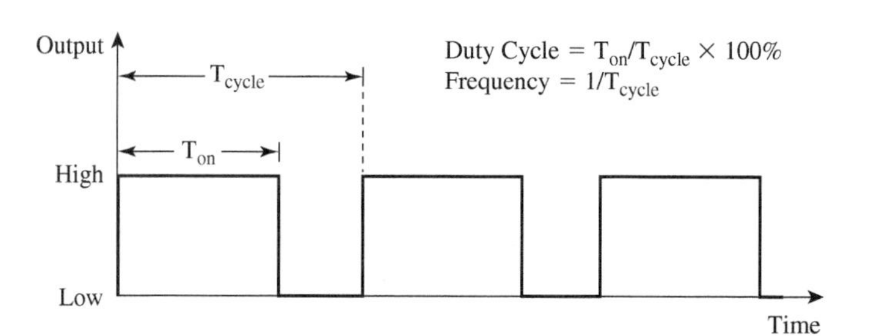

Figure 4.14

Illustration of different PWM duty cycles

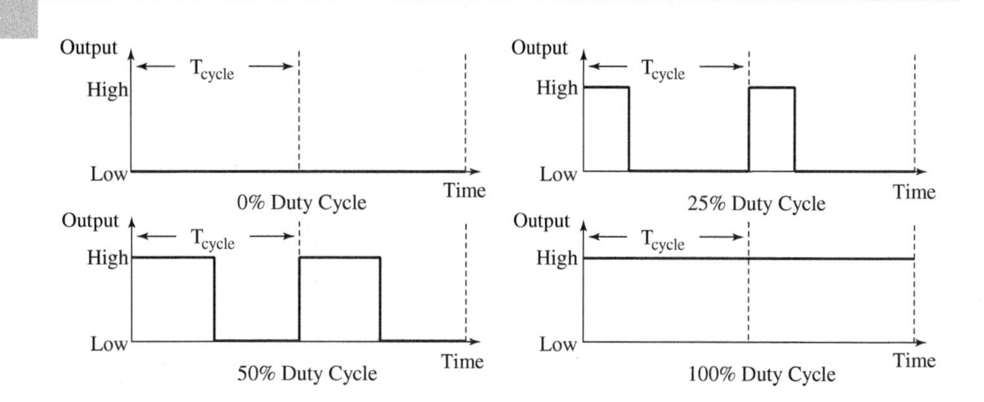

On **AVR MCUs**, the PWM output mode is one of the modes of operation of the timers provided on the chip.

4.6.1 PWM Generation and Modes

A PWM (Pulse Width Modulation) signal is a digital signal possessing only two values, 0 volts and V_{DD} (or V_{CC}), typically generated within an MCU through the combined utilization of a counter and a comparator. The process of PWM generation can be summarized as follows:

1. **Counter Initialization**: The counter value begins at zero.
2. **Incrementation**: The counter value is incremented by one at a rate defined by the timer linked with that particular counter.
3. **Maximum Value Check**: Once the counter value reaches the maximum count value, it is either reset to 0 or counts back down to zero, initiating the cycle anew.
4. **Comparison with Set Value**: Simultaneously, a comparator constantly examines the current counter value against a predetermined set value.
5. **PWM Output Determination**: If the current counter value is less than the set value, the PWM output will register as high. Conversely, if the current counter value equals or exceeds the set value, the PWM output will register as low.

Crucially, the maximum count value establishes the PWM frequency, while the set value dictates the duty cycle of the PWM signal. This combination of elements allows for precise control over the signal's modulation.

MCUs can generate several PWM modes (see Figure 4.15). These are:

- Fast PWM
- Phase Correct PWM
- Phase and Frequency Correct PWM

Fast PWM mode, as its name implies, is capable of generating high-frequency PWM signals. In this mode, the counter is immediately reset to 0 once it reaches the maximum count value, thereby enabling the generation of faster signals. This PWM mode is particularly suitable for applications such as power regulation and rectification. The high frequency of Fast PWM mode allows for the use of physically smaller external components like coils and capacitors. Consequently, this can lead to a reduction in the overall size of the system and, in turn, a decrease in total system cost [14]. Unlike the Fast PWM mode, in **Phase Correct PWM**, or **Phase and Frequency Correct PWM** mode, the counter gradually counts back to zero after the counter reaches the maximum count value, but the PWM frequency in these modes will be roughly half that of the Fast PWM mode. These two modes result in a more symmetrical PWM output signal when the duty cycle is changed. This is illustrated in the lower part of Figure 4.15, which shows two different duty cycle PWM signals corresponding to two different counter-set values. Due to its symmetric signal generation, these two modes should be used for motion control applications which ensures consistent torque and smoother motor operations. Unlike fast PWM, which can produce electrical noise from rapid changes, Phase Correct PWM offers a more gradual transition, minimizing harmonics and electrical interference. This results in enhanced low-speed performance, less mechanical stress due to reduced vibrations, and overall improved efficiency. The Phase and Frequency Correct PWM mode is like the Phase Correct PWM mode but allows more control of the maximum count value resulting in a large range of achievable PWM frequency values and not just the specific frequencies obtained by changing the timer prescaler values. The last mode is mostly for sound generation as it allows a large range of frequencies.

Figure 4.15

PWM modes: (a) Fast PWM and (b) Phase Correct PWM

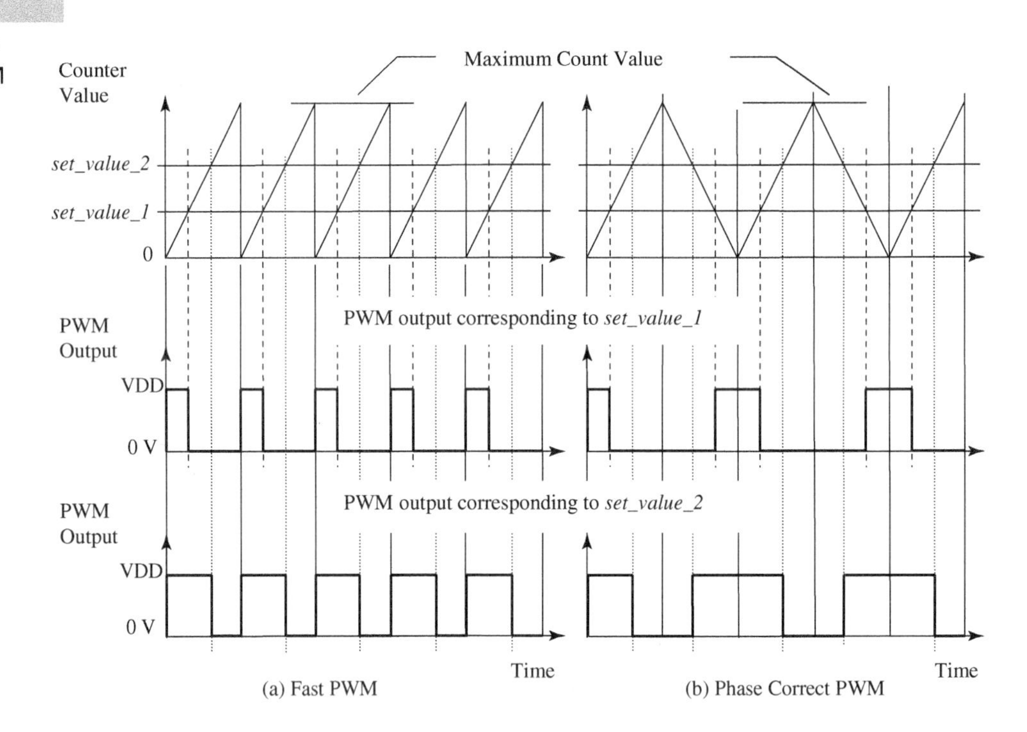

4.6.2 PWM Details in ATmega328P MCU

The ATmega328P MCU has three independent PWM generators each controlling two PWM output pins. A separate timer controls each generator. The previously mentioned three PWM modes can be obtained using this MCU. Pins 5 and 6 are controlled by *Timer0*, Pins 9 and 10 by *Timer1*, and Pins 3 and 11 by *Timer2*. Table 4.7 lists information about these pins. When using the Arduino development environment, the PWM signal is controlled using the *analogWrite()* function. The function takes two arguments, one is the PWM pin number and the other is the PWM duty cycle (0 to 255), where 0 corresponds to 0% duty cycle (always off), and 255 to 100% duty cycle (always on). Note that this function does not allow you to change the PWM frequency but uses a set default frequency (shown in bold in Table 4.7). To change the PWM frequency, one needs to change the prescaler or divisor value for the clock. This is done by the code statement shown below:

TCCR** = (TCCR** & 0xF8) | setting_value

where *TCCR*** is the timer control register associated with the particular PWM pin, and *setting_value* is the value needed to get the correct prescaler value. Table 4.7 also shows the PWM frequencies for different setting values, and Example 4.4 illustrates the selection of a different frequency than the default one. The table assumes that the MCU has a 16 MHz clock. Note that *Timer0* is used in the *millis()*/*micros()* timing functions. Thus, if the frequency is changed from the default value, it will affect the operation of these functions.

Table 4.7

PWM details in ATmega328P MCU

	PWM PIN # on Arduino Uno Board					
	5	**6**	**9**	**10**	**3**	**11**
Pin Designation on ATmega328P Chip	PD5	PD6	PB1	PB2	PD3	PB3
Timer Controlling PWM	0		1		2	
Counter Size	8-bit		16-bit		8-bit	
PWM Type	Fast PWM		Phase-Correct PWM		Phase-Correct PWM	
Timer Control Register	TCCR0B		TCCR1B		TCCR2B	
Setting_Value to Adjust Frequency	Prescaler or Divisor	Freq. (Hz)	Prescaler or Divisor	Freq. (Hz)	Prescaler or Divisor	Freq. (Hz)
0x01	1	62500	1	31372.549	1	31372.549
0x02	8	7812.5	8	3921.57	8	3921.57
0x03	64	**976.563**	64	**490.196**	32	980.392
0x04	256	244.141	256	122.549	64	**490.196**
0x05	1024	61.035	1024	30.6373	128	245.098
0x06	N/A		N/A		256	122.549
0x07	N/A		N/A		1024	30.6373

As seen in this table, the Fast PWM mode gives higher PWM frequencies than the Phase Correct Mode for the same prescaler value. Note the PWM frequency is given by the formulas shown below.

$$PWM_{freq}(\text{Fast PWM Mode}) = \text{Clock\_Freq.}/(256 \times \text{Prescaler}) \quad \textbf{(4.3)}$$

$$PWM_{freq}(\text{Phase Correct PWM Mode}) = \text{Clock\_Freq.}/(510 \times \text{Prescaler}) \quad \textbf{(4.4)}$$

Example 4.4 PWM Frequency

Show how pin 3 can be set to generate a PWM frequency of 980.392 Hz.

Solution:

From Table 4.7, we can see that pin 3 is controlled by *Timer2*. Thus, to change the PWM frequency from the default value of 490.196 Hz, we need to write to the timer control register associated with that timer. That register is TCCR2B. From Table 4.7 for a desired PWM frequency of 980.392 Hz, the setting_value for that register is 0x03. The code statement to set this frequency is shown below. Note that this statement can be added to the Arduino sketch as is with no need to include any header files.

```
TCCR2B = (TCCR2B & 0xF8) | 0x03
```

The first three bits of the register are cleared using the bitwise AND operation (&), and then the same bits are set to the setting_value using the bitwise OR operation (|).

To use the Phase and Frequency Correct mode, the user has to change the maximum count value by writing to registers OCR1A or ICR1 (see data sheet for more details). In this mode, the PWM frequency is given by the following formula:

$$PWM_{freq}(\text{Phase and Frequency Correct PWM Mode}) = \text{Clock\_Freq.}/(2 * \text{top} * \text{Prescaler}) \quad \textbf{(4.5)}$$

where *top* is limited to 255 for 8-bit timers and 65535 for 16 bits timers. The number of available duty cycles is affected by the value of *top*.

4.7 AVR MCU Components and Features

In this section, we will discuss several components and features of AVR MCUs. These include EEPROM data, delays and timers, watchdog timer, and reset operations.

4.7.1 EEPROM Data

The EEPROM (Electrically Erasable Programmable Read-Only Memory) is a type of non-volatile memory that can be used to store data in embedded systems, including Arduino-based platforms. Unlike RAM (Random Access Memory), where data is lost when power is removed, data stored in EEPROM remains intact even when the device is turned off. The data in EEPROM can be written either during the programming phase or while the program is executing. Writing data to EEPROM during program execution takes longer than writing to a RAM location, but the advantage is that the data is permanently stored.

The Arduino development environment provides a standard library, known as 'EEPROM,' with functions that enable easy access to the EEPROM memory. These functions include:

EEPROM.read(address): This function reads the data stored in EEPROM at the specified memory address. It is used to retrieve data that has been previously stored, allowing for persistent data storage between power cycles or resets.

EEPROM.write(address, value): This function erases the existing data at the specified memory address and writes the new value to the EEPROM. This enables the modification of stored data, and since EEPROM has a finite number of write cycles (often around 100,000), the erasing and rewriting should be done judiciously.

In the context of real-world applications, EEPROM is often utilized to store configuration settings, calibration data, user preferences, or any information that needs to be retained after a power loss. It provides a small but reliable storage solution where data persistence is essential, even though the access times are longer compared to RAM.

4.7.2 Timing Delays and Timers

The Arduino development environment provides two functions for timing delays. These are *delay()* and *delayMicroseconds()*. The maximum delay interval that can be set with the *delay()* function is around 50 days (set in millisecond units), but this value may vary depending on the Arduino board being used. The *delayMicroseconds()* function allows more precise delays as the delay is set in microsecond units, but the maximum delay interval for this function is about 70 minutes. These functions are typically used for creating a simple time delay between different parts of a program, and they work by executing a precise number of instructions to cause the requested delay. They do not use any timers. An important consideration when using these delay functions is that other code execution (except for interrupt service routines) is blocked during the delay. This means that if you have tasks that need continuous monitoring or frequent updates, using these functions may lead to unresponsive behavior. For more advanced timing control, such as measuring time intervals between events, tracking the occurrence of an event relative to another point in the code, or creating non-blocking delays, it may be necessary to access timers directly. This is discussed in Chapter 6.

The ATmega328P has three timers referred to as *Timer0*, *Timer1*, and *Timer2*. *Timer1* uses a 16-bit counter while *Timer0* and *Timer2* use 8-bit counters. Table 4.7 has information about these timers. Some of these timers can operate as either timers or counters. In the timing mode, the count value is incremented every clock cycle or at a multiple of it if a prescaler is used, where a **prescalar** is a user-set factor that can divide the input clock frequency for that timer. In the counter mode, the timer module will increment on every rising or falling edge of the external signal connected to the microcontroller. The user has a choice of setting the clock source for these timers, as well as setting the maximum overflow interval. All the above timers can use a prescaler (see Table 4.7). Having a prescale factor higher than 1:1 increases the maximum counting interval before the counter overflow. Example 4.5 illustrates the use of the prescale factor. More information about these timers is included in Chapter 6.

Example 4.5 Timer Counting Interval

Using an 8 MHz internal clock, and 1:8 prescale factor, determine the maximum counting interval for *Timer1* on the ATmega328P.

Solution:
The input clock frequency to *Timer1* would be 1 MHz (1/8 $\times$ 8) or a counter resolution of 1 μsec. Since *Timer1* is associated with a 16-bit counter, the maximum counting interval is $2^{16} \times 1$ μsec = 65536 μsec or 65.53 ms. Thus, with a 8:1 prescale factor, the *Timer1* counter will overflow once every 65.53 ms.

4.7.3 Watchdog Timer

The watchdog timer (WDT) is a special counter that resets the processor if it overflows. The purpose of the WDT is to cause the processor to reset if it times out in a lengthy operation. To prevent overflow, the program needs to periodically reset the counter associated with the watchdog timer. The MCU's hardware handles the WDT, independent of any program running on the MCU. In AVR MCU, the WDT is controlled by the WDTCSR (Watchdog Timer Control) register. The watchdog timer uses a dedicated internal low-frequency clock running at 128 kHz for its timing functions which is independent of the internal high-frequency oscillator clock. The overflow period is processor dependent, and on the ATmega328P, the overflow period can range from 16 ms to 8 seconds. In addition to resetting the processor, the WDT on the AVR MCU can also cause an interrupt or both actions simultaneously. If the WDT is enabled in the code, then the user program should include a call to periodically reset the counter associated with the watchdog timer. If the code was stuck in a code segment for an interval longer than the WDT overflow period, there is no chance to restart the WDT before it overflows, and WDT will keep resetting (i.e., causing the program to start from the beginning).

4.7.4 Reset Operations

There are different reset operations in AVR MCUs. These include Power-on Reset (POR), Brown-out Reset (BOR), Watchdog Timer Reset, and External Reset. These are discussed below. A **POR** occurs whenever the power (Vcc line) is turned off and then on to the chip. After a POR, the code on the chip starts executing from the instruction placed at the reset vector address, and the I/O lines and some of the registers on the chip will reset to their 'Reset' state. A **BOR** (if enabled) produces the same result as a POR and occurs whenever the V_{CC} voltage level falls below the rated voltage (between 1.8 and 5.5 volts for most chips). The BOR does not occur unless certain registers on the chip were set to detect this condition. This feature is useful in battery-powered applications to detect low-voltage conditions. The **Watchdog Timer Reset** occurs whenever the counter associated with the WDT overflows, while an **External Reset** occurs whenever the reset pin (the RESET pin on AVR MCUs) goes to low for a certain minimum interval. Note that the MCU has a special register that can indicate which type of reset has occurred. In AVR MCUs, it is the called MCUSR (MCU Status Register).

4.8 Chapter Summary

This chapter focused on microcontrollers, which are versatile, single-chip devices integrating a microprocessor, memory, and peripheral devices all on the same chip. This consolidation of components results in a compact yet powerful tool that has become indispensable in digital electronic design, from simple home projects to complex industrial systems. In the preliminary sections, the chapter introduces readers to essential numbering systems used in programming and interfacing microcontrollers, namely, binary and hexadecimal systems, as well as binary addition and subtraction operations. It emphasizes that understanding these systems is crucial, as data in microcontrollers and microprocessors is processed, stored, and transferred in binary form, while hexadecimal notation often offers a more readable format for humans, especially when dealing with larger binary numbers. The chapter then dives into an in-depth exploration of the structure and workings of microprocessors and microcontrollers, making a distinction between them. While a microprocessor is essentially the 'brain' of a computer, executing instructions and performing computations, a microcontroller is a self-contained system with a microprocessor, memory, and peripherals. The discussion spans topics like different types of memory (including RAM, ROM, and flash), buses for data and instruction transport, and design architectures.

After providing this foundational knowledge, the chapter zeros in on a specific microcontroller: the ATmega328P, which is the heart of the widely used Arduino Uno board. The fundamental operations performed by the ATmega328P microcontroller are thoroughly examined, starting with digital input and output. This part of the discussion covers how the microcontroller interacts with digital signals, explaining how it reads the state (high or low) of a digital input pin, and how it can set the state of a digital output pin. Next, the chapter explores the analog-to-digital conversion (ADC) capability of the ATmega328P. This feature allows the microcontroller to interpret analog signals (like those from a temperature sensor or potentiometer) and convert them into digital values that it can process. The section on Pulse Width Modulation (PWM) output explains how the ATmega328P can simulate analog output effects by rapidly switching digital output pins on and off, with varying 'on' to 'off' ratios. This can be used, for example, to control the brightness of an LED or the speed of a motor. The last part of this chapter covers other significant features and devices integrated into AVR microcontrollers, including EEPROM, the timing and counter modules, the watchdog timer, and reset operations.

Questions

4.1 What is the base in the hexadecimal numbering system?

4.2 What is meant by MSB in a binary number?

4.3 What is the 2's complement representation?

4.4 What distinguishes a microcontroller from a microcomputer?

4.5 List some differences between a microcontroller and FPGA.

4.6 Where are program instructions stored in a microcontroller?

4.7 Which bus handles the transfer of data between CPU and memory?

4.8 What is Flash memory?

4.9 What is the difference between the Von Neumann and the Harvard architectures?

4.10 What is the difference between the CISC and RISC design approaches?

4.11 What function does the program counter perform?

4.12 What is the purpose of the stack memory?

4.13 How many bytes can the ATmega328P MCU store in program memory?

4.14 Name several pin layouts for MCUs.

4.15 Name three external clock sources.

4.16 What are the minimum connections needed for an MCU to operate?

4.17 How is an MCU 'programmed'?

4.18 What is a boot-loading application?

4.19 What is the advantage of pull-up mode in digital input operations?

4.20 What are the different PWM modes available on the ATmega328P?

4.21 Which PWM mode offers the highest frequencies?

4.22 What is the limitation of the use of the *delay()* function?

4.23 What is the advantage of storing data in EEPROM Data memory?

4.24 What is the purpose of a watchdog timer?

4.25 What happens after a power reset?

Problems

Section 4.2 Numbering Systems

P4.1 Convert the following decimal numbers to hexadecimal and binary. Do not use a built-in programming function to do the conversion.

a. 22

b. 184

c. 630

P4.2 Convert the following decimal numbers to hexadecimal and binary. Do not use a built-in programming function to do the conversion.

a. 37

b. 215

c. 822

P4.3 Determine how many bytes are in the following memory sizes.

a. 2K

b. 4M

c. 1G

P4.4 Convert the following hexadecimal decimal numbers to binary and decimal. Do not use a built-in programming function to do the conversion.

a. 0x31

b. 0x26B

c. 0xA4F7

P4.5 Find the complement of the following decimal numbers assuming an 8-bit binary field.

a. 5

b. 12

c. 68

P4.6 Find the 2's complement representation for the following numbers using an 8-bit field.

a. −1

b. −43

c. −121

P4.7 Find the 2's complement representation for the following numbers using an 8-bit field.

a. −11

b. −65

c. −115

P4.8 Perform the following binary operations.

a. 1011 + 0010

b. 00011101 + 01001111

c. 1010 − 0011

d. 00101001 − 00010101

P4.9 Perform the following binary operations.

a. 1001 + 0101

b. 01010110 + 01101011

c. 1110 − 0101

d. 101010 − 001101

P4.10 Find the binary representation for the following numbers using the IEEE 742 standard.

a. 0.078125

b. − 0.5

c. 10.5

Section 4.3 Microprocessors and Microcontrollers

P4.11 Consider a simple 8-bit microprocessor with a program memory of 256 bytes and a data memory of 64 bytes. The microprocessor has an 8-bit address bus and an 8-bit data bus. Each instruction is one byte long.

a. Calculate the addressable memory space for program and data memory in bytes.

b. How many different instructions can this microprocessor support?

Section 4.4 AVR Microcontrollers

P4.12 Using the Microchip website, select one or more 8-bit AVR MCUs with a small number of pins that are suitable for the following applications.

a. Monitoring of four digital input lines and writing to five digital output lines.

b. Monitoring and update of 10 digital I/O lines, reading four analog inputs, and communicating with a PC using RS-232 protocol.

c. Same as part b, but also using 3 PWM lines.

P4.13 Using the Arduino Uno board, draw a connections diagram for an application that requires the following.

a. Reads 4 digital I/O lines

b. Writes to 4 digital I/O lines

c. Reads 2 analog signals

Be sure to identify and label all the pins on the board that need to be used.

P4.14 Using the Arduino Uno board, draw a connections diagram for an application that requires the following.

a. Reads 2 digital I/O lines

b. Writes to 4 digital I/O lines

c. Reads 2 analog signals

d. Sends 2 PWM signals

e. Provides 5 volts and ground signals to an external circuit

Be sure to identify and label all the pins on the chip that need to be used.

P4.15 If one needs an Arduino board that can support a much better timing accuracy than the Uno board, then one should look for a board that has a built-in real-time clock module (RTC). Research and identify an Arduino board that has this module and compare the timing accuracy of the identified board with that of the Uno board.

Section 4.5 Digital Input/Output and Analog-to-Digital Conversion Operations

P4.16 Map the following input voltages to low or high logic levels using Figure 4.11: 0.1, 1.2, 2.0, 3.5, and 4.5 volts.

P4.17 Map the following output voltages to low or high logic levels using Figure 4.11: 0.3, 0.7, 2.0, 3.0, and 4.9 volts.

P4.18 Draw a circuit for interfacing a digital I/O on an AVR MCU to a MOSFET transistor that switches a small 12-volt motor on/and off.

P4.19 Draw a circuit for interfacing the Arduino Uno with the following components.

a. An LED that is turned on/off by the MCU

b. An N.O. push-button switch that is read by the MCU

c. A rotary potentiometer that is used to set the desired operating value for a control system

Section 4.6 PWM Operation

P4.20 Determine which PWM pin on the ATmega328P shall be used to enable a PWM operation at a frequency of close to 1 kHz and a duty cycle of 25%.

P4.21 Determine the duty cycle argument for the *analogWrite()* function and the setting_value (in Table 4.7) to produce a PWM frequency at 62,500 Hz and with a duty cycle of 30%. Assume the MCU has a 16 MHz clock.

P4.22 *Timer0* and *Timer2* are both 8-bit timers. Explain why the use of the same prescale factor such as 1, 8, 64, or 1024 results in different PWM frequencies.

Section 4.7 AVR MCU Components and Features

P4.23 Determine the maximum counting interval for *Timer1* on the ATmega328P using a 1:8 prescale factor and a 16 MHz external clock.

P4.24 Determine the maximum counting interval for *Timer2* on the ATmega328P using a 1:256 prescale factor and a 16 MHz external clock.

P4.25 Research and identify three features of some AVR microcontrollers for saving power. Explain how each feature saves power.

Laboratory/Programming Exercises

L/P4.1 Write code to read one of the A/D converter channels on the Arduino Uno or a similar board. Connect the output of a rotary potentiometer to the A/D channel, and display on the Serial Monitor the values read from the A/D channel as you rotate the potentiometer.

L/P4.2 Write an Arduino code to read inputs from three push buttons and control two LEDs using the Arduino Uno or a similar board. Pressing the first button should turn on the first LED, pressing the second button should turn on the second LED, and pressing the third button should turn on both LEDs.

L/P4.3 Develop and download a program for Arduino Uno or a similar board that does the following.

a. Turn on and off several LEDs on a breadboard in a particular pattern (one on and the next off, or two on and the next two off ...). You can create any pattern you like.

b. Use the A/D reading from a rotary potentiometer that is connected to one of the A/D channels to vary the timing speed of the pattern. Turning the potentiometer CW (as seen from above) should cause the pattern to turn on and off more rapidly. Use a timing delay function to create the timing delay.

L/P4.4 Write an Arduino code to read an analog input from a light sensor, such as a photoresistor, and control the brightness of an LED using an Arduino Uno or a similar board. The LED should be dimmer when the light intensity is higher and brighter when the light intensity is lower.

L/P4.5 Write code to display the effect of changing the PWM duty cycle on the Arduino Uno or similar board. Add a loop in the code to generate PWM signals with increasing or decreasing duty cycles. Connect the output of the PWM to an oscilloscope and monitor the output.

L/P4.6 Write code to display the effect of changing the PWM frequency on the Arduino Uno or a similar board. Add a loop in the code to generate PWM signals with several different frequencies selected from Table 4.7. Use a delay of 10 seconds before displaying the next frequency. Connect the output of the PWM to an oscilloscope and monitor the output.

L/P 4.7 Design and build a circuit to interface an Arduino Uno or similar board to an LED, a push-button switch, and a rotary potentiometer. Specify any resistors needed plus wiring of all needed pins. Test your circuit by writing a simple program to turn on the LED if the value read by the potentiometer exceeds a specified value. Your code should turn off the LED whenever the push-button is pressed.

5

Chapter

Data Acquisition and Microcontroller/PC Interfacing

Chapter Objectives:

When you have finished this chapter, you should be able to:

- Explain the difference between analog and digital signals
- Determine the requirements for proper sampling of analog signals
- Determine the voltage resolution, digitizing accuracy, and input and output values of an analog-to-digital converter
- Determine the voltage resolution, digitizing accuracy, and input and output values of a digital-to-analog converter
- Explain data acquisition boards
- Explain how to set or read a particular bit in a parallel digital input/output port, and know the tools available in MATLAB for data acquisition
- Explain RS-232 communication
- Outline the I2C and SPI interfacing methods

5.1 Introduction

The heart of any mechatronics system is typically a computer or an embedded processor. This processor interfaces with the actuators and sensors integral to the system, facilitating communication and coordination among these components. For the system to function effectively, data transfer between the processor and these components is essential. Data can be categorized as either analog or digital signals. **Analog signals** are continuous, meaning they can take any value within a specified range. The term 'analog' stems from 'analogous' or 'similar.' For instance, a pressure transducer with an analog output produces a time-varying voltage signal that is analogous to the time-varying pressure it measures. Examples of such analog voltage signals include the voltage supplied by power companies or the voltage used to power DC electric motors (refer to Chapter 9). Sensors like thermocouples and tachometers also produce analog voltage outputs (see Chapter 8).

On the other hand, **digital** or **discrete signals** are discontinuous and can only assume specific, predefined values, often represented as '0's and '1's in binary code. An encoder, a type of digital displacement measurement sensor (refer to Chapter 8), is an example that produces digital output signals. Inherently, all microprocessor circuits use digital signals. A notable advantage of digital signals is their resilience against noise, which can compromise the resolution of analog signals. Converting a signal from the analog domain to the digital domain requires the use of an analog-to-digital converter (A/D). Similarly, converting a signal from the digital domain to the analog domain requires the use of a digital-to-analog (D/A) converter.

This chapter discusses sampling as well as techniques to interface a processor to the outside world using different interface devices (such as analog-to-digital converters, digital-to-analog converters, parallel digital input/output ports, and asynchronous and synchronous serial ports). Chapter 4 discussed interface devices on AVR microcontrollers (such as the A/D converter). In this chapter, more information is given on the operation and programming of these devices. Interfacing is important for the operation of control systems because a control system interacts with sensors and actuators through these interface devices. For further reading on the topics covered in this chapter, see [5] and [15].

5.2 Sampling

5.2.1 Sampling Theory

In converting an analog signal to a digital signal, the analog signal is 'sampled' to obtain the digital or discrete values. By sampling, we mean that the analog signal is read at defined time instances and the continuous-time signal is replaced by a sequence of numbers [16]. Thus, if we have a continuous-time signal, $y(t)$, and the signal is sampled once every T_s seconds, where T_s is the sampling period, then the digital samples are given by

$$y[n] = y(t = nT_s), \quad n = 0, 1, 2, \ldots \tag{5.1}$$

The sampling period T_S is the inverse of the sampling frequency f_S ($T_S = 1/f_S$). As an example, if we have a 1 Hz continuous time sinusoidal signal written as $y(t) = \sin(2\pi 1\, t)$, and the sampling frequency is 10 samples/s, then the sampled or discrete-time signal is given by

$$y[n] = \sin\left(2\pi 1 \frac{n}{10}\right) = \sin\left(\frac{\pi n}{5}\right) \tag{5.2}$$

Figure 5.1 shows a plot of this continuous time signal and its sampled version.

Figure 5.1

Plot of a continuous time 1 Hz sinusoidal signal and its sampled version at a sampling frequency of 10 samples/s

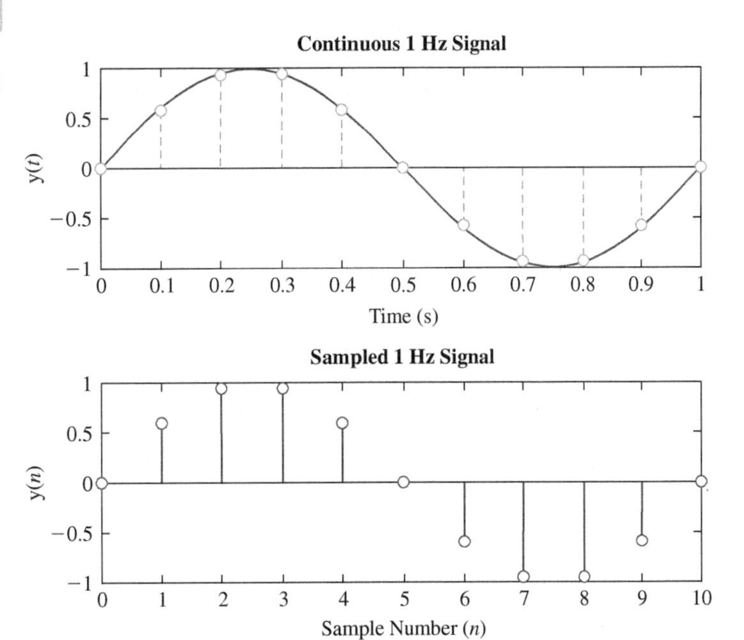

The discrete time signal is just a sequence of numbers and carries no time information. Thus, it is possible for a discrete signal to represent more than one continuous-time signal. Example 5.1 illustrates this point.

Example 5.1 showed that the 600 Hz signal when sampled at 1000 Hz gave the same discrete sequence as the 400 Hz signal. Here the 600 Hz signal was 'aliased' to 400 Hz, so the sampling operation was not able to capture the frequency content of the 600 Hz signal. **Aliasing** (▶ available—V5.1) occurs when

Example 5.1 Sampling of Continuous Time Signals

Show that the continuous time signals $y_1(t) = \cos(2\pi 400\ t)$ and $y_2(t) = \cos(2\pi 600\ t)$ produce the same sampled sequence when sampled at a frequency of 1000 samples/s.

Solution:

The digital samples for the 400 Hz signal can be written as

$$y_1[n] = \cos\left(2\pi 400 \frac{n}{1000}\right) = \cos\left(\frac{4\pi n}{5}\right)$$

For the 600 Hz signal, the digital samples are

$$y_2[n] = \cos\left(2\pi 600 \frac{n}{1000}\right) = \cos\left(\frac{6\pi n}{5}\right)$$

But note that $\cos\left(\frac{4\pi n}{5}\right) = \cos\left(\frac{6\pi n}{5}\right)$ for any n. Thus, $y_1[n] = y_2[n]$, and the sampled sequence for both signals will be the same. Figure 5.2 shows a plot of the two continuous signals and their sampled sequences.

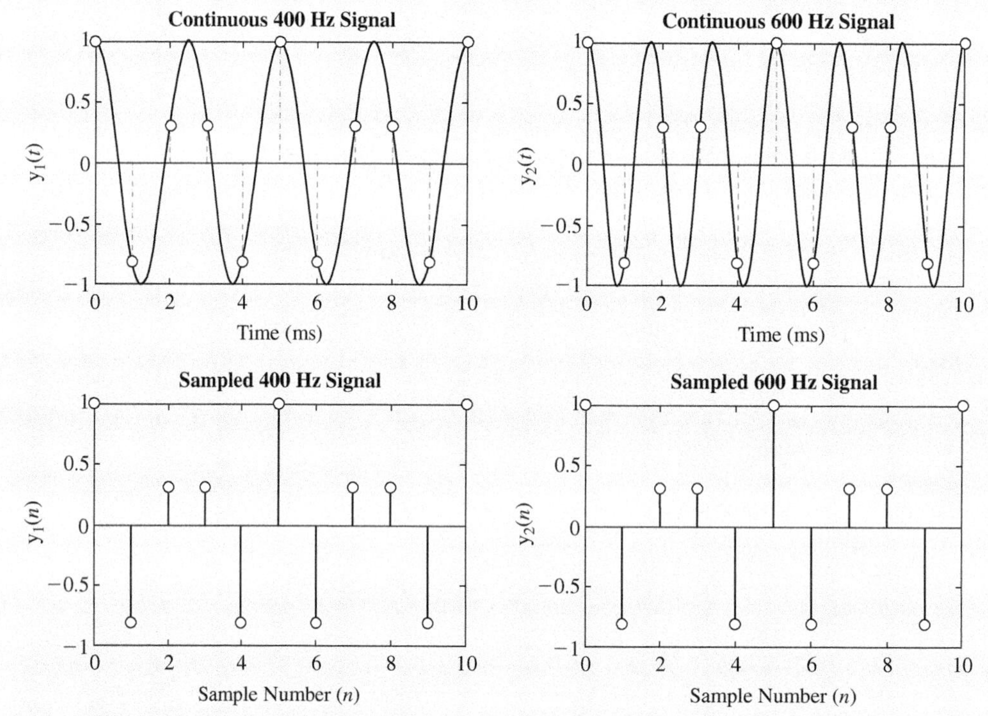

Figure 5.2 A plot of two continuous signals and their sampled sequences

If each signal was a sine function instead of a cosine function, then each of the two sampled signals will be the negative of the other signal since $\sin(\theta) = -\sin(2\pi n - \theta) = \sin(\theta - 2\pi n)$, while $\cos(\theta) = \cos(2\pi n - \theta) = \cos(\theta - 2\pi n)$.

higher frequency components in a signal are incorrectly represented as lower frequencies due to insufficient sampling. So, this brings the question: What should be the sampling rate of a given continuous-time signal, so its frequency content does not get aliased? The requirement for the proper sampling rate is given by **Shannon's Sampling Theory** [17], which states that the sampling frequency should be at least twice that of the highest frequency in the signal ($f_{\max}$). Thus, a 1000-Hz sinusoidal analog signal should be sampled at a frequency of 2000 Hz or higher. The minimum sampling rate, or $2\,f_{\max}$, is called the **Nyquist rate**. Alternatively, the **Nyquist frequency** (▶ available—V5.2), which is half the sampling frequency f_s, is defined as the highest frequency that a signal can have so it can be sampled at a frequency of f_s without creating aliasing in the sampling process.

So, if we sample a signal at a frequency below the Nyquist rate (twice the highest frequency in the signal), a phenomenon known as **under-sampling** occurs, leading to aliasing. When this happens, the signal gets aliased to another frequency, f_a, called the aliased frequency. A simplified expression for the **aliased frequency** of a signal f sampled at a frequency f_s is given by Equation (5.3)

$$f_a = |f - f_s| \tag{5.3}$$

However, it is important to note that this equation is a basic representation and might not cover all aliasing scenarios, such as when multiple high frequencies are present. Applying this equation to Example 5.1, we see that the 600 Hz signal will appear as a 400 Hz signal due to aliasing when sampled at 1000 Hz since the 1000 Hz sampling frequency is below the Nyquist rate (1200 Hz) for this signal.

As the sampling rate increases above the Nyquist rate, the sampled waveform looks more like the original signal. In practice, a sampling rate of at least five times higher than the highest frequency in the signal is typically used. To illustrate signal aliasing, Figure 5.3 shows a 1-Hz cosine signal and the corresponding

Figure 5.3

Illustration of signal aliasing. The dashed line shows the aliased signal.

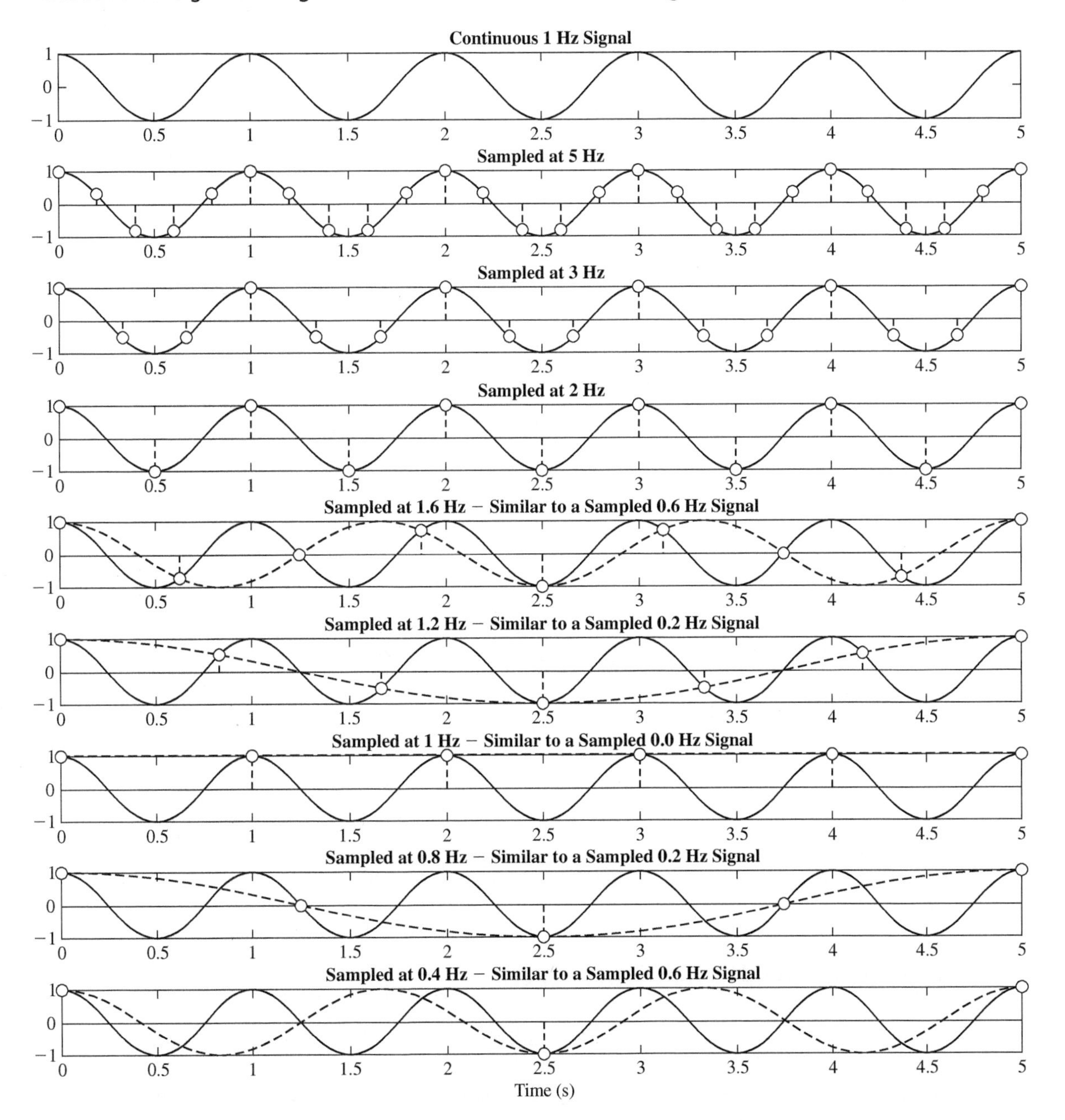

sampled signals at different frequencies ranging from 0.4 Hz to 5 Hz. Note the distortion in the signal when sampled below 2 Hz, the minimum frequency specified by Shannon's sampling theory, and how it is aliased to other frequencies. Depending on the characteristics of the original signal and the sampling process, filtering may be required to remove or reduce the effects of aliasing and noise. Low-pass filtering techniques (see Chapter 7) can be utilized to limit the frequencies beyond the Nyquist frequency.

5.2.2 Signal Reconstruction

While the Shannon Sampling Theorem specifies the minimum sampling rate required to reconstruct the continuous-time signal from its samples without aliasing, it does not provide explicit guidance on the reconstruction process itself. It serves as a theoretical foundation, highlighting the importance of a sufficiently high sampling rate to capture the essential information contained in the continuous-time signal. To practically reconstruct a continuous-time signal from its discrete samples, various methods and techniques have been developed based on signal processing principles. One widely used approach is known as the **zero-order hold (ZOH)**, which is commonly employed in the digital-to-analog conversion process (see Section 5.4).

The ZOH (▶ available—V5.3) is a simple and intuitive method for signal reconstruction that assumes that each sample value remains constant until the next sample arrives. It is termed 'zero-order' because it employs a zeroth-degree polynomial (i.e., a constant) to represent the signal between samples. This technique effectively converts the discrete samples into a piecewise constant signal, which can be seen as a staircase-like approximation of the original continuous-time signal. Figure 5.4 shows a 1-Hz sinusoidal signal and its zero-order hold reconstruction at two different sampling frequencies.

Figure 5.4

A 1-Hz sinusoidal signal and its zero-hold reconstruction at two different sampling frequencies

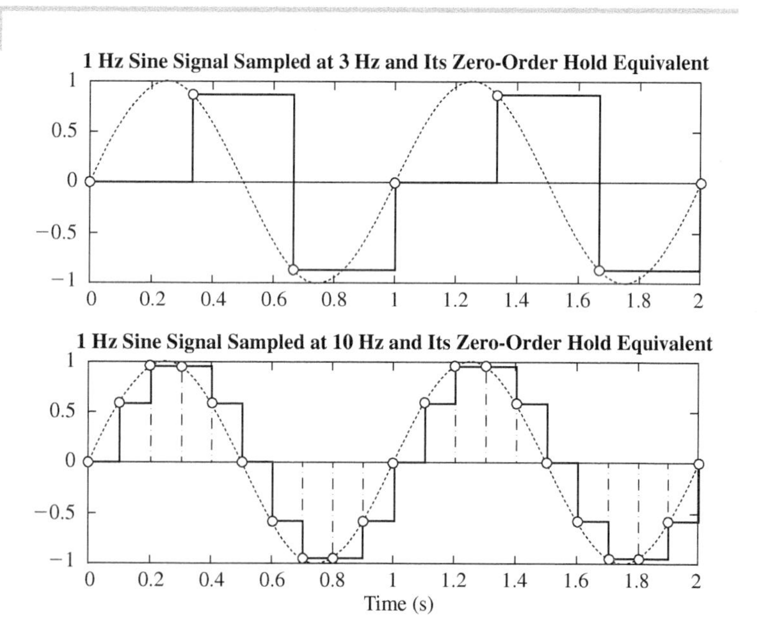

While zero-order hold provides a simple and practical way to reconstruct a continuous-time signal from its samples, it introduces some limitations. The piecewise constant nature of the approximation leads to a loss of high-frequency information and potential distortion of the signal.

5.3 Analog-to-Digital Converter

An analog-to-digital (A/D) converter (▶ available—V5.4) is a hardware device for converting continuous-time analog signals to digital signals. An A/D converter operates as follows. The continuous-time analog signal is first sampled at regular intervals, which involves measuring the instantaneous value of the signal at discrete points in time. Once a sample is acquired, the sample-and-hold circuit of the A/D converter (similar in function to the ZOH discussed in the previous section but operates here at the input stage) becomes active. This circuit maintains the sampled value constant throughout the sampling interval. Its primary role is to provide a steady value to the quantizer, ensuring that variations in the analog signal during the quantization process do not introduce errors. The held value is then quantized to represent it digitally. Quantization involves approximating the continuous range of the analog signal into a finite set of discrete levels. The outcome of the A/D conversion process is a series of discrete digital values, which effectively represents the original continuous-time analog signal in a discrete format.

5.3.1 A/D Characteristics

Among the critical characteristics of an A/D converter are its conversion rate, voltage range, bit resolution, and quantization error. The **conversion rate** indicates how many analog-to-digital conversions are performed within a specified time unit, typically a second. On a microcontroller, this rate hinges on the selected clock signal, its speed, and the chosen bit range. A/D devices embedded in PC data acquisition cards typically exhibit conversion rates of under 100,000 conversions per second for entry-level models. In contrast, advanced models can achieve rates exceeding one million conversions per second. It is essential to understand that these figures represent the peak performance capabilities of the standalone A/D device. When integrating an A/D converter into a digital feedback control system, the effective conversion rate tends to decrease, primarily influenced by the processing speed of the system. The **voltage range** of an A/D converter refers to the analog voltage range that the device can handle. On a microcontroller, the range is 0 to V_{DD} (the supply voltage) unless an external reference voltage (*AREF*) is used, in which case the range is 0 to *AREF*. On a PC data acquisition card, most A/D devices allow both unipolar and bipolar ranges ranging from 0.05 to 10 V but can tolerate an overload voltage of up to 30 V. The voltage conversion range is normally set by a software call to the device controller. The **bit resolution** of the A/D device is quoted as the number of bits that the converted analog signal is mapped into. Common bit sizes are 12 to 16 bits, but many microcontrollers have A/D devices that have only a 10-bit range. The bit resolution affects the **voltage resolution** or increment of the device, which is shown in the formula

$$\text{Voltage Resolution} = \text{Range}/2^n \tag{5.4}$$

where n is the bit resolution of the A/D converter. To understand the relationship between voltage range and voltage resolution, let us consider a 12-bit A/D device with a –10 to 10 V range. In this case, a 20-V range (–10 to 10 V) is mapped into 2^{12} binary combinations. Thus, the device can map voltage values to discrete values at increments of

$$\text{Range}/2^n = 20/4096 = 4.88 \text{ mV}$$

This means that the input voltage can increase by up to 4.88 mV without changing the output value of the A/D converter. For example, if the input voltage that was measured by this device was the output of an analog temperature sensor with a sensitivity of 10 degrees per volt (°/V), then the device will not be able to measure temperature changes that are less than 0.0488°. The discrete output of an A/D converter subjected to an input voltage V_{in} is given by

$$\text{Digital output} = \text{ceiling}\left((\text{V}_{\text{in}} - \text{V}_{\text{ref}} - \text{voltage resolution})/\text{voltage resolution}\right) \tag{5.5}$$

Where V_{ref} is the reference voltage for the A/D which is typically zero volts. Along with the voltage resolution of the device, the **quantization error** (or digitization accuracy) of the A/D converter is also an important performance parameter and is directly related to the bit range of the device. The digitization accuracy refers to the uncertainty in the discretized voltage value, and it is +/– one-half of a bit. To further understand the concept of digitization accuracy, let us consider a contrived 2-bit A/D device with a range of 0 to 5 V. This A/D device maps a 5 V analog range (0 to 5 V) into four different binary values (2^2 or 0, 1, 2, and 3). As seen in Figure 5.5, the output of the A/D converter will be 0 if the input analog voltage happens to be in the range of 0 to 1.25 V, 1 if the input analog voltage happens to be in the range of 1.25 to 2.5 V, and so forth. Notice that the maximum digital output level (3 in this case) is reached before the input reaches full scale or 5 V. Nominally, we say that if the A/D converter outputs a value of 1, then this corresponds to a nominal analog voltage of 1.875 V. The input voltage can change up to +/–0.625 V (or +/– one-half of a bit) without any change in the output of the A/D. Thus, at any of the discrete output values of the A/D converter, we say that we have an uncertainty of +/– half the voltage increment of the A/D device. Some A/D converters are built with an intentional offset of –1/2 bit. The staircase output curve of such a device will be shifted to the left and will start at 0.625 V for the 2-bit A/D example. The input/output relationship of an A/D converter is further illustrated in Example 5.2.

Figure 5.5

Two-bit A/D Mapping

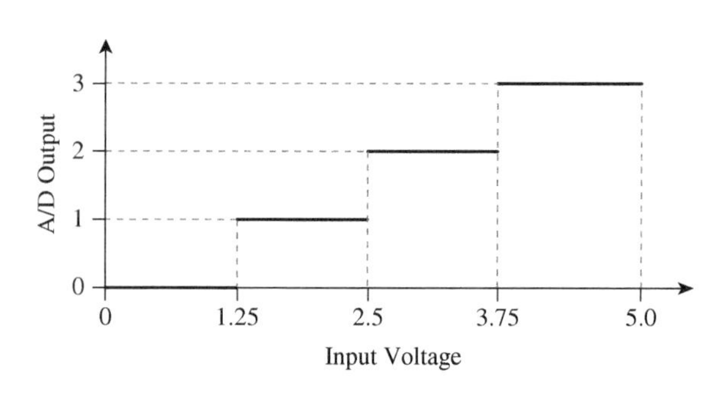

Example 5.2 A/D Converter

Determine the voltage resolution and digitization accuracy of a 12-bit A/D converter with a 0 to 10 V range. Determine the output level if the input voltage is 6.5 V. Also, determine the corresponding analog input voltage at the following digital output values: 0 and 1000.

Solution:

From Equation (5.4), the voltage resolution of this A/D device is $10/2^{12} = 2.441$ mV. Thus, the digitization accuracy of the conversion is +/– 1.220 mV.

From Equation (5.5), the discrete output level is given as the integer ceiling of ((6.5 V – 2.441 mV) /2.441 mV) or 2662.

This 12-bit A/D converter maps the 0 to 10 analog voltage range into 0 to 4095 digital values. When the A/D converter outputs a value of 0, the analog input voltage of this A/D converter is in the range of 0×2.441 mV to 1×2.441 mV or 0 to 2.441 mV. Similarly, when the A/D converter outputs a value of 1000, the analog input voltage is in the range of 1000×2.441 mV to 1001×2.441 mV or 2.441 to 2.443 V.

5.3.2 A/D Operation

Most A/D converters are built to operate on the principle of successive approximation. In the **successive approximation method,** an internal digital-to-analog (D/A) converter (which converts digital signals to analog signals—see Section 5.4) and a comparator circuit are used to converge on the digital signal that is closest to the sampled analog signal. Starting with the MSB, the bits in the D/A converter are set/reset one at a time until the sampled analog signal matches the output signal from the D/A converter to within the least significant bit. The binary pattern of the D/A converter is then the digital input signal. This conversion technique is a good compromise between speed, resolution, and cost. Example 5.3 illustrates the operation of a successive approximation 3-bit A/D converter with a 0 to 10 V analog input range. Other types of A/D converters include flash/parallel, integrating, and digital ramp.

Most microcontrollers have several A/D channels. On a PC, the A/D converter is packaged with a digital-to-analog converter and a parallel port to form a data acquisition card. The card is placed in one of the available computer slots, and a cable then is used to connect the card to an interface board commonly known as a distribution panel or screw terminal. To set up an A/D converter in a PC data acquisition board, the user has to make certain important selections, including the input range and the input signal configuration. The input range is normally set using software.

Example 5.3 Successive Approximation A/D

Illustrate the operation of a 3-bit, 0 to 10 V successive approximation A/D subjected to an analog input voltage of 8 V.

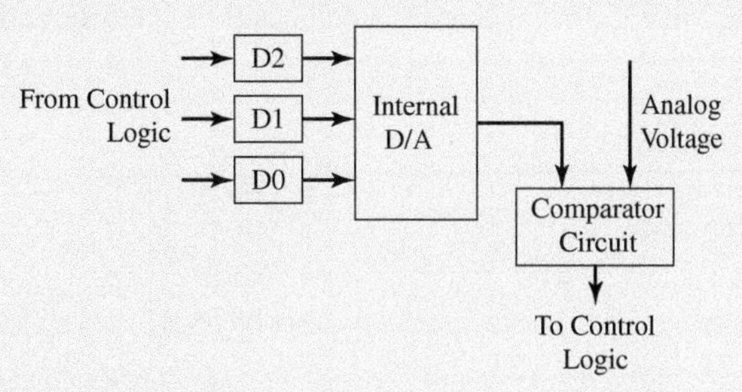

Figure 5.6 Three-bit successive approximation A/D

Solution:

With reference to Figure 5.6, the control logic of this A/D device will first turn the most significant bit (bit D2) of the internal D/A device associated with this A/D. The analog output voltage corresponding to this bit is 5 V (see table in Example 5.4). The comparator circuit will then compare the analog output of the internal D/A with the supplied analog input (8 V). Since the output of the internal D/A is smaller than the supplied voltage, bit D2 remains on, and the next bit (bit D1) is turned ON. The analog output (7.5 V) of the internal D/A is still less than the supplied voltage, so bit D1 remains ON. When bit D0 is turned ON, the analog output is 8.75 V, which is greater than the supplied voltage. Thus, bit D0 is turned OFF, and the output of the A/D will be 0x06, which is the same digital pattern on the internal D/A.

5.3.3 A/D Input Signal Configuration

The configuration of an A/D input signal is classified into two types: single-ended input and differential input. In the **single-ended input** mode, the input signal is referenced to the A/D board's signal ground. As illustrated in Figure 5.7(a), the signal is transmitted using two wires. The wire carrying the signal is connected to any of the input channel terminals, while the other wire is linked to the board's signal ground, often referred to as low-level ground or LLGND. In this setup, the A/D converter determines the difference between the signal and the board's ground. However, a single-ended connection is prone to noise since the signal wire can inadvertently function as an antenna, capturing electrical interference. It is essential to note that the single-ended configuration is suitable only for floating signal sources to prevent ground loops (as explained in Section 2.10). A floating signal source is not connected to any common ground or fixed potential, and lacks any ground connection at the signal source. Battery-powered devices are typical examples of floating signal sources. Being independent of an external power source or ground, their signal outputs remain unanchored to a reference point.

In the **differential input mode**, the high (or positive) input signal's value is ascertained relative to the low input (or negative) signal. Typically, three wires are used for connection to the board. The wire conveying the high-input signal is connected to one of the A/D input channel terminals, for instance, 0, 1, 2, ..., 7 (assuming 16 single-input channels exist). The low signal connects to an adjacent channel, like channel 8 in Figure 5.7(b) if the main signal connects to channel 0. This low signal wire also joins the board's signal ground via a resistor. When set to the differential input mode, the total number of usable input channels is reduced by half. Thus, a board with 16 single-ended input channels will offer only eight channels in differential mode. Differential input mode boasts superior noise handling compared to its single-ended counterpart. Electromagnetic interference typically affects both leads of the signal. Since the A/D converter in differential mode gauges the discrepancy between the high and low inputs, voltages common to both inputs are effectively negated. This mode is particularly recommended for reading outputs from analog sensors prone to noise, such as thermocouples and strain gauges (discussed in Chapter 8). A measure of the ability of an A/D converter used in differential input mode to eliminate the common voltage is called the **common mode rejection ratio (CMRR),** which was previously discussed in the context of op-amps (see

Section 2.9.5). In an ideal A/D converter, any voltage common to both signal wires will be completely canceled. In a real A/D converter, a perfect cancellation does not occur, and a fraction of the common voltage will show. The CMRR, which is expressed in decibels (dB), is the reciprocal of the voltage fraction that is passed. It is desirable to have an A/D converter with a high CMRR ratio. Note that most MCUs do not support differential mode A/D input, but Microchip manufactures special A/D ICs that support differential A/D input. An example is the MCP3301 IC, which is a dedicated 13-bit differential input A/D converter IC.

Figure 5.7

Data acquisition board wiring for single-ended and differential input mode

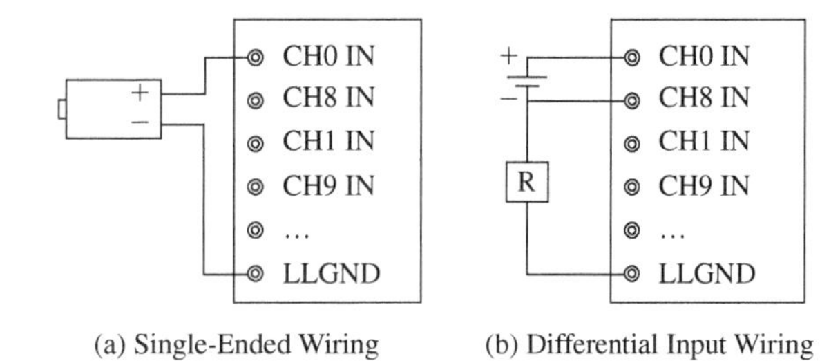

(a) Single-Ended Wiring (b) Differential Input Wiring

5.4 Digital-to-Analog Converter

A digital-to-analog (D/A) converter (▶ available—V5.5) is a device that converts digital signals, represented as discrete binary values, into continuous analog signals. The discrete values could represent audio samples, sensor readings, control values, or any other form of digitally encoded information. The D/A converter takes the digital input and converts it into an analog signal that varies continuously based on the digital value. Most microcontrollers do not have a D/A converter, but the function of the D/A converter is approximated on microcontrollers using the PWM output feature (see Section 4.6).

5.4.1 D/A Characteristics

All of the performance parameters of an A/D (such as conversion rate, voltage range, bit resolution, and digitization accuracy) that were discussed before are similarly applied here for the D/A converter, so they will not be repeated, but there is an example (Example 5.4) that discusses the mapping between digital input values and analog output values.

Example 5.4 D/A Mapping

A 3-bit D/A converter is set for 0 to 10 V output range. Map all the possible digital input values to their corresponding analog output values.

Solution:

A 3-bit D/A converter has 2^3 possible digital input values (0 to 7 decimal or 000 to 111 binary). The voltage resolution is $10/2^3 = 1.25$ V. The corresponding analog output values for each possible digital input value are shown here:

Binary Input	Analog Output (V)	Binary Input	Analog Output (V)
000	0.00	100	5.00
001	1.25	101	6.25
010	2.50	110	7.50
011	3.75	111	8.75

Note that while this D/A converter's nominal range is 0 to 10 V, the maximum output analog value is only 8.75 V due to the coarseness of its resolution. If the bit resolution was 10 instead of 3, then the maximum analog output voltage would be 9.990 V, or to generalize

$$\text{Maximum output} = \text{range} - \text{resolution} = 10 - 10/2^{10} = 9.990 \text{ V}$$

For a D/A converter, the digital input that gives a certain analog output voltage V_{out} is given by

$$\text{Digital input} = \text{ceiling}\left((V_{out} - V_{min} - \text{voltage resolution})/\text{voltage resolution}\right) \qquad \textbf{(5.6)}$$

where V_{min} is the minimum voltage supplied by the D/A. For example, using the data from Example 5.4, the digital input needs to be 6 for an output of 8 V. Due to the coarse resolution of this A/D converter, the analog output will only be 7.5 V (6 × 1.25). If this was a 10-bit D/A converter instead, then the output would be 7.998 V at a digital input of 819. Most commercial D/A converters have a 12-bit output resolution.

5.4.2 D/A Operation

To illustrate the operation of a D/A converter, let us consider the **weighted resistor summing amplifier circuit** shown in Figure 5.8. The digital input acts as an electronic switch in this circuit, providing a connection between V_R and the respective resistor if the binary value is 1. For the digital input 1011 shown in the figure, the output of this circuit is $-11\ V_R$. This circuit is not used in practice, because it requires resistances of certain ratios which are difficult to satisfy with good accuracy. This is especially true for a D/A converter with more than 4 bits. For example, a 12-bit D/A would require that the 11th-bit resistor have a resistance of 1/2048 of the 0th-bit resistor.

Figure 5.8

Weighted resistor summing amplifier circuit

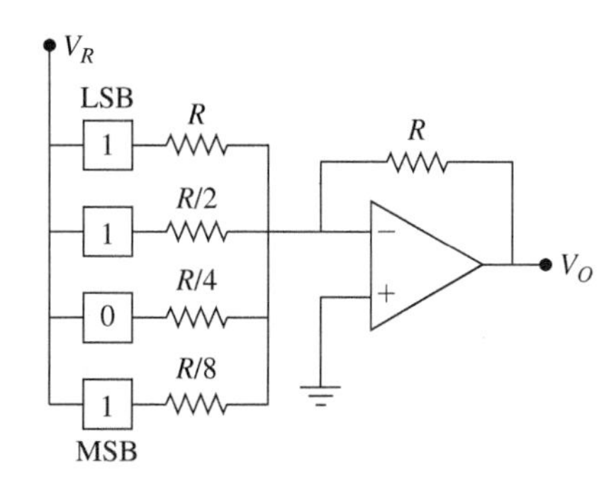

A commonly used circuit for D/A conversion is the ***R/2R* ladder resistor network** [18]. Unlike the weighted resistor summing amplifier circuit, the *R/2R* ladder circuit requires only two resistor values *R* and *2R* regardless of the number of bits. The *R/2R* ladder circuit is shown in Figure 5.9. It consists of a repeating pattern of *2R* and *R* resistors arranged in a ladder form. The *2R* termination resistor is connected to the ground and is used to make the Thevenin resistance of the network at each ladder leg equal to *R* when all the bits are grounded (see Figure 5.9).

Figure 5.9

R/2R ladder resistor network

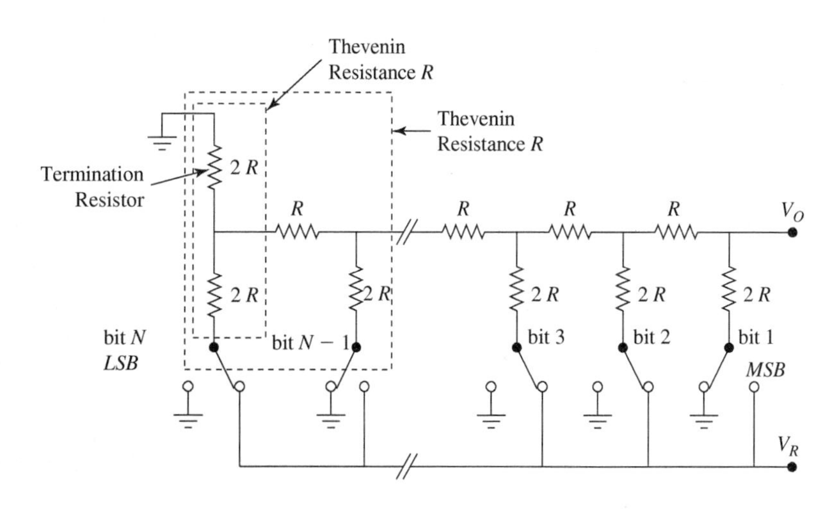

The output voltage V_O when a voltage V_R is connected to a bit $i (1 \leq i \leq N)$ with all the other bits grounded is given as

$$V_O = \frac{V_R}{2^i} \qquad \textbf{(5.7)}$$

If more than one bit is connected to V_R, then the principle of superposition applies. Thus, for example, if we have a 3-bit D/A with all the bits connected to V_R, then the analog output is equal to $V_R/2 + V_R/4 + V_R/8 = 7/8\ V_R$. If V_R is equal to 10 V, then the output is 8.75 V, as was the case in Example 5.4. The *R/2R* network provides the most accurate method of digital-to-analog conversion.

Practical D/A converters employ a **ZOH** (see Section 5.2.2). After a digital value is converted to an analog value by the D/A converter, the ZOH ensures this analog value remains constant until the next digital value is received and converted. This means that the current output of the D/A device will remain constant until a new output is sent to the device.

5.5 Data Acquisition

5.5.1 Data Acquisition Boards

Data Acquisition boards and modules, commonly known as **DAQ boards** (available—V5.6), are important interfaces bridging the analog world to the digital computational domain. They are widely employed across scientific, engineering, and industrial domains. These boards capture, measure, and relay environmental data for immediate or later digital analysis or for sending control output signals. Many of these boards come equipped with analog input and output, digital I/O, and counter/timers. Figure 5.10 illustrates two commercial multifunction DAQ boards. Figure 5.10(a) displays a board designed to operate with the PCI Express (PCIe) bus on PCs, whereas Figure 5.10(b) features one that uses USB for interfacing. While older models were designed for traditional computer slots like PCI, newer models have adopted interfaces such as PCIe and USB, with USB-based DAQ devices gaining popularity due to their portability and user-friendly nature.

Figure 5.10

Multifunction DAQ:
(a) multifunction PCI Express board (PCIe-DAS1602/16) and
(b) multifunction USB device (USB-1608G). (Courtesy of Digilent)

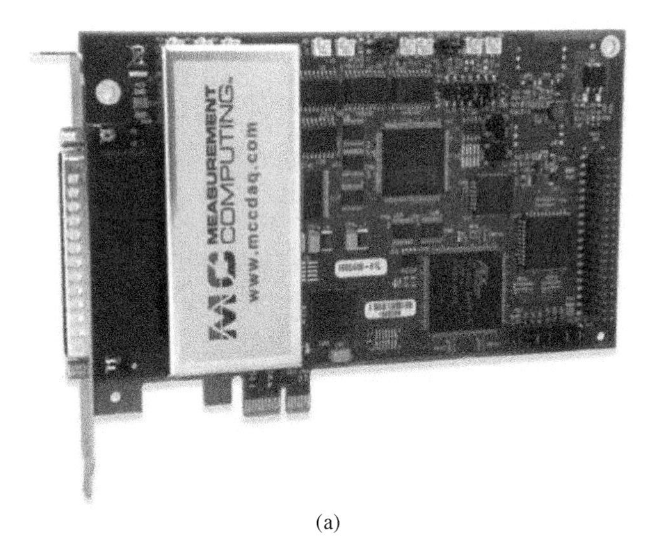

(a)

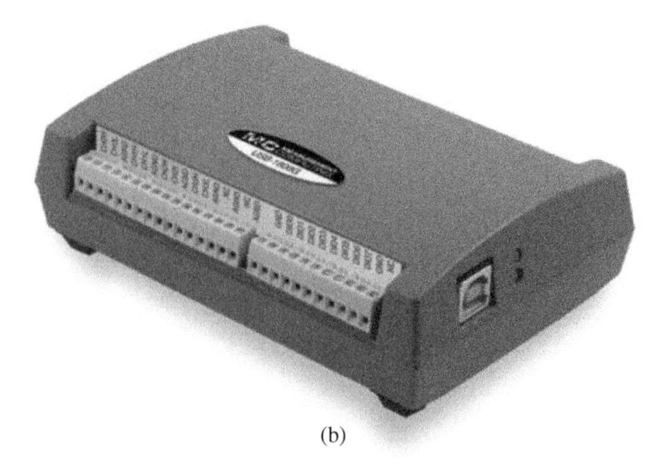

(b)

Modern DAQ boards often feature multiplexers (see Section 3.6.3), enabling a single A/D converter to read from multiple input channels connected to various sensors. This feature is advantageous as it reduces the need for multiple A/D converters, thus reducing cost and board space while still allowing multiple channels of data acquisition. Some DAQ boards incorporate signal conditioning circuits to optimize incoming signals through amplification, filtering, or isolation. Onboard clocks manage precise timing, a crucial element for high-speed data applications. Additionally, built-in memory buffers minimize the risk of data loss during fast acquisitions or computational delays. These boards are also designed to seamlessly integrate with popular software platforms and are bundled with dedicated drivers and software development kits compatible with platforms like LabVIEW, MATLAB, or Python.

A/D and D/A converters on data acquisition boards utilize different data transfer modes, including **direct memory access** (DMA) and **programmed I/O**. With DMA, data moves directly between the memory and the DAQ board without tasking the system processor. In contrast, in programmed I/O, the system processor orchestrates the data transfer. DMA facilitates rapid data transfer rates and, while it requires more setup, its efficiency shines when transferring large datasets, allowing the CPU to multitask effectively.

Some DAQ boards, like those from Measurement Computing (now Digilent), feature digital I/O pins that operate like a parallel port, often with configurations such as four 8-bit ports. The term '**parallel port**' refers to an interface where data is sent simultaneously across multiple lines. This name derives from the fact that all data presented to the device is transmitted in parallel. An example of this is the printer port found on older PCs. These ports can be configured through software to function as input, output, or a combination of both. Most of these parallel interfaces utilize transistor-transistor logic (TTL) family chips, with specific voltage levels for representing logical states. These voltage ranges are listed in Table 5.1.

Table 5.1

TTL input and output levels

Operation	Low State Voltage Range	High State Voltage Range
Input	0–0.8 V	2.0–5.0 V
Output	0–0.5 V	2.7–5.0 V

The input parallel port is commonly used to read data from switches and on/off type sensors, such as proximity sensors and limit switches. The output parallel port is normally used to activate lights, solenoids, and relays. While the software that comes with most DAQ boards provides functions to access the parallel digital input/output port, it does not provide means to read a single input bit on the port or set a particular output bit without disturbing the rest of the bits. These operations can be done by using the bitwise logical operators that are available with programming languages such as C. Example 5.5 illustrates this using C syntax.

Example 5.5 Setting a Bit on a Digital Output Parallel Port

Illustrate how bit #5 of an 8-bit digital output parallel port can be set either high or low without changing the current output on the port.

Solution:

To enable us to perform this operation, we need to have a variable that stores whatever was sent to the port. Let us call this variable PortValue. To set bit #5 to high, we simply perform a bitwise OR operation between the variable PortValue and the value 0x20. We can write it as

PortValue = PortValue | 0x20

Notice that in using this operation, the value of each bit, other than bit #5, is not changed from its current value, since 'oring' a bit with 0 does not change its value. To set bit #5 to low, we need to perform a bitwise AND operation with the value 0xDF, which has a value of 1 in all its bits except bit #5, which has a value of 0.

PortValue = PortValue & 0xDF

The updated value of PortValue is then sent to the port.

5.5.2 MATLAB/Simulink Data Acquisition

MATLAB is widely recognized for its capabilities in numerical computations and data analysis. Beyond these core capabilities, MATLAB also supports data acquisition through its **Data Acquisition Toolbox** (▶ available—V5.7). This toolbox facilitates the acquisition of analog data from A/D converters, dispatch of digital data via D/A converters, and interaction with digital input/output devices across diverse DAQ hardware. Primarily operating within the MATLAB environment, it provides a suite of commands to directly funnel data into the MATLAB workspace. Once in the workspace, users can leverage MATLAB's extensive functions and toolboxes for data analysis and visualization. However, it is worth noting that while the toolbox excels in data acquisition and allows for integration with Simulink, it does not guarantee real-time execution. This makes it less suitable for time-sensitive applications.

For tasks requiring feedback control (see Chapter 10) or any other time-critical operations, the **Simulink Desktop Real-Time** (SDRT) Toolbox emerges as a more fitting choice. As implied by its name, SDRT is explicitly crafted for real-time applications (see Chapter 6), ensuring that Simulink models execute in real-time, a critical requirement for numerous controls, testing, and hardware-in-the-loop simulation scenarios. Integrated Case Study II.B discusses the application of the SDRT Toolbox in interfacing and governing the linear motion slide system.

Integrated Case Study II.B: Data Acquisition Using MATLAB/Simulink

In Integrated Case Study II (DC Motor-Driven Linear Motion Slide—see Section 11.3), the slide's motion was controlled using the MATLAB Simulink Desktop Real-Time software package. Figure 5.11 illustrates the interface between MATLAB-Simulink and the associated hardware. The Measurement Computing DAS 6036 DAQ board is equipped with two 16-bit D/A analog output channels. One of these channels was employed to transmit the voltage signal to the servo drive connected to the motor. The motor's position was captured using a 24-bit counter board. This board processes signals from the incremental encoder (see Section 8.3.3) attached to the motor. Notably, the Desktop Real-Time toolbox operated in Kernel mode. This mode is a privileged operating state where software can directly interact with hardware resources, bypassing typical user-space restrictions, thus offering reduced latency and rapid response times—essentials for real-time applications. Furthermore, the Desktop Real-Time toolbox necessitates a plugin specific to the hardware being used for data acquisition.

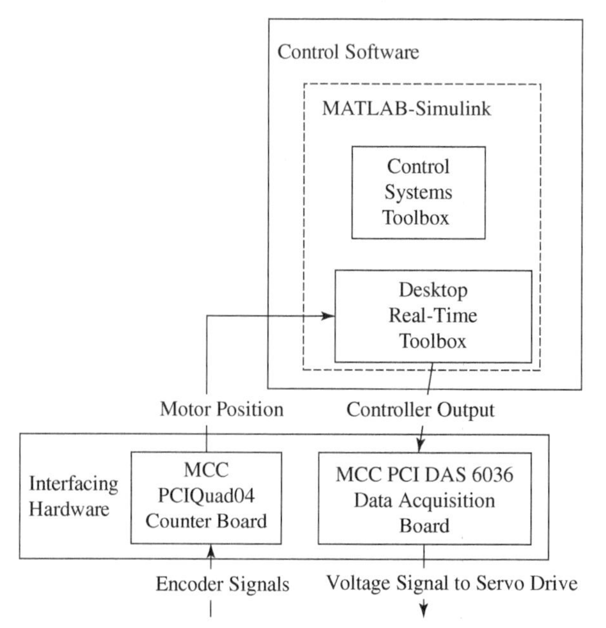

Figure 5.11 Interface between MATLAB-Simulink and the hardware

MATLAB was used to perform tests to identify the dynamic response (see Chapter 7) of the slide. A voltage signal was sent from the PC and the position of the motor was recorded by the software. A screenshot of the MATLAB screen to perform this identification is shown in Figure 5.12. The MATLAB discrete derivative was used to differentiate the position signal to obtain the velocity output. Since the encoder is a sampled signal, the discrete derivative provides a cleaner derivative of the signal. The Desktop system runs at a sampling frequency of 1000 Hz.

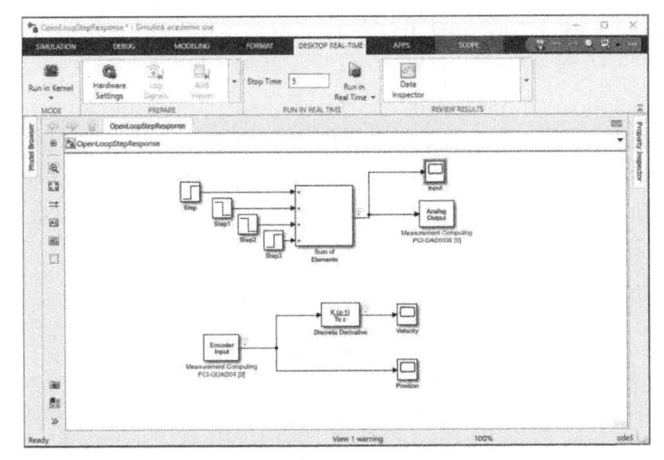

Figure 5.12 Screenshot of the MATLAB/Simulink screen to perform dynamic response tests of the system

5.6 Serial Communication

A parallel digital input/output port is optimal for transmitting data over short distances due to its simultaneous multi-channel design. However, for longer distances, **serial communication** is typically more suitable because of techniques often associated with it, such as differential signaling, that bolster noise resilience. Additionally, serial connections tend to use fewer wires than parallel ones. Serial ports were commonly used in devices like modems, mice, and keyboards to transmit data one bit at a time, especially before the widespread adoption of USB. This mode of communication involves breaking data into smaller units, which, for simplicity, we'll refer to as 'packets' here, though they are not packets in the traditional network communication sense. Each bit is sent sequentially over a single channel.

Serial data is transmitted in either asynchronous or synchronous modes. In **asynchronous communication**, devices rely on start and stop bits to signal the beginning and end of a packet, allowing the transmitter and receiver to operate independently without a shared clock. In contrast, **synchronous communication** (see next two sections) involves continuous streams of data that are coordinated with a common clock signal between devices. Traditional serial ports on PCs typically use asynchronous communication, as they often lack the requisite hardware for synchronous transmission. The term **baud rate** in serial communication usually signifies the number of signal changes or symbols per second. In many serial protocols, this corresponds to the number of bits transmitted per second, and this rate can range from as low as 300 to well over a million, contingent on the specific technology and medium employed.

In asynchronous transmission (available—V5.8), means must be provided to inform the receiver of the start and end of a data packet, since the timing of the transmission of the packet is not known in advance. This is accomplished by structuring the **data packet** to include a **start bit** at the beginning of the packet to inform the receiver of the start of the packet and a **stop bit** at the end of the packet to indicate its completion. The number of bits in a packet can vary, but 10-bit serial data packets are common and are discussed here. These packets are structured to include one start bit (the first bit that is transmitted), 7 or 8 bits of data representing the character to be transmitted, an optional **parity bit** for 7-bit data, and one stop bit (or two stop bits for 7-bit data with no parity bit) at the end of the packet. The start bit is always low (space) while the stop bit is always high (mark). Figure 5.13 shows the packet structure for one start bit, seven data bits, one parity bit, and one stop bit. The data is structured with the LSB first. The practice of transmitting the LSB first in serial communication is quite common, although it depends on the specific communication protocol or system being used. This approach is known as 'LSB-first' or **'little-endian'** bit ordering. It is one of the two main methods for ordering bits; the other is 'MSB-first' or **'big-endian'** bit ordering, where the MSB is transmitted first.

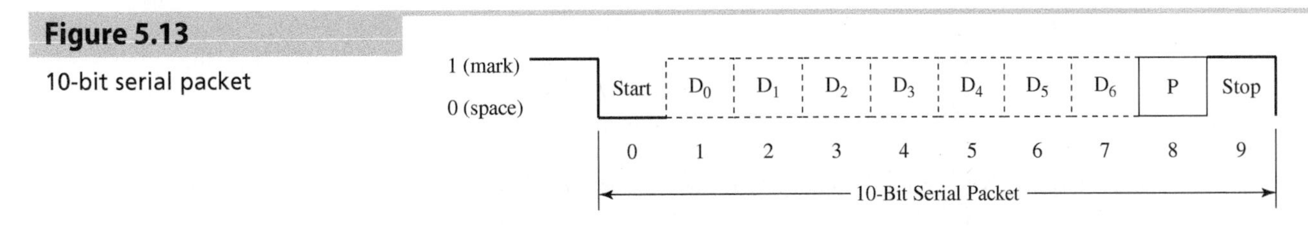

Figure 5.13
10-bit serial packet

Since the receiving end does not know ahead what data is transmitted, the parity bit can be used to provide a crude method of error checking. The different **parity methods** are listed here.

Even: This means that the total number of one bits in the packet (excluding stop bit) is even. Thus, the value of the parity bit is set to 1 or 0 to make the total number of one bits even.

Odd: This means that the total number of one bits in the packet (excluding stop bit) is odd. Thus, the value of the parity bit is set to 1 or 0 to make the total number of one bits odd.

Mark: Parity bit is always set to logical 1 (mark).

Space: The parity bit is always set to logical 0 (space).

None: No parity bit is used.

For even/odd parity, the parity method of error checking works by counting the total number of one bits in the packet (excluding the stop bit) that is received. If even (odd) parity was selected, then a transmission error has occurred if that number is not even (odd). For mark or space parity, the parity bit in the received packet is checked to see if it matches the parity set mode. The parity method can detect a single-bit error but is not guaranteed to detect multibit errors. The seven or eight bits of data are converted using an encoding scheme such as the ASCII code (see Appendix B for a list of the codes). Example 5.6 illustrates the serial packet structure.

Example 5.6 Serial Packet Structure

Show the 10-bit serial packet for the letter 'B' using a 7-bit data, one start bit, one stop bit, and a parity bit. Illustrate for both even and odd parity.

Solution:

The 7-bit ASCII code for the letter 'B' is 0x42 or b100 0010 (see Appendix B). Thus, for even parity, the 10-bit serial packet is

0 + 0100001 + 0 + 1
Start + 7-bit ASCII + parity + stop bit

or '0010000101'.

For odd parity, the 10-bit serial packet is

0 + 0100001 + 1 + 1
Start + 7-bit ASCII + parity + stop bit

or '0010000111'. Note that the start of the packet is from the left end, and the ASCII code for the letter is written in the packet with the least significant bit first.

Serial ports installed on PCs and some laptops take the form of a 9-pin male 'D' connector, known as DE-9 (older PCs have a 25-pin male connector, termed DB-25, but it is common for many of these pins to be unused in standard serial communication). In a serial connection, the data can be sent in full-duplex mode or half-duplex mode. **Full-duplex mode** means that the data between the two devices in communication can be transmitted simultaneously in both directions. In **half-duplex mode** (which is outdated and not commonly used), data is transmitted in one direction at a time, but the direction can be changed. For full-duplex mode, a serial connection between two ports requires (physically) a minimum of three wires if no hardware flow control is used. One wire is used for sending the data, the second wire is used for receiving the data, and the third is used as ground. The remaining pins in the connector are used for control purposes.

There are several protocols for serial interfaces. **RS-232C**, now referred to as EIA-232, stands out as the most prevalent, especially in PC serial ports. The 'RS' in 'RS-232' denotes 'Recommended Standard,' and the appended 'C' marks the latest revision of this standard. While it is a popular protocol, the RS-232 has inherent limitations concerning speed, cable length, and device compatibility. Another notable protocol is **RS-422**, which, by leveraging differential signaling (using two wires for each signal), offers improved noise immunity and can support longer cable runs. Furthermore, it might be worth noting the RS-485 in discussions about serial protocols. Like RS-422, **RS-485** utilizes differential signaling but has the added advantage of supporting multi-point communications. This means multiple devices can be connected on the same serial bus, allowing for one-to-many and many-to-many communications.

To ensure data integrity and prevent buffer overflow in the receiving device, various **handshaking** or **flow control methods** can be employed. These can be software-based, hardware-based, or a combination of both. One common software flow control method is XOnXOff. In this method, special control characters are sent between the receiver and the transmitter to regulate data flow. Specifically, 'XOn' and 'XOff' stand for 'transmit on' and 'transmit off' respectively. When the receiver is ready to accept more characters, it sends an XON character (often represented as 0x11) to the transmitter. Conversely, when its buffer approaches full, it sends an XOFF character (often represented as 0x13) to temporarily halt data transmission. Being software-based, this method can be implemented even with a simple three-wire serial connection.

An alternative, hardware-based handshaking method employs the **Request-to-Send** (RTS) and **Clear-to-Send** (CTS) signals. Typically, the transmitting device asserts or activates its RTS signal to request permission to send data. In response, if the receiving device is ready to accept data and its buffer is not full, it asserts its CTS signal, granting the transmitter permission. If the receiver's buffer nears capacity, it will de-assert the CTS signal, signaling the transmitter to stop sending data. It is worth noting that both RTS/CTS (hardware) and XON/XOFF (software) flow controls can be utilized simultaneously for enhanced communication reliability.

A common application of USART (Universal Synchronous/Asynchronous Receiver-Transmitter) in microcontrollers is to facilitate full-duplex asynchronous serial communication, often adhering to the RS-232 protocol standards. It is important to differentiate between the USART hardware capability in a microcontroller and the RS-232 communication standard, with the former being a versatile on-chip communication module and the latter specifying the characteristics of serial communication. It is worth noting that with the increasing prevalence of USB ports, the use of traditional serial ports on PCs and laptops has become less common. However, USB-to-serial adapters can be used to interface with devices that still rely on serial communication. The basic principles and steps of serial communication remain the same, regardless of whether the physical interface is a native serial port or a USB-to-serial adapter.

The Arduino platform offers built-in functions for RS-232 serial communication, commonly used to exchange data between an Arduino and computers, sensors, or other microcontrollers. The serial communication process is initiated by calling the *Serial.begin()* function in Arduino's *setup()* function, where the baud rate for the communication is specified as a parameter. After initializing the serial port, data can be sent from the Arduino to another device using the *Serial.print()* or *Serial.println()* functions. While *Serial.print()* sends data as is, *Serial.println()* appends a newline and a carriage return characters at the end of the data. To receive data, the *Serial.available()* function is used to check if data is present in the receive buffer. If data is available, it can be read using the *Serial.read()* function, which reads a single byte of data from the receive buffer. The *Serial.write()* function can also be used to send binary data or a series of bytes. In situations where serial communication needs to be terminated, the *Serial.end()* function can be called to disable the serial port, freeing up the hardware resources for other purposes.

5.7 Serial Peripheral Interface

To facilitate rapid and deterministic communication, many AVR MCUs come equipped with built-in synchronous serial port modules. The ATmega328P MCU, for example, supports both the Serial Peripheral Interface (SPI™) and the 2-wire Serial Interface (commonly known as the Inter-Integrated Circuit (I^2C™) interface). SPI and I^2C communication methods have become prominent due to their suitability for compact and power-efficient devices. They require fewer pins than parallel interfaces, making them ideal for smaller devices. Their design supports multiple device connections, with I^2C using unique addresses and SPI using a 'chip select' mechanism. Additionally, many modern sensors and peripherals come standard with SPI or I^2C interfaces, allowing for easy integration.

The SPI (▶ available—V5.9), operates in full duplex mode, can achieve speeds of 1 Mbps or even higher, and follows a leader-follower configuration. This interface is simple to implement, typically requiring only four wires.

These four wires are given here.

Serial Clock Signal (SCK pin): This is the clock pulse signal that the leader sends to the follower. One bit of data is transmitted for each clock pulse.

Leader Out Follower In (Serial Data Out or SDO pin): Output data from leader to follower.

Follower Out Leader In (Serial Data In or SDI pin): Output data from follower to leader.

Follower Select (SS pin): This signal is used to select the particular follower in the case of one leader and several followers.

In the Arduino Uno board (see Table 4.6), the SCK pin is digital pin #13, the SDO pin is digital pin #11, the SDI pin is digital pin #12, and the SS pin is digital pin #10. In establishing an SPI interface between two devices, one device must be designated as the leader and the other as the follower. The leader generates the clock pulses while the follower receives them. The leader determines the clock rate, with many devices supporting speeds up to or exceeding 10.0 Mbps. Additionally, the phase and polarity of the clock, which dictate whether data is sent on the rising or falling edge of the clock signal, must be defined. To visualize the SPI communication mechanism, consider Figure 5.14. In it, each SPI device is shown to have both a buffer and a shift register. The shift register, a series of interconnected flip-flops that can store binary numbers (see Section 3.7), plays a pivotal role in data transmission and reception during SPI communication. The primary interaction with the SPI for data transfer occurs through these shift registers. When a data transfer is initiated, the relevant data is loaded from the buffer into the shift register. With each clock pulse generated by the leader, data is transmitted bit-by-bit between devices. During this transmission, as the leader shifts out its most significant bit (MSB), it can simultaneously receive the MSB from the follower into its least significant bit (LSB) position, and vice versa. This simultaneous transmission and reception signify the full-duplex nature of SPI. Once the data transfer completes, which could be of various lengths and not strictly 8 bits, the shift registers' content can be moved to their respective buffer registers. This transfer prepares the devices for the next data interaction, which could be either a read or a write operation.

Figure 5.14

Illustration of SPI interface

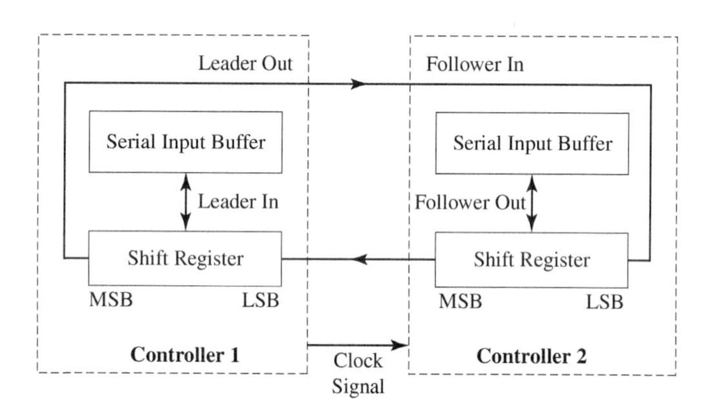

The Arduino IDE platform has built-in functions to set and access the SPI interface through the SPI library. They include the *SPI.begin()* function that initializes the SPI bus interface. Before engaging in any data transfer, the SPI settings, such as clock speed, bit order, and data mode, are configured using the *SPI.beginTransaction()* method. This is crucial, especially when interfacing with multiple SPI devices that may have varied configurations. The data exchange between the Arduino and another SPI device takes place via the *SPI.transfer()* function. This is a full-duplex function, meaning that as data is sent out, data is also simultaneously received. Upon completing a specific SPI data transaction, the *SPI.endTransaction()* function is invoked to reset the settings, preventing any conflicts with subsequent SPI operations or devices. If the SPI functionality is no longer required during the Arduino's operation, the SPI can be disabled with the *SPI.end()* function.

5.8 Inter-Integrated Circuit Interface

The **I^{2}C interface** (pronounced I-Squared-C), also known as **Two Wire Interface** (TWI) in some contexts, is a synchronous serial communication protocol developed by Philips Semiconductor, now known as NXP Semiconductors. The I^{2}C interface requires just two wires for communication: SDA for data transmission and SCL for the clock signal. On the Arduino UNO, which employs the ATmega328P, the SDA line is typically on pin A4 and the SCL line on pin A5 (see Table 4.6). Figure 5.15 illustrates the connection between an I²C leader (e.g., an AVR MCU) and an I²C follower (such as a serial EEPROM). It is essential to recognize that the SDA and SCL lines are open-collector types, necessitating a pull-up resistor on each line.

The I²C protocol (▶ available—V5.10) supports configurations with one leader and multiple followers. The leader dominates the communication by initiating with a start bit, then transmitting the follower's address followed by a bit indicating a write or read operation. If the sent address matches the internal address of the follower, then the follower will send back an acknowledgment bit to the leader. In a write operation, the leader dispatches a data byte, waiting for the follower's acknowledgment. Conversely, in a read operation, the follower sends a data byte and then waits for an acknowledgment from the leader. Communication concludes when the leader transmits a stop bit. Standard I²C speeds are 100 kHz, but it can operate at elevated speeds like 400 kHz, 1 MHz (Fast-mode Plus), and even up to 3.4 MHz (High-speed mode). However, it is worth noting that some AVR microcontrollers might not adhere entirely to the high-speed I²C specifications. When compared to the I²C, the SPI interface often boasts faster speeds, contingent on the specific microcontroller and connection quality.

Figure 5.15

I²C wiring

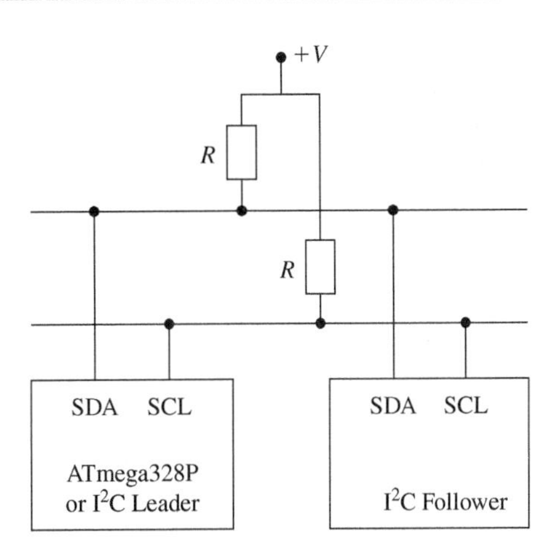

The Arduino platform has several functions for accessing the I²C interface through the Wire library. For instance, the *Wire.begin()* function initializes I2C communication, acting as either a follower or leader based on whether an address is provided or not. To initiate a transmission to a specific device, *Wire.beginTransmission(address)* is used where *address* is the 7-bit address of the device to transmit to. The *Wire.write()* function serves to dispatch data, either as a singular byte or as byte arrays. For receiving data, a request for bytes from a specific device is placed using *Wire.requestFrom(address, quantity)*, after which *Wire.read()* can be employed to retrieve the bytes. The *Wire.available()* function provides information on the number of bytes ready to be read. Subsequently, the *Wire.endTransmission()* function concludes the data transmission process.

5.9 Chapter Summary

This chapter discussed the means for interfacing a processor with external devices and peripherals. It commences with an exploration of Shannon's Sampling Theory, which postulates that for accurate signal representation without aliasing, the sampling frequency should be no less than double the highest frequency present in the analog signal. This foundational theorem is of paramount importance, especially when converting continuous analog signals into discrete digital values. The narrative then shifts its focus to the intrinsic operating principles of two critical components: the analog-to-digital converter (A/D) and the digital-to-analog converter (D/A). With regard to A/D and D/A converters, the chapter covers essential characteristics such as sampling rate, voltage range and resolution, bit resolution, and quantization error. Additionally, the chapter contrasts two distinct input modes of A/D converters: the single and differential. Here, it is highlighted that the differential input mode, owing to its inherent design, offers superior noise immunity compared to the single-ended mode. The subsequent section introduces readers to data acquisition boards, and the detail of manipulating individual bits on a parallel digital input/output port, especially when this port is an integral component of a data acquisition board. It also discussed data acquisition in MATLAB and Simulink. Transitioning from parallel to serial interfacing techniques, the chapter covered the details of the legacy RS-232 communication protocol. Concluding the chapter is an in-depth discussion of two pivotal serial communication protocols: the Serial Peripheral Interface (SPI) and the Inter-Integrated Circuit (I^2C^{TM}) interface. Both these interfaces underscore the flexibility and versatility of modern communication between a processor and its peripherals.

Questions

5.1 Explain what is meant by signal aliasing.

5.2 What is the aliasing frequency of a 20 Hz cosine signal sampled at 15 Hz?

5.3 What is the Nyquist frequency for a signal sampled at 20 Hz?

5.4 What is the Nyquist rate for a 100 Hz signal?

5.5 What is a ZOH?

5.6 What affects the voltage resolution of an A/D converter?

5.7 What is the quantization error?

5.8 What is the purpose of differential wiring in A/D reading?

5.9 Does the ATmega328P MCU have a D/A converter?

5.10 What is the limitation of using the weighted resistor summing amplifier circuit for a D/A converter?

5.11 What is the advantage of using the *R*/2*R* ladder resistor network for D/A conversion?

5.12 What advantages does serial interfacing have over parallel interfacing?

5.13 What type of operators are used to set or reset a single bit on a parallel port without disturbing the rest of the bits?

5.14 Explain the different types of parity methods used in the RS-232 protocol.

5.15 What is meant by full-duplex and half-duplex modes?

5.16 List the differences and similarities between RS-232 interfacing and SPI/I^2C interfacing.

5.17 Which serial communication method has the fastest data transfer rate?

Problems

Section 5.2 Sampling

P5.1 What is the minimum sampling frequency, or the Nyquist rate, needed to sample the following signals to prevent aliasing?

a. $f(t) = \sin(\pi t)$

b. $f(t) = \sin(5\pi t)$

c. $f(t) = \cos(8\pi t)$

P5.2 What is the minimum sampling frequency, or the Nyquist rate, needed to sample the following signals to prevent aliasing?

a. $f(t) = \cos(5\pi t)$

b. $f(t) = 3\sin(2\pi t) + 3\cos(2\pi t)$

c. $f(t) = 3\sin(2\pi t) + 3\cos(3\pi t)$

P5.3 Determine the Nyquist frequency if the signal was sampled at the following frequencies.

a. 20 Hz

b. 100 Hz

c. 200 Hz

P5.4 Determine the aliasing frequency if the signal $\cos(2\pi 10\,t)$ was sampled at the following frequencies.

a. 8 Hz

b. 13 Hz

c. 15 Hz

P5.5 Determine the aliasing frequency if the signal $\cos(2\pi 50\, t)$ was sampled at the following frequencies.

a. 20 Hz

b. 50 Hz

c. 80 Hz

Section 5.3 Analog-to-Digital Converter

P5.6 Determine the digital output of a 3-bit A/D converter with 0 to 10 V analog voltage range if subjected to the following analog inputs.

a. 1 V

b. 3.5 V

c. 8 V

P5.7 Determine the digital output of a 10-bit A/D converter with 0 to 5 V analog voltage range if subjected to the following analog inputs.

a. 1 V

b. 2.5 V

c. 5 V

P5.8 An ATmega328P MCU with a 10-bit A/D has its A/D converter set with $AREF = 3$ V. Determine the digital output of the A/D converter if subjected to the following analog inputs.

a. 1 V

b. 0 V

c. 2.5 V

P5.9 Determine the minimum number of bits required for an A/D converter that has a -10 to 10 V range to achieve a voltage resolution of 0.02 V.

P5.10 Determine the minimum number of bits required to digitize an analog signal with a resolution of:

a. 0.1%

b. 1%

P5.11 Consider a 12-bit analog-to-digital (A/D) converter. The input voltage range of the A/D converter is 0 to 4 V. Determine the following.

a. The smallest voltage increment that can be resolved by this A/D converter.

b. The quantization error of this A/D converter.

c. The maximum input voltage that can be measured without saturating the A/D converter.

P5.12 For the A/D converter in Problem P5.11, assume that the A/D converter has a maximum sample rate of 100 kHz and a total conversion time of 10 microseconds. Determine the following.

a. The minimum time required to sample and convert an analog signal with a frequency of 10 kHz.

b. The maximum input frequency that can be accurately sampled by this A/D converter.

P5.13 A temperature sensor was connected to a 16-bit A/D converter with a 0 to 5 V analog range. The sensor sensitivity is 10 mV/°C and the sensor output is 0 volts at zero degrees. Determine the following.

a. The temperature reading if the A/D converter output is 1000.

b. The measurement uncertainty due to the quantization error of the A/D.

P5.14 The A/D converter on the Arduino board is not designed to measure negative voltages and will be damaged if a negative voltage is directly connected to it. Describe how using a voltage divider circuit would allow the A/D converter on the Arduino Uno board to read negative voltages.

Section 5.4 Digital-to-Analog Converter

P5.15 A 4-bit D/A converter is set for a 0 to 5 V output range. Map all the possible digital input values to their corresponding analog output values.

P5.16 For an *R/2R* ladder resistor network similar to that shown in Figure 5.9 with $N = 4$, determine the voltage output of the network if all the bits were connected to V_R.

P5.17 For an *R/2R* ladder resistor network similar to that shown in Figure 5.9 with $N = 8$, determine the voltage output of the network if bits 1 to 5 were connected to V_R.

P5.18 Research the advantages and critical considerations of using matched resistor networks in A/D and D/A converter circuits, emphasizing their impact on performance and precision.

Section 5.5 Data Acquisition

P5.19 With reference to Table 5.1, map the following analog input voltages into their respective logic values if using the TTL interface: 0.1, 0.6, 1.5, 3.5, and 4.7 V.

P5.20 Illustrate how bit #3 of an 8-bit output parallel port can be set either high or low without changing the current output on the port.

Section 5.6 Serial Port

P5.21 Estimate the time it takes to send a file that has 20,000 characters using RS232-serial interfacing if the baud rate is set at 38,400 bps using a 10-bit data packet.

P5.22 Show the 10-bit serial packet for the following characters using 7-bit data, one start bit, one stop bit, and a parity bit. Illustrate for both even and odd parity.

a. 5

b. L

Section 5.7 Serial Peripheral Interface

P5.23 Perform research to identify two different sensors that use SPI interfacing. For each sensor, list the manufacturer and part number, and the sensor details.

Section 5.8 I^2C

P5.24 Perform research to identify two different sensors that use I^2C interfacing. For each sensor, list the manufacturer and part number, and the sensor details.

Laboratory/Programming Exercises

L/P5.1 Use MATLAB or an equivalent program to show the effects of signal aliasing. Assume the input signal is a cosine function with a frequency of 5 Hz. Plot the sampled signal if the signal is sampled at the following sampling frequencies: 4, 8, and 20 Hz. Redo this exercise but replace the input cosine function with a sine function.

L/P5.2 Use MATLAB or an equivalent program to show the effects of signal aliasing. Assume the input signal is a cosine function with a frequency of 2 Hz. Plot the sampled signal if the signal is sampled at the following sampling frequencies: 10, 8, 4, 3, 2, 1, and 0.5 Hz. For sampling frequencies below the Nyquist rate, plot also the aliased signal (see Figure 5.3 as an example).

L/P5.3 Use MATLAB or an equivalent program to show a 1-Hz sinusoidal signal and its zero-hold reconstruction at 5 and 8 Hz sampling frequencies.

L/P5.4 Using the Arduino Uno or similar board, write a program that allows the MCU to receive and transmit characters to the Serial Monitor in Arduino.

L/P5.5 With reference to the datasheet of a digital potentiometer (such as Microchip MCP42050-I/P that uses SPI interfacing), do each of the following.

a. Build a circuit on a breadboard to interface this potentiometer to an Arduino board.

b. Develop a program to set the resistance of the potentiometer. Note that the resistance is set by sending a byte to the chip. Use a multimeter to verify the set resistance.

L/P5.6 With reference to the datasheet of a digital temperature sensor (such as the MAXIM DS1631+ sensor that uses I^2C interfacing), do each of the following.

a. Build a circuit on a breadboard to interface this sensor to an Arduino board.

b. Develop a program to read the temperature measured by the sensor.

c. Display the read temperature to the Serial Monitor in Arduino.

Chapter 6

Control Software

Chapter Objectives:

When you have finished this chapter, you should be able to:

- Explain the concept of time and timers
- Outline how timers are implemented in different computing platforms
- Differentiate between traditional and event-driven programming
- Organize the operation and control of physical systems into tasks and states
- Develop state-transition diagrams for control of physical systems
- Explain code organization in state-transition diagrams
- Implement state-transition diagrams in different computing platforms

6.1 Introduction

The previous two chapters discussed microcontrollers and the means of interfacing microcontrollers or PCs. This chapter focuses on software development issues when using a microcontroller (and to a lesser extent, the personal computer) as the controller in a mechatronic system. Key issues include how to incorporate time into a control program, how to structure the operation and control of physical systems into tasks and states, and how to write control code that is responsive and non-blocking. Incorporating time is essential in many software situations, especially in the digital implementation of closed-loop control systems, since the sampling rate, which is the frequency at which the control signals are sent to the system, affects the response of the system.

Software development is a very important piece in the development of a mechatronic system. Without software, an MCU or PC cannot function. A software-based control system offers flexibility over a hardware-based one since the controller structure and control logic can be changed by simply changing the code in the program. Due to the different types of control activities that need to be performed (such as feedback control or discrete event control), it is advantageous to develop a uniform control software structure that can handle a variety of control applications. This chapter presents such a structure that is based on the task/state software structure. According to Lyshevski [19], developing a control software structure is one of the most challenging problems in mechatronics system design. In many situations, more than one control task needs to be controlled at the same time, and this chapter addresses the issue of multitasking. For further reading on the topics covered in this chapter, see [2] and [20].

6.2 Time and Timers

Time is a very important element in the software used in mechatronic applications. Recording the precise moment of an event's occurrence, known as event time-stamping, is indispensable for diagnostics, debugging, and system monitoring. For instance, pinpointing the exact moment a motor malfunctions can correlate this anomaly with other concurrent system activities. Additionally, intentional time delays are essential in many mechatronic systems. For instance, after sending a command to a motor to start, the system might wait for a brief period to ensure the motor has spun up before sending further instructions. Furthermore, in many mechatronic systems, there is a need to repeatedly execute certain pieces of code at precise intervals. This could be for monitoring sensor values, implementing feedback control loops, or periodically sending or receiving data.

We will consider both **absolute** and **relative** timing modes and explore their implementation in software. Certain applications, such as alarm monitoring and scheduling, necessitate absolute time, encompassing both date and specific time details. Take, for instance, a programmable thermostat designed to regulate a home's temperature. It allows users to designate desired temperatures based on the time of day (e.g., morning or evening) and the specific day of the week. Therefore, this thermostat functioning as a binary or ON-OFF controller not only synchronizes with absolute time but also continuously references it for effective operation. Absolute time also plays a crucial role in security or alarm monitoring systems where recording the exact timestamp of events, like a door being opened, is paramount. On the other hand, relative time is also known as **interval timing** and concerns the duration or elapsed time between events rather than their precise timestamps. It becomes vital when setting specific sampling rates in feedback control systems or when creating time delays. With interval timing, the focus is on controlling the time gap between the present moment and the forthcoming operation, regardless of the exact hour or day it occurs.

Regardless of the mode of timing, a timer is implemented in a computer system using a combination of a clock and a counter. The clock is any device (like a stable crystal oscillator) that produces a regular train of pulses at a specific frequency. These pulses are fed to a counter, which tallies them. To retrieve the counter's value, the program might read it directly from a specific memory-mapped register or through a function call, depending on the system's architecture. Counters can operate in two primary modes: count-up or count-down. In **count-up mode**, the counter initiates at 0 (or another value), incrementing by 1 with each pulse. Conversely, in **count-down mode**, the counter begins with a preset value, decrementing by 1 for every incoming pulse. Given the finite size of counters, such as 16-bit or 32-bit, a count-up counter will overflow after reaching its maximum value, wrapping around to 0. A count-down counter, however, underflows upon reaching 0, wrapping back to its maximum possible value. When a counter overflows or underflows, it typically reloads with the starting value, and the counting sequence resumes.

An important characteristic of a timer is its **resolution**. This refers to the smallest time change that can be measured by a timer. For a clock that generates pulses at a frequency of f pulses per second, the timer resolution in seconds is

(6.1)
$$\text{Timer resolution} = 1/f$$

The maximum time interval that can be measured by an n-bit counter before it overflows is given by

(6.2)
$$\text{Maximum interval} = 2^n/f$$

For example, a clock that operates at a frequency of 12,000 Hz and a 16-bit counter has a maximum time interval of 65,536/12,000 = 5.461s.

In **relative-timing** mode, time intervals are obtained by dividing the difference between two counter readings by the clock frequency. For example, a counter operating in count-up mode has the time interval given by

(6.3)
$$\Delta T = (C_2 - C_1)/f$$

for $C_2 > C_1$, and

$$\Delta T = (2^n - C_1 + C_2)/f$$

for $C_2 < C_1$, where C_2 and C_1 are the counter readings at the end and start of the interval, respectively, and n is the bit size of the counter. C_2 can be less than C_1 if the second reading of the counter was obtained after the counter has overflowed.

To determine time in absolute timing mode, the starting time is first synchronized with a reference time (such as the current time and date information). The absolute time is then determined by adding the interval between the current time and the last time the counter was read to the starting time. To have accurate absolute timing information, the counter has to be read at least once during an overflow period.

To illustrate problems with **counter overflow** (available—V6.1), let us consider a counter that overflows every 24 hours. If this counter was read after 1 hour from the last time it overflowed, and then read again after 6 hours from the last time it overflowed, then by taking the time difference between these two readings, we say that 5 hours have elapsed. However, if the second reading of this counter was done the next day instead (29 hours after the first reading), we still get a difference of 5 hours between the two readings, although the two readings are separated by 29 hours. This is because when the counter overflows, it resets itself back to zero.

For accurate timekeeping over periods longer than the counter overflow period, the counter has to be read at least once during each overflow period to prevent timing errors resulting from overflow of the counter. The successive time intervals between the current and the last time the counter was read are added to obtain the current time. The challenge is ensuring the counter is read before it overflows. The answer is to set a maximum read interval for the counter that is less than the overflow interval, and to call any reading in which the time difference exceeds the maximum read interval overflow. This scheme, which was proposed in [2], may not be able to detect all overflow instances, but should be able to detect the overflow error over many readings of the counter.

To further illustrate, let us revisit our counter-example, which overflows every 24 hours, and set the maximum read interval to 16 hours. Assume the counter readings were 2 hours, 19 hours, and 21 hours (see Figure 6.1). Then according to this scheme, we get an overflow at the second reading, since the difference between the second and the first reading is 17 hours, which is greater than the maximum read interval that we set (16 hours). On the other hand, at the third reading of 21 hours, this scheme does not detect an overflow error, since the difference between the third and the second counter readings is only 2 hours, while in reality the third reading was obtained on the second day. Nevertheless, this scheme will be able to detect some overflow errors over many readings of the counter. Detection of such errors indicates the need to revise the code to read the timer more often.

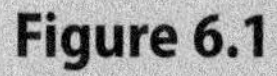

Figure 6.1

Illustration of counter overflow

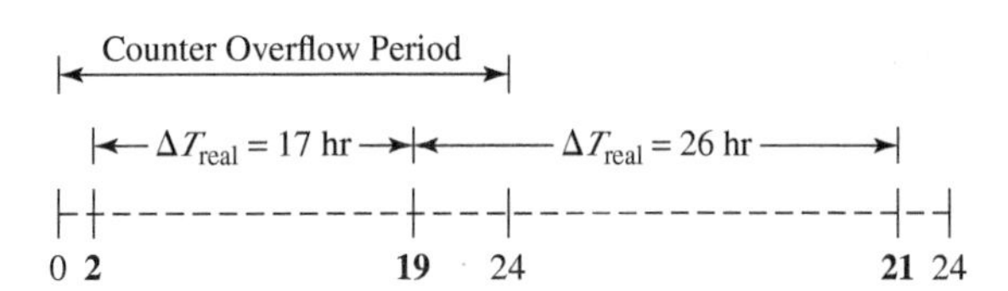

6.3 Timing Functions

Most programming languages provide functions for accessing time information that make use of the hardware timer that is available on the microcontroller or the PC. In this section, we will discuss timer implementation in PCs and AVR microcontrollers. In PCs, we discuss timer implementation in MATLAB.

6.3.1 Timer Implementation in MATLAB

MATLAB has several functions to obtain timing information (available—V6.2). In this section, we will discuss

- TIC and TOC functions
- CLOCK and ETIME functions
- TIMER object

The **TIC** and **TOC** built-in functions are designed to be used together and use the PC clock. The TIC function starts a stopwatch timer, while the TOC function reads the current time in seconds from when the

TIC function was called. To get timing information, the TIC function is called first. A subsequent call to the TOC function returns the elapsed time since the TIC function was called. For example, if we type in the command window:

tic

toc

Then MATLAB will print:

Elapsed time is 2.647394 seconds.

The TIC and TOC functions prove useful as an interval timer. For example, the following MATLAB code (see Figure 6.2) can be used to check if a certain time interval (such as 10 s) has elapsed.

Figure 6.2

MATLAB code listing for implementation of an interval timer

```
tic;
start_time = toc;
while (toc - start_time) < 10

end
disp ('10s seconds has elapsed')
```

The TIC/TOC functions have a sub-millisecond time resolution, but the actual resolution obtained in a given application is dependent on the computer hardware used and on what other applications are running at the same time.

The **CLOCK** function returns a six-element vector that has the current date and time information in decimal format. Typing CLOCK in the command window gives

clock

ans =

*1.0e+003 **

2.0230 0.0010 0.0140 0.0120 0.0020 0.0128

where the first element is the year information and the remaining five elements are respectively: month, day, hour, minute, and second. To access only one element of the date and time information, the CLOCK function is assigned to a variable, and the corresponding element of that variable is used. To get the information in integer format, the clock function is called using the FIX function or *fix(clock).* The **ETIME** function along with the CLOCK function can be used to determine the time (in seconds) that has elapsed between two values of the clock vector. Thus, the following code will give the elapsed time:

t1 = clock;

etime (clock,t1);

MATLAB does not recommend the CLOCK and ETIME functions for accurate timing, since they are based on the system time, which the operating system can adjust periodically. The TIC/TOC functions give a more accurate event or interval timing.

MATLAB allows the creation of a **TIMER** object that can be used to automatically time the execution of a special function called the timer callback function. The TIMER object is created using the command:

T = timer('PropertyName1', PropertyValue1, 'PropertyName2', PropertValue2, ...)

where *T* (or any other name) is the TIMER object, and the property name/value pairs are used to specify the operating characteristics of the TIMER object. Some of the available properties and their description are listed in Table 6.1.

Table 6.1

TIMER object properties

Property Name	Property Description
AveragePeriod	The average time between *TimerFcn* executions since the timer started
ExecutionMode	Determines how the TIMER object schedules execution of the *TimerFcn*
Period	The delay in seconds between the execution of the *TimerFcn*
TasksToExecute	The number of times the timer should execute the *TimerFcn* if the *ExecutionMode* is not a single shot
TasksExecuted	The number of times the timer has called the *TimerFcn* since the timer was started
TimerFcn	Timer callback function

The timer *Period* property cannot be less than 1 ms. If unspecified, the default setting is then 1.0 s. The *ExecutionMode* property of the timer defines how the timer is run. It can run once (if the property is *singleShot*, which is the default setting) or it can do multiple executions (if the property is set to *fixedDelay*, *fixedRate*, or *fixedSpacing*). Figure 6.3 shows the different execution modes. In this figure the timer period is the same in all cases, but the point of time at which the execution begins is different. In the *fixedRate* mode, the timer period starts at the point where the timer callback function is added to the queue. In the *fixedDelay* mode, the timer period starts at the beginning of the execution of the timer callback function, while in the *fixedSpacing* mode, the timer period begins at the point where the timer callback function finishes executing. Note how the absolute timing of the execution of the *TimerFcn* is different in each mode.

Once the TIMER object has been created, it needs to be started. It can be started immediately by calling the function *start(T)*, or it can be started to run at a specified time using the function *startat(T, start_time)*. The timer can be stopped by calling the *stop(T)* function.

Figure 6.3

Illustration of the different execution modes of the TIMER object

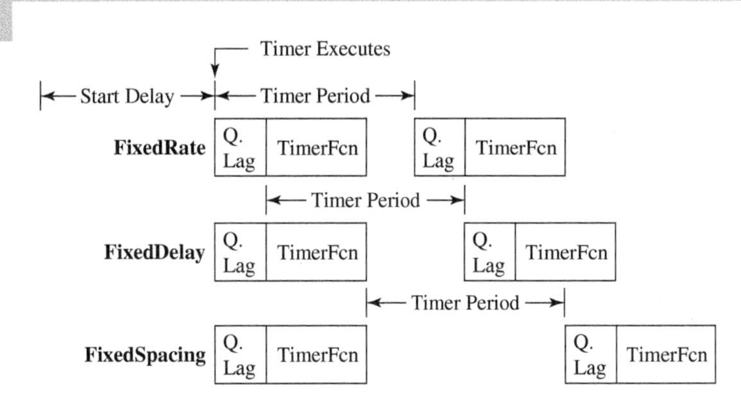

Note: Q. Lag means Queue Lag

As an example, Figure 6.4 shows the code listing to create and start a TIMER object with a period of 0.5 s. The timer has a callback function that needs to be executed 20 times in *fixedRate* mode. Included in this code is a listing of the *timer_callback_fcn.*

Once the timer has completed its execution, the timer should be removed from memory using the *delete* command. The TIMER object will be utilized later in this chapter for the implementation of state-transition diagrams in MATLAB.

6.3.2 Timing in AVR Microcontrollers

Similar to PCs, microcontrollers also have timers (available—V6.3) that can be used for timing purposes. We will discuss the timing features of the ATmega328P, the microcontroller used in the popular Arduino Uno board (details of this microcontroller were covered in Chapter 4). The ATmega328P has three timers called *Timer0*, *Timer1*, and *Timer2*. Table 6.2 lists information about these timers.

Figure 6.4

MATLAB code listing demonstrating the TIMER object

```
% Demo of TIMER Object
% File: DemoTimer.m

function DemoTimer
% Create the TIMER object T
T = timer('TimerFcn', @timer_callback_fcn, 'period', 0.5,'TasksToExecute', 20,'ExecutionMode',
'fixedRate');
% Start the timer
start(T);
% Wait for the timer to complete the tasks
while(get(T,'TasksExecuted') < 20)

% do nothing
end
disp('Tasks Execution is done');
% Remove the timer from memory
delete(T);

% Listing of the timer callback function
function timer_callback_fcn(obj,event)

disp('In timer call back function \n')
```

Table 6.2

Timers in ATmega328P microcontroller

	Timer0	Timer1	Timer2
Bit size	8-bit	16-bit	8-bit
Operate as a counter?	Yes	Yes	No
Programmable prescaler?	Yes	Yes	Yes
Prescaler values	1:1, 1:8, 1:64, 1:256, 1:1024	1:1, 1:8, 1:64, 1:256, 1:1024	1:1, 1:8, 1:32, 1:64, 1:128, 1:256, 1:1024
Maximum timer interval at 16 MHz Clock (with maximum prescale)	$0.016 \times 1024 = 16.384$ ms	$4.096 \times 1024 = 4194.304$ ms	$0.016 \times 1024 = 16.384$ ms
Timer overflow interrupts	Yes	Yes	Yes

Timers 0 and 1 can operate as either timers or counters. In the timing mode, the count value is incremented every instruction cycle (or at a multiple of it if a prescaler is used). In the counter mode, the *Timer0* or *Timer1* module will increment on every rising or falling edge of the external signal connected to a specific pin on the microcontroller. Using a programmable prescaler reduces the instruction cycle rate, as seen by the timer allowing for longer timing intervals to be counted by the timer before overflow. For example, if the 16-bit Timer1 was used as a timer with no prescaler (or a prescale value of 1:1), then at a clock rate of 16 MHz, this timer will overflow every 4.096 ms. This is obtained from Equation (6.2) or

$$\text{Maximum interval} = \frac{2^{16}}{16 \times 10^6} = 4.096 \text{ ms}$$

With a 1:8 prescaler, the overflow period will be eight times longer or 32.768 ms.

Any of the three timers can be used to keep time as long as they are read often to prevent timer overflow from affecting the accuracy of the reading. When any of the three timers overflow, a flag is set up. The program can poll this flag to check for this overflow condition. Alternatively, the overflow of the timer can be set to generate an interrupt. Arduino provides two timing functions that use Timer0 for timekeeping. These timing functions are:

millis()—Return the time in milliseconds since the Arduino program started running. This function overflows after approximately 50 days.

micros()—Return the time in microseconds since the Arduino program started running. This function overflows approximately after 70 minutes. The *micros()* function has a finer time resolution (about 4 microseconds but it is dependent on which Arduino board one is using) than the *millis()* function.

These timing functions do not block the execution of other code while running, unlike the delay functions (*delay()* and *delayMicroseconds()*) that are provided with Arduino which block the execution of other code while they are running. The way these timing functions work is that they access the built-in hardware *Timer0* whose operation does not interfere with the code running in Arduino and whose value gets incremented at a specific counting frequency starting from a count of zero. When the counter reaches its maximum counting value (such as approximately 50 days in ms units for the counter associated with the *millis()* function), it overflows. Overflow means that the counter resets itself to zero and starts counting again. Overflow is not an issue if these timing functions are used to determine a time interval and if the time interval is less than the overflow interval. Figure 6.5 shows an Arduino code listing that uses the *millis()* function to implement a one-time 10 second time delay but also allows other code to run at the same time.

Figure 6.5

Arduino code listing that implements a non-blocking time delay

```
const unsigned long interval = 10000;  // Delay interval in milliseconds
unsigned long lastTime;                // Variable to store the last time the action was performed
byte Flag = 0;                         // Flag to run the delay one time

void setup() {
 Serial.begin(38400);
 lastTime = millis();                  // Store the starting time
}

void loop() {

 if (millis() - lastTime >= interval and Flag == 0) {
   Serial.println("10 Seconds have elapsed");
   Flag = 1;
                                       // 10 seconds have passed
                                       // Add code here to perform the desired action
 }
                                       // Place other non-blocking code here
}
```

6.4 Events and Event-Driven Programming

Control software, often referred to as a control program, is vital for the efficient functioning of a mechatronics system. This software serves as a bridge, coordinating and connecting various components of the system, such as sensors and actuators. Put simply, it enables these components to communicate and collaborate effectively. The hallmark of a proficient control program is its agility—its capacity to swiftly handle and respond to unexpected events throughout its operation.

So, what exactly are 'events'? **Events** can be viewed as triggers or disturbances, either spontaneous or anticipated. They might arise from human actions, like the basic task of pressing a button on a graphical user interface (GUI) or adjusting a control parameter. On the other hand, they could be generated by the system itself due to changes in data, such as a sudden shift in a temperature sensor or an anomaly in a pressure reading. Some events are time-based, initiating actions at specific moments or intervals, like a task set to execute every few seconds. Example 6.1 discusses some of the events that can occur during the execution of a program to control the operation of an elevator.

Example 6.1 Events in an Elevator Operation

Name several events that could occur during the execution of a program to control the operation of an elevator.

Solution:

Below are some potential events that can occur:

Button Press Inside the Elevator Car: If a passenger inside the elevator selects a floor, the elevator schedules a stop for the chosen level.

External Floor Selection Buttons: Elevators are equipped with call buttons on each level: an upward arrow and a downward arrow. If a user presses the up button on a particular floor, it indicates they want to go to a floor above their current one. Similarly, pressing the down button indicates a desire to go to a lower floor.

Reaching Desired Floor: Upon reaching a designated floor, the elevator engages its doors to open.

Safety Mechanism Activation: Elevators have safety sensors on the doors to prevent them from closing if there's an obstruction (like a person or object). If the sensor is triggered, the elevator will re-open its doors.

Automatic Door Closure: If the doors have been open beyond a certain time limit without any obstruction (to allow passengers to enter or exit), the elevator should close its doors.

Emergency Halt Activation: Elevator cabins have an emergency button. When activated, it prompts the elevator to cease all operations instantly.

Excess Weight Alert: Elevators contain weight monitoring systems. If a cabin's weight surpasses the safety threshold, an alert activates, and the elevator remains stationary until the overload is addressed.

Fire Alert Protocol: Upon a fire alert in many structures, elevators return to a default floor (usually the ground) and may be rendered inactive, directing occupants to use stairways for evacuation.

Travel Limits Triggered: Elevators have limit switches at the top and bottom of their travel to ensure they do not exceed their intended range of motion.

System Error or Fault Detected: The elevator's control system can detect various types of errors or faults, such as mechanical issues, software errors, or communication breakdowns. When detected, the elevator might enter a safe mode or halt operation.

Traditionally, programming methodologies leaned heavily toward a **sequential paradigm**. Here, operations are executed in a strict sequence, moving predictably from one step to the next. Each operation occupies a defined position, constrained by either time or particular conditions. This step-by-step approach is often visualized using flowcharts, which depict the progression of operations. For instance, Figure 6.6(a) illustrates this method.

Figure 6.6

(a) Traditional programming and (b) event-driven programming

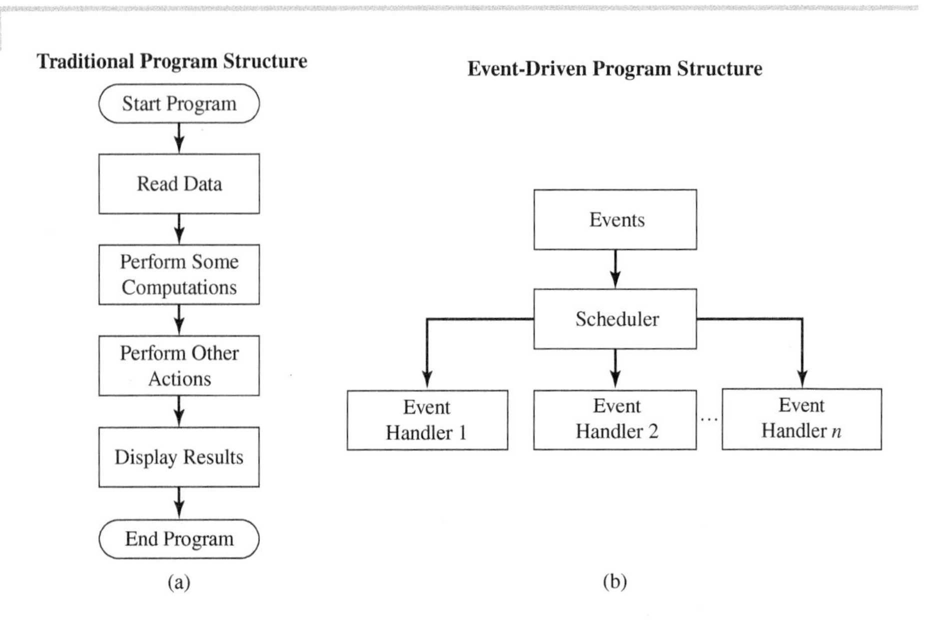

Traditional sequential programming, where tasks are performed in a set order, struggles when faced with unpredictable events. Events can occur at any moment and are not constrained to a predetermined sequence. This dynamic nature inherent in mechatronics systems calls for a more adaptable strategy: **event-driven programming** (▶ available—V6.4). Unlike the linear, step-by-step flow of sequential programming, event-driven approaches are adaptable and responsive. Instead of following a fixed sequence, the system continuously monitors for external triggers. In many systems, when an event is recognized, it is passed to a scheduler or event loop, which then invokes the appropriate event handler to address the event. Without events to act upon, the relevant code segments remain inactive. Figure 6.6(b) illustrates this approach. However, the unpredictable nature of events poses complexities. Events can appear sporadically and even seem simultaneous, demanding an ever-alert software framework. State-transition diagrams, introduced in the following section, provide a structured blueprint for designing event-driven control programs. While event-driven programming is the paradigm that facilitates responsiveness to external triggers, state-transition diagrams provide a systematic way to navigate, manage, and design within this paradigm.

The next question is, how promptly should a control program react to these events, and what is the acceptable **latency**—the delay between recognizing an event and addressing it? This largely hinges on the application of the mechatronics system. For systems fundamental to self-driving cars, advanced machinery, flight control systems, or military technology, delays are unacceptable. In contrast, systems with less time-sensitive demands can tolerate longer latency periods.

Control programs that handle time-sensitive applications are categorized as real-time programs. In **a real-time program**, the timing of the output is as critical as the accuracy of the output itself [20]. This implies that the correctness of a real-time program is not solely about producing the right result; it is also about ensuring the result is produced within a stipulated time frame. If it is not, the output may lose its value or relevance.

For instance, consider a program designed to compute the roots of a quadratic equation. Here, the operation and the output are not necessarily time-bound. Whether beginning the program now or 1 minute later, or whether it takes a few milliseconds or a few seconds to compute the roots, as long as the solution is obtained within a time frame that does not impact the desired outcome or user experience, it remains valid. However, when we look at more dynamic applications, timing becomes crucial. For example, take a control program that manages anti-lock braking in vehicles. The moment a wheel skid is detected, the control program must respond almost instantaneously. Any significant delay might lead to ineffective braking, resulting in a potential accident. Another example can be seen in the realm of manufacturing. When cutting a material with a heat source like a laser, the duration for which the laser interacts with the material is pivotal. This duration, determined by the relative speed of the laser head to the material, dictates the cutting outcome. If the material is not exposed to the laser long enough, it might not be cut adequately due to insufficient heating. Conversely, prolonged exposure could cause the material to overheat and burn. In this laser-cutting scenario, the precise timing of the control output is of the essence.

Real-time control programs are typically classified into two main categories**: hard** and **soft real-time systems**. In hard real-time systems, missing a deadline is deemed a system failure. Conversely, in soft real-time systems, while meeting deadlines is the objective, occasional misses are undesirable but not always catastrophic. For both hard and soft real-time applications, a real-time operating system (RTOS) is often recommended. An **RTOS** (▶ available—V6.5) is a specialized operating system tailored to meet the rigorous requirements of real-time applications. It provides critical mechanisms for task scheduling, priority management, and interrupt handling, ensuring high-priority tasks meet their specified deadlines. Examples of RTOS include FreeRTOS, VxWorks, and QNX, among others. While real-time programming involves developing software to meet stringent time-sensitive requirements, an RTOS offers the underlying infrastructure and tools essential for the efficient execution of this software.

In the realm of task management and scheduling, there is also the concept of **cooperative multitasking**. This chapter delves into the cooperative control mode program structure, which is based on the concept of cooperative multitasking. When paired with non-blocking code and fast processors, it offers a simple methodology for designing control programs that can operate in real time. It is essential to clarify that cooperative multitasking is about how tasks yield control to one another, usually voluntarily, whereas soft real-time systems focus on meeting task execution deadlines with some level of flexibility. Although they both pertain to task scheduling and management in computing, their primary objectives and mechanisms differ.

6.5 Task/State Control Software Structure

The first step in setting up a control program in software is to organize the control actions that need to be performed into separate groups. Each group will be called a **task**. Depending on the nature of the control job, the grouping will be done differently. For example, if the control job is to control the operation of an assembly machine that consists of several modules that work together, such as a part feeder, a robot, and an indexing table, then we need to set up a separate task to control each module of this assembly system. Thus, the grouping of control actions in this example is set based on the physical structure of the machine to be controlled. Depending on the complexity of the control actions that need to be performed for each module, more than one task can be designed to control that module too. The grouping also can be done based on the software activities to be performed (such as handling Internet communication or sending pulse and direction signals to a stepper motor). Within a control task, the code should be organized into separate states. Each **state** signifies a distinct condition of operation. A machine or a process can be in only one state at a time. An example of the state structure is the state of a traffic light. These states are red, yellow, and green. At any point in time, the traffic light can be in only one of these three states, and there are distinct conditions or events (such as the elapse of a time interval) that cause the transition from one state to the other.

The states of a task and their transitions can be represented graphically using **state-transition diagrams** (▶ available—V6.6). A **state machine** is a conceptual model that captures the behavior of a system (be it a machine, software program, or other entities) in terms of its states and the transitions between those states, driven by certain conditions or events. State-transition diagrams can be used to visually represent the behavior of event-driven programs. Each state in the diagram can represent a particular configuration or condition of the program, and the transitions represent events or triggers that move the system from one state to another.

State-transition diagrams are invaluable tools used in various domains, including the design of sequential logic circuits, modeling character behavior and game states in game development, and protocol design in networking. At its core, a state machine describes the possible states of a system and the rules or events that dictate how it transitions from one state to another. There are two primary models of state machines: Moore and Mealy (▶ available—V6.7). The **Moore machine** generates its output based solely on its current state, meaning that the output is a function of the state only. On the other hand, the **Mealy machine** produces its output based on both its current state and the current input, meaning the output is a function of the state and the input. As a result, Mealy machines can be more responsive to changes since they consider the current input, but they can also be more complex. In practical software design, especially in more complex systems, one might not strictly adhere to just one model. Instead, a combination or a higher-level abstraction may be used to best match the problem's needs. The state-transition diagrams that are illustrated in this text follow the Moore model, and the content covered in this section draws inspiration from concepts presented in [2] and [21].

As an illustration of task/state organization, let us create a state-transition diagram for a program that needs to control the operation of a single-axis positioning stage driven by an electric motor as shown in Figure 6.7.

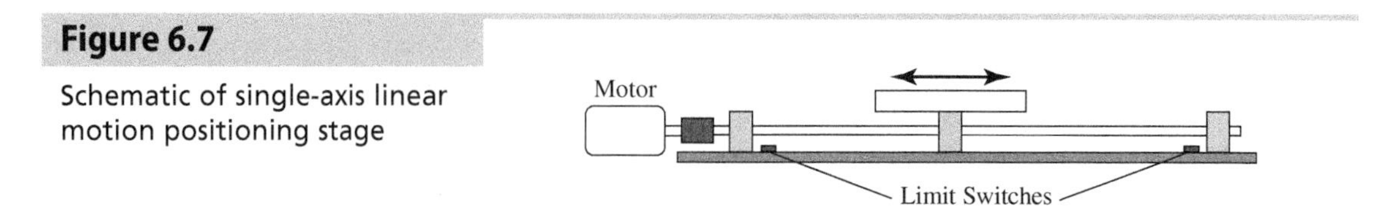

Figure 6.7

Schematic of single-axis linear motion positioning stage

The positioning stage is controlled by three commands issued by the user. These are:

Right: Causes the stage to move to the right

Left: Causes the stage to move to the left

Stop: Causes the stage to stop

Furthermore, the stage is equipped with two limit sensors, one at each end of its travel. If the stage hits either of these sensors, it should stop moving. The state-transition diagram that controls the operation of the stage is shown in Figure 6.8. The diagram has three states labeled *Stop Motion*, *Right Motion*, and *Left Motion*. Notice how these states are mutually exclusive and the stage can only be in one state at any point in time. Before motion begins, the stage is in the *Stop Motion* state. It also goes to this state if either the limit switch was activated while it is moving, or the *Stop* command was issued. To transition from the *Stop Motion* to the *Right Motion* state, the user must issue the *Right* command and the stage should not also be at the right limit switch position since there should be no motion to the right if the stage has hit the right limit switch. A similar transition condition is needed to go to the *Left Motion* state from the *Stop Motion* state. A *Left* command while the stage is moving to the right causes the state to change from the *Right Motion* state to the *Left Motion* state. An opposite action happens if a *Right* command is issued while the stage is in the *Left Motion* state.

Figure 6.8

State-transition diagram for the stage motion

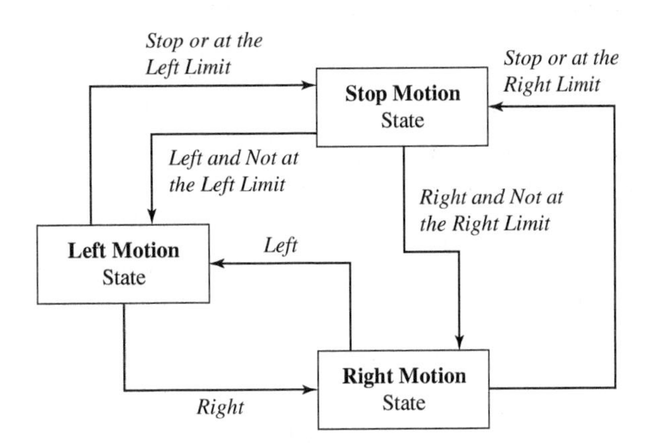

The state-transition diagram example shows that in a proper state-transition diagram, there should be no transition from one state to another unless some event has occurred such as a new user command or a change in the value of a sensor that is being monitored.

In this section, we will apply the state-transition diagram concepts to the control of discrete-event tasks and feedback control tasks.

6.5.1 Discrete-Event Control Tasks

A **discrete-event control task** controls a sequence of actions in response to specific events, rather than continuously monitoring changes. Typical actions in such systems might include the opening or closing of a valve, the starting or stopping of a motor, or the activation of devices like heaters, fans, or pumps. Examples of discrete-event systems abound, spanning automated cutting machines, assembly lines, item dispensers, entry/exit systems, and many process control systems. These actions can be instigated by time-based triggers (e.g., a valve being open for precisely 3 seconds) or event-based triggers (e.g., a robot arm lifting when a sensor detects the presence of a part). The task/state control software structure is particularly apt for managing discrete-event systems. To effectively control such a system, it is essential to model its operations using state-transition diagrams. These diagrams provide an intuitive visual representation of system behavior. In this section, we delve into a more intricate discrete-event task: the operation of automated sliding entry doors (▶ available—V6.8) commonly found in public spaces (refer to Figure 6.9). Example 6.2 discusses this control task. Five mutually exclusive states are used in the example with the transitions between the states being either sensor-based or time-based. Each of the five states represents a unique state of operation of the automated entry door. The control system for this door can be in only one of these states at any point in time. For this automated entry door, a single task is sufficient to represent the operation of the system.

Figure 6.9

A schematic of an automated sliding entry door

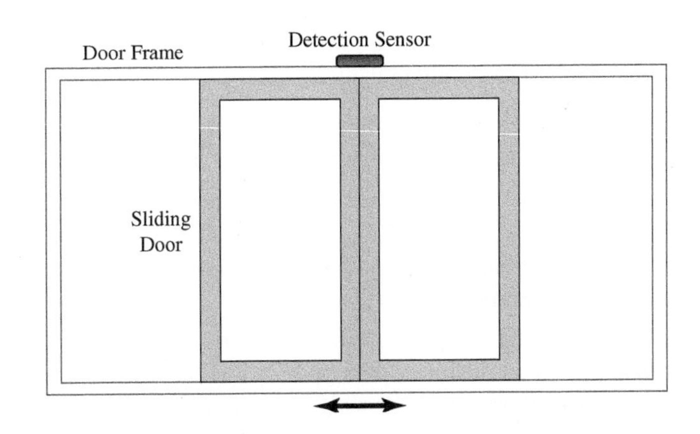

Example 6.2 Automated Entry Door State-Transition Diagram

Develop a state-transition diagram for the operation of an automated sliding entry door that is commonly available in public places. Assume that the control system will open the door when signaled from the sensor attached to the door. Assume also that the door will stay open while the detection sensor is ON and for a specified interval after the detection sensor switches to OFF.

Solution:

The state-transition diagram for the operation of the automated entry door is shown in Figure 6.10. The control system starts in the *Start* state, at which point the door is closed and the detection sensor is OFF. When the detection sensor turns on, the control system switches to the *OpenDoor* state. When this state first runs, control signals are sent to the actuators to open the door. The control system will stay in this state until the proximity sensors (or limit switches) that indicate that the door is fully open are trigged on. The control system then switches to the *DoorIsOpen* state and stays in that state until the detection sensor is no longer ON. For safety reasons, rather than closing the door right away, the control system switches to a Wait state to wait for the elapse of an interval timer before issuing the command to close the door. Since it is possible that the detection sensor is triggered again while the control system is waiting for the timer interval to elapse, the control system switches back to the *DoorIsOpen state* if this occurs. When the wait interval is over, the control system switches to the *CloseDoor* state. Like the *OpenDoor* state, the control system waits for the door to fully close before switching from the *CloseDoor* state to the Start state. Similarly, the control system can go back from the *CloseDoor* state to the *OpenDoor* state if the detection sensor was triggered while the door was closing.

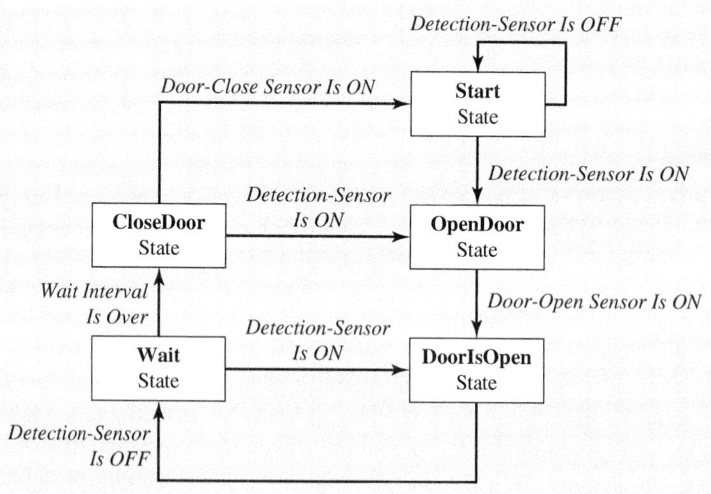

Figure 6.10 State-transition diagram for the operation of the automated entry door example

To further illustrate this topic, consider the state-transition diagram modeling the operation of a **digital heating thermostat** (see Figure 6.11). The thermostat has a switch that turns it ON and OFF. The thermostat should operate according to the following rules:

a. The thermostat operates when the switch is in the ON position. The thermostat does not do any control if the switch is in the OFF position.

b. The thermostat turns ON the heating equipment when the room temperature is 1° below the desired temperature.

c. The thermostat turns OFF the heating equipment when the room temperature is 1° higher than the desired temperature.

d. When the heating equipment is turned ON, it cannot be turned off unless a user-specified delay interval (such as 2 minutes) has elapsed since it was turned ON. This is done to protect the equipment from rapid turn ON/turn OFF.

Figure 6.11

Schematic of a digital heating thermostat

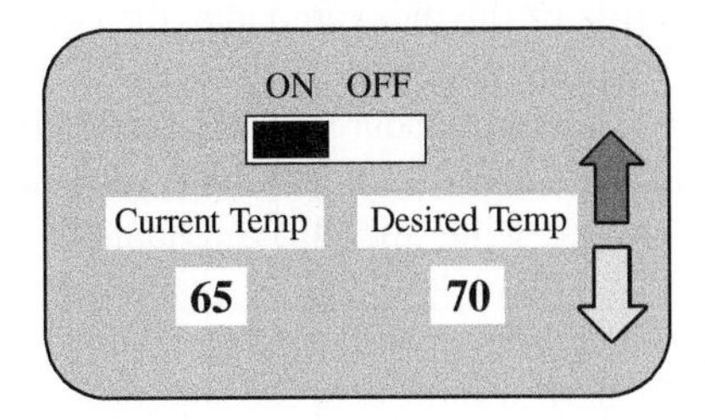

The state-transition diagram for the operation of a thermostat that follows the above rules is shown in Figure 6.12. It has four states. The state diagram starts in the *SwitchOffHeaterOff* state where the ON/OFF switch is OFF and the heater is OFF. When the user moves the switch to the ON position, and the current temperature is 1° or more below the desired temperature, then the state switches to the *HeaterOn* state. If the current temperature is not 1° or more below the desired temperature when the switch was moved to the ON position, then the state switches to the *HeaterOff* state.

In the *HeaterOn* state, the heater is turned ON and the current time is recorded. Both of these actions should be done in a section of the code that is only executed once (more on this in the next section). From the *HeaterOn* state, the state switches to the *HeaterOff* state (where the heater is turned OFF) only if the temperature exceeds the desired temperature by at least 1° and the waiting interval since the heater was started is over. The other possible transitions from the *HeaterOn* State are activated by the user turning the switch to the OFF position. Since the switch can be turned OFF before the time delay interval is over, the state switches to the *SwitchOffHeaterOn* state (where the heater is still on) to wait for the interval to be completed before switching to the *SwitchOffHeaterOff* state where the heater is OFF. Note that at a steady state, the state diagram switches between the *HeaterOn* and *HeaterOff* states based on the relationship between the desired and current temperature. Similar to the automated entry door example, a single task is used to model the operation of the thermostat.

Figure 6.12

State-transition diagram for the operation of a heating thermostat

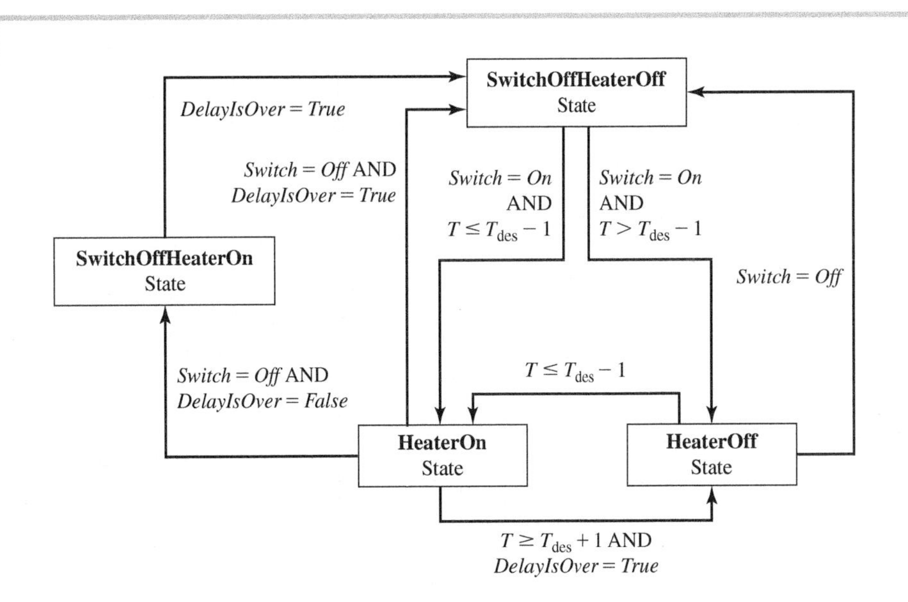

In **more complicated systems**, more than one task is needed to organize the operation of the system. This is the case for the operation of **automated assembly systems** (▶ available—V6.9) such as the one shown in Figure 6.13, which is a simplified version of automated multiple-part industrial assembly systems. This assembly system assembles two components. A circular disk is placed on a rectangular base via the use of a four-station rotary indexing table. The rectangular and circular parts are fed using a conveyor belt. The assembly operation proceeds as follows. First, the rectangular part is transferred to station 1 on the indexing table, using pick and place robot #1. After the indexing table rotates 90° clockwise, the circular disk is placed on the rectangular base at station 2, using pick-and-place robot #2. After another 90° rotation of the indexing table, the two-part assembly of the circular and rectangular parts is inspected at station 3. After the third motion of the indexing table, the completed assembly (if passed inspection) is transferred to conveyor belt *C*, using pick-and-place robot #3. If the assembled part fails inspection, then either the machine operator is alerted to address this situation, or the failed assembly is picked and dropped onto a rejection bin (not shown here). At steady state, each station of the indexing table will be performing its part of the assembly process while the indexing table is stationary. When all of these actions are completed, the indexing table will rotate 90°, and the process repeats. As a first step in designing a control system for this machine, the operation of each module in this assembly machine should be structured as a separate task. Each task will have several states depending on the operations done on that task. Example 6.3, which follows, shows the state-transition diagram for the operation of pick-and-place robot #1 of this system.

Figure 6.13

A simplified automated assembly system

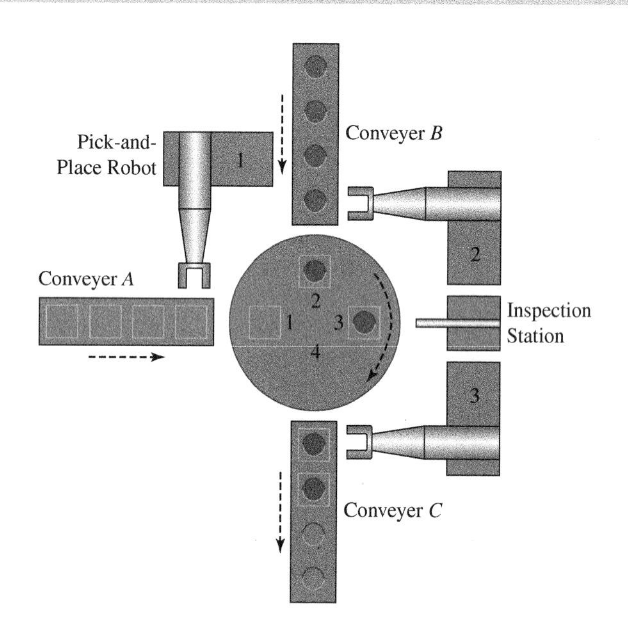

Example 6.3 Assembly Robot #1 State-Transition Diagram

Draw a state-transition diagram for the operation of pick-and-place robot #1 in the assembly system shown in Figure 6.13. Assume that the robot used in this setup is a pneumatically driven one. Such robots are assembled from linear or rotary motion axes with two possible locations for each axis. Thus, moving the robot to the *Pick-Up Location* entails sending a digital signal to the solenoid valve(s) that controls the air supply to the robot axis (axes) that causes the desired motion. To determine if the robot has completed its travel, assume that end-of-travel proximity sensors (see Section 8.4), which are mounted at either end of the motion, are used. Picking or dropping a part corresponds to closing or opening the gripper that is attached to the end of the robot. Assume that a pneumatically actuated gripper is used here, and the opening and closing actions of the gripper are similarly accomplished by actuating the solenoid valve that controls the air supply to the gripper.

Solution:

The state-transition diagram for pick-and-place robot #1 is shown in Figure 6.14. The diagram consists of eight states, and the robot will be waiting in the *Start* state until it receives a signal from the cell controller to start its sequence and a part is ready for pick-up at the end of conveyer belt *A*. At this point, the robot will be in the *Move-to-Pick-Up-Location* state where it will be moving to the part pick-up location. When the robot reaches the part-pick-up location, as indicated by feedback signal(s) from an end of travel sensor(s), the robot changes state to the *Pick-and-Lift-Part* state. If the pick-up was successful, the robot switches states to the *Transfer-Part-to-Indexing-Table* state, where the carried part is moved to station 1 on the indexing table. If the pick-up was not successful, the robot switches states to the *Wait-for-Operator-1* state. In this state, the operator is alerted to check the system, and on receiving a *Resume* signal from the operator, the pick-up is attempted again. Similarly, when the part has reached the drop-off location, the state changes to the *Place-Part-on-Indexing-Table* state. After the part was successfully transferred to the indexing table, the robot changes state to the *Move-to-Starting-Location* state. Similarly, if the placement was not successful, the robot switches states and waits for a *Resume* signal from the operator before switching to the *Move-to-Starting-Location* state. When the robot reaches the starting location, the state switches back to the *Start* state. As seen from this example, each state of this diagram represents a distinct phase of operation for the robot. Transition between these phases is initiated by the appropriate sensor signals or operator input.

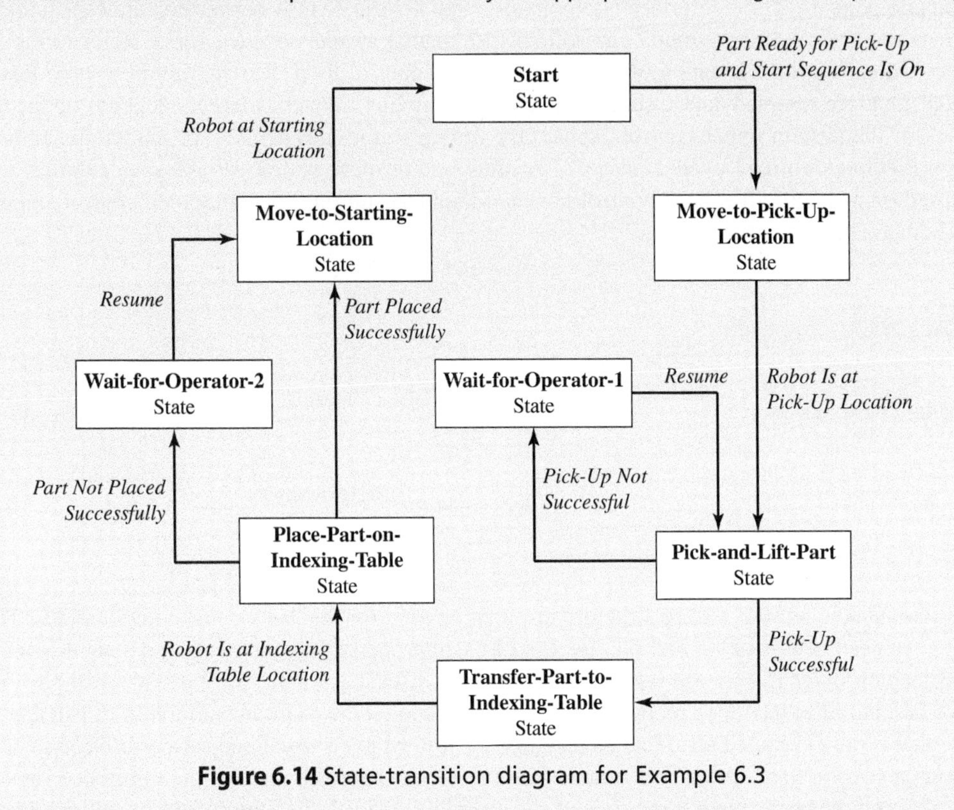

Figure 6.14 State-transition diagram for Example 6.3

The state-transition diagrams for the other pick-and-place robots are similar in structure. The state-transition diagrams for the other components can be similarly developed, but each will have different states specific to that component. For example, part of the state-transition code for the **indexing table** includes a homing sequence to determine a starting position for the table. This homing sequence is executed once during the initialization phase. Thus, to control the operation of this assembly machine, we should have nine tasks: one for each component of the machine plus one task to schedule the operation of the machine. The function of the scheduling task is to decide which components of the machine will be active during a particular cycle of the machine. In this assembly system, not all components of the machine will be active during the start and end phases of the assembly. For example, in the first step of making the first assembly, all of the operations will take place at station #1 on the indexing table, and all of the components on the other stations will be idle. Similarly, during the last step of making the last assembly, all of the operations will take place on station #4, and all of the other components will be idle.

In this assembly example, some of the components could require feedback control action for their proper operation (e.g., if the robots were servo-driven). In this case, a feedback control task (to be discussed next) will be needed, but from the point of view of the task/state methodology, the structure would remain the same. We then replace the state that checks if the robot has reached its end of travel by monitoring a proximity sensor with a state that waits for a flag that is set by the feedback controller when the desired motion is completed.

6.5.2 Feedback Control Tasks

Feedback control systems are instrumental in regulation and tracking control applications, as detailed in Chapter 10. One primary advantage of these systems is their ability to maintain adherence to a designated set point or reference value, even when faced with external disturbances. Furthermore, the intrinsic nature of feedback negates the need for calibration since the system's input is dynamically adjusted based on feedback control action. Numerous real-world applications underscore these advantages. For instance, feedback control tasks can be observed in the speed and position control of electric motors, temperature regulation of ovens, and comprehensive environmental control systems. The essential components of a digital feedback control system encompass the controller, the controlled element, feedback devices, and interface components. These interface components, such as A/D or D/A converters, are especially pivotal when there's a need to bridge digital controllers with analog systems or vice versa. Refer to Figure 6.15 for a block diagram representation of a digital feedback control loop. It is important to note that digital control systems are sampled data systems. The chosen sampling frequency largely depends on the inherent dynamics of the system under control. Thus, task timing in these systems is of paramount importance. A typical **feedback control cycle** consists of reading one or more output variables, calculating a control input based on a specific feedback control law, and then transmitting this computed control signal to the system being managed.

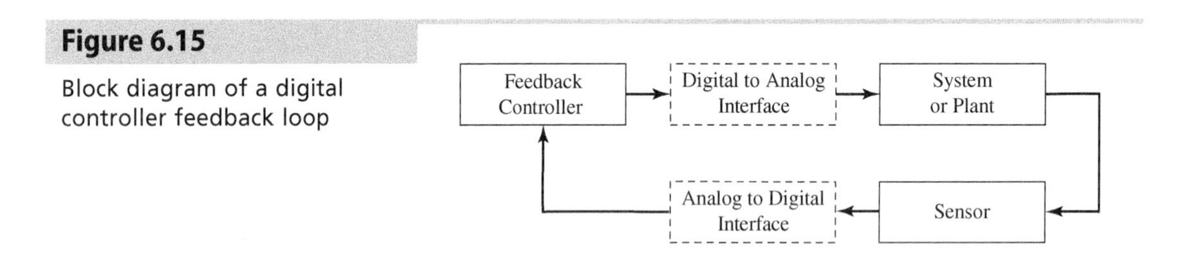

Figure 6.15

Block diagram of a digital controller feedback loop

The task/state control software structure also can be used for feedback control applications. The only thing that we need is to have access to a timer, so we can properly time the control actions. Figure 6.16 is a state-transition diagram that can be used to implement a feedback control loop. This task has only one state. We first initialize the timer, record the starting time, and set the counting variable *K* to 1 in a section of the code that is only executed once (called an entry section, which is discussed further in Section 6.6.1). In the action part of the state, the current time is read in every scan of this state. If the difference between the current time and the start time is a multiple of the sampling interval, then the software calls the *DoControl* function in which the control action is performed. The *DoControl* routine will have the code for the feedback controller used such as a PID controller (see Chapter 10). After the controller completes its job, the counting index *K* is incremented, and the task goes back to waiting for the elapse of the next sampling interval to perform the control action again. As seen here, the control of a feedback task has an identical structure to a discrete-event control task.

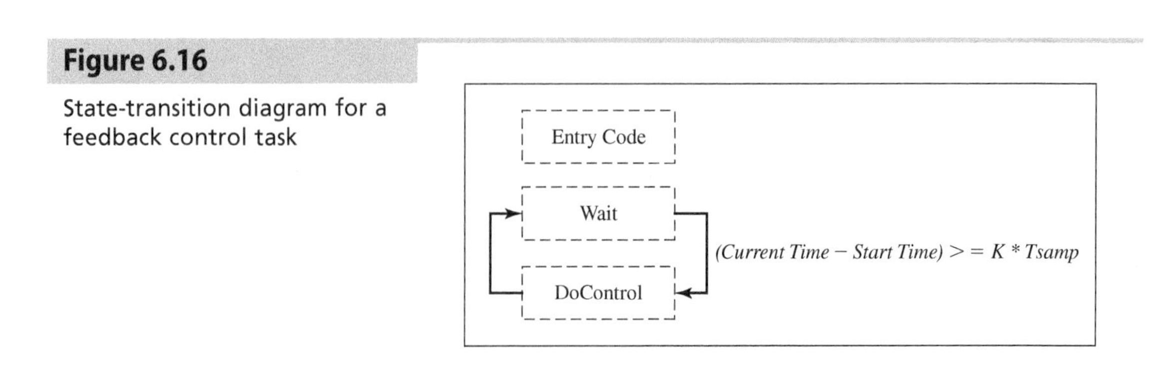

Figure 6.16

State-transition diagram for a feedback control task

6.5.3 State-Transition Diagrams for Integrated Case Studies

The following example state-transition diagrams are relevant in three out of the four Integrated Case Studies featured in this book. Integrated Case Study I.B delves into the development of the state-transition diagram for the operation of a stepper motor-driven rotary table. Integrated Case Study III.B discusses the two state-transition diagrams that are used to structure the operation of the temperature-controlled heating system. Lastly, Integrated Case Study IV.B covers the development of the state-transition diagram related to the motion of a mobile robot.

Integrated Case Study I.B: A State-Transition Diagram for Table Operation

For a mechatronics system such as the rotary table that is considered in Integrated Case Study I (Stepper Motor-Driven Rotary Table—see Section 11.2), a state-transition diagram should be created to guide the development of the software that controls the motion of this event-driven system. A state-transition diagram should be designed based on how the system needs to operate, and for this rotary table system, the motion of the table is controlled by three commands issued by the user. An explanation of the desired function of each of these commands is given here.

CW: This command should cause the table to rotate clockwise (as seen from above the disk) and should cause the table to automatically stop when it reaches the optical sensor. If the table was originally at the home position, this command should cause the table to make one complete revolution.

CCW: This command should cause the table to rotate counterclockwise (as seen from above the disk) and should cause the table to automatically stop when it reaches the optical sensor. If the table was originally at the home position, this command should cause the table to make one complete revolution.

Stop: This command should stop the table if it is moving, but the program should still be running.

At startup, the program should rotate the disk until the notch on the disk is aligned with the optical sensor (home position). If the disk is already aligned with the sensor, then there is no need to home the table. We can enter any of the three commands using a single character interface scheme (c for CW, w for CCW, and s for Stop) in which the commands are sent from the Serial Monitor on Arduino IDE.

The desired operation commands can be translated into seven states as shown in the **state-transition diagram** in Figure 6.17. At startup, the program starts in the *Start* state. If the disk happens to be at the home position, then the state transitions to the *AtHome* state. If the disk is not at the home position, then the disk is set to rotate clockwise until it hits the home position. In the *AtHome* state, the table can transition to either the *Initial CW Motion* or *Initial CCW Motion* state, depending on the user input. These initial motion states are used for a short time while the disk clears the home sensor zone. No monitoring of the home sensor is done in either state to prevent a transition based on false triggering. When the short time interval elapses, the table is in the *CW Motion* or *CCW Motion* state. Monitoring is done in each of these states for a home sensor signal, a *Stop* command, or a reversal of the motion direction command, and the corresponding transition is taken. In the *Stopped* state, the table waits for a *CW* or *CCW* command to start moving again. While the table is stopped in both the *AtHome* state and the *Stopped* state, two states are used here to distinguish the fact that in the *AtHome* state, the table is at the home position, which is not the case with the *Stopped* state.

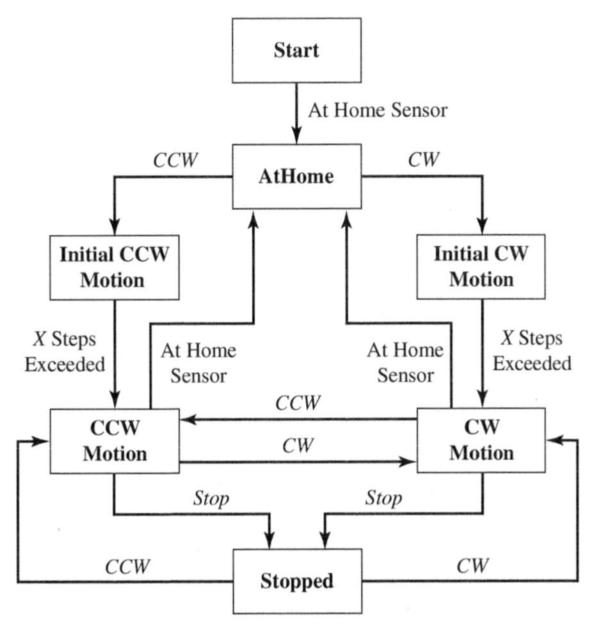

Figure 6.17 State-transition diagram for the operation of the stepper motor-driven rotary table

The code that implements this state-transition diagram is listed in Section 11.2.4.

Integrated Case Study III.B: Control Software for System Operation

For Integrated Case Study III (Temperature-Controlled Heating System—see Section 11.4), the heated plate is controlled by an Arduino Uno which manages the operations of this system. The user sends single-letter commands to the microcontroller through the Serial Monitor. These commands include the controller gains (K_p and K_i), the mode of operation (open- or closed-loop control), the control duration interval, the desired temperature for closed-loop operation or the voltage step input value for open-loop operation, and the start/stop command. Open-loop control is used to identify the dynamics of the system by sending a step input voltage signal to the heater (see Chapter 7) and recording the temperature of the plate. For both control modes, the control operations are synchronized with time, and the Arduino *millis()* function is used for timekeeping. Figure 6.18 shows the control software structure for the operation of this system.

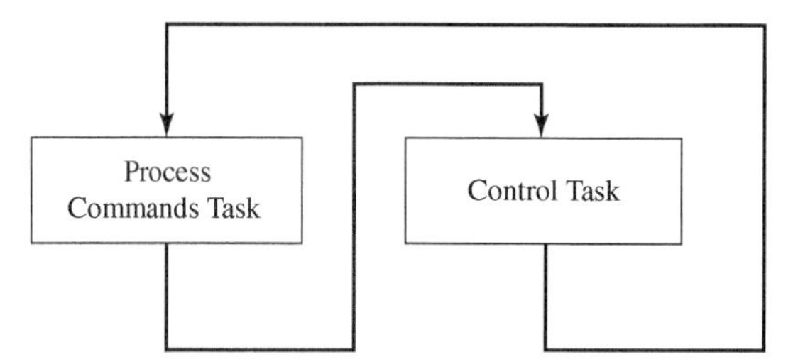

Figure 6.18 Software control structure for the heated plate system

The software is structured as two tasks that are continuously scanned (see Section 6.6.2). The *Process Command Task* is used to check for any serial commands received from the user and then it processes these commands. The *Control Task* performs either open- or closed-loop feedback control of the plate temperature. The use of two separate tasks gives modularity to the control system as the details of either task could be changed without affecting the other task. Each of these tasks is implemented in code using a single state in each task, and the code for this case study is shown in Section 11.4.4.

Integrated Case Study IV.C: State-Transition Diagram for Robot Motion

For Integrated Case Study IV (Mobile Robot—see Section 11.5), an intriguing application involves enabling the mobile robot to trace the shape of an *n*-sided polygon with equal sides (see Figure 6.19), such as a square or hexagon, using closed-loop wheel odometers. For greater flexibility, the user should be able to specify both the number of sides and the length of one side of the polygon. To trace this shape, the robot will alternate between moving in straight lines along each polygon side and executing in-place rotations between straight segments.

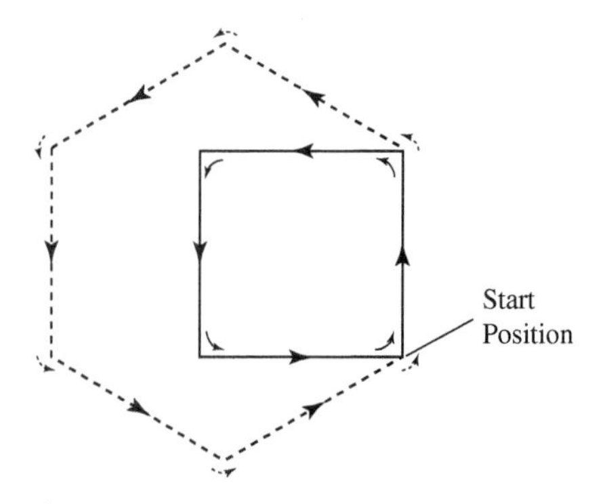

Figure 6.19 Two examples of polygon tracing: a square and a hexagon

Feedback control techniques, as discussed in Chapter 10, will ensure that each linear or rotational segment is completed within a specified error band before starting the next one. A state-transition diagram illustrating this control operation is shown in Figure 6.20.

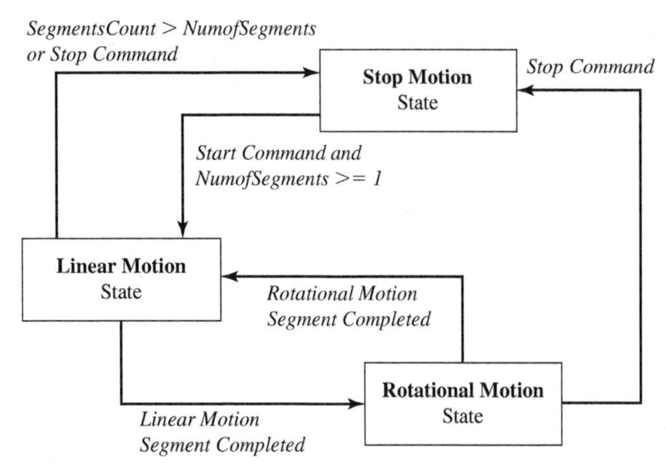

Figure 6.20 State-transition diagram for the mobile robot to perform polygon tracing

The diagram has three states, and the code starts in the *Stop Motion* state and waits for the user to issue a *Start* command. Upon receiving this command, the code transitions to the *Linear Motion* state where the robot is moved in a straight-line fashion (i.e., each wheel is commanded the same trajectory). Upon completion of the straight move, the code transitions to the *Rotation Motion* state, where the robot is rotated in place (each wheel is commanded equal but opposite sign displacements). Depending on the number of segments that the shape has, the code will alternate between these states until the path is traced. Note that in the entry section of the *Linear* or *Rotary Motion* states, the code calls a function to generate a trapezoidal velocity profile move for each wheel motion to result in controlled acceleration of the mobile robot. The code that implements this transition diagram is shown in Section 11.5.5.

6.6 Task and State Structure in Code

A state-transition diagram acts as a blueprint for the development of the control code. After creating a state-transition diagram for a control task, the next step is to translate this diagram into code using a programming language. Typically, a task is coded as a function, and each state within the task is represented as a conditional statement within that function. If the programming language supports the Switch/Case statement, then each state can be coded as one of the cases within that statement. Figure 6.21 depicts the fundamental structure of a task. Whenever the task is executed, the code checks to identify the active state. In subsequent invocations of the task function, the active state continues to be visited until the state transitions to another state.

Figure 6.21

Basic code structure of a task

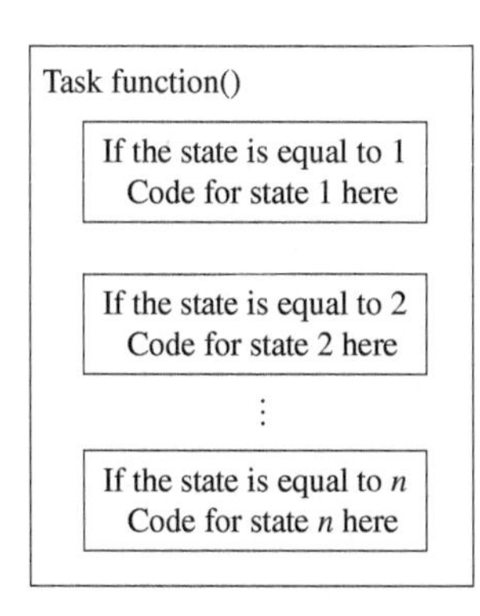

6.6.1 Details of States in a Task

Within a state, the code should be organized into one of four different types depending on the actions that need to be done:

Entry Code: Executed only once on entry to the state

Action Code: Executed on every scan of the state

Test Code: Executed to check the condition for transition (to go to another state)

Exit Code: Executed if the associated transition is taken

An *Entry Code* section is the part of the state code that needs to be executed only once on entry to that state. An example is to record the initial value of a timer or to send a signal to open a valve. On subsequent visits to the same state, the code in the *Entry Code* section should not be executed again. An *Action Code* is executed in every scan of the state. An example is to increment an index for the implementation of a timing mechanism for a feedback controller. Some states may not need an *Action Code* section if, for example, the state is just waiting for a sensor signal to change its value. A *Test Code* section is used to check if the transition condition that causes the transition to another state in the task is satisfied. On satisfaction of the transition condition, the code in the *Exit Code* section is executed. Every state should have *Test* and *Exit Code* sections. If the code transitions from one state to another automatically without a transition condition, then these two states are not set correctly since they can be combined into a single state.

As an illustration of this coding structure, consider the *OpenDoor* state in Example 6.2. Refer to Figure 6.22, which gives an implementation of the state code for this state in C language.

Figure 6.22

State coding organization for the *OpenDoor* state of Example 6.2

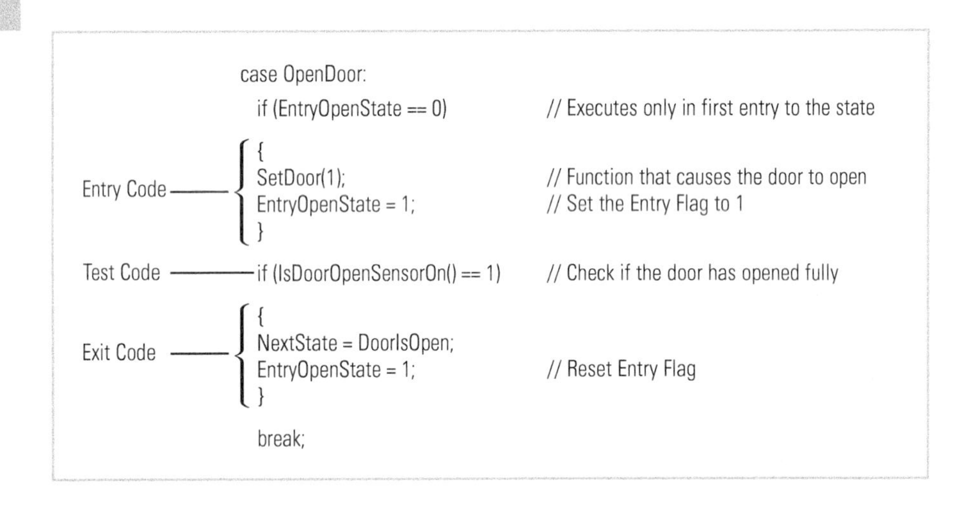

For this state, the *Entry Code* section includes the code that causes the door to open (such as actuating the mechanism that opens the door by sending a signal to a relay). This code should be executed once on the first entry to this state. On subsequent entries to this state, the code should not be executed again. This easily can be accomplished in the software through the use of a **flag variable** (*EntryOpenState*) whose value indicates whether the code has been executed or not. There is no *Action Code* here for this state. The *Test Code* for this state is the code that checks if the door is fully open. This is a logical check that returns either a true or a false condition. This code should be executed in every scan of this state. The *Exit Code*, which only gets executed if the test condition is true (i.e., the door is fully open as indicated by the appropriate sensor), should include code that resets the entry flag to its initial value, so that the next time this state becomes active, the code executes similarly. The *Exit Code* should also cause the *DoorIsOpen* state to be scanned in the next scan of this task. This can be easily accomplished using a variable named, for example, *NextState*, whose value indicates which state needs to be the active state in the next scan. The value of the *NextState* variable is only updated when we need to transition from one state to another. On the next scan of the task, the program will go directly to the state that is currently indicated by the *CurrentState* variable, which is assigned the contents of the *NextState* variable. Figure 6.23 shows the code structure for selecting the active state.

Figure 6.23

Code structure for selecting the active state in a task

```
void Door_Task(void)
{

    CurrentState = NextState;
    switch (CurrentState)
    {

    case Start:
      ...
    case OpenDoor:
      ...

    }
}
```

Example 6.4 illustrates the state coding organization for another state in the automated entry door state-transition diagram.

Example 6.4 State Coding Organization for the *Wait* State

Show the state coding organization for the *Wait* state in Example 6.2.

Solution:

Figure 6.24 shows the state coding organization for this state implemented in C language. In the *Entry Code* section, the start time is read from the timing function *ReadTimer* and stored in a variable called *StartTime*. Since there are two possible transitions from this state, two test codes are included here. The first *Test Code* checks if the detection sensor returns a true value. In this case, the next state should be the *DoorIsOpen* state. The other *Test Code* checks if the delay interval (3 s here) has elapsed. If that is the case, then the code should go to the *CloseDoor* State in the next scan. The entry flag (*EntryWaitState*) is reset to its initial value in the *Exit Code* section.

```
                case Wait:
               {  if (EntryWaitState == 0) {
Entry Code ----{     StartTime = ReadTimer();              // Record the start time
               {     EntryWaitState = 1;                   // Set Entry Flag
               {  }
               {  if (DetectionSensor() == 1) {
               {     NextState = DoorIsOpen;
               {     EntryWaitState = 0;                   // Reset Entry Flag
               {  }
Test and   ----{  if ((ReadTimer() - StartTime) >= 3)      // Check if 3 seconds wait interval has elapsed
Exit Code      {  {
               {     NextState = CloseDoor;
               {     EntryWaitState = 0;                   // Reset Entry Flag
               {  }
                  break;
```

Figure 6.24 State coding organization for the *Wait* state

6.6.2 Cooperative Control Mode

After the control job is split into tasks, the next step is to decide on a mechanism to execute these tasks in a programming language. There are several ways to do this. The simplest mechanism is called the **cooperative control mode** or the cooperative multitasking structure (see Figure 6.25 which shows a visualization of this structure). In this structure, each of the *M* tasks in the control system is placed in a loop that executes repeatedly and the tasks are called (through a function call for example) in the order of their placement in the loop. When a task is called (i.e., the function associated with that task is called), the scanning mechanism goes through the states in that task (i.e., scans the states), identifying and executing the current state. Once the code for the current state is executed, the software exits the task. This process is repeated on the next scan through the task. Only one state (the current or active state) from each task is executed in a single scan through the loop.

Figure 6.25

Scanning of multiple tasks in the cooperative control mode; the underlined states indicate active or current states

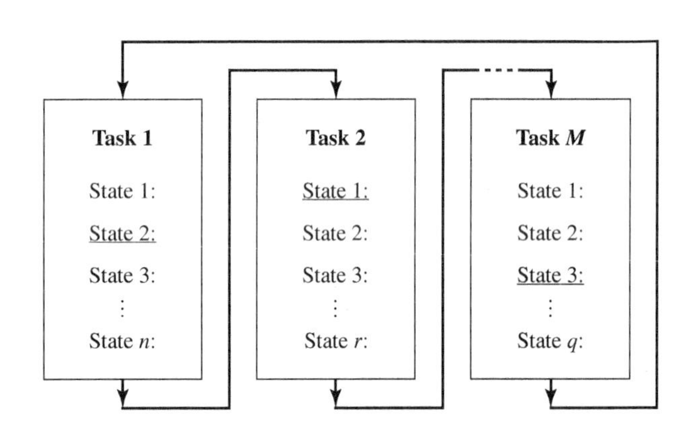

To effectively implement this scanning mechanism in a PC or a microcontroller, two critical conditions must be met. First, the **execution time** of the active state should be kept minimal. This ensures each task has its opportunity swiftly, providing an illusion of concurrent or near-parallel execution. While modern high-speed processors can expedite computations, it is essential to understand that the actual execution time is influenced not just by the number of lines of code but also by their complexity and nature. Second, the code associated with any state should be **non-blocking** (▶ available—V6.10), meaning it does not pause or wait for an external event or condition before continuing. For instance, operations that wait for user inputs or changes in signal levels would be considered blocking. An example of blocking code in C is shown in Figure 6.26.

Figure 6.26

Example of a blocking code

```
while (Input1 <= 5)
  Input1 = Read_AD(1);
```

In this example, the use of the *while* statement halts the execution of the program until the variable *Input1* has a value greater than 5. Since it is not known when the value of *Input1* will be greater than 5, this code is blocking. While waiting for the input to change value, no other code in the program can be executed. This can cause serious problems if we have other tasks that need to be scanned and if these other tasks include time-sensitive actions (such as counting, motion tracking, or screen updates).

Other examples of blocking code include the use of the *scanf()* statement in C-language programs to read user input and the use of the *input()* function in Python programs to gather input from the user. The *scanf()* statement is blocking because it waits for the user to hit the carriage return (or Enter key) before the next statement is allowed to execute. Similarly, the *input()* function in Python is blocking because no subsequent code can execute until the user provides input and hits the Enter key. These characteristics are essential assumptions for implementing control systems based on this model.

6.6.3 Cooperative Control Mode Implementation

The cooperative control mode, a mechanism where tasks voluntarily yield control, can be easily implemented in any programming language, as shown in the pseudocode in Figure 6.27.

Figure 6.27

Pseudocode for implementing the cooperative control mode in software

```
While (Stop not equal to 1)
{
Increment Scan Counter
Call Task1()
Call Task2()
Call Task3()
Every number of scans
        Call function to allow background processing
}
```

The figure showcases an infinite loop that perpetually operates unless the *Stop* variable equals 1. This loop encompasses functions that invoke tasks; for this instance, three tasks are considered. On invoking a task, its active state code is executed. Upon completion, the subsequent task is called upon, thus perpetuating the cycle. The final lines within our loop, pertinent to Figure 6.27, facilitate background processing, a crucial component if deployed within systems like Windows OS. Such processing ensures the handling of background events like user interactions or system notifications. Lacking this, the application might seem unresponsive or 'frozen' since it does not react to user inputs. It is vital to highlight the flexibility in checking frequency for background processes. For instance, assessing after every 1000 task invocations may suffice, contingent on acceptable latency for event responses. The primary takeaway is that it is unnecessary, and possibly counterproductive, to manage background processes within every iteration, as this could potentially reduce the task processing rate.

Note that in the cooperative control mode, all tasks have the same **priority**, and we cannot preempt the operation of one task and start another. The execution of the next task in the loop cannot be started until the previous task has completed its operation. With the use of fast processors that provide high scanning rates (over 1,000,000 task scans per second) and the requirement of no blocking code, the tasks operate in nearly parallel fashion in this mode. It should be emphasized that the 'near parallel' operation in this mode is an illusion of parallelism due to the high speed of execution, not actual parallel execution. One major advantage of the cooperative control mode is that data can be easily exchanged between tasks without concern about data corruption for cases where more than one task uses the same data. This is because each task completes its operation before another task starts. A disadvantage of the cooperative control mode is that control actions (such as activating a relay or recording the position signal from an encoder) cannot be guaranteed to occur within a particular time interval relative to the event that requires these control actions. We cannot stop the execution of one task and start another. Each task has to wait for its turn to execute. This control mode or scheduling method is sometimes referred to as 'round-robin.' Nevertheless, the simplicity of this control mode makes it very attractive to use in many applications.

If we wish to have the capability to halt the execution of one task and initiate another, we employ what is known as preemptive scheduling or **preemptive control mode** (▶ available—V6.11). This mode permits one task to be interrupted so that another, often more urgent, task can begin execution. This is particularly crucial in scenarios such as alarm processing where a specific task must be executed immediately in response to an alarm signal. To realize this mode, tasks are typically assigned priority levels. In a preemptive environment, when a task of higher priority is ready to run, it can preempt a currently executing task of lower priority. Once the higher-priority task is completed, the preempted lower-priority task can then resume. Moreover, if two tasks are awaiting execution and one has a higher priority than the other, the preemptive scheduler ensures the higher priority task is executed first. Only after its completion will the lower-priority task commence.

It is worth noting that while preemptive scheduling introduces complexities such as potential race conditions and the need for synchronization mechanisms, it is not solely a feature of RTOS. Many general-purpose operating systems like Linux and Windows also utilize preemptive scheduling. However, as mentioned before RTOS are specially designed to meet stringent timing requirements, making them apt for real-time applications. The implementation of a preemptive scheduler, particularly in an RTOS setting, is non-trivial. Although many RTOS come equipped with preemptive scheduling, developing such systems and ensuring their deterministic behavior requires considerable expertise.

6.7 Examples of Control Task Implementation in Software

In this section, we will illustrate the software implementation of the heating thermostat state-transition diagram using two computing platforms: MATLAB and AVR microcontrollers. The state-transition diagram will be implemented as a single task using the cooperative control mode. One feature of the heating thermostat is that it can easily be simulated in the software without the need for any actual hardware. With proper software structuring, code for simulation can easily be transferred into actual implementation by replacing a few functions in the code.

6.7.1 Implementation in MATLAB

MATLAB offers two ways to implement a state-transition diagram. One way is to implement each task as a function that is called by the TIMER object callback function. The timer callback function offers a built-in mechanism for repeatedly calling one or more software tasks. Coupled with the use of the graphical user interface (GUI) that is available in **MATLAB App Designer**, one can build an event-driven type application. The other way to implement a state-transition diagram is to use the **Stateflow** toolbox in conjunction with Simulink. This approach is not covered in this text.

The user interface for the thermostat problem is shown in Figure 6.28(a), and a snapshot of the code while operating is shown in Figure 6.28(b). The user interface uses two *push-buttons*, one for the *START* command and the other for the *EXIT* command. It also uses a *panel* on which several controls are placed. These controls include a *toggle button* for the thermostat on/off switch, a *drop-down menu* for a list of desired temperatures, and five *labels* to display information.

Figure 6.28

(a) User interface created using MATLAB App Designer and (b) snapshot of the interface while code is running

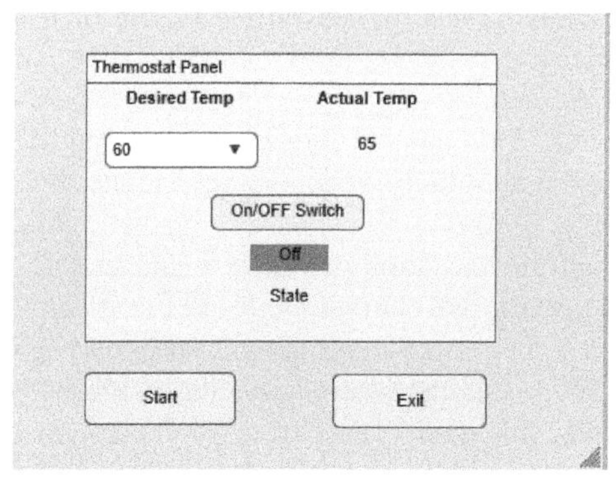

(a)

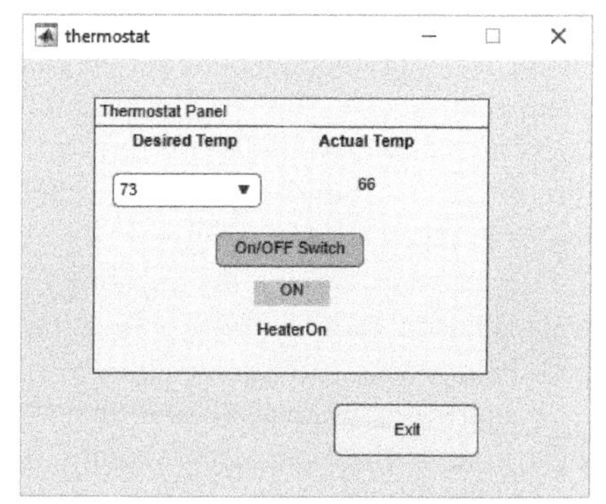

(b)

The code works by pressing the *START* push-button (which disappears after this action). The *Startbutton* callback function (see code listing in Figure 6.29) associated with the *START* button performs initialization as well as setting two timer objects, *TIMER*1 and *TIMER*2. The callback function associated with *TIMER*1 runs every 0.5 s and handles the *Thermostat task* state-transition diagram. The callback function associated with *TIMER*2 runs every 5 s and is used for code that simulates heating/cooling action. The *Startbutton* callback function also calls the function *TIC*, which will be used later with the function *TOC* to implement the heater delay.

Figure 6.29

MATLAB code listing for the *Startbutton* callback function

```
% Button pushed function: Startbutton
    function Startbutton_Callback(app, event)
      % Create GUIDE-style callback args - Added by Migration Tool
      [hObject, eventdata, handles] = convertToGUIDECallbackArguments
      (app, event); %#ok

      % hObject    handle to Startbutton (see GCBO)
      % eventdata  reserved - to be defined in a future version of MATLAB
      % handles    structure with handles and user data (see GUIDATA)
      tic; % reads the hardware timer
      global TIMER1 TIMER2 Temp SwitchOn HeaterOn HeaterDelay NextState DesTemp
      % Perform Initialization
      Temp = 65;
      SwitchOn = 0;
      HeaterOn = 0;
      HeaterDelay = 0;
      NextState = 'SwitchOffHeaterOff';
      % Read the timer
      tic;
      %Read the desired temperature
      valuepop = get(handles.Desiredpopupmenu,'Value');
      stringpop = (get(handles.Desiredpopupmenu,'String'));
      DesTemp = str2num(stringpop{valuepop});
      %Setup the two timer objects
      TIMER1 = timer('TimerFcn', @app.timer_callback_fcn1, 'period', 0.5,'ExecutionMode',
      'fixedRate');
      start(TIMER1);
      TIMER2 =  timer('TimerFcn', @app.timer_callback_fcn2, 'period', 5.0,'ExecutionMode',
      'fixedRate');
      start(TIMER2);
      set(handles.Startbutton,'Visible','Off');
    end
```

The code for the Thermostat task, which implements the state-transition diagram of Figure 6.12, is shown in Figure 6.30. The code has four states with names similar to those used in the state-transition diagram and is implemented using a *switch-case* structure. Note the use of the *NextState* variable for transition between states, as well as the use of the *global* variables which are shared among the functions that make use of these variables. Also, note the use of *persistent* declaration for variables that are local to this function but need to keep their values in each call of the function.

The elapse of the timing delay for the heater is implemented by setting the variable *DelayIsOver* to 1 whenever the current time (obtained by reading the *TOC* function) minus the *StartTime* is larger than the *HeaterDelay*. Note that *HeaterDelay* is set to 0 during initialization (see Figure 6.29) and is set to the desired amount in the entry code section of the *HeaterOn* state. Simulation of the heater operation is done by setting the global variable *HeaterOn* to either 1 or 0. This variable is used in the callback function associated with *TIMER2* (see Figure 6.31) to update the temperature.

Figure 6.30

Thermostat task implemented inside the TIMER1 callback function in MATLAB

```
function timer_callback_fcn1(app, obj, event)
    % global lists variables that can be accessed from any function
     global HANDLESVAR Temp SwitchOn HeaterOn DesTemp HeaterDelay NextState
    % persistent stores variables that retain their values between calls to the function
    persistent StartTime DelayIsOver EntryHeaterOnState
     if isempty(NextState)                    % Start in this state at the beginning
       NextState = 'SwitchOffHeaterOff';
    end
    if isempty(EntryHeaterOnState)
       EntryHeaterOnState = 0;
    end
    if (toc - StartTime) >= HeaterDelay       % read time and update DelayIsOver flag

       DelayIsOver = 1;
    else
       DelayIsOver = 0;
    end

    State = NextState;                        % Set the active state
    switch State

       case 'SwitchOffHeaterOff'
          if SwitchOn == 1 && Temp <= (DesTemp - 1)
             NextState = 'HeaterOn';
          elseif SwitchOn == 1
             NextState = 'HeaterOff';
          end
      case 'HeaterOn'
          if EntryHeaterOnState == 0
             StartTime = toc;                 % Record the start time
             HeaterOn = 1;                    % Turn on heater
             HeaterDelay = 10;                % Set heater delay time
             DelayIsOver = 0;
             EntryHeaterOnState = 1;
          end
          if (Temp >= (DesTemp + 1)) && DelayIsOver
             NextState = 'HeaterOff';
             HeaterOn = 0;
             EntryHeaterOnState = 0;
          elseif SwitchOn == 0
             if DelayIsOver
                HeaterOn = 0;
                NextState = 'SwitchOffHeaterOff';
             else
                NextState = 'SwitchOffHeaterOn';
             end
             EntryHeaterOnState = 0;
          end
      case 'HeaterOff'
          if (Temp <= (DesTemp - 1))
             NextState = 'HeaterOn';
          elseif SwitchOn == 0
             NextState = 'SwitchOffHeaterOff';
          end
      case 'SwitchOffHeaterOn'
          if DelayIsOver
             HeaterOn = 0;
             NextState = 'SwitchOffHeaterOff';
          end
    end
    set(HANDLESVAR.StateText,'string',State);
    set(HANDLESVAR.ActualTempText,'string',int2str(Temp));
 end
```

The code for simulating the heating/cooling action is shown in Figure 6.31. A simple scheme is used here in which the temperature is increased by one unit if *HeaterOn* is set to 1 and is decreased by one unit every four calls of the *TIMER2* callback function (controlled by the *count* variable) if *HeaterOn* is set to 0. The decrease in temperature when the heater is OFF is to simulate the heat loss that occurs when the current temperature is higher than the ambient temperature.

Figure 6.31

MATLAB code listing for simulating heating/cooling action which is implemented inside the TIMER2 callback function

```
function timer_callback_fcn2(app, obj, event)
      global HeaterOn Temp
      persistent count
      if isempty(count)
        count = 0;
      end

      if HeaterOn                % Simulate heating action
        Temp = Temp + 1;
      else
        count = count + 1;       % Simulate cooling action
        if count == 4
          Temp = Temp - 1;
          count = 0;
        end
      end

    end
```

It is recommended to dispose of the timer objects once the program completes its execution. This is done inside the code of the callback function associated with the *ExitButton*. The code listing is shown in Figure 6.32, and it stops the two timer objects and then deletes them.

Figure 6.32

MATLAB code listing for the *ExitButton* callback function

```
% Button pushed function: ExitButton
function ExitButton_Callback(app, event)
% Create GUIDE-style callback args - Added by Migration Tool
[hObject, eventdata, handles] = convertToGUIDECallbackArguments(app, event); %#ok

 % hObject    handle to ExitButton (see GCBO)
 % eventdata  reserved - to be defined in a future version of MATLAB
 % handles    structure with handles and user data (see GUIDATA)
 global TIMER1 TIMER2
 stop(TIMER1);
 delete(TIMER1);
 stop(TIMER2);
 delete(TIMER2);
 close;
end
```

6.7.2 Implementation in AVR Microcontroller

In the previous section, we have illustrated the implementation of the thermostat control problem using a PC computing platform. In this section, we illustrate it using the Arduino Uno control board which uses an ATmega328P microcontroller. While an MCU is normally not used to just implement simulation code, it was done here to illustrate coding using an MCU. The Serial Monitor in Arduino IDE is used to display the current and desired temperatures as well as the current state information. For specifying the desired temperature, this implementation uses a rotary potentiometer that is connected to analog input channel 0 on the board. The output of the 10-bit A/D on the Arduino analog input channel is mapped to give desired temperatures in the range of 20 to 40 degrees. One of the digital input pins set to operate in pull-up input mode (see Section 4.5) is used to simulate the operation of the thermostat ON/OFF switch.

Figure 6.33 shows the variable declaration section for the Arduino program. The *#define* statement is used to define each of the four states in the state-transition diagram.

Figure 6.33

Variable declarations and *setup()* function for the thermostat implementation on the Arduino Uno

```
int analogPin = 0;              // Potentiometer to change temperature
int OnOffSwitchPin = 6;         // Digital pin to simulate on/off switch

#define SwitchOffHeaterOff 1    // Define the states
#define HeaterOn 2
#define HeaterOff 3
#define SwitchOffHeaterOn 4

unsigned long Time;             // Variable to record time using Timer1
unsigned long Tupdate;          // Update interval for heater simulation

int count = 0;                  // Variable used in heater simulation
int EntryHeaterOnState = 0;     // HeaterOnState Entry variable
int DelayIsOver;                // Variable to indicate heater delay is over
long StartTime1, StartTime2;    // Variables used for interval timing
int State, NextState;           // State and NextState of transition diagram
int HeaterDelay = 0;            // Heater delay variable with initialization
int Temp = 25;                  // Actual temperature
int DesTemp;                    // Desired temperature
int HeatOn;                     // Variable to indicate heater status

void setup()
{
 Serial.begin(38400);           // Initialize serial communication at 38400 baud rate

 pinMode(OnOffSwitchPin, INPUT_PULLUP);  // Set digital switch to input with pull-up mode
 NextState = 1;                 // Set the starting state in the state-transition function
}
```

The *loop()* function as well as the function to simulate the heater operation and the timing function are shown in Figure 6.34. In the *loop() function,* the two timers, one used for the delay interval and the other for simulating heating and cooling, are initialized to the current time. Afterward, an infinite loop is run in which the *Thermostat_Task()* and the *heater()* functions are repeatedly called. The *heater()* function updates the current temperature every *Tupdate* interval using a simple scheme. If the variable *HeatOn* is set to 1, then the temperature is increased by one unit every *Tupdate* interval. If *HeatOn* is set to zero, then the temperature is decreased by one unit every four *Tupdate* intervals. The *heater()* function also displays the current and desired temperatures and current state to the Serial Monitor.

Figure 6.34

loop(), *heater()*, and *GetTimeNow()* functions for the thermostat implementation on the Arduino Uno

```
void loop()
{

 Tupdate = 5000;                // Update interval in msec
 StartTime1 = GetTimeNow();     // Record initial time
 StartTime2 = GetTimeNow();
 NextState = SwitchOffHeaterOff;
```

Figure 6.34

loop(), heater(), and *GetTimeNow()* functions for the thermostat implementation on the Arduino Uno (*Continued*)

```
 while ( 2 > 1)                // Start infinite loop
 {
  Thermostat_Task();
  heater();
 }
}

void heater(void)
{
 if ((GetTimeNow() - StartTime2) >= Tupdate)
 {
  if (HeatOn == 1)
  {
   Temp = Temp + 1;          // Heating action
  }
  else
  {
   count = count + 1;
   if (count == 4)
   {
    count = 0;
    Temp = Temp - 1;         // Cooling action
   }
  }
  StartTime2 = GetTimeNow();
  Serial.print("D ");
  Serial.print(DesTemp);
  Serial.print(" A ");
  Serial.print(Temp);          // Display current temperature
  Serial.print(" State ");
  Serial.println(State);
 }
}

long GetTimeNow(void)        // Returns time in units of ms
{
 return (millis());
}
```

Figure 6.35

The state-transition diagram for the thermostat task implemented on the Arduino Uno

```
void Thermostat_Task(void)
{
 if ((GetTimeNow() - StartTime1) >= HeaterDelay )
 {
  DelayIsOver = 1;
 }
 else
 {
  DelayIsOver = 0;
 }

 DesTemp = 20 + analogRead(analogPin) * 20. / 1023.; // Read desired temp from pot
                                                      // DesTemp will range from 20 to 40
 State = NextState;
 switch (State)
 {
  case SwitchOffHeaterOff:
```

Figure 6.35

The state-transition diagram for the thermostat task implemented on the Arduino Uno (*Continued*)

```
      if ((digitalRead(OnOffSwitchPin) == 0) && (Temp <= (DesTemp - 1)))
      { // Is OnOffSwitch on and Temp <= (DesTemp - 1)
       NextState = HeaterOn;
      }
      else if (digitalRead(OnOffSwitchPin) == 0)
       NextState = HeaterOff;
      break;

     case HeaterOn:
      if (EntryHeaterOnState == 0)
      {
       StartTime1 = GetTimeNow(); // Record the start time
       HeatOn = 1;                // Turn On Heater
       HeaterDelay = 10 * 1000;   // Heater delay in ms
       DelayIsOver = 0;
       EntryHeaterOnState = 1;
      }
      if ((Temp >= (DesTemp + 1)) && (DelayIsOver == 1))
      {
       NextState = HeaterOff;
       HeatOn = 0;
       EntryHeaterOnState = 0;
      }
      else if (digitalRead(OnOffSwitchPin) == 1 )
      {
       if (DelayIsOver == 1)
       {
        HeatOn = 0;
        NextState = SwitchOffHeaterOff;
       }
       else
       {
        NextState = SwitchOffHeaterOn;
       }
       EntryHeaterOnState = 0;
      }
      break;

   case HeaterOff:
      if (Temp <= (DesTemp - 1))
       NextState = HeaterOn;
      else if (digitalRead(OnOffSwitchPin) == 1 )
       NextState = SwitchOffHeaterOff;
      break;

     case SwitchOffHeaterOn:
      if (DelayIsOver == 1)
      {
       HeatOn = 0;
       NextState = SwitchOffHeaterOff;
      }
      break;

   }
  }
```

The implementation of the state-transition diagram for the thermostat task is shown in Figure 6.35. The coding is very similar to that used in the MATLAB version. The *switch-case* statement is used to implement the state-transition diagram. The delay before turning off the heating action is controlled by the variable *DelayIsOver* which is set to 1 if the time interval since the heater was turned on has elapsed.

6.8 Chapter Summary

This chapter addresses timing, events, and control software structures, focusing on software challenges that emerge when using a microcontroller or a PC as the controller in a mechatronics system. We began by exploring how timers are implemented using a combination of a clock source and a counter, delving into both absolute and relative timing modes. Subsequently, timer-related issues such as resolution and overflow were discussed. This was complemented by an examination of timing functions in both MATLAB and AVR microcontrollers. The chapter then shifted its attention to events and event-driven programming, explaining the distinctions between event-driven and traditional programming. The emphasis then turned to the task/state software structure for controlling mechanical systems. Here, a 'task' refers to a set of control activities, whereas a 'state' denotes a unique operational condition. Notably, states are mutually exclusive, meaning a machine or process can only exist in one state at any given moment. The intricacies of state-transition diagrams were thoroughly examined, showcasing their utility in representing control activities for discrete-event and feedback control applications. While discrete-event control pertains to the regulation of a sequence of events or actions, feedback control is leveraged for regulation or tracking purposes.

The chapter also provided a detailed explanation of how state-transition diagrams are converted into code. It offered guidance on organizing code for a specific state into one of four categories, depending on the actions required within that state. A detailed discourse on the cooperative control mode followed, highlighting it as a potent tool for executing control tasks in programming. This method's success is contingent upon two pivotal factors: non-blocking code and short code duration. To round off the chapter, coding examples illustrating state-transition diagrams in MATLAB and AVR microcontrollers were provided.

Questions

6.1 List several timing needs in mechatronic systems.

6.2 How is a timer implemented in a processor?

6.3 For what applications is absolute timing mode needed?

6.4 Provide examples where relative timing is used.

6.5 Explain what is meant by timer overflow.

6.6 How can timer overflow be detected?

6.7 List the available timing functions in MATLAB.

6.8 Name two timing functions in Arduino.

6.9 What is an event?

6.10 Give several examples of events.

6.11 How is event-driven programming different from traditional programming?

6.12 What is the difference between soft and hard real-time systems?

6.13 Provide several examples of time-sensitive mechatronic applications.

6.14 How can control actions be organized into tasks?

6.15 What is a state-transition diagram?

6.16 Why are state-transition diagrams important?

6.17 Name the types of control tasks.

6.18 List examples of conditions or events that cause transitions between states.

6.19 Name the code sections in a particular state.

6.20 List several examples of blocking code.

6.21 Is a '*For-Loop*' inside a state considered a blocking code?

6.22 What control mode allows a higher priority task to interrupt a currently executing lower priority task?

Problems

Section 6.2 Time and Timers

P6.1 Identify the type of timer–interval or absolute–required for the following operations and provide a justification for your choice.

a. Alarm monitoring system

b. Climate control system

c. Elevator door opening/closing

d. Screen saver program

e. Feedback control system

P6.2 Determine the resolution and the maximum counting interval of a timer that uses a 1 MHz clock that feeds into a 16-bit counter.

P6.3 Determine the time interval measured by an 8-bit counter operating at a frequency of 25 kHz if the first and second counter readings were:

a. 45 and 218

b. 130 and 12

Section 6.3 Timing Functions

P6.4 Write pseudocode (syntax is not important) that allows one to determine the execution time of a certain function or code segment. Assume you have access to the function that returns the current time information.

P6.5 Write pseudocode (syntax is not important) to check how long the timing function itself takes to run. Assume you have access to the function that returns the current time information.

Section 6.4 Events and Event-Driven Programming

P6.6 Name several events that could occur during the operation of a residential garage door opener.

P6.7 Name several events that could occur during the operation of a clothes washing machine.

P6.8 Name several events that could occur during the operation of an autonomously operated vehicle or self-driving car.

Section 6.5 Task/State Control Software Structure

P6.9 Develop a state-transition diagram for a basic video player that has three states: Stopped, Playing, and Paused. The user can press *PLAY*, *PAUSE*, and *STOP* buttons. Upon launch, it is in the Stopped state. Pressing *PLAY* transitions it to Playing, while *PAUSE* shifts it to Paused. Pressing *PLAY* from Paused resumes playback. A *STOP* button returns it to the Stopped state regardless of its current state.

P6.10 Develop a state-transition diagram for a basic elevator in a three-story building. Each morning, the elevator starts at the ground floor. Users can request any floor, and the elevator moves to that floor and waits; if already on the requested floor, it stays.

P6.11 Develop a state-transition diagram for a software counter. The counter should count up when the user presses the *UP* command, and count down when the user presses the *DOWN* command. The counting should stop when the user presses the *STOP* command or when the count reaches a user-specified limit such as 100 or −100.

P6.12 Develop a state-transition diagram for a traffic light system that includes a protected left-turn signal. Consider the following sequence.

i. The main light turns green, allowing straight-through traffic. This lasts for 30 seconds.

ii. The main light turns yellow for 5 seconds.

iii. The main light turns red. Simultaneously, the left-turn signal turns green, allowing protected left turns for 15 seconds.

iv. The left-turn signal turns yellow for 4 seconds.

v. All lights are red for a 5 second clearance interval.

Use the above sequence to guide the transitions between the states in your diagram.

P6.13 Develop a state-transition diagram for the operation of a garage entry system that operates as follows. The user presses a button to get a ticket or swipes a card in a card scanner. Once the ticket is picked up by the user or the card is validated, the gate arm rotates upward. The gate arm remains in a raised position until the vehicle has completely cleared the gate or a waiting interval has elapsed, at which point the gate drops down. The system has a proximity sensor to prevent the gate from striking people and vehicles, and the gate rotates upward if an object is detected while moving downward.

P6.14 Develop a state-transition diagram for the operation of the four-position rotary indexing table shown in Figure 6.13 and discussed in Example 6.3. Assume that the indexing table needs to be first homed to determine a starting position for the table.

P6.15 Develop a state-transition diagram for the operation of a vending machine. The vending machine operates as follows. The user enters the required money and then selects an item to be bought. The machine dispenses the selected item and returns the change to the user if needed. The machine displays a message if the selected item is not available. The user can cancel the transaction before an item is selected.

P6.16 Modify the state-transition diagram for the heating thermostat to one that can handle cooling instead of heating.

Section 6.6 Task and State Structure in Code

P6.17 Using pseudocode, write and identify the entry, active, test, and exit code for the three states in the state-transition diagram shown in Figure 6.8.

P6.18 Using pseudocode, write and identify the entry, active, test, and exit codes for the *Start* and *CloseDoor* states in the state-transition diagram shown in Figure 6.10.

P6.19 Using pseudocode, write and identify the entry, active, test, and exit codes for the *Start* and the *Move-to-Pick-Up-Location* states in the state-transition diagram shown in Figure 6.14.

P6.20 Using pseudocode, write and identify the entry, active, test, and exit code for the *Pick-and-Lift-Part* and *Wait-for-Operator-1* states in the state-transition diagram shown in Figure 6.14.

Laboratory/Programming Exercises

L/P6.1 Using the TIMER object in MATLAB, implement MATLAB code that displays a message to the command window every 10 s.

L/P6.2 Use the Arduino *millis()* function to implement code that displays a message to the Serial Monitor every 10 s. Do not use the *delay()* function.

L/P6.3 Use Arduino *millis()* function to implement code to inform the user when a push-button, connected to one of the digital input ports on the Arduino Uno or a similar board, was pressed since the start of a program operation.

L/P6.4 Use the *millis()* function in Arduino to write and implement code to determine the execution time of a for-loop. Try different index values for the loop and determine the relationship between the execution time and the index value.

L/P6.5 Use the *millis()* function in Arduino to write and implement code to turn ON and OFF two LEDs that are connected to the Arduino Uno or a similar board at two different rates (e.g., one every 1 second while the other every 2 seconds). Do not use the *delay()* function.

L/P6.6 Implement, using the Arduino Uno or a similar board, the software counter discussed in Problem P6.11. The count value should be displayed to the user using the Serial Monitor. The *UP*, *DOWN*, and *STOP* commands should be implemented through push-button switches that are connected to the board or through single-letter commands (such as *u* for *UP*) that are sent through the Serial Monitor. Use a rotary potentiometer connected to one of the analog input channels to vary the counting rate.

L/P6.7 Implement, using the Arduino Uno or a similar board, the state-transition diagram for the operation of the positioning system whose state-transition diagram is shown in Figure 6.8. Use a counter to keep track of the stage position and display that to the user using the Serial Monitor. The *Left*, *Right*, and *Stop* commands should be implemented through push-button switches that are connected to the board or through single-letter commands (such as *l* for *Left*) that are sent through the Serial Monitor.

L/P6.8 Use the TIMER object in MATLAB to implement the software counter discussed in Problem P6.11. The count value should be displayed to the user using a label. The user should be able to change the counting rate while the code is running. Use the MATLAB App Designer to develop the user interface for this system.

L/P6.9 Use the TIMER object in MATLAB to implement the state-transition diagram for the operation of the positioning system whose state-transition diagram is shown in Figure 6.8. Use the MATLAB App Designer to develop the user interface for this system.

7

Chapter

System Response

Chapter Objectives:

When you have finished this chapter, you should be able to:

- Explain the time response of first- and second-order systems subjected to a step input
- Explain transfer functions and how transfer functions are obtained
- Explain the frequency response of dynamic systems
- Interpret a frequency response plot
- Explain the phenomena of resonance
- Explain how to obtain the bandwidth of a system
- Explain the different types of filters and their use in signal processing
- Explain the several ways available in MATLAB to obtain the solution for the set of differential equations that represent the model of a dynamic system

7.1 Introduction

In mechatronics system design, studying the system's response is crucial as it offers evaluative data on the system's performance. By understanding this response, design optimizations can be made, enhancing performance and reliability. Chapter 8 discusses sensors, emphasizing the importance of selecting a sensor based on its response rate to changes in the measured quantity. Likewise, when amplifying a sensor output signal using an amplifier, understanding how the amplifier processes signals of different frequencies is crucial to maintaining signal fidelity. System response is equally pivotal in actuator selection (Chapter 9) and control system design (Chapter 10).

Analysis of the system response can be approached from either the time domain or the frequency domain. In the time domain, responses are assessed using time-related units, like the time constant measured in seconds. In contrast, the frequency domain evaluates responses using units like bandwidth, measured in Hz.

Effective system response analysis hinges on the availability of mathematical models for physical systems, whether they're sensors, actuators, or an entire mechatronics system. Often in mechatronics applications, these models are represented by either first- or second-order differential equations or by employing transfer functions. This chapter covers both the time and frequency responses of these systems, signal filtering, and the simulation of dynamic systems in MATLAB. For a more in-depth exploration of system response, see [22–24].

7.2 Time Response of First-Order Systems

A **first-order system** is one whose dynamic model is described by a first-order differential equation. Examples of first-order systems include the model of the speed of a mass subjected to a force input, the model of the speed of a rotational load subjected to a torque input, the model of the voltage in an RC circuit, the model of the height of fluid in a tank with flow input, or the model of the temperature of a heated plate. Examples 7.1–7.3 illustrate the derivation of three such models. Appendix D includes additional examples.

Example 7.1 Block Subjected to a Force Input

Develop a model for the velocity of the block shown in Figure 7.1 which has a mass m. The viscous damping coefficient is b, and the block is subjected to a force F applied by an actuator.

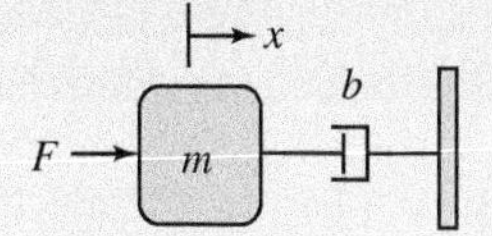

Figure 7.1 Block with viscous damping

Solution:

From application of Newton's Second Law ($\sum F = m\,\ddot{x}$) for linear motion, the force F applied by the actuator minus the viscous frictional resistance force is equal to the acceleration force in the block. Putting this in equation form, we can write

$$F - b\dot{x} = m\ddot{x} \tag{1}$$

Since the velocity of the block v is the derivative of the block's position x (or $v = \dot{x}$), we can write Equation (1) in terms of the velocity v. Doing this we get

$$F - bv = m\dot{v} \tag{2}$$

Placing the v and $\dot{v}$ terms on the same side, we get

$$F = m\dot{v} + bv \tag{3}$$

This is a first-order differential equation that relates the force applied by the actuator to the speed of the block.

Example 7.2 Voltage in an RC Circuit

Develop a model for the voltage across the capacitor for the circuit shown in Figure 7.2, consisting of a resistor R and a capacitor C connected in series. Assume that the voltage input to the circuit is v_S.

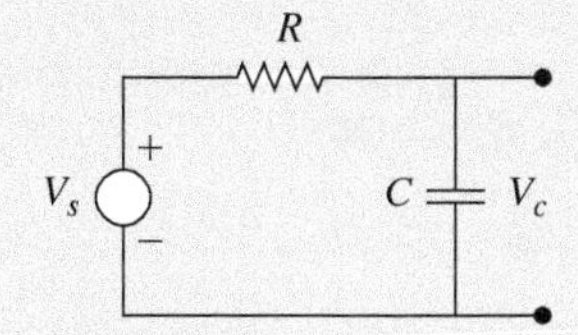

Figure 7.2 RC circuit model

Solution:

Let us apply KVL (see Section 2.4) to the RC circuit. We get

$$v_S - iR - v_C = 0 \tag{1}$$

But from Equation (2.3), the voltage across the capacitor is related to the current in the capacitor by

$$\frac{dv_C}{dt} = \frac{1}{C}i \tag{2}$$

Solving for the current in Equation (1), and substituting it in Equation (2), we get

$$\frac{dv_c}{dt} = \frac{1}{C}\left(\frac{v_s - v_c}{R}\right) = \frac{1}{RC}(v_s - v_c) \tag{3}$$

By rearranging terms, Equation (3) can be written in the form

$$RC\frac{dv_c}{dt} + v_c = v_s \tag{4}$$

This first-order differential equation relates the voltage applied to the circuit to the voltage across the capacitor. Notice that this equation has the same form as the model for the mechanical system discussed in Example 7.1.

Example 7.3 Modeling of a Mechanical System Driven by an Electric Motor

Obtain the dynamic model that relates the torque applied by the electric motor to the motor speed for the drive system shown in Figure 7.3, where a load disk is actuated by a motor through a gear drive system with a speed reduction ratio of N. Let the viscous damping coefficient at the input shaft be b_1 and at the output shaft be b_2. Assume that the shafts are rigid and let I_1 represent the combined inertia of the motor shaft, input shaft, coupling, and pinion; and I_2 represent the combined inertia of the gear, the output shaft, and the load disk.

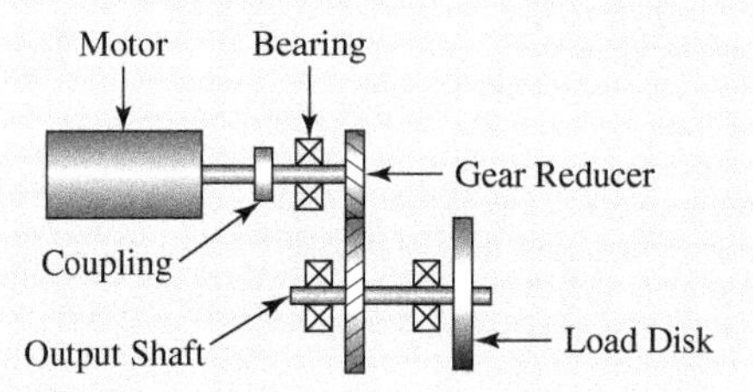

Figure 7.3 Motor-driven gear system

Solution:

With the shafts assumed to be rigid and no external torques acting on the system, the torque (T_m) supplied by the motor is used to overcome the inertia and damping in the system. Thus, by application of Newton's Second Law $\left(\sum M = I\ddot{\theta}\right)$ for rotational motion, we can write

$$T_m = I_1\ddot{\theta}_1 + b_1\dot{\theta}_1 + \left(I_2\,\ddot{\theta}_2 + b_2\dot{\theta}_2\right)/N \tag{1}$$

where θ_1 is the angular position of the input shaft and θ_2 is the angular position of the output shaft. Note that in a geared system, the torque on the output side of a gear system is reduced by a factor of N when reflected to the input side as shown in the expression for the last term in Equation (1). Due to gearing,

$$\dot{\theta}_2/\dot{\theta}_1 = 1/N \tag{2}$$

Using Equation (2) and its derivative to express the velocity and acceleration, respectively, of the output shaft in terms of the input shaft, we obtain

$$T_m = I_1\,\ddot{\theta}_1 + b_1\,\dot{\theta}_1 + \frac{\left(I_2\,\ddot{\theta}_1 + b_2\dot{\theta}_1\right)}{N^2} = I_{eff}\,\ddot{\theta}_1 + b_{eff}\,\dot{\theta}_1 \tag{3}$$

where $I_{eff} = I_1 + \frac{I_2}{N^2}$, and $b_{eff} = b_1 + \frac{b_2}{N^2}$. Much like inertia, the damping coefficient is reduced by the factor N^2 when reflected to the input shaft. Equation (3) is a model for this system that relates the input torque to the angular acceleration of the input shaft. Since $\omega = \dot{\theta}$, Equation (3) can be written using the angular velocity of the input or motor shaft as the output variable. The model is then

$$T_m = I_{eff}\,\dot{\omega}_1 + b_{eff}\,\omega_1 \tag{4}$$

This is a first-order differential equation that relates the torque applied by the motor to the motor's angular speed.

The **time response** of a system (▶ available—V7.1) is the output of the system as a function of time when subjected to an input signal. In general, the response of a system is the sum of the free and forced responses. The **free response** is the response of the system due to initial conditions, while the **forced response** is the response due to the input or forcing function. As an example, let us assume that our first-order system is represented by the differential equation

$$\dot{v} + a\,v = f(t) \tag{7.1}$$

This is a generic equation that can represent any first-order system. For positive a, the response of the system $v(t)$ due to an initial condition $v(0) \neq 0$ and a forcing function $f(t)$ in the form of a step input of magnitude b is given by

$$v(t) = v(0)e^{-at} + \frac{b}{a}(1 - e^{-at}) \tag{7.2}$$

A **step input** in an input that at time $t = 0$ s changes its value from zero to some fixed value b instantly (see Figure 7.4). It is commonly used to represent certain inputs applied to physical systems such as the step voltage signal applied to a circuit from the closing of the power supply switch in the circuit, or the sudden force applied to a stationary mass in a mechanical system.

Figure 7.4

Step input signal

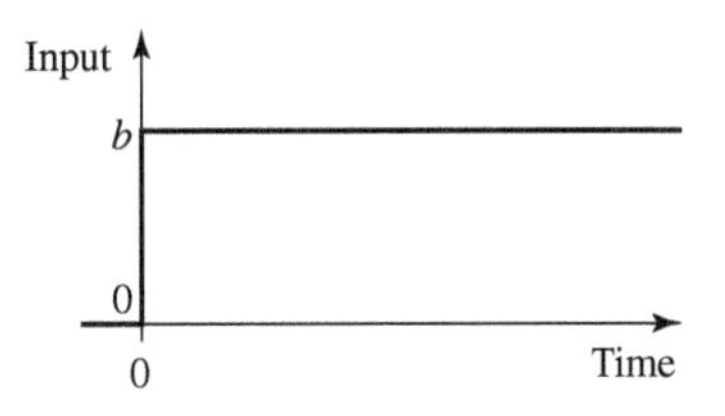

The first term in Equation (7.2) is the free response, while the second term is the forced response. If a is negative in Equation (7.1), then its response is unbounded or unstable, and this discussion does not apply to it. The free and the forced responses are shown in Figures 7.5 and 7.6, respectively for the case where $v(0)$, a, and b all have a value of 1.

Figure 7.5

Free response of the model given by Equation (7.1)

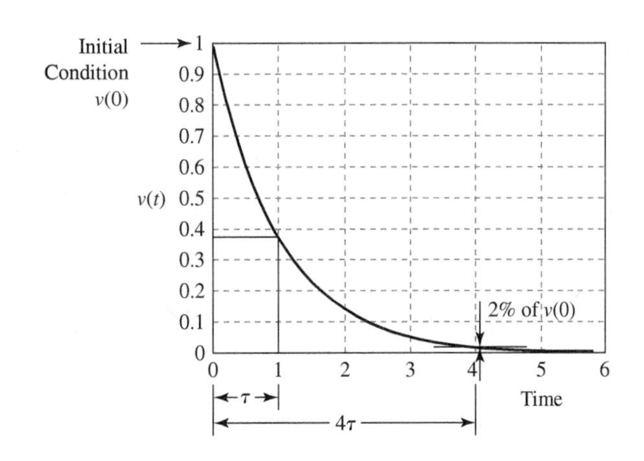

As seen in Figure 7.5, the free response takes the form of an exponential decay with a time constant $\tau = 1/a$. The **time constant** is an inherent property of a system and is defined as the time that it takes the output to reach 63.2% of the difference between the initial and final values. The response v starts at the initial condition $v(0)$ (or 1 in this case), and after one time constant, the output has completed 63.2% of the difference between the initial and final values (or the output has decayed to 36.8% of the initial value). After four time constants, the free response is only about 2% of the initial value. As time increases further, the free response goes to zero. Similarly, the forced response starts at zero (see Figure 7.6) and increases exponentially. After one time constant, the output has reached 63.2% of the final value. After four time constants, the output is within 2% of the final steady value (or the output has precisely reached 98.2% of the final value). Customarily, four time constants are used to measure the system's responsiveness. Thus, if we know the time constant for a system, we can approximately predict when its output will reach its final value using the four time constant approximation. From this, we see that for a first-order system, the time constant of the system solely determines the responsiveness of the system. Systems that have a small time constant (such as DC motor-driven systems) have a fast response, while those that have a large time constant (such as thermal systems) have a slow response. For linear systems, the time constant is independent of the magnitude of the input. Thus, the time constant for a step input of magnitude 1 or 10 is the same.

Figure 7.6

Forced response of the model given by Equation (7.1)

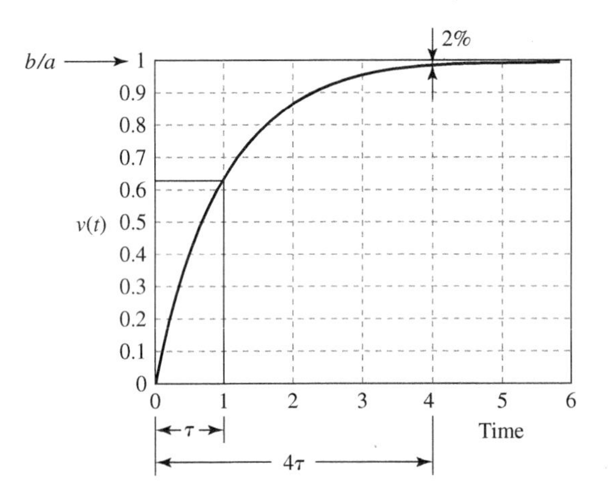

Instead of looking at the time response in terms of what caused it (i.e., free and forced parts), we can also view it in terms of its transient and steady state time characteristics. If we group the two exponential terms in Equation (7.2) together, we get the following form for the response.

(7.3)
$$v(t) = \frac{b}{a} + \left(v(0) - \frac{b}{a}\right)e^{-at}$$

The first term in Equation (7.3) is the **steady state solution**, which is the part of the response that stays indefinitely. The second term in Equation (7.3) is the **transient solution**, which is the part of the response that tends to disappear as $t \to \infty$. Thus, at steady state, the response tends to the constant value of b/a. Example 7.4 illustrates the response of a first-order system.

Example 7.4 Response of a First-Order System

Plot the response of the first-order system $\dot{v} + av = f(t)$ subjected to a step input of magnitude b and with an initial value of $v(0)$ under the following conditions: (a) $b/a > v(0)$, (b) $b/a < v(0)$, and (c) $b/a = v(0)$.

Solution:

(a) The response of a first-order system subjected to a step input always starts at the initial value $v(0)$ and tends to b/a at steady state. Since b/a is greater than $v(0)$, the response increases exponentially between $v(0)$ and b/a, as shown in Figure 7.7.

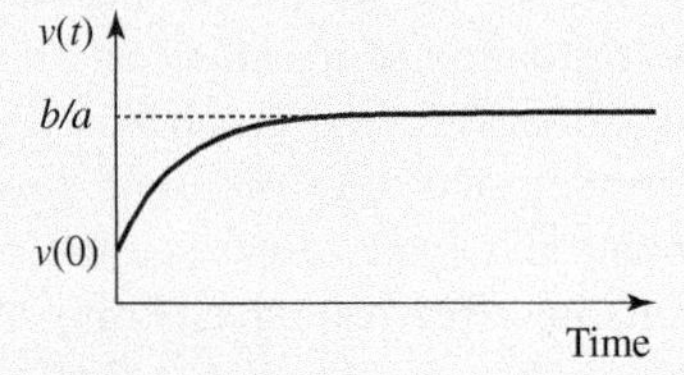

Figure 7.7 Response with $b/a > v(0)$

(b) Since b/a is smaller than $v(0)$, the response decreases in exponential fashion between $v(0)$ and b/a as shown in Figure 7.8.

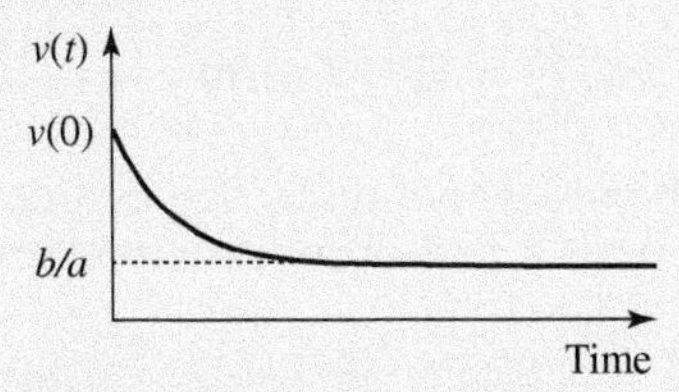

Figure 7.8 Response with $b/a < v(0)$

(c) For b/a equal to $v(0)$, the response is a constant value of b/a as shown in Figure 7.9, with no transient part. Note that in practice it is a rare situation to have the initial value $v(0)$ precisely match the ratio b/a.

Figure 7.9 Response with $b/a = v(0)$

Figure 7.10 shows the thermal response of a commercial IC temperature sensor (LM35C) that is discussed in the next chapter (see Section 8.10.1). The plot shows that the sensor output has first-order system response characteristics, and it takes the output several minutes to teach 100% of its final value. Integrated Case Study III.C discusses the determination of dynamic model parameters from step response data.

Figure 7.10

Thermal response of a commercial IC temperature sensor—LM35C

(Courtesy of Texas Instruments Incorporated, Dallas, TX)

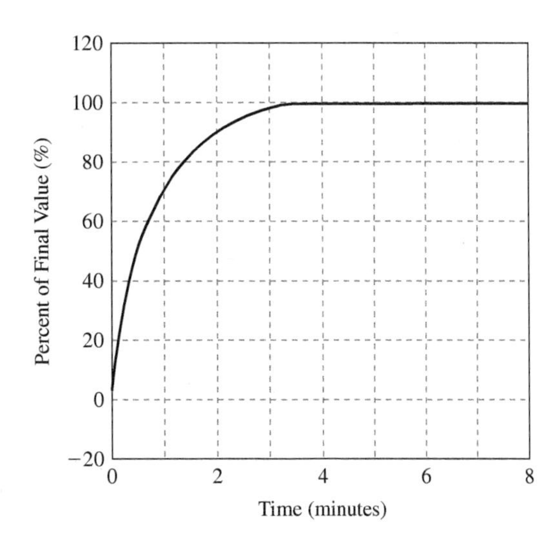

Integrated Case Study III.C: System Model Parameters from Step Response Data

For Integrated Case Study III (Temperature-Controlled Heating System—see Section 11.4), the system uses a copper plate that is heated by a resistance-type heater whose heating output is controlled by the magnitude of the DC input voltage supplied to it (12 V generates 10 W). An IC temperature sensor (LM35C—see Section 8.10.1) is used to measure the temperature of the plate. A step input voltage (or heat input) is used to identify the dynamic model of the plate. First, using the Conservation of Heat Energy principle and Newton's law of cooling [22], a basic model of the copper plate (excluding radiation effects) is

(1)
$$RC\frac{dT}{dt} = T_a - T + Rq$$

where T = plate temperature, T_a = ambient temperature, q = heater output (W), C = thermal capacitance, and R = convective resistance. This is a first-order system whose time constant is given by RC.

The solution is (assuming that $T(0) = T_a$)

(2)
$$T(t) = T_a + Rq\left(1 - e^{-t/RC}\right)$$

The parameters R and C can be determined experimentally by analyzing the temperature response of the plate to a given heat input. The dashed line in Figure 7.11 shows the temperature response of the plate over 1 hour when subjected to a step input of 9 V or $q = 7.5$ W. Using Equation (2), and noting that the exponential term is zero at steady state, the parameter R is obtained from

$$R = \frac{T_{SS} - T_a}{q} = \frac{88 - 29}{7.5} = 7.9\ °\text{C/W}$$

where T_{SS} is the steady state temperature. The time constant τ (or RC) is obtained from differentiating Equation (2) and solving for the derivative at $t = 0$. The derivative is also the same as the slope of the temperature versus time plot at $t = 0$. By equating the derivative (Rq/RC) to the slope value ((80 − 30)/(900 − 0)), the time constant is found to be 1.1×10^3 s. The solid line in Figure 7.11 shows a plot of Equation (2) using these values. The model shows a close agreement with the data.

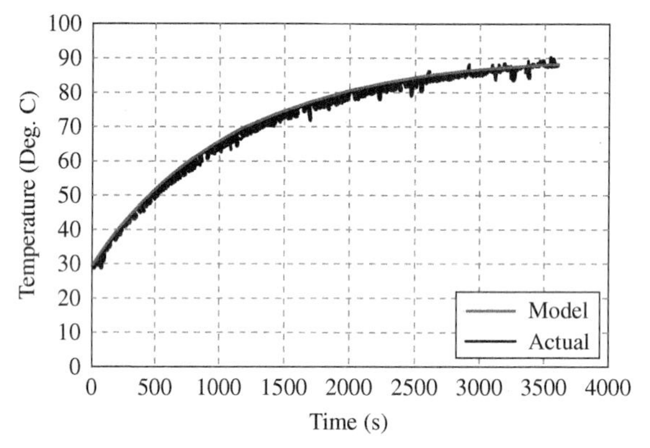

Figure 7.11 Plate response to a step input

7.3 Time Response of Second-Order Systems

A second-order system is one whose dynamic model is described by a second-order differential equation. Examples of **second-order systems** include the model of the position of a mass subjected to a force input in a mass-spring-damper system, the model of a voltage in an RLC circuit, or the model of the height of a fluid in a coupled two-tank system. Example 7.5 illustrates the derivation of a model of a second-order system.

Example 7.5 Model of an RLC Circuit

Obtain a dynamic model of the voltage across the capacitor for the RLC circuit shown in Figure 7.12, which consists of a resistor R, an inductor L, and a capacitor C connected in series. Assume that the voltage input to the circuit is v_S. Note that RLC circuits represent many devices including radio receivers, induction heaters, and transmission lines. Let the voltage across the capacitor be denoted by v_C and the current through the circuit be denoted by i.

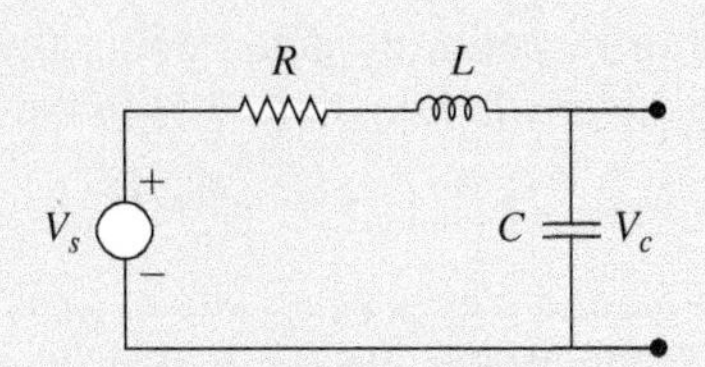

Figure 7.12 RLC Circuit

Solution:

Applying Kirchhoff's voltage law (KVL) (see Section 2.4) around the circuit and using Equation (2.4) for the voltage across an inductor gives

$$v_s - iR - L\frac{di}{dt} - v_c = 0 \quad (1)$$

But from Equation (2.3), the current in the capacitor and the voltage across it are related by

$$i = C\frac{dv_c}{dt} \quad (2)$$

Substituting the current *i* from Equation (2) and its derivative in Equation (1), we get

$$v_s - RC\frac{dv_c}{dt} - LC\frac{d^2v_c}{dt^2} - v_c = 0 \quad (3)$$

By rearranging terms, and dividing every term by C, we get

$$L\frac{d^2v_c}{dt^2} + R\frac{dv_c}{dt} + \frac{v_c}{C} = \frac{v_s}{C} \quad (4)$$

This is a second-order differential equation that relates the voltage across the capacitor to the supply voltage v_s.

Like a first-order system, a second-order system response has free and forced components, and the details of its response are dependent on the roots of the characteristic equation of the system model. The **characteristic equation** is an algebraic equation derived from a given linear differential equation in which the coefficient and power of the variable in each term correspond to the coefficient and order of a derivative in the original equation. A first-order system has just one root, while a second-order system has two roots. Equation (7.4) is the differential equation model for the dynamics of a system that has a mass *m*, a spring with a stiffness *k*, and a viscous damper with a viscous damping coefficient *c* subjected to a force input *f* (see Figure 7.13), but any second-order system can be represented by such a model (see the model for the RLC circuit in Example 7.5 which has the same structure).

Figure 7.13

Mass-spring-damper system subjected to a force input

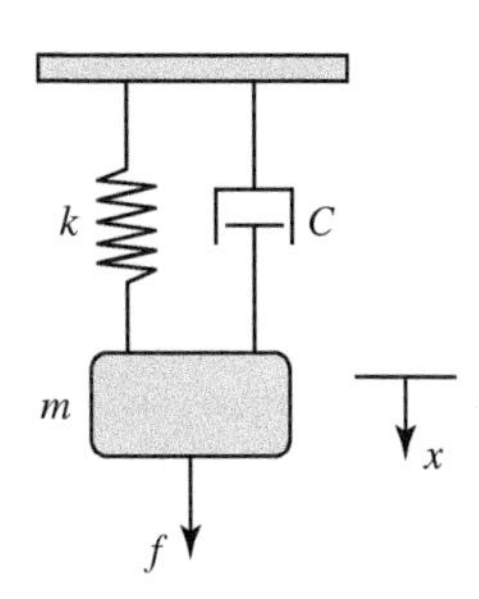

$$m\ddot{x} + c\dot{x} + kx = f(t) \quad \textbf{(7.4)}$$

The **characteristic equation** of this system is

$$ms^2 + cs + k = 0 \quad \textbf{(7.5)}$$

Equation (7.5) has two roots. These are

$$s_{1,2} = \frac{-c \pm \sqrt{c^2 - 4mk}}{2m} \quad \textbf{(7.6)}$$

The **roots** can be real or imaginary, depending on the sign of the quantity under the square root. They also can have positive or negative real parts. When the root has a positive real part, the response of the system is unbounded or unstable, while a non-zero imaginary part means that the response of the system will be oscillatory. For the case of **stable response** (all the roots lie in the negative half of the real-complex plane), we can characterize the type of response by determining the damping ratio of the characteristic equation. To do this, we write the characteristic Equation (7.5) in the form shown in Equation (7.7).

(7.7)
$$s^2 + 2\zeta\omega_n s + \omega_n^2 = 0$$

This is obtained by letting $k/m = \omega_n^2$ and $c/m = 2\zeta\omega_n$, where ω_n is defined as the natural frequency and ζ is the damping ratio. The **natural frequency** is the oscillation frequency at which the system naturally oscillates or vibrates when disturbed from its equilibrium position under the absence of damping. It is an inherent property of the system. The **damping ratio** is defined as the ratio of c/c_c, where c_c is the critical damping, which is the value of the damping that causes the roots of the characteristic Equation (7.5) to have two repeated real roots or the term under the square root in Equation (7.6) to be exactly zero.

Using the damping ratio ζ to characterize the response (available—V7.2) (or $x(t)$ in this case), we can have three cases.

Underdamped Case ($0 < \zeta < 1$) In this case, the roots will have an imaginary part and the free response, or the forced step-input response, will be oscillatory. Figure 7.14 shows the free response, and Figure 7.15 shows the forced step response, respectively, for various values of ζ. The plots are shown for the case $\omega_n = 1$ rad/s, $m = 1$ kg, and $F = 1$ N.

Critically Damped Case ($\zeta = 1$) The roots in this case will be two repeated roots with no imaginary component. The response will not be oscillatory. This case gives the fastest response without any oscillation, and it is very desirable for a system to have such a response.

Overdamped Case ($\zeta > 1$) The roots will also be real in this case, but not repeated. The response will not be oscillatory and will be slower than the critically damped case.

Another case is when $\zeta = 0$ or there is no damping in the system. The roots are purely imaginary in this case and have no real component. The free response in this case will be oscillatory but will not decay with time, unlike the three previous cases. The system in this case is called **neutrally stable**.

Figure 7.14

Free response of a second-order system for various values of ζ

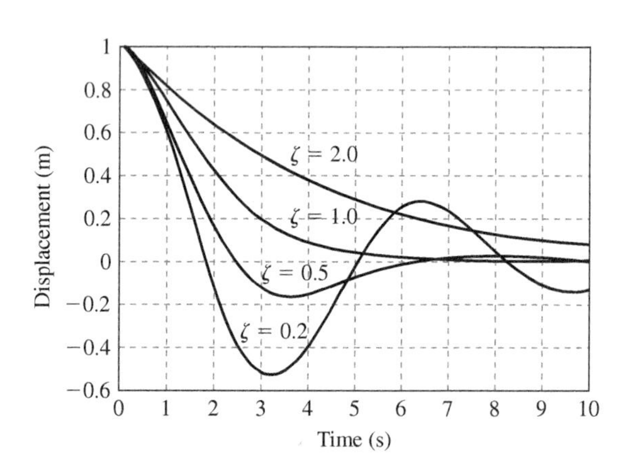

Figure 7.15

Forced response of a second-order system under a unit step input for various values of ζ

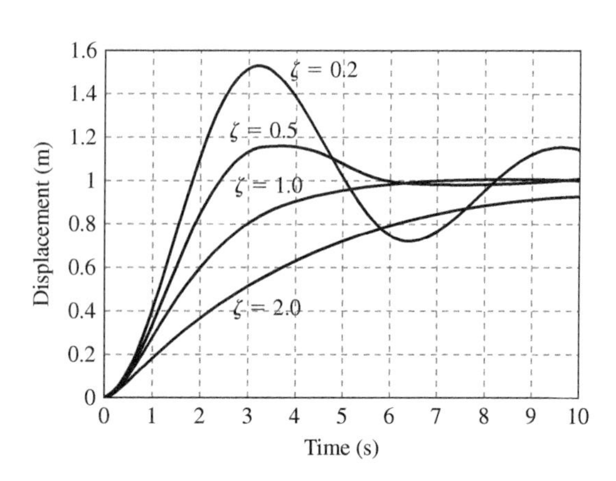

Figure 7.16 gives a graphical interpretation of the location of the root of a second-order system. For $0 < \zeta < 1$, the root has real and imaginary components. The vector from the origin to the root location has a length of ω_n and makes an angle of θ with the negative real axis. The real component magnitude is given by $\zeta\omega_n$. Note that $\zeta = \cos(\theta)$. Thus, when ζ is 1, θ is zero and the root has no imaginary component; when $\zeta = 0$, θ is 90° and the root has no real component. The imaginary component magnitude is equal to $\omega_d = \omega_n\sqrt{1-\zeta^2}$, where ω_d is the damped natural frequency. The **damped natural frequency** ω_d is smaller than the natural frequency ω_n, and it is the frequency that one can observe for a real system that undergoes oscillation, as the natural frequency represents the frequency at which a system would oscillate if there were no damping and is not observable for real systems which have damping. Note that the further away the root is from the real axis, the higher the damped frequency of oscillation will be. Similar to first-order systems, the time constant is also defined for the response of a second-order system subjected to a step input. The time constant is determined from the reciprocal of the real root location. For under-damped and critically damped systems, there is only one time constant, since for both cases, there is one value for the real location of the roots. For over-damped systems, we have two different time constants, each corresponding to the two distinct real roots of the system. The further the root is away from the imaginary axis, the smaller the time constant is, and the faster the response would be due to that root.

Figure 7.16

Graphical interpretation of root location

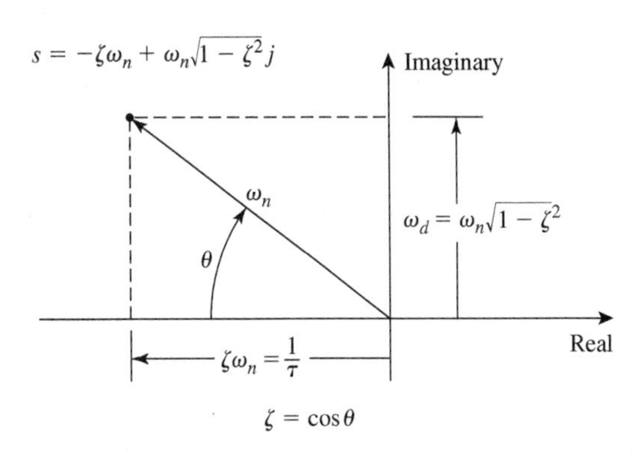

Table 7.1 lists the time response parameters of first- and second-order systems.

Table 7.1

Time response parameters for first and second-order systems

First-Order System	
Model	$\dot{v} + av = f(t)$
Characteristic equation	$s + a = 0$
Time constant (s)	$\tau = 1/a$
Second-Order System	
Model	$m\ddot{x} + c\dot{x} + kx = f(t)$
Characteristic equation	$ms^2 + cs + k = 0$ or $s^2 + 2\zeta\omega_n s + \omega_n^2 = 0$
Damping ratio	$\zeta = \frac{c}{2\sqrt{km}}$
Natural frequency (rad/s)	$\omega_n = \sqrt{k/m}$
Damped natural frequency (rad/s)	$\omega_d = \omega_n\sqrt{1-\zeta^2}$
Time constant (s)	$\tau = \frac{1}{\zeta\omega_n}$ for $\zeta \leq 1$ For $\zeta > 1$: $\tau_1 = \frac{2m}{c - \sqrt{c^2 - 4mk}}$ $\tau_2 = \frac{2m}{c + \sqrt{c^2 - 4mk}}$

While our discussion has been limited to first and second-order systems, it should be noted that for any linear time-invariant dynamic system, the overall response of the system is influenced by the individual contributions of each pole or root in the characteristic equation. For the system to be stable, it is crucial that all poles are located in the left half of the s-plane (i.e., all poles have negative real parts). Note that the response of a first-order system under a step input does not exhibit oscillation, while an underdamped second-order system exhibits oscillation under a step input. If we have a plot of the response of such a second-order system, its parameters can be obtained from analyzing the response plot of the system. Figure 7.17 shows a typical plot and a few of the performance characteristics of the plot. These include the **percentage overshoot,** the **peak time,** and the **2% settling time**; these are defined on the right side of Figure 7.17. Example 7.6 illustrates how to obtain model parameters from the step response plot.

Figure 7.17

Performance characteristics of a second-order system

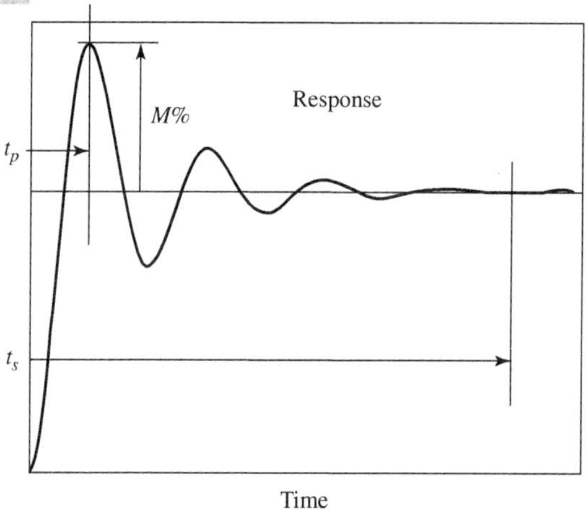

Maximum percent overshoot:

$$M\% = 100 e^{-\pi\zeta/\sqrt{1-\zeta^2}}$$

$$A = \ln \frac{100}{M\%}, \ \zeta = \frac{A}{\sqrt{\pi^2 + A^2}}$$

Peak time:

$$t_p = \frac{\pi}{\omega_n \sqrt{1-\zeta^2}}$$

Settling time (2%):

$$t_s = \frac{4}{\zeta \omega_n}$$

Example 7.6 Obtaining Model Parameters from the Step Response Plot

Determine the dynamic model of a second-order system whose response to a step input of magnitude equal to 10 N is shown in Figure 7.18.

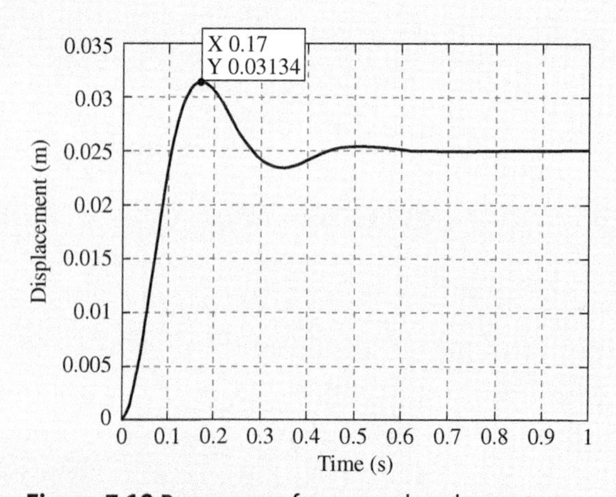

Figure 7.18 Response of a second-order system

Solution:

We will first determine the damping ratio for this system using the maximum overshoot data from the plot. Using the peak and the steady displacement values of 0.0313 m and 0.025 m, respectively, the maximum percent overshoot is

$$M\% = \frac{0.0313 - 0.025}{0.025} \times 100\% = 25.2\%$$

Using $M\%$, we will next compute A and ζ using the expressions shown in Figure 7.17. This gives A to be 1.38 and ζ to be 0.402. Using the peak time of 0.17 second, we then compute ω_n using the expression for t_p given in Figure 7.17. This gives

$$\omega_n = \frac{\pi}{0.17\sqrt{1 - 0.402^2}} = 20.2 \text{ rad/s}$$

From Equation (7.4), the steady state displacement is given by F/k since the acceleration and velocity are zero at steady state. Since F is 10 N and the steady state displacement is 0.025 m, k will be 400 N/m. From $\omega_n = \sqrt{k/m}$, we obtain m as 0.98 kg. Finally, the damping coefficient c is obtained from the expression for the damping ratio shown in Table 7.1. This gives c to be 15.9 Ns/m. Note that if we are not given the magnitude of the input that generated this response, then we can only determine the damping ratio and the natural frequency for the system, but not the parameters m, k, and c.

7.4 Transfer Functions

In analyzing the response of dynamic systems, especially for frequency response analysis (which is covered in the next section), the transfer function form of the dynamic model of a system is used instead of the differential model. The **transfer function** (▶ available—V7.3) of a system is defined as the ratio of the output to the input of the system under zero initial conditions and is expressed as the ratio of two polynomials in the parameter s. Thus, the transfer function $G(s)$ between the input signal U and the output signal Y is written as:

$$G(s) = \frac{Y(s)}{U(s)} = \frac{b_0 s^m + b_1 s^{m-1} + \ldots + b_m}{a_0 s^n + a_1 s^{n-1} + \ldots + a_n} \tag{7.8}$$

where $m \leq n$. The transfer function relationship can be graphically represented using block diagram concepts. A block diagram of the transfer function given by Equation (7.8) is shown in Figure 7.19. In a **block diagram**, the dynamics of the system are represented by the transfer function block, and the input and output of the system are indicated by the direction in which the arrows point.

Figure 7.19

A block diagram representation of a transfer function

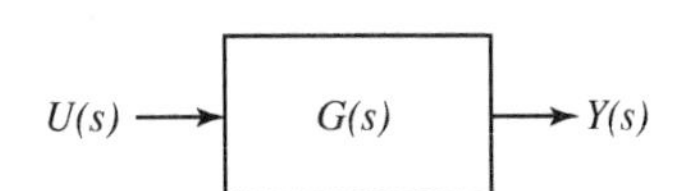

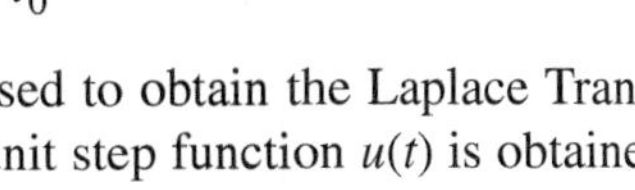

For linear systems with zero initial conditions, the transfer function of a differential equation model is obtained using the Laplace Transform, where the **Laplace transform of a function $f(t)$** is defined as:

$$F(s) = \mathcal{L}(f(t)) = \int_0^\infty f(t)\, e^{-st}\, dt \tag{7.9}$$

where s is a complex number. This definition can be used to obtain the Laplace Transform of given time functions. For example, the Laplace Transform of the unit step function $u(t)$ is obtained as:

$$F(s) = \int_0^\infty f(t)\, e^{-st}\, dt = \int_0^\infty 1e^{-st}\, dt = \left[-\frac{1}{s}e^{-st}\right]_0^\infty = \frac{1}{s} \tag{7.10}$$

The Laplace Transform (▶ available—V7.4) is usually covered in detail in textbooks on system dynamics or feedback control (see for example [22] or [25]), so the details of it will not be covered here. Appendix C has a brief listing of some of the important properties and tables for the Laplace Transform. We will only point out certain aspects of it as it relates to converting differential equations into algebraic equations. One such aspect is the **derivative property**, which gives an expression for the Laplace Transform of the time derivative of a function of time. For the first time derivative, the property is:

$$\mathcal{L}\left(\frac{df(t)}{dt}\right) = s\mathcal{L}(f(t)) - f(0) = sF(s) - f(0) \tag{7.11}$$

and for the second time derivative, the property is:

(7.12)
$$\mathcal{L}\left(\frac{d^2f(t)}{dt^2}\right) = s^2F(s) - sf(0) - \dot{f}(0)$$

As an illustration of this method, let us consider the second-order differential model given by Equation (7.4). Applying the Laplace Transform to Equation (7.4), and using Equations (7.11) and (7.12) with the assumption of zero initial conditions, we obtain the following equation:

(7.13)
$$F(s) = ms^2\,X(s) + cs\,X(s) + k\,X(s)$$

Factoring $X(s)$ from the right part of Equation (7.13), the transfer function between $X(s)$ and $F(s)$ is given as

(7.14)
$$\frac{X(s)}{F(s)} = \frac{1}{ms^2 + cs + k}$$

Example 7.7 illustrates the derivation of the transfer function for a first-order system.

Example 7.7 Transfer Function for a System

For the system model given in Example 7.2,

(a) Derive the transfer function for the model under the assumption of zero initial conditions.
(b) If the applied voltage $v_s(t)$ is a step type of input, obtain an expression for the voltage across the capacitor in terms of the Laplace variable s.

Solution:

(a) We obtained the following dynamic model in Example 7.2

$$RC\frac{dv_c}{dt} + v_c = v_s \quad (1)$$

Applying the Laplace Transform to both sides of Equation (1), we get

$$RCs\,V_c(s) - RC\,v(0) + V_c(s) = V_s(s) \quad (2)$$

Since transfer functions are obtained under the assumption of zero initial conditions, we set the second term in Equation (2) to zero. Solving for the ratio of $V_c(s)$ to $V_s(s)$, we get

$$\frac{V_c(s)}{V_s(s)} = \frac{1}{RCs + 1} \quad (3)$$

Note that for a first-order system, the transfer function denominator polynomial has a power of 1, while for a second-order system, the denominator polynomial will have a power of 2, and so forth.

(b) From Equation (7.10), the Laplace Transform for a step input of magnitude V is given by

$$\mathcal{L}(v_s(t)) = V_s(s) = \frac{V}{s} \quad (4)$$

Substituting this result in Equation (3), and solving for $V_c(s)$, we get

$$V_c(s) = \frac{V}{s(RCs + 1)} \quad (5)$$

7.5 Frequency Response

7.5.1 Frequency Response Plots

In addition to the time response, one can study the response of a system as a function of the frequency of the input signal. This is called the **frequency response** (▶ available—V7.5) of a system, and it is defined as the response of the system to a sinusoidal input signal in which the amplitude of the input signal remains constant, but its frequency is varied. The frequency response is important in many applications including system dynamics, control systems, and communications. An example application that will be discussed later in this chapter is the design of filters that allow the passing of signals with desired frequencies, but block or reduce the effect of signals of unwanted frequencies. The frequency response is normally shown as two semilog plots: a magnitude plot and a phase plot. They are commonly known as the **Bode plots**. In the **magnitude plot,** the ratio of the output to the input is plotted as a function of frequency using a logarithmic scale for the frequency axis. The magnitude plot is displayed in units of **dB** (decibels) where 1 dB $= 20 \log_{10}$ (output/input). In the **phase plot,** the angle between the output and the input signals is displayed as a function of frequency using a logarithmic scale for the frequency axis. A logarithmic scale is used instead of a linear scale for the frequency axis to allow a wide range of frequencies to be displayed in the plots. Table 7.2 shows the relationship between the output/input ratio or gain and their decibel values. The table shows that the decibel value for a given gain is the same as the negative decibel value of the inverse of that gain.

Table 7.2

Relationship between the output/input ratio or gain and their decibel values

Output/Input ratio or gain	$20 \log_{10}$(Output/Input ratio) in dB units
0.001	−60
0.01	−40
0.1	−20
$1/\sqrt{2}$ or 0.707	−3
1	0
$\sqrt{2}$ or 1.41	3
10	20
100	40
1000	60

To illustrate the process of generating the frequency response plots, let us consider the transfer function of a low-pass filter (see Section 7.6), which is given by $G(s) = 1/(1 + \tau s)$, where τ is the time constant of the filter. We replace the variable s in the transfer function with the complex term $j\omega$, where ω is the angular frequency in rad/s and j is $\sqrt{-1}$. Performing this, we obtain

$$G(j\omega) = \frac{1}{1 + j\omega\tau} \quad \textbf{(7.15)}$$

To get rid of the complex quantity in the denominator, we multiply the numerator and the denominator of Equation (7.15) by the complex conjugate of $1 + j\omega\tau$, which is $1 - j\omega\tau$. This gives

$$G(j\omega) = \left(\frac{1}{1 + j\omega\tau}\right)\frac{1 - j\omega\tau}{1 - j\omega\tau} = \frac{1 - j\omega\tau}{1 + \omega^2\tau^2} = \frac{1}{1 + \omega^2\tau^2} - \frac{j\omega\tau}{1 + \omega^2\tau^2} \quad \textbf{(7.16)}$$

Since $G(j\omega)$ is a complex quantity, it has a magnitude as well as a phase angle. **The magnitude** or **magnitude ratio** (M) of the transfer function is obtained by taking the square root of the sum of the squares of the real and imaginary parts in Equation (7.16). This gives

$$|M(j\omega)| = \sqrt{\left(\frac{1}{1 + \omega^2\tau^2}\right)^2 + \left(\frac{\omega\tau}{1 + \omega^2\tau^2}\right)^2} = \frac{1}{\sqrt{1 + \omega^2\tau^2}} \quad \textbf{(7.17)}$$

The **magnitude in dB units** (commonly called small m) is given by Equation (7.18)

$$m(j\omega) = 20 \log M = 20 \log\left(\frac{1}{\sqrt{1 + \omega^2\tau^2}}\right) \quad \textbf{(7.18)}$$

The phase angle is obtained from evaluating the inverse tangent of the ratio of the imaginary part to the real part or

$$\varphi = \tan^{-1}\left(\frac{-\omega\tau}{1}\right) = -\tan^{-1}(\omega\tau) \quad \textbf{(7.19)}$$

Equations (7.18) and (7.19) can be plotted as functions of the angular frequency ω to obtain the frequency response plot of the transfer function. These plots are given in Figures 7.20 and 7.21, respectively, where the magnitude plot is displayed in dB units, and τ is set to 1.

Figure 7.20

Magnitude plot of a first-order low-pass filter

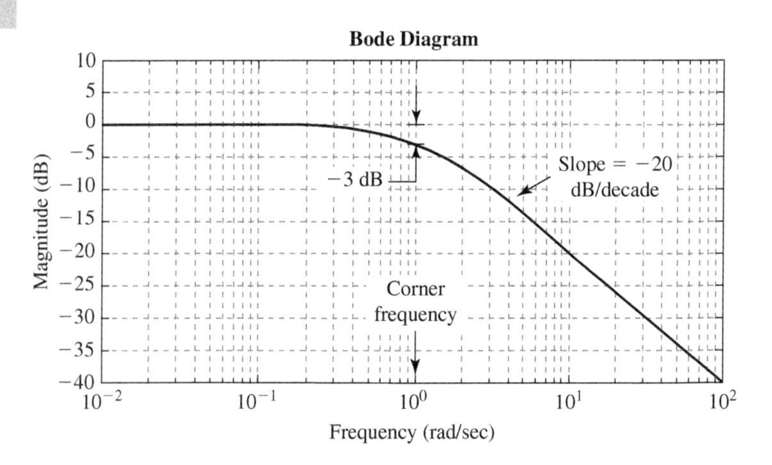

Figure 7.21

Phase plot of a first-order low-pass filter

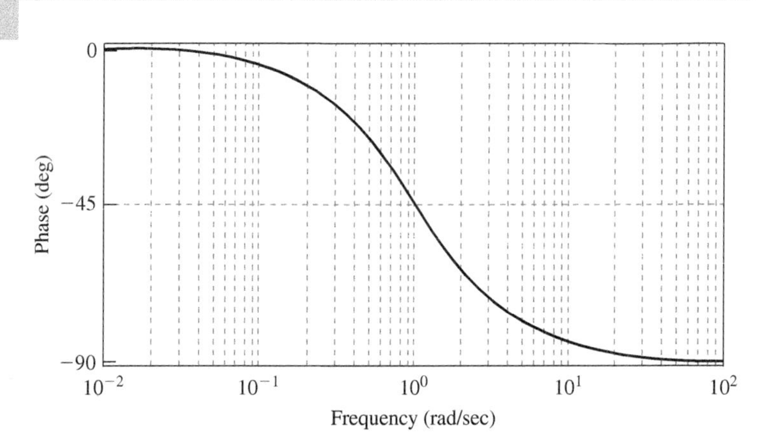

From Equation (7.18), we see that when $\omega\tau$ is very small; the magnitude is approximately equal to 1. Since the logarithm of 1 is zero, the magnitude in dB units is equal to 0 for small values of $\omega\tau$. Similarly, when $\omega\tau$ is very large, Equation (7.18) can be approximated by $(\omega\tau)^{-1}$. Hence, the magnitude in dB is equal to $-20 \log_{10} \omega - 20 \log_{10} \tau$. This is the equation of a straight line if we think of $\log_{10} \omega$ as the independent variable and $\log_{10} \tau$ as a constant. Thus, the magnitude plot in dB will be asymptotic to a straight line with a slope of -20 dB per decade, where a decade is 10 times the increase or decrease in the frequency. When $\omega = 1/\tau$, or is at the **corner frequency**, the magnitude has a value of -3 dB. From Table 7.2, we see that at a gain of -3 dB, the output is approximately 30% less than the input or the output/input ratio is 0.7. Similarly, for the phase plot, we can see from Equation (7.19) that when $\omega\tau$ is very small, its inverse tangent is close to 0°, while when $\omega\tau$ is very large, its inverse tangent is close to $-90°$. A negative **phase angle** means that the output signal lags the input signal, while a **positive phase angle** means that the output signal leads the input signal. The relationship between the phase angle (φ in degrees) and the **lead/lag time** (Δt in seconds) at any particular frequency (ω in rad/s) is given by

(7.20)
$$\Delta t = \frac{\varphi\pi}{180\omega}$$

As an example, at a frequency of 10 rad/s, the output signal from the low-pass filter whose frequency response is shown in Figures 7.20 and 7.21 will lag the input signal by 0.147 seconds.

MATLAB has two commands for generating the frequency response plots, the ***Bode*** and ***Bodemag*** commands. The *Bode* command generates both the magnitude and phase plots, while the *Bodemag* command generates the magnitude plot. Figure 7.22 shows the Bode plot for the second-order system given by Equation (7.14). Notice the hump in the magnitude plot, which shows up in underdamped systems with a damping ratio of less than 0.707. For this second-order system, the magnitude decreases at the rate of 40 dB per decade after the corner frequency (versus 20 dB per decade for a first-order system), and the phase plot approaches $-180°$ for large frequencies (versus $-90°$ for a first-order system). One property of linear systems is that the frequency response of the product of several transfer functions is the same as that obtained by summing the frequency response of each transfer function. This follows from the fact that $\log(ab) = \log(a) + \log(b)$.

Figure 7.22

Frequency response plot for an underdamped second-order system

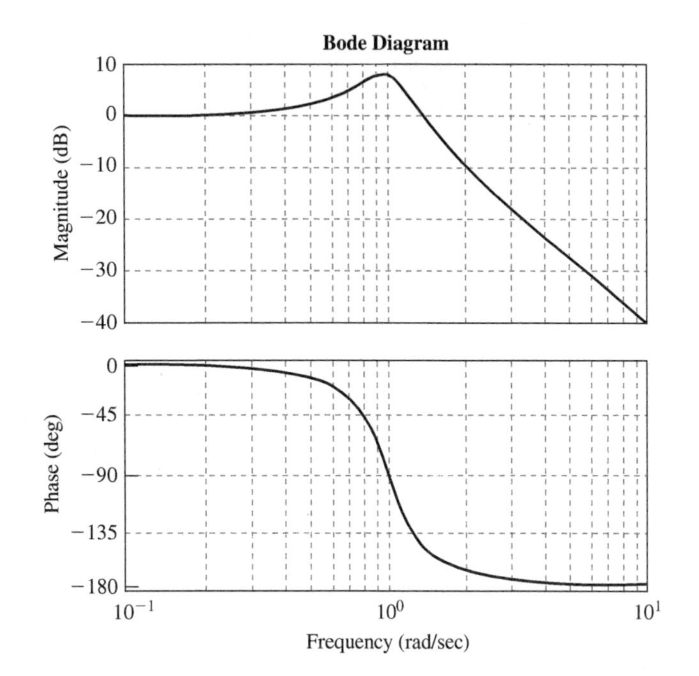

One characteristic of a linear system is that if the input to the system is a sinusoidal signal expressed as $A \sin \omega t$ with a frequency ω and an amplitude A, then the output is also sinusoidal with an amplitude B with the same frequency ω but with a phase shift φ. The relationship between the amplitude of the output and input signals is indicated by the magnitude ratio $M = B/A$, which as discussed before, is the magnitude of the transfer function between the output and the input signals.

As an example of the frequency response plot for a commercial device, Figure 7.23 shows the magnitude frequency response curves for the instrumentation amplifier (IA128) that was discussed in Section 2.9.5. Figure 7.23 shows that as the gain of the amplifier increases, the range of frequencies at which the device can maintain these gains at a constant value decrease. Example 7.8 illustrates the use of frequency response data in determining the system output.

Figure 7.23

Instrumentation amplifier gain versus frequency plot (Courtesy of Texas Instruments Incorporated, Dallas, TX)

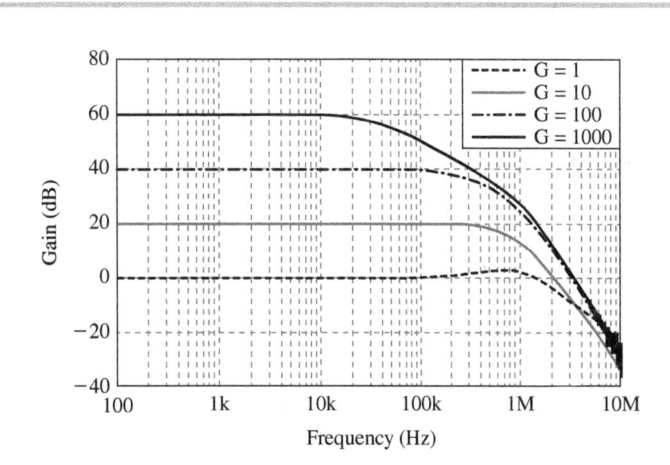

Example 7.8 System Output from Frequency Response Plot

For the second-order system given by the transfer function shown in Equation (1), determine the output of the system if the system is subjected to a sinusoidal input with a frequency of 10 rad/s with an amplitude of 10 N. Let m = 1 kg, $\zeta = 0.4$, and $\omega_n = 6$ rad/s.

$$\frac{X(s)}{F(s)} = \frac{1/m}{s^2 + 2\zeta\omega_n s + {\omega_n}^2} \tag{1}$$

Solution:

A bode plot for this transfer function is shown in Figure 7.24. At the specified frequency of 10 rad/s, the magnitude in dB (small m) is −38.1. Using the definition of m, we have

$$m = -38.1 = 20 \log M \quad (2)$$

Solving Equation (2) for M, we get $M = 0.0124$. But M is the ratio of the output to the input. Since the input amplitude is 10 N, the output will be $0.0124 \times 10 = 0.124$ m. Alternatively, one could have obtained the answer without obtaining a frequency plot by substituting $s = j\omega$ in Equation (1) and evaluating the magnitude of the complex ratio at $\omega = 10$ rad/s.

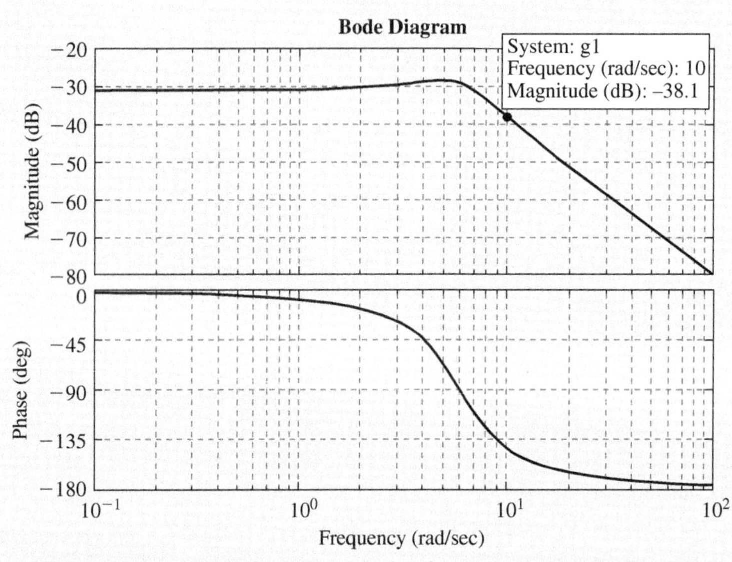

Figure 7.24 Bode plot for the transfer function given by Equation (1)

7.5.2 Resonance

In the magnitude plot shown in Figure 7.22, we see that the magnitude increases as the frequency is increased from 0.1 rad/s, reaching a peak at a frequency close to 1 rad/s. This behavior is called resonance and it shows up in systems with low damping. **Resonance** (available—V7.6) is a phenomenon that occurs when a dynamic system such as a mass-spring-damper system is driven at its natural frequency or a close multiple of it. It is characterized by a significant increase in the system's response, typically resulting in a large amplitude or magnitude. To quantify this resonance behavior, let us start with the transfer function between displacement and force for the mass-spring-damper system given by Equation (7.14). If we multiply the numerator by k, we obtain a dimensionless transfer function $T(s)$ given by Equation (7.21)

(7.21)
$$T(s) = \frac{kX(s)}{F(s)} = \frac{k}{ms^2 + cs + k}$$

This transfer function can be written in terms of the damping ratio ζ and the natural frequency ω_n as

(7.22)
$$T(s) = \frac{kX(s)}{F(s)} = \frac{k/m}{s^2 + \frac{c}{m}s + k/m} = \frac{kX(s)}{F(s)} = \frac{\omega_n^2}{s^2 + 2\zeta\omega_n s + \omega_n^2}$$

Substituting $s = j\omega$ into Equation (7.22) and plotting the magnitude ratio m in dB as a function of the frequency ratio $r = \omega/\omega_n$ gives the plot shown in Figure 7.25. The plot shows that the magnitude ratio m becomes very large when the frequency ratio r approaches 1, or the forcing frequency is close to the natural frequency, for cases where the damping ratio is small. This resonance behavior is not just limited to mechanical systems; resonance can occur in various physical systems including mechanical, electrical, and acoustic systems. In electrical systems, resonance occurs, for example, in series-tuned and parallel-tuned circuits comprised of inductors and capacitors, while in acoustic systems, it occurs in acoustic instruments. In general, resonance is not desirable to have in many physical systems, and in designing mechanical systems, it is essential to ensure that the system's natural frequency does not match the frequency of expected external disturbances. For example, motor-driven systems can introduce at resonance large vibrations to a support structure if they are unbalanced and are operated at a speed close to the natural frequency of the structure. However, there are several applications where resonance is useful. One of these is the design of oscillators (see Section 4.4.4) that provide timing signals for microcontrollers and similar devices.

Figure 7.25

Frequency response as a function of the frequency ratio $r = \omega/\omega_n$

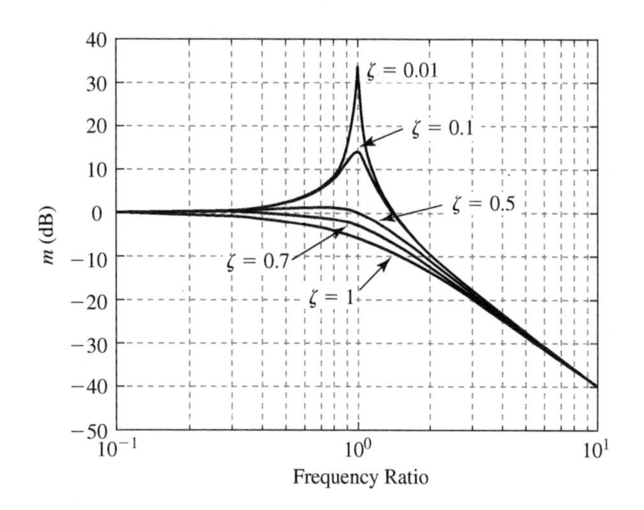

The resonance frequency and the magnitude ratio at resonance can be determined from the formulas given in Table 7.3 for the mass-spring-damper system. It should be noted that these formulas are obtained when the transfer function of the system is written in the form of Equation (7.22).

Table 7.3

Resonance frequency and magnitude ratio for a mass-spring-damper system

System transfer function	$\frac{k}{ms^2 + cs + k} = \frac{\omega_n^2}{s^2 + 2\zeta\omega_n s + \omega_n^2}$
Natural frequency	$\omega_n = \sqrt{k/m}$
Damping ratio	$\zeta = \frac{c}{2\sqrt{km}}$
Resonance frequency	$\omega_r = \omega_n\sqrt{1 - 2\zeta^2}$ *for* $0 \le \zeta \le 0.707$
Magnitude ratio at resonance	$M_r = \frac{1}{2\zeta\sqrt{1 - \zeta^2}}$ *for* $0 \le \zeta \le 0.707$

7.5.3 Bandwidth

We can use the frequency response plot to obtain a measure of the range of frequencies to which a system such as a sensor or actuator is especially responsive. This measure is called **bandwidth** (▶ available—V7.7), and it is formally defined as the range of frequencies over which the power transmitted or dissipated by a system is no less than one-half of the peak power in the system [22]. In practice, the bandwidth is obtained by drawing a horizontal line through the magnitude frequency response plot that is 3 dB below the peak magnitude ratio. The intersection points of this line with the plot define the lower and upper bounds of the bandwidth frequency range of the system as shown in Figure 7.26.

Figure 7.26

Illustration of the definition of bandwidth

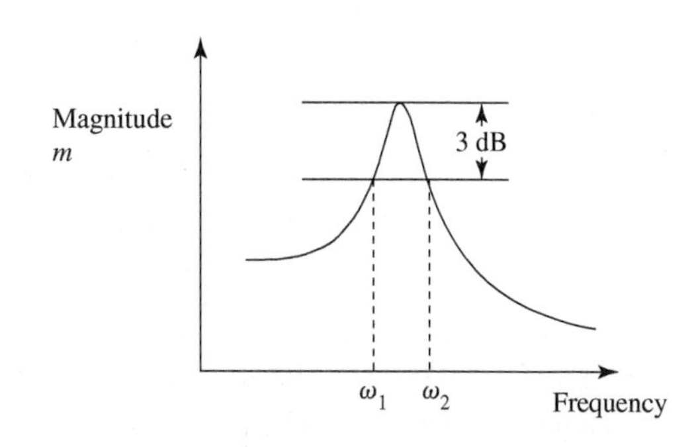

Designate the lower frequency as ω_1 and the higher frequency as ω_2. The bandwidth B is

(7.23)
$$B = \omega_2 - \omega_1$$

The frequencies ω_1 and ω_2 are known as the **half-power points** of the system since the power transmitted by the system is only half the peak power in the system. Note at a -3 dB magnitude ratio, the output of the system is only 70.7% of the input signal magnitude (see Table 7.2). Many measurement systems and low-pass filters (see Section 7.6) do not have a magnitude frequency response that looks like Figure 7.26. In many cases, such as the response shown in Figure 7.20, the lower frequency ω_1 cannot be determined since the peak magnitude occurs at low frequencies. For this case, the bandwidth is just the frequency ω_2 or the frequency at which the response drops by -3 dB from the low-frequency value. For the plot in Figure 7.20, the bandwidth is 1 rad/sec.

A system with a wider bandwidth can respond to a wider range of frequencies, while a system with a narrower bandwidth is more restricted in the frequencies it responds to. Bandwidth also affects the time response of a system, and in general, there is an inverse relationship between the time constant and bandwidth of a system. For a simple first-order system or filter, such as the low-pass filter discussed in Section 7.5.1, the bandwidth (B in rad/s) is related to the time constant (τ in s) by the formula

(7.24)
$$B = 1/\tau$$

It is important to note that this relationship holds for simple first-order systems or filters. In more complex systems, such as higher-order systems or systems with resonant behavior, the relationship between the time constant and bandwidth becomes more intricate and may not follow the same formula. Note also that when two systems are connected in series, their effective bandwidth is governed by the one with lower bandwidth.

Many manufacturers of sensors and amplifiers provide the bandwidth data for their products. Typically, the frequency response plot for the sensor or the amplifier looks like that of the first-order filter shown in Figure 7.20. For example, the driver (Model CPL190/CPL290) for a capacitive-type non-contact displacement sensor made by Lion Precision has a user-selectable bandwidth that ranges from 100 Hz to 15 kHz. For 15 kHz bandwidth, the frequency response for the driver looks similar to that shown in Figure 7.27. While at this setting, the sensor driver has a bandwidth of 15 kHz. However, it is not recommended to use the sensor up to the bandwidth limit, since at the bandwidth frequency the sensor output will only be 70.7% of the output at low frequency. The input frequency range in which the sensor output will be 100% of the measured quantity is smaller than the bandwidth limit as seen in the plot.

Figure 7.27

Typical sensor driver frequency response plot

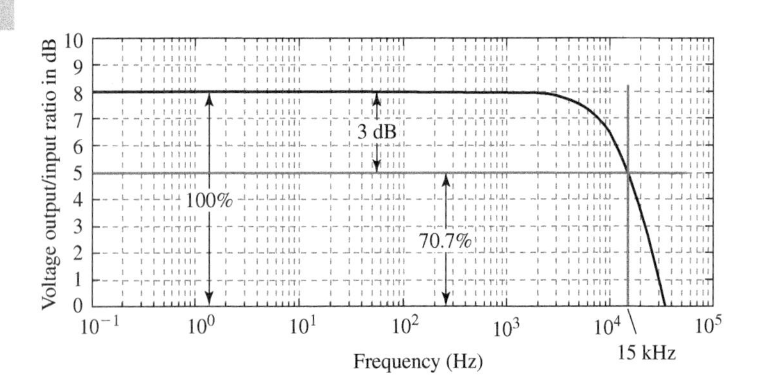

In computer networking, the term bandwidth is also used, but it means a different thing. In **computer networking**, bandwidth refers to the maximum amount of data that can be transmitted over a network connection in a given period. The unit for bandwidth for networking is the number of bits transmitted per second, typically shown as megabits per second (Mbps) or gigabits per second (Gbps).

7.6 Filtering

Signal filtering is a signal processing technique used to modify, extract, or remove specific components or characteristics from a signal. It involves the application of filters to alter the frequency content or other properties of the signal, resulting in a desired output. Filtering can be applied to various types of signals, including audio signals, biomedical signals, image signals, and sensor data, among others. The goal of signal filtering is to enhance the desired information, remove unwanted noise or interference, and/or extract specific features from the signal. This section discusses different types of filters. These include low-pass, high-pass, notch, and bandpass. The names of these filters reflect the specific frequency ranges they are designed to pass.

A filter that is implemented using analog circuit components such as op-amps and capacitors (see Chapter 2) is referred to as an **analog filter**. If a filter is implemented using computer code in a digital signal processor chip, it is referred to as a **digital filter**. Digital filters offer flexibility over analog filters since changing the filter characteristics involves only code change. However, analog filters are inexpensive and more robust. Analog filters can be classified as passive or active. A **passive filter** does not require any external power to operate. An example of a passive analog filter is the first-order RC-circuit filter. An **active filter**, on the other hand, requires external power to operate. Active filters use components such as op-amps.

The filtering action is best described using frequency response plots and transfer function concepts. The ideal magnitude frequency response characteristics for the low-pass, high-pass, bandpass, and notch filters are shown in Figure 7.28, where f_c is the filter corner frequency. Figure 7.28 shows that over the frequencies at which the filtering is active, the output signal is zero. As such, these ideal frequency characteristics cannot be achieved. First, filters with zero passing are difficult to construct. Second, the sharp transitions shown in the figure cannot be achieved.

Figure 7.28

Ideal magnitude frequency response characteristics of various filters

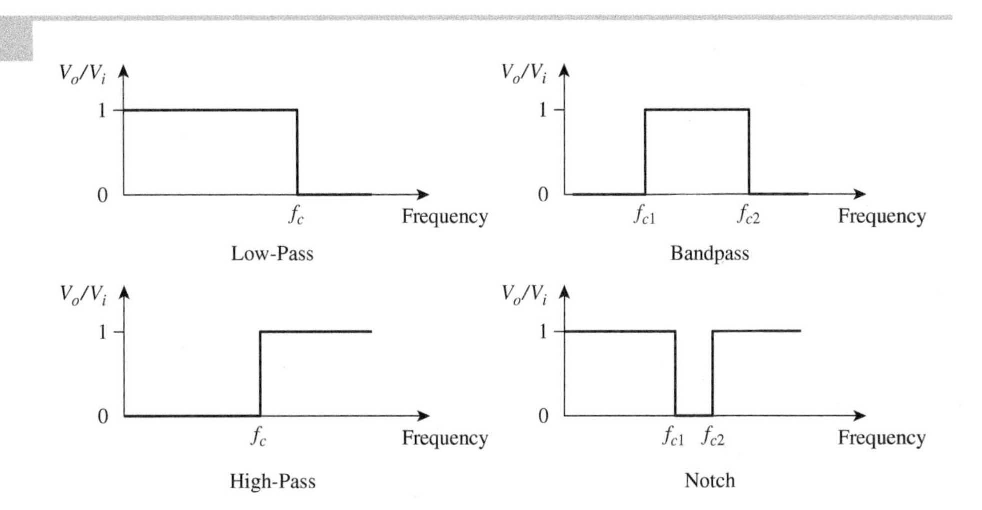

7.6.1 Low-Pass Filters

Low-pass filters (available—V7.8) are used to attenuate frequency signals above the corner frequency f_c such as the 60 Hz interference signals from AC power-operated equipment. These filters are commonly used to process signals from tachometer and temperature sensors (see Chapter 8). There are many forms of low-pass filters, so we will restrict our discussion to the first-order, low-pass filter given by the transfer function

$$H(s) = \frac{1}{\tau s + 1} = \frac{1}{\frac{s}{2\pi f_c} + 1} \quad \textbf{(7.25)}$$

where τ is the time constant of the filter. The **time constant** τ in seconds and the **corner frequency** f_c in Hz are related by

$$f_c = \frac{1}{2\pi\tau} \quad \textbf{(7.26)}$$

The order of the filter refers to the highest power in the denominator of the filter transfer function. A first-order filter is also known as a **single-pole filter** since the characteristic equation of the filter has only one pole or one root. The magnitude and phase frequency response plots for a first-order low-pass filter were discussed in Section 7.5.1 (see Figures 7.20 and 7.21). The magnitude plot shows that the filter output-to-input magnitude has a unity gain before the corner frequency at ω_c, where ($\omega_c = 1/\tau$), but the gain decreases slowly as the frequency approaches the corner frequency. At the corner frequency, the output magnitude is -3 dB with the output at 70.7% of the input. After the corner frequency, the magnitude starts dropping much faster at the rate of 20 dB/decade. The phase plot shows that the filter has nearly zero phase lag at low frequencies. At the corner frequency, the phase angle is -45 degrees, and the phase angle approaches -90 degrees for frequencies much higher than the corner frequency. Example 7.9 illustrates the computation of the filter output at any frequency.

A simple passive low-pass filter can be easily constructed using a resistor and a capacitor circuit as shown in Figure 7.29(a). The filter time constant is the product of the resistance and the capacitance or $\tau = RC$. An active low-pass filter is implemented using op-amps as shown in Figure 7.29(b), where the filter time constant is given as $\tau = R_2C$. Note for the active filter, the filter gain is given by $(-R_2/R_1)$. The attenuation rate of the first-order low-pass filter is, however, not high. It is desirable to have high attenuation rates such as −60 dB/decade. Such an attenuation rate can be achieved by other types of filters such as Butterworth and Chebyshev. The details of these and other filter types can be found in many signal-processing texts such as [17].

Figure 7.29

Circuit for (a) passive low-pass filter and (b) active low-pass filter

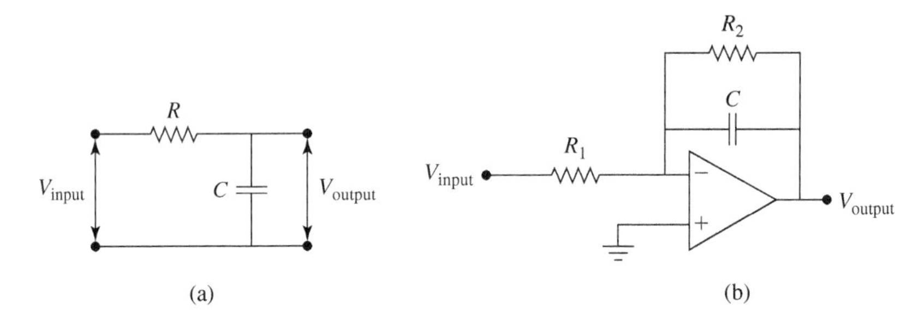

If the sensor data is available digitally, then one can use a digital filter instead. A **digital low-pass** filter can be implemented using Equation (7.27).

(7.27)
$$y(k) = (1 - \alpha)\,x(k) + \alpha\,y(k - 1)$$

where $y(k)$ is the output sequence, $x(k)$ is the input sequence, k is the current index, and α is a factor that is dependent on the filter corner frequency f_c and the sampling time T. Alpha is given by:

(7.28)
$$\alpha = e^{(-2\pi f_c T)}$$

For example, if $f_c = 10$ Hz, and T is 1 ms, then α is 0.9391, and the digital filter equation is

(7.29)
$$y(k) = 0.0609\,x(k) + 0.9391\,y(k - 1)$$

This iterative equation can be easily implemented in code using a *For-Loop* or similar construct. Figure 7.30 shows the filter output for a 10 Hz sinusoidal input signal for three corner frequencies: 1, 5, and 10 Hz, and a sampling time of 1 ms. Notice how the output magnitude is smaller than the input magnitude for all three cases. Moreover, a more significant reduction in the output magnitude is achieved as the ratio of the signal frequency to the corner frequency increases.

Figure 7.30

Digital filter output for different corner frequencies

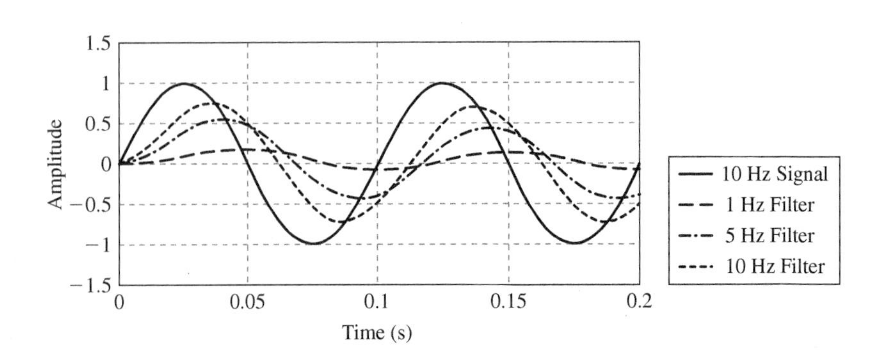

Example 7.9 Low-Pass Filter Output

If the corner frequency of the low-pass, first-order filter is 100 Hz, determine the filter output at 500 Hz.

Solution:

The output is obtainable using two methods. The first is an exact method that is based on plugging the value of the frequency into the transfer function. The second is based on using the slope information.

From Equation (7.25), if we substitute $s = j\omega$, where ω is the circular frequency, we obtain

$$H(s) = \frac{1}{\tau j\omega + 1} = \frac{1 - \tau j\omega}{1 + \tau^2\omega^2}$$

and the magnitude of this complex expression is given by

$$\left|\frac{V_o}{V_{in}}\right| = \frac{1}{\sqrt{1 + \omega^2\tau^2}}$$

Replacing ω with $2\pi(500)$ and τ with $1/(2\pi 100)$ in this expression, we obtain an output-to-input ratio of 0.196, or the output is 19.6% of the input.

Using the slope information, we first determine the number of decades between the 500 Hz and the 100 Hz frequency by taking the logarithm of the ratio of the two frequencies or $\log_{10}(500/100) = 0.699$. A first-order filter has a slope of −20 dB/decade, so at 500 Hz, the filter output is −13.98 dB (0.699 × −20 dB) or 20% of the input. Thus, there is an 80% attenuation at a frequency of 500 Hz. The result is very close to that using the transfer function method.

7.6.2 High-Pass Filters

High-pass filters (▶ available—V7.9) are used to attenuate frequency signals below the corner frequency, f_c. These filters are used in instrumentation, audio systems, or communication systems to attenuate DC and low-frequency components from a signal so the high-frequency components of the signal become more distinguished. Similar to the low-pass case, there are many forms of high-pass filters, so we will restrict our discussion to the first-order, high-pass filter given by the transfer function

$$H(s) = \frac{\tau s}{\tau s + 1} \tag{7.30}$$

where τ is the time constant of the filter. The magnitude and phase frequency response plots of this filter are shown in Figure 7.31, and they are opposite to those of the low-pass filter. The plot shows that the filter has nearly a unity gain above the corner frequency ω_c. Before this frequency, the magnitude increases at the rate of 20 dB/decade.

Figure 7.31

Magnitude and phase for first-order, high-pass filter ($\omega_c = 100$ rad/s)

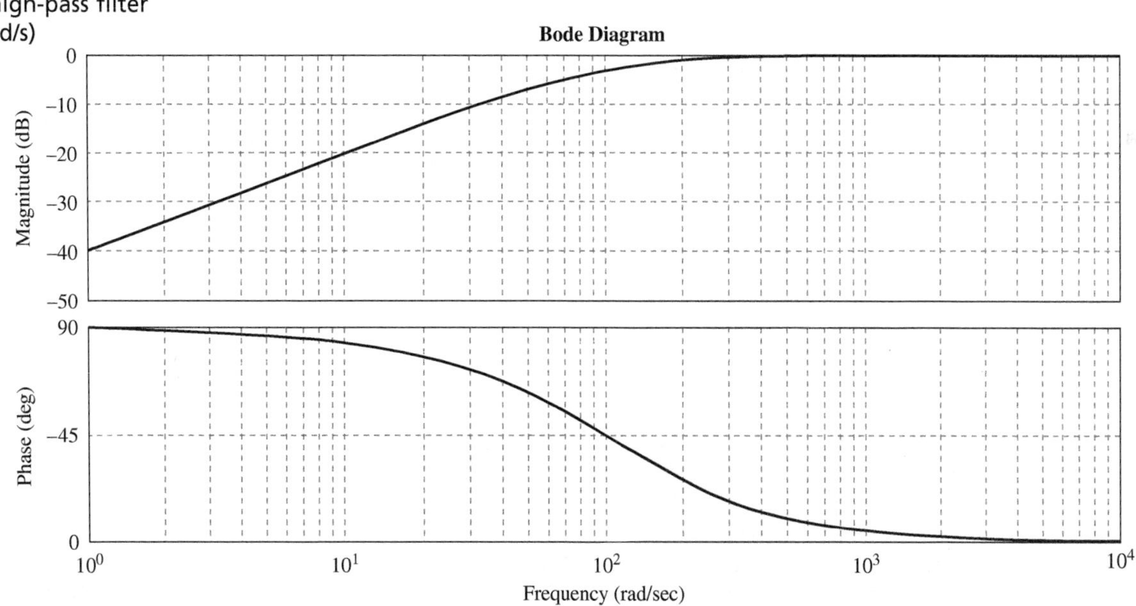

A simple passive high-pass filter can also be constructed using a resistor and capacitor circuit as shown in Figure 7.32. The circuit is similar to the low-pass filter, but the resistor and the capacitor locations are interchanged.

Figure 7.32

Circuit for a first-order, high-pass filter

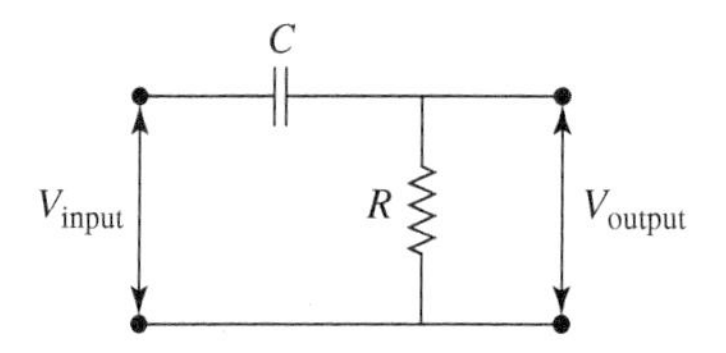

7.6.3 Bandpass Filters

Bandpass filters (▶ available—V7.10) are used to allow signals in a certain frequency range and to attenuate all signals outside of this range. The two corner frequencies that define this frequency range are f_{cl} and f_{ch}. A bandpass filter is a cascade combination of a low-pass and a high-pass filter, where a high-pass filter is followed by a low-pass filter. The transfer function of the filter is

(7.31)
$$H(s) = \frac{\tau_1 s}{(\tau_1 s + 1)(\tau_2 s + 1)}$$

where $\tau_1 = 1/(2\pi f_{cl}) = 1/\omega_{cl}$ and $\tau_2 = 1/(2\pi f_{ch}) = 1/\omega_{ch}$. A passive bandpass filter is shown in Figure 7.33.

Figure 7.33

A passive bandpass filter

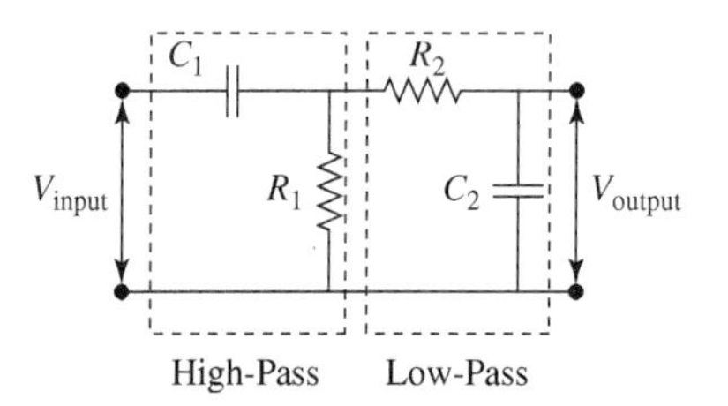

The magnitude and phase frequency response plots of this filter are shown in Figure 7.34. The magnitude plot shows that the filter has nearly a unity gain inside the desired frequency band. Outside this band, the magnitude increases or drops at the rate of 20 dB/decade. The phase angle is close to zero around the center of the pass band, but approaches +/−90 degrees away from the frequency band. The **center frequency** of a bandpass filter is defined as the frequency at which the output gain of the filter is maximum. It is given in Equation (7.32) as the geometric mean value of the lower and upper cutoff frequencies and not the average of the two frequencies. For the filter shown in Figure 7.34, the center frequency is 100 rad/s.

(7.32)
$$\omega_{center} = \sqrt{\omega_{cl}\,\omega_{ch}}$$

Figure 7.34

Magnitude and phase for bandpass filter (ω_{cl} = 10 rad/s, ω_{ch} = 1000 rad/s)

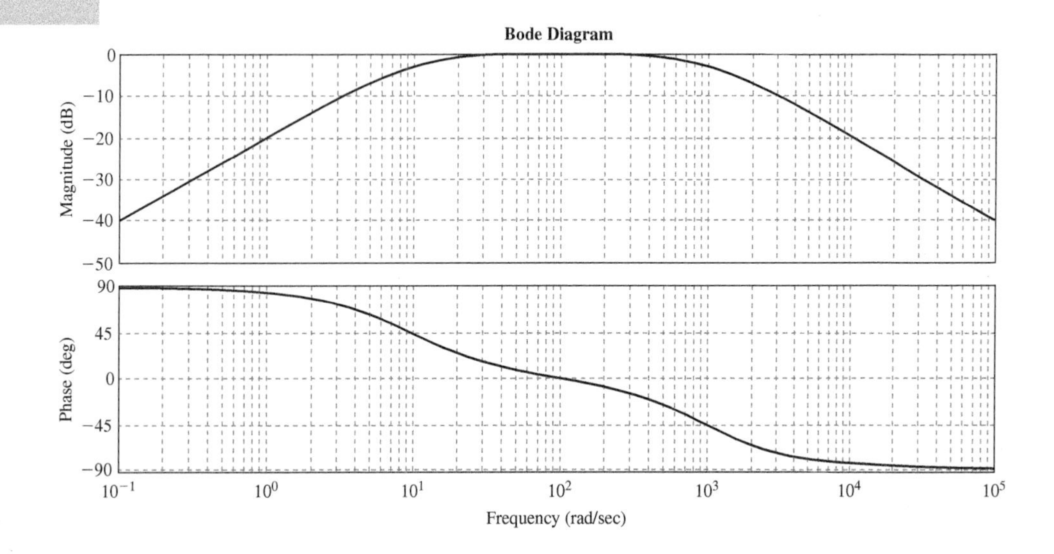

7.6.4 Notch Filters

Notch filters ( available—V7.11) are used to attenuate a narrow band of frequencies from a signal. For example, if we know that the noise is originating at a particular frequency such as 60 Hz, then we can design a filter to eliminate this noise. A notch filter is a second-order system and has the following transfer function:

$$H(s) = \frac{\tau^2 s^2 + 1}{\tau^2 s^2 + 4\tau s + 1} \tag{7.33}$$

where, as before, the time constant τ and the notch frequency f_c are related by Equation (7.26). The magnitude and phase frequency response plots of this filter are shown in Figure 7.35. The magnitude plot shows that the filter has zero output at the notch frequency (10π rad/s), with a sharp roll down and roll up at either side of the notch frequency. The phase plot shows that away from the notch frequency, the phase angle is zero. If the disturbance frequency is different from f_c, then the disturbance frequency is not eliminated. Passive and active-type notch-filters are available. Example 7.10 discusses the use of an RLC circuit as a passive notch filter.

Figure 7.35

Magnitude and phase for a notch filter

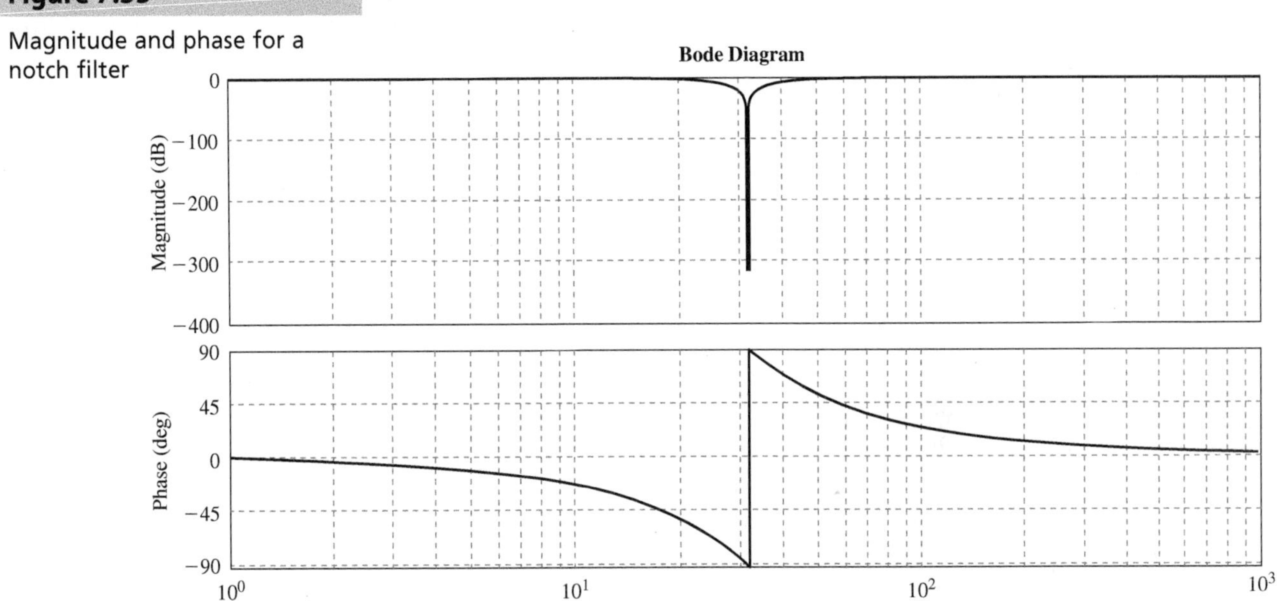

Example 7.10 Notch Filter

For the RLC circuit shown in Figure 7.36, if the combined voltage across the capacitor and the inductor is considered as the filter output, then this circuit acts as a notch filter. For this circuit, (a) show that the transfer function of this filter's output to input is like that given by Equation (7.33), and (b) determine the *R*, *L*, and *C* values to create a filter with a notch frequency close to 360 Hz.

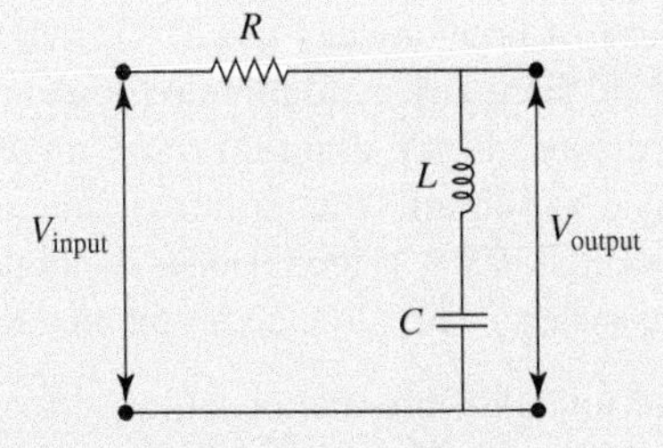

Figure 7.36 RLC circuit used as a notch filter

Solution:

(a) To obtain the filter transfer function, obtain an expression for the current in the circuit, and then use that expression to obtain the output voltage of the filter as a function of the input voltage. Applying Kirchhoff's voltage law (KVL) (see Section 2.4) around the circuit and using Equation (2.4) for the voltage across an inductor gives

$$v_{input} = iR + L\frac{di}{dt} + v_c \tag{1}$$

Differentiating Equation (1) with respect to time, substituting the expression given in Equation (2.3) for dv_c/dt, and applying the Laplace Transform to the resulting equation gives

$$s\,V_{input}(s) = Rs\,I(s) + Ls^2 I(s) + \frac{1}{C}I(s) \tag{2}$$

From Equation (2), the current $I(s)$ is then

$$I(s) = \frac{s\,V_{input}(s)}{Ls^2 + Rs + 1/C} \tag{3}$$

The filter output is given as

$$v_{output} = L\frac{di}{dt} + v_c \tag{4}$$

Performing similar steps to that done to obtain the current, the output voltage in Laplace Transform is then given as

$$V_{output}(s) = (Ls^2 + 1/C)\frac{I(s)}{s} \tag{5}$$

Substituting Equation (3) into Equation (5) gives the transfer function for the filter output as

$$\frac{V_{output}(s)}{V_{input}(s)} = \frac{LCs^2 + 1}{LCs^2 + RCs + 1} \tag{6}$$

This is the same form as Equation (7.33) with $\tau^2 = LC$ and $4\tau = RC$.

(b) From Equation (7.26), setting the notch frequency at 360 Hz gives τ as 4.42×10^{-4} s. Since $\tau^2 = LC$, and selecting C as 10 μF, which is a commonly available ceramic capacitor value, gives L as 19.5 mH. Since 19.5 mH is not a commonly available inductor value, we choose L as 20 mH, which is a commonly available inductor value. Using these values of L and C gives us a modified notch frequency of 355.9 Hz or $\tau = 4.47 \times 10^{-4}$ s. From $4\tau = RC$, we get R as 178.8 Ω. We choose R as 178 Ω which is a standard resistor value in the E96 series (see Appendix A). A MATLAB Bode plot of this filter using the chosen R, L, and C values gave a notch frequency of 356 Hz. Note that to realize a notch filter using this circuit with a much smaller notch frequency such as 20 or 50 Hz requires the availability of inductors and capacitors with large values that are not commonly available.

7.7 MATLAB Simulation of Dynamic Systems

MATLAB is a software package used for modeling and simulating a variety of linear and nonlinear systems in both the continuous and discrete domains. The program is widely used in engineering programs throughout the world, so most of the readers of this textbook should be familiar with it. Here we will provide only a brief overview of how one can use MATLAB to obtain a solution for a dynamic model. Readers should consult any of the many available textbooks on this subject for further reading; see for example [26–27]. We also discuss block diagram representation and simulation in MATLAB.

MATLAB offers several ways to obtain the solution for the set of differential equations (▶ available—V7.12) that are obtained when a model of a dynamic system is obtained. The method depends on the characteristics of the system. The three common methods are listed:

- State space solution methods for linear differential equation systems
- Direct integration using ODE solvers for nonlinear differential equation systems
- Transfer function methods for linear differential equation systems with zero initial condition

A brief outline of each method is included. In addition, block diagram representation and simulation using Simulink are also discussed.

7.7.1 State Space Solution Method

This method applies to linear differential equation systems. The obtained differential equation(s) are represented as a set of n first-order differential equations in the form

$$\dot{x} = Ax + Bu \tag{7.34}$$

and the input-output relationship is represented as

$$y = Cx + Du \tag{7.35}$$

where x is the $n \times 1$ state space vector, u is the $m \times 1$ input vector, A is the $n \times n$ model coefficient matrix, B is the $n \times m$ input coefficient matrix, y is the $q \times 1$ output vector, C is the $q \times n$ output matrix, and D is the $q \times m$ output-input matrix. For single-input single-output systems, $m = q = 1$. With the A, B, C, and D matrices specified, the **state space model** is created with the MATLAB command:

$$sys1 = ss(A,B,C,D)$$

where *sys1* is a user-defined name for the system. The response of the system can then be obtained in several ways. If we are interested in obtaining the response of the system to a pre-defined input signal such as a step signal, then the ***step*** command can be used with the typical calling format:

$$step(sys1)$$

If we would like to obtain the response for a user-defined input vector *uv*, then the ***lsim*** command can be used with the calling format:

$$lsim(sys1,\ uv,\ t)$$

where *uv* is a user-defined input vector specified over the time interval t.

As an illustration of this method, let us consider the differential equation model for the motor-driven geared system obtained in Example 7.3. The equation is:

$$T_m = I_{eff}\,\ddot{\theta}_1 + b_{eff}\,\dot{\theta}_1 \tag{7.36}$$

This is a second-order linear differential equation, and thus we need two state variables to represent this model. Let x_1 be the angular position θ_1, and x_2 be the angular velocity $\dot{\theta}_1$. Then the following two first-order differential equations (Equations (7.37) and (7.38)) are equivalent to Equation (7.36).

$$\dot{x}_1 = x_2 \tag{7.37}$$

and

$$\dot{x}_2 = \frac{1}{I_{eff}}(T_m - b_{eff}x_2) \tag{7.38}$$

If the output of the system is the angular position θ_1, then the state space matrices for this system are:

$$A = \begin{bmatrix} 0 & 1 \\ 0 & \dfrac{-b_{eff}}{I_{eff}} \end{bmatrix},\ B = \begin{bmatrix} 0 \\ \dfrac{1}{I_{eff}} \end{bmatrix},\ C = [1 \quad 0],\ D = [0] \tag{7.39}$$

Letting $I_{eff} = 0.1$ kg m^2, $b_{eff} = 0.2$ N·m s/rad, and $T_m = 1$ N·m, Figure 7.37(a) shows the unit step response obtained by using the command:

$$step(sys1,2)$$

for the state space system defined by Equation (7.39), where 2 is the time duration of the simulation in seconds. Figure 7.37(b) shows the corresponding step response when the angular speed was made as the output of the system or C, in this case, is equal to [0 1].

Figure 7.37

Step response for system defined by Equation (7.39) (a) angular position response and (b) angular velocity response

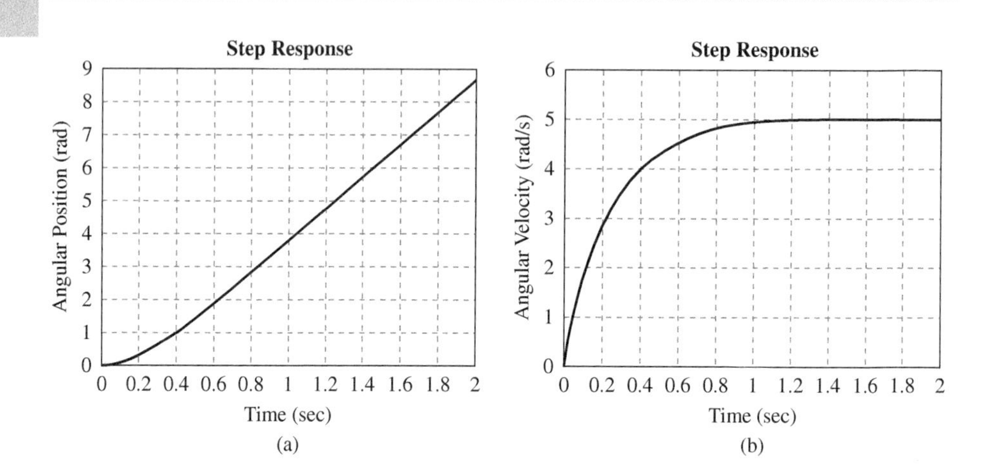

7.7.2 Direct Integration Using ODE Solvers

For nonlinear differential equations, we can obtain the solution by direct integration using any of the ODE solvers in MATLAB. The user needs to define the differential equation model, the period to perform the integration, and the initial conditions before the ODE solver is called. The differential equation model is defined using the *function* script in MATLAB. As an illustration of this process, let us consider the following nonlinear differential equation (Equation (7.40)) which represents the motion of a pendulum.

(7.40)
$$I_O\ddot{\theta} + k\sin\theta = 0$$

This differential equation is represented in the MATLAB function $f(t,y)$ listed in Figure 7.38.

Figure 7.38

MATLAB function for Equation (7.40)

```
function dydt = f(t,y)
dydt = [y(2); -k/IO * sin(y(1))];
```

The function $f(t,y)$ is dependent on time and the vector y. The angular position θ is $y(1)$ and the angular velocity $\dot{\theta}$ is $y(2)$. Note the output argument *dydt* defines a 2×1 vector which gives expressions for the derivatives of y, i.e.,

(7.41)
$$\begin{bmatrix} \dot{y}(1) \\ \dot{y}(2) \end{bmatrix} = \begin{bmatrix} \dot{\theta} \\ \ddot{\theta} \end{bmatrix} = \begin{bmatrix} y(2) \\ -k/I_O \sin(y1) \end{bmatrix}$$

To integrate this problem, the following statements are typed in MATLAB:

```
tspan = [0, 2.5];
y0 = [pi()/4; 0];
[t,y] = ode45(@f, tspan, y0);
```

Where *y*0 is the initial conditions vector, and *ode45* is a differential equations solver based on the use of an explicit Runge-Kutta formula. The *ode45* solver is used to solve ordinary differential equations with high accuracy. MATLAB has other differential equation solvers such as *ode23*, but *ode45* is the preferred solver for most problems since it offers higher accuracy than *ode23*.

Figure 7.39 shows the solution of Equation (7.40) for the following set of parameters:

$$I_O = 0.775 \text{ kg}\cdot\text{m}^2 \text{ and } k = 15.696 \text{ N}\cdot\text{m}$$

and the initial condition $\theta_0 = \pi/4$ and $\dot{\theta}_0 = 0$. Note that due to the lack of friction, the pendulum keeps oscillating between $\pi/4$ and $-\pi/4$. Note also how the pendulum angular velocity is maximum when the pendulum goes through the zero position and is zero at the extremes of motion.

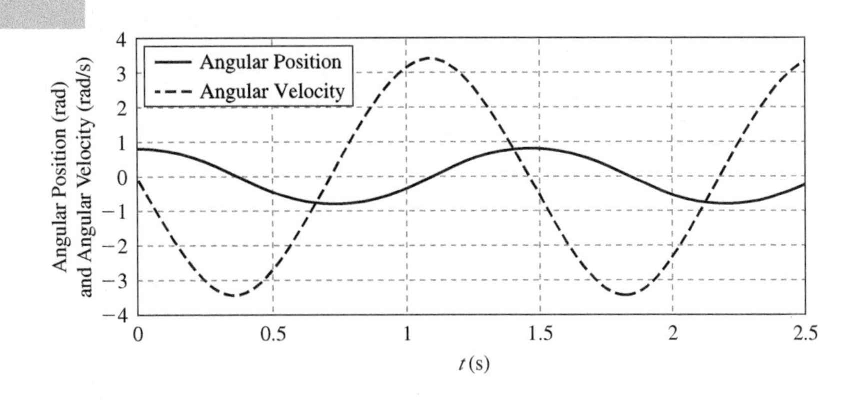

Figure 7.39
Pendulum simulation

7.7.3 Transfer Function Methods

For linear systems with zero initial conditions, the differential equation model can be transformed into a transfer function form using the Laplace Transform, and the transfer function commands in MATLAB can then be used to obtain a solution.

As an illustration of this method, let us consider again the differential equation obtained for the dynamics of the motor-driven geared system obtained in Example (7.3). Applying the Laplace Transform to Equation (7.36), and using the derivative property (see Equations (7.11) and (7.12)) with the assumption of zero initial conditions, we obtain the following equation:

$$T_m(s) = I_{eff}\, s^2\, \Theta_1(s) + b_{eff}\, s\, \Theta_1(s) \tag{7.42}$$

This equation can be written in transfer function form as:

$$\frac{\Theta_1(s)}{T_m(s)} = \frac{1}{I_{eff}\, s^2 + b_{eff}\, s} \tag{7.43}$$

The transfer function in Equation (7.43) can be defined in MATLAB using the ***Transfer Function* command** or *TF*(*NUM*, *DEN*) function which creates a continuous time transfer function. *NUM* and *DEN* are row vectors that specify the numerator and denominator coefficients, respectively, of the transfer function in a descending order. Thus, the transfer function in Equation (7.43) is specified with the following command:

$$G_1 = tf([1],[I_{eff}\ b_{eff}\ 0])$$

If we are interested in obtaining the response of the system to a pre-defined input signal such as a step signal, then the *step* command can be used with the typical calling format:

$$step(G_1)$$

While multi-input, multi-output systems can be represented in transfer function form, the transfer function method is better suited for single-input, single-output systems.

7.7.4 Block Diagram Representation and Simulation in Simulink

An alternative method to the three previous ways to solve and simulate dynamic models in MATLAB is to use a graphical method to perform this task. This is done using a block diagram representation of a dynamic system, which offers a graphical representation of the interaction of the different elements in the system. MATLAB offers through the Simulink package a mechanism for performing this representation in software. Simulink is an add-on to MATLAB and cannot run without having MATLAB installed.

Simulink offers a library of many pre-defined elements (called blocks in software) that one can choose from (see Figure 7.40). These blocks are grouped into categories. For the simulation of dynamic systems in the continuous time domain, blocks from the *Continuous* category are used. This category includes many blocks including a transfer function block, a state space block, a derivative block, and an integrator block.

Figure 7.40

Simulink 'Continuous' category blocks

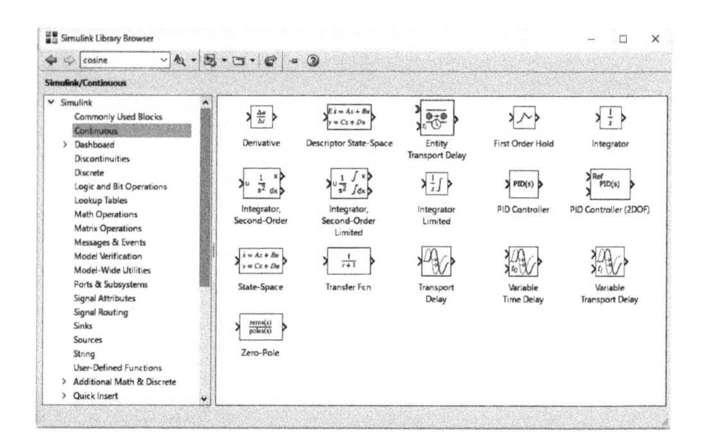

To build a dynamic model using Simulink, the user selects and drops the needed blocks into the Simulink model sheet. The blocks are then joined together to represent the desired signal flow between the blocks. Before the simulation is started, the **block parameters** for each block need to be defined so that the Simulink model corresponds to the dynamic system that is modeled. The block parameters can be set to either numeric values or left as variables that need to be defined in a script or *.m* file that is run before the simulation is started.

To illustrate the use of Simulink, let us build a model of the dynamics of the motor-driven geared system considered before. The transfer function of this system is given in Equation (7.43). As seen in Figure 7.41, we represent this transfer function in Simulink using two cascaded blocks, a *Transfer Function* block and an *Integrator* block. While we could have used a single transfer function block, we chose this representation so we can access the velocity of the system which is obtained from the output of the transfer function block. Also, the given transfer function allows this representation because we can factor out an s term from the denominator. The model parameters I_{eff} and B_{eff} are set as $I_{eff} = 0.1$ and $B_{eff} = 0.2$ in this simulation. The Simulink model library includes many blocks to supply inputs. We have chosen the *Step* block here which supplies a unit step torque input to the system. To capture the speed and position response of this model, two *Scope* blocks are included in the diagram. Once the model is built, and the model parameters are defined, the simulation is started by pressing the arrow button in the *Simulation* tab.

Figure 7.41

Simulink representation of the model given by Equation (7.43)

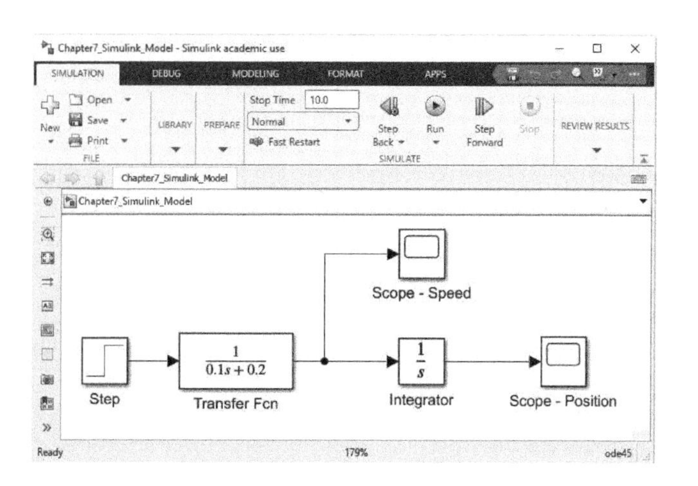

The speed and position response with $I_{eff} = 0.1$ and $B_{eff} = 0.2$ are shown in Figure 7.42.

Figure 7.42

Speed and position response for the model in Figure 7.41

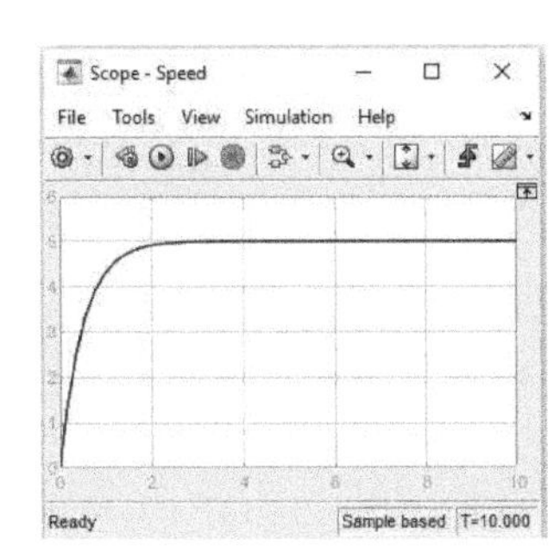

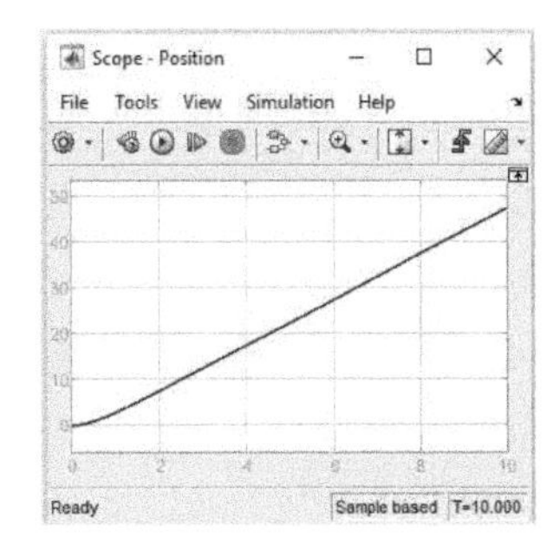

7.8 Chapter Summary

This chapter provides an in-depth exploration of time and frequency response analysis of first- and second-order systems, resonance phenomena, bandwidth, filtering techniques, and MATLAB simulation of dynamic systems. The chapter began by introducing the concept of time response, which examines how a system responds to changes in the input signal over time. First, the analysis focused on first-order systems whose model is represented by a first-order differential equation. The concept of the time constant of these systems was introduced, which plays a crucial role in determining their response speed. Next, the discussion extended to second-order systems such as an RLC circuit or a mass-spring-damper system. The chapter covered the time response analysis of second-order systems, considering both underdamped, critically damped, and overdamped cases. The concept of natural frequency and damping ratio was introduced to describe the behavior of these systems. Moving on, the chapter covered the concept of transfer functions and the frequency response of systems, exploring how systems behave in the frequency domain. Frequency response plots, called Bode plots, were introduced. Resonance phenomena, where systems exhibit peak response at a specific frequency, were explained in detail. The concept of bandwidth, which defines the range of frequencies over which a system operates effectively, was discussed. Filtering techniques were then covered, with an emphasis on low-pass, high-pass, bandpass, and notch filters. Finally, the chapter concluded with the different methods available in MATLAB to simulate and solve models of dynamic systems.

Questions

7.1 What is the free response of a system?

7.2 What is the forced response of a system?

7.3 What is the transient response of a system?

7.4 What is the steady state response of a system?

7.5 Provide examples of step input in mechanical and fluid systems.

7.6 A first-order system has a time constant of 2 seconds. How long does it take for the output of this system to approximately reach a steady state when subjected to a step input?

7.7 Is the time constant defined for an unstable system?

7.8 Can the response of a first-order system subjected to a step input be oscillatory?

7.9 What is the natural frequency of the system?

7.10 Is the damped natural frequency always lower than the natural frequency of the system?

7.11 What type of characteristic roots does a critically damped system have?

7.12 What should be the damping ratio of a second-order system for the system to not exhibit oscillation when subjected to a step input?

7.13 Under what conditions is the transfer function for a system obtained?

7.14 What is the Bode plot?

7.15 Define what is meant by sensor bandwidth.

7.16 What affects the bandwidth of a mechanical system?

7.17 What is resonance?

7.18 Which type of a second-order system exhibits resonance?

7.19 What happens to the resonance response of a system as the damping ratio becomes smaller?

7.20 Give an example of a passive low-pass filter.

7.21 For what applications is a low-pass filter used?

7.22 What is the order of a bandpass filter?

7.23 What type of filter is used to attenuate a narrow band of frequencies from a signal?

7.24 What is the difference between an active and a passive filter?

7.25 What MATLAB method should be used to solve a system of nonlinear differential equations?

7.26 What is the state space solution method in MATLAB?

7.27 Which solution method in MATLAB uses a block diagram representation of a dynamic system?

Problems

Section 7.2 Time Response of First-Order Systems

P7.1 A first-order system has a differential model given by $\dot{x} + 4x = 2$, and an initial condition of $x(0) = 2$. Determine expressions for (a) the free response and (b) the forced response of this system.

P7.2 For the model given in Problem P7.1, determine expressions for (a) transient response and (b) steady state response of the system.

P7.3 A first-order system has a differential model given by $2\dot{x} + 5x = 3$, and an initial condition of $x(0) = 5$. Determine expressions for (a) the free response and (b) the forced response of this system.

P7.4 For the model given in Problem P7.3, determine expressions for (a) transient response and (b) steady state response of the system.

P7.5 Determine the time constant, if it exists, for the following first-order systems.

a. $\dot{x} + 4x = 0$

b. $2\dot{x} + 4x = 0$

c. $2\dot{x} + 4x = 4$

d. $2\dot{x} - 5x = 0$

P7.6 For the differential model given in Example 7.3, determine an expression for the time constant of the system.

Section 7.3 Time Response of Second-Order Systems

P7.7 Determine the type of response (oscillatory or non-oscillatory, and stable or unstable) for the following second-order systems, whose characteristic equations are included, when subjected to a step input.

a. $s^2 + 5s + 6 = 0$

b. $s^2 + 2s + 1 = 0$

c. $s^2 - 2s - 3 = 0$

d. $s^2 + 2s + 3 = 0$

P7.8 Determine the type of response (underdamped, critically damped, or overdamped) for the following second-order systems, whose characteristic equations are included, when subjected to a step input.

a. $s^2 + 4s + 4 = 0$

b. $s^2 + 4s + 2 = 0$

c. $s^2 + 5s + 4 = 0$

P7.9 Determine, if applicable, the natural frequency, damping ratio, damped natural frequency, and time constant for the following systems whose characteristic equations are:

a. $s^2 + 6s + 9 = 0$

b. $s^2 + 4s + 16 = 0$

c. $s^2 - 4s + 16 = 0$

P7.10 A second-order system with a damping ratio of 0.5 and a natural frequency of 10 rad/s is subjected to a step input. Determine the peak time, settling time, and maximum percent overshoot for this system.

P7.11 An experimentally recorded response of a second-order system subjected to a step input showed an overshoot of 20% and a peak time of 1 second. Using these values, determine the natural frequency and damping ratio of this system.

P7.12 A mass-spring-damper system subjected to a step input of magnitude 50 N shows a response that has a peak value of 0.060 m, a steady state value of 0.050 m, and a peak response time response of 0.156 seconds. Using these values, determine the parameters m, c, and k of this system.

Section 7.4 Transfer Functions

P7.13 Determine the transfer function between the output $X(s)$ and the input $Y(s)$ for the following differential models.

a. $2\dot{x} + 3x = 3y(t)$

b. $4\ddot{x} + 5\dot{x} + 2x = 5y(t)$

c. $3\ddot{x} + 5x = 2y(t)$

P7.14 For the RLC circuit considered in Example 7.5, (a) determine the transfer function for this system and (b) the relationship between the circuit parameters that will cause the system to be critically damped.

P7.15 For the mass-spring-damper system whose transfer function is given by Equation (7.14) and is:

$$\frac{X(s)}{F(s)} = \frac{1}{ms^2 + cs + k}$$

Determine what happens to the natural frequency of the system if (a) the stiffness was doubled, (b) the mass was doubled, or (c) the damping was doubled.

Section 7.5 Frequency Response

P7.16 Determine expressions for the magnitude and phase of the following transfer functions.

a. $\dfrac{2}{3s + 5}$

b. $\dfrac{10}{2s + 6}$

c. $\dfrac{s}{s + 1}$

P7.17 Determine expressions for the magnitude and phase of the following transfer functions.

a. $\dfrac{5}{s^2 + 3s + 6}$

b. $\dfrac{2s}{20s^2 + 12s + 1}$

c. $\dfrac{4s^2 + 1}{4s^2 + 8s + 1}$

P7.18 Figure P7.18 shows the frequency response plot for a first-order system. Determine the magnitude ratio M for this system at the following frequencies: 1, 10, and 100 rad/s.

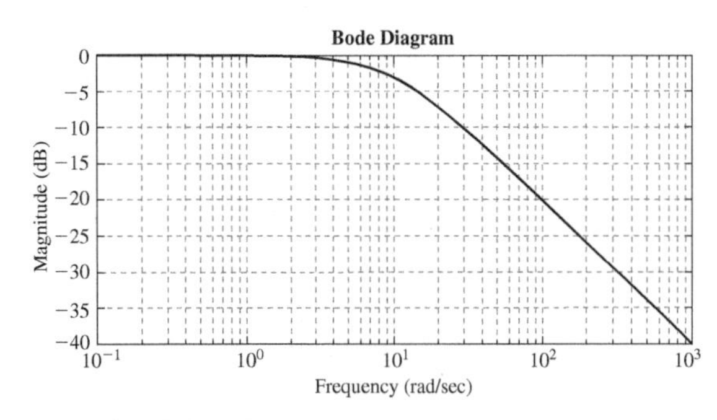

Figure P7.18

P7.19 A system with the transfer function $3/(s + 2)$ was subjected to a sinusoidal input given by $y(t) = 2\sin(3t)$. Provide values for the output amplitude B and the phase shift φ of the sinusoidal steady state output of this system under this input signal.

P7.20 A system with the transfer function $2/(s^2 + 6s + 8)$ was subjected to a sinusoidal input given by $y(t) = 4\sin(2t)$. Provide values for the output amplitude B and the phase shift φ of the sinusoidal steady state output of this system under this input signal.

P7.21 Figure P7.21 shows the frequency response plot for a mass-spring-damper system. Determine:

a. The resonance frequency and the magnitude ratio M at resonance.

b. If the system is subjected to a sinusoidal force of amplitude of 10 N at a frequency of 20 rad/s, determine the displacement amplitude of the system under this input force.

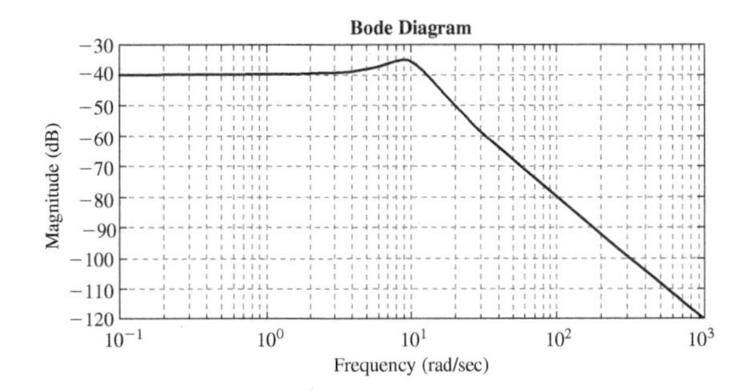

Figure P7.21

P7.22 A mass-spring-damper system has a damping ratio of 0.5 and a natural frequency of 1 rad/s. Determine (a) the resonance frequency and (b) the magnitude ratio at resonance for this system.

P7.23 Determine the bandwidth of the systems shown in (a) Figure P7.18 and (b) Figure P7.21.

P7.24 A continuous-time signal has a bandwidth of 400 Hz. You are tasked with digitizing this signal using an A/D converter. However, the only A/D converters available to you have fixed sampling rates of either 600 Hz or 1500 Hz.

a. Can either of these A/D converters be used without introducing aliasing? Explain your answer.

b. If aliasing will occur with one of the available sampling rates, what is the minimum sampling rate you would need to avoid aliasing?

P7.25 What is the effective bandwidth as seen by a signal when it passes through an amplifier with 1 MHz bandwidth connected in series with a low-pass filter with a bandwidth of 10 kHz?

Section 7.6 Filtering

P7.26 Determine the corner frequency (in Hz) of the following low-pass filters whose transfer function is given by:

a. $\frac{1}{0.01s + 1}$

b. $\frac{1}{0.002s + 1}$

P7.27 An A/D converter is sampling signals at a frequency of 1000 Hz. Determine the corner frequency of a low-pass filter that should be used to prevent aliasing.

P7.28 Determine the parameters of a low-pass passive filter constructed using an RC circuit to have a corner frequency of 100 Hz and a low-frequency gain of 1.

P7.29 Determine the parameters of a low-pass active filter that uses op-amps to have a corner frequency of 100 Hz and a low-frequency gain of -3.

P7.30 Determine the transfer function of a bandpass filter that has a frequency range between 10 to 1000 Hz.

P7.31 Determine R_1 and R_2 using resistors from the E92 series (see Appendix A) for a passive bandpass filter with $f_{cl} = 1$ Hz, $f_{ch} = 100$ Hz, and $C_1 = C_2 = 0.1\ \mu$F.

P7.32 Determine the transfer function of a notch filter that can be used to attenuate a 50 Hz signal.

P7.33 Determine the notch frequency of a passive notch filter implemented using an RLC circuit if $L = 10$ mH and $C = 1$ uF. What should be the value of the resistor for this filter?

P7.34 Determine the parameters of a low-pass digital filter with a corner frequency of 5 Hz and a sampling frequency of 1 ms.

Laboratory/Programming Exercises

L/P7.1 Use MATLAB, Excel, or similar software to simulate the step response of the first-order system given in Problem P7.1. From the response plot, determine the time constant of the system.

L/P7.2 Build an RC circuit with $R = 5.1\ \text{k}\Omega$ and $C = 0.1\ \mu$F. Use the function generator to apply a square wave signal with an amplitude of 4 volts at a frequency of 100 Hz. Use the oscilloscope to view the input and output signals and take a picture of the plot displayed on the scope. How long does it take the output to reach a steady state after each transition in the input signal?

L/P7.3 Use MATLAB or similar software to obtain the Bode plots for the transfer functions given in Problem P7.16.

L/P7.4 Use MATLAB/Simulink or similar software to simulate the operation of a low-pass filter with a 100 Hz corner frequency. Generate plots to show the filtering effects for sinusoidal signals with a frequency of 20, 200, and 2000 Hz.

L/P7.5 Build an RC filter circuit with $R = 5.1\ \text{k}\Omega$ and $C = 0.1\ \mu$F. Use the function generator to apply a sinusoidal signal with frequencies ranging from 10 to 3000 Hz. Record the amplitude of the input and output signals as well as the phase shift as a function of frequency. Plot the filter response data (magnitude and phase) and compare it to that expected from the theory.

L/P7.6 Use MATLAB, Excel, or similar software to simulate the operation of a low-pass digital filter (Equation 7.27) with a corner frequency of 5 Hz and a sampling frequency of 1 ms. Generate plots to show the filtering effects for sinusoidal signals with a frequency of 1, 10, and 20 Hz.

L/P7.7 Use MATLAB or similar software to create a notch filter with a notch frequency of 60 Hz. Apply sinusoidal signals with the same amplitude but with the following frequencies: 54, 59, 60, and 61 Hz. For each signal, record the steady state output amplitude of the notch filter.

L/P7.8 Use MATLAB or similar software to simulate the response of the state space model given by Equation (7.39) to a step input of magnitude 2 N·m over 2 seconds. Use the following values for the parameters of the system: $I_{eff} = 0.05$ kg m^2 and $b_{eff} = 0.2$ N·m s/rad.

L/P7.9 Use the Arduino Uno or a similar board to capture the open-loop step speed response of a motor or any other dynamic system. The user specifies the sampling interval, the test duration, and the desired step magnitude. The program applies the step input to the system at time zero and then displays the desired number of samples at the desired sampling rate to the serial monitor. The output from the serial monitor can be copied using the CTRL+C command and pasted to a plotting software such as EXCEL to plot the response.

Chapter 8

Sensors

Chapter Objectives:

When you have finished this chapter, you should be able to:

- Interpret a sensor performance specification
- Select a specific sensor for a given measurement application
- Explain the different types of displacement, proximity, speed, force, torque, temperature, and vibration sensors
- Predict the output of strain gauge-based sensors
- Explain the principle of operation of many of the different sensors covered in this chapter
- Analyze a bridge circuit to process the output of resistance-type sensors

8.1 Introduction

Sensors play a pivotal role in a mechatronics system, providing vital information essential for the monitoring and control of the system. Without sensory input, automated systems cannot operate optimally. A sensor is a device that generates an output corresponding to changes in physical quantities, such as temperature, force, or displacement. In many sensors, the component directly responsible for converting a physical quantity to another form (often an electrical signal) is called a **transducer**. This conversion is achieved through various means, such as modifying a property of the transducer (e.g., its resistance, capacitance, or inductance), leveraging piezoelectric effects, or other mechanisms. Electronic circuits then process these changes, translating them into discernible voltage or current signals. While the terms 'sensor' and 'transducer' are frequently used interchangeably, especially when referring to devices that produce electrical outputs, it is essential to recognize that not all sensors yield an electrical signal. Some examples of this include the mercury bulb thermometer, which displays temperature mechanically, and the spring scale force sensor, which indicates force without an electrical output. The distinction lies in understanding that while all transducers can be sensors, not all sensors function as transducers in the strict sense of energy conversion.

Figure 8.1 shows a block diagram of the process of measurement using a sensor with a transducer. Normally, the output signal from the transducer is not in a form suitable to be read by a display device or meter, and signal conditioning operations (such as filtering or amplification) are needed to process the sensor output.

There are a variety of sensors available that are commonly used. These include sensors that measure motion-related information (such as strain, speed, displacement, and acceleration). Also, sensors are available to measure process parameters (such as temperature, level, and pressure). This chapter will focus more on sensors that measure motion-related information. The next chapter will discuss actuators. Both sensors and actuators are key to implementing feedback control of motion-control systems. We start this chapter by discussing some of the performance parameters of sensors. For further reading on sensors and measurements, see [28–30].

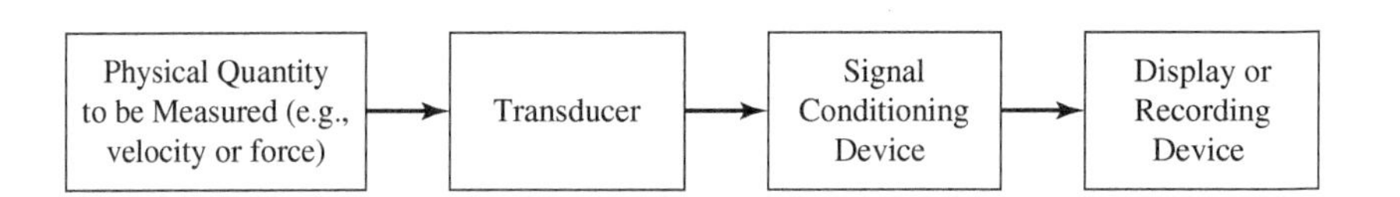

Figure 8.1

Measurement process

8.2 Sensor Performance Terminology

Several parameters characterize sensors' performance (video available—V8.1). The time-independent characteristics are called the static characteristics, while the time-dependent characteristics are called the dynamic characteristics. The **static characteristics** characterize the sensor output after it has settled due to changes in the physical quantity being measured. The **dynamic characteristics** describe the sensor characteristics from the time the physical quantity has changed to the time before the output has settled.

8.2.1 Static Characteristics

Range The minimum to maximum value that can be measured is the range. The range defines the allowable range of the physical quantity that can be detected by the sensor.

Accuracy The difference between the true and actual measured values is the accuracy. It is commonly expressed as a percentage of full-scale value. For example, if a temperature sensor has a range of 0 to 200°C and an accuracy of $+/-0.5\%$ the full-scale value, then the temperature read by the sensor is off from the true actual temperature by $+/-1°$. Note that accuracy is also referred to as bias, and the accuracy error can be improved by calibration.

Sensitivity The relationship between the measured input and the output of the sensor is its sensitivity. If the sensor has a linear input–output relationship, then the sensitivity is the slope of this curve. Sometimes, this parameter is used to indicate the sensitivity of the sensor to non-measured input (response due to transverse motion when the sensor is designed to measure axial motion) or the environment (temperature).

Resolution The smallest change in input value that will produce an observable change in the output is the resolution. The inherent resolution should be distinguished from the display device resolution.

Hysteresis The maximum difference in sensor output for the same input quantity is the hysteresis, with one measurement taken while the input is increasing from the minimum input and the other by decreasing the input from the maximum input. A sensor with hysteresis will have a different output value that is a function of whether the input quantity was increasing or decreasing when the measurement was made. The hysteresis error is illustrated in Figure 8.2.

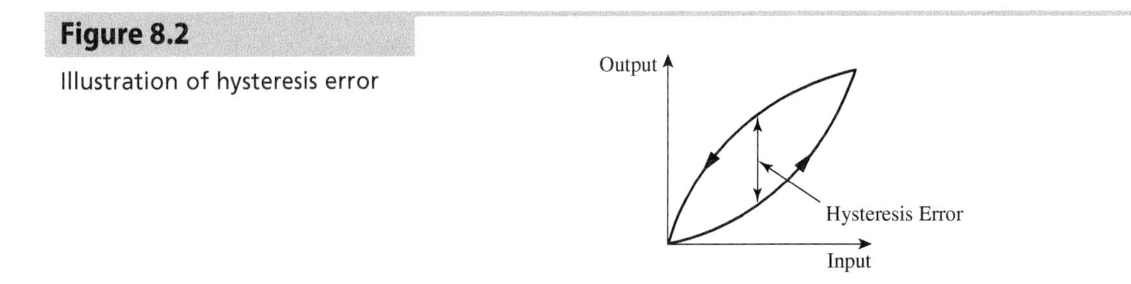

Figure 8.2

Illustration of hysteresis error

Repeatability Repeatability refers to the consistency of output values when the same measurement is repeated under identical conditions. The closer the output values are to each other for repeated measurements, the higher the measurement's repeatability and, by extension, the precision. Factors like signal interference, vibration, temperature fluctuation, instrument drift, and user variability can introduce variability and affect repeatability. While calibration primarily addresses systematic biases in measurements and does not directly improve repeatability, techniques like taking multiple measurements and using statistical methods can help manage and potentially reduce repeatability errors.

Non-Linearity Error Most sensors are designed to have a linear output, but their output is not perfectly linear. The nonlinearity error is a measure of the maximum difference between the sensor's actual output and a straight line fit to the sensor input–output data and is usually specified as a percentage of the full-scale output. There is no unique way to obtain the straight-line fit. The straight line can connect the minimum and maximum output values that define the sensor range, or it can be obtained from a least-square fit to the entire input–output data or from a least-square fit to the input–output data with one end of the line passing through the origin. Figure 8.3 illustrates these cases and shows that the magnitude of this error is dependent on how this error is defined.

Figure 8.3 Illustration of nonlinearity error

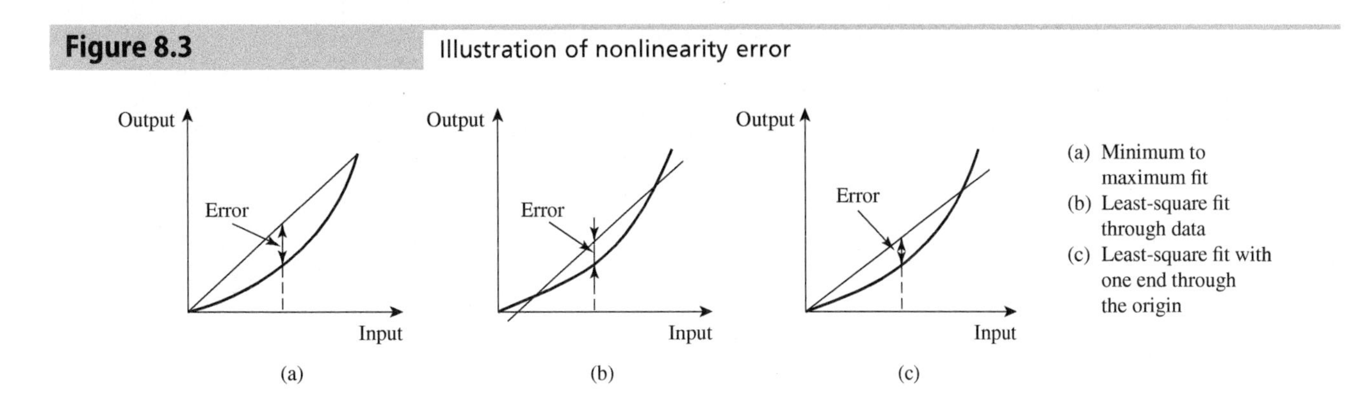

Stability Stability or drift refers to the variation of the output with time when the input quantity is not changing. The output variation is called the zero drift when no input is applied to the sensor. Stability affects the repeatability of the measurement.

8.2.2 Dynamic Characteristics

The details of many of the dynamic characteristics listed below were covered in the previous chapter, so they will not be repeated here.

Rise Time The time it takes the output to change from 10 to 90% of its final steady-state value is the rise time. It provides a measure of how quickly the sensor output reaches its steady state.

Time Constant This is defined as the time it takes the output to reach 63.2% of the difference between the final and the initial output. A large time constant implies a sluggish sensor, while one with a small value indicates a rapidly responding sensor.

Settling Time The time it takes the output to reach within a certain percentage of the final steady–state value is the settling time. A common value is the 2% settling time. The rise time, time constant, and settling time are illustrated in Figure 8.4 for a sensor with first-order response characteristics.

Figure 8.4

Illustration of basic dynamic response characteristics

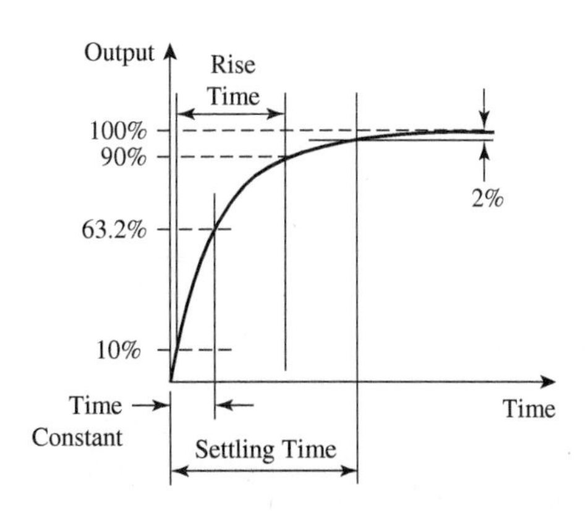

Bandwidth The bandwidth of a sensor defines the frequency range over which it is designed to operate effectively. When utilizing a sensor for feedback in a closed-loop control system, the sensor's bandwidth needs to exceed that of the controller to ensure that the sensor can accurately measure and relay system responses to the controller without introducing phase lags or distortions that might destabilize the control system.

Values for each sensor performance characteristic are found in the manufacturer datasheet for the particular sensor. Normally, the specification in the datasheet is grouped into categories, including dynamic or performance, electrical, mechanical, environmental, and physical. An example of some of the characteristics of a compression load cell is shown in Table 8.1, including a description of each specification.

Table 8.1

An example of the specifications for a load cell sensor

Item	Value	Explanation
Rated capacity	10 lbs	The maximum weight that the cell is rated to handle
Excitation	10 VDC	The cell is designed for an excitation voltage of up to 10 VDC, but it can also operate with lower voltages
Rated output	2 mV/V nominal	The nominal cell output at the rated load (10 lbs) and the maximum excitation voltage (10 V) will be 20 mV (2 mV/V × 10 V)
Linearity	+/−0.25% FS	With a 5 lb load applied, the cell output will indicate a load of 5 lbs +/−0.025 lb
Hysteresis	+/−0.15% FS	Due to hysteresis, the cell output can vary by +/−0.015 lb
Maximum load (safe overload)	150% of rated capacity	The cell can handle up to 15 lbs without risk of damage, but accuracy may be affected
Bridge resistance	350 Ω	The resistance of the strain gauge inside the load cell

Example 8.1 illustrates the computation of the maximum error in a sensor measurement.

Example 8.1 Computation of Total Sensor Error

A linear displacement sensor with a measuring range of 0 to 50 mm has the following specifications: a resolution of 0.01 mm, an accuracy of 0.5% of full-scale reading, and a linearity error of 0.2% of full-scale reading. Determine the maximum possible error (in mm) one could expect.

Solution:

The total error will be the sum of the sensor resolution error, the accuracy error, and the linearity error.

The accuracy error = accuracy (as %) × full-scale reading/100 = 0.5 × 50/100 = 0.25 mm.

The linearity error = linearity error (as %) × full-scale reading/100 = 0.2 × 50/100 = 0.1 mm.

Summing the three errors (0.01 + 0.25 + 0.1) gives the total error as 0.36 mm. This is the maximum possible error when taking a reading with this displacement sensor.

This solution assumes that all the sources of error are additive, which represents a worst-case scenario. This assumption often provides a conservative estimate of error that ensures safety and robustness in various applications, particularly in engineering. Some errors might be systematic (always in one direction) while others might be random. In such cases, the errors might not necessarily add up to a worst-case scenario.

8.3 Displacement Measurement

Displacement sensors provide information about the change in the position of a rigid body. The sensors can be classified as those that provide **analog output** (such as potentiometers and resolvers) and those that provide **digital output** (such as encoders). Displacement sensors also can be classified as contact or non-contact, depending on whether the sensor contacts the object during measurement. **Contact-type displacement** sensors include strain gauges and potentiometers, while **non-contact displacement** sensors include optical encoders as well as inductive and capacitive-type displacement sensors. This section will discuss different types of displacement sensors.

8.3.1 Potentiometers

A potentiometer (▶ available—V8.2) is a contact-type sensor that provides displacement information by measuring the voltage drop across a resistor. Potentiometers can be of the linear or rotary type. A **linear potentiometer** is designed to measure linear displacement. The sensor has a linear slider arm that is attached to the object whose displacement needs to be measured (see Figure 8.5). The displacement of the slider changes the electrical resistance between nodes *a* and *b*, which then can be used as a measure of displacement. The *b* node is commonly called the wiper. In normal operation, a DC voltage is applied between nodes *a* and *c*, and the voltage output between nodes *a* and *b* is used as a measure of displacement. The linear potentiometer acts as a voltage divider as the wiper is moved between nodes *a* and *c*. If node *a* is connected to the ground, then the *b-node* voltage will increase as the slider travels from *a* to c and will be equal to the supply voltage as the wiper reaches node *c*. On the other hand, if the node *a* is connected to the supply voltage and node *c* to ground, then the *b-node* voltage will decrease as the slider travels from *a* to *c* and will be equal to 0 volts as the wiper reaches node *c*.

Figure 8.5

Model of a linear potentiometer

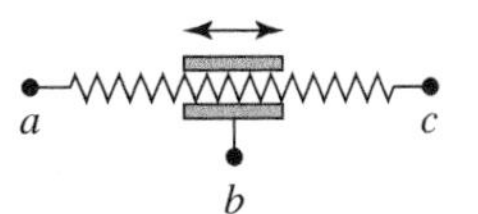

A **rotary-type potentiometer** (see Figure 8.6) is designed to measure angular displacement. The sensor has a rotary knob that is coupled to the shaft of the object whose angular displacement needs to be measured. Similar to a linear potentiometer, the rotation of the knob changes the electrical resistance between the leads of two nodes on the potentiometer. Rotary potentiometers are available as **single-turn** or **multi-turn** devices. Multi-turn devices can measure several shaft revolutions, while single-turn devices are designed to measure a rotation of up to one revolution. Some single-turn devices cannot measure a complete revolution due to the construction of the potentiometer with a dead zone, which prevents the wiper on the potentiometer from making a complete turn. The contact element in a potentiometer is constructed either from a wound wire or from conductive plastic. Wire-wound elements provide better stability and linearity than conductive plastic, but conductive plastic offers better resolution and longer life. The use of a wound wire results in a step change in the output voltage (and hence a coarser resolution than conductive plastic) as the slider moves from one turn in the coil to the next turn. Potentiometers have the advantage that they are easy to use, but because they are contact-type devices, they have a frictional resistance which affects the motion of the measured object. Because potentiometers provide an analog voltage as their output, they also need to be interfaced with an A/D converter before the signal is read by a PC or a microcontroller.

Figure 8.6

A commercial rotary potentiometer

(© Wayne Higgins/Alamy Stock Photo)

The resistance of a potentiometer is important. A high resistance results in a smaller current and hence less heat loss through the potentiometer while it is in operation, but it also worsens the loading error since, in practice, the output voltage of the potentiometer is read by a device that does not have infinite impedance (see Section 2.6). Loading introduces **nonlinearities** into the potentiometer output. To see this, refer to

Figure 8.7, and assume that the potentiometer has a resistance R_P, the measuring device has a resistance R_L, and the supply voltage is V_S. Then the voltage output at any position x ($0 \leq x \leq 1$) is given by the relationship

(8.1)

$$V_O = \frac{xV_S}{1 + x(1 - x)\frac{R_P}{R_L}}$$

Figure 8.7

Model of a potentiometer interfaced with a measuring device with load resistance (R_L)

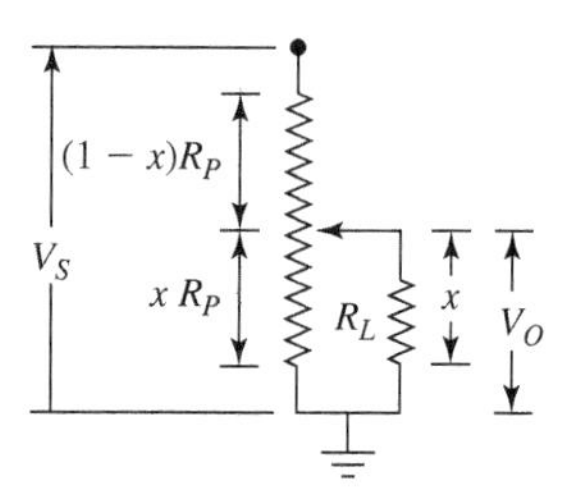

This relationship is obtained by using the voltage dividing rule to compute the output voltage (V_O) and noting that the load resistor (R_L) and the potentiometer resistor ($x\,R_P$) are two resistors in parallel. Note that if the load resistance (R_L) is infinite, then R_P/R_L is zero and Equation (8.1) reduces to $V_O = xV_S$, where the output voltage is directly proportional to the slider position x. If R_L is not infinite, then the output voltage varies nonlinearly with the slider position x. The nonlinearity worsens as the ratio of R_P/R_L increases.

Figure 8.8 shows a plot of Equation (8.1) for three values of the ratio R_P/R_L: one for $R_P/R_L = 0$, another for $R_P/R_L = 0.1$, and the third for $R_P/R_L = 1.0$. Note how the output voltage varies nonlinearly with x, especially for $R_P/R_L = 1$. To eliminate this problem, one should select a potentiometer with as small a resistance as possible, and one should also consider the power dissipation capacity of the potentiometer to ensure safe operation. Example 8.2 illustrates computations for a potentiometer.

Figure 8.8

A plot of Equation (8.1)

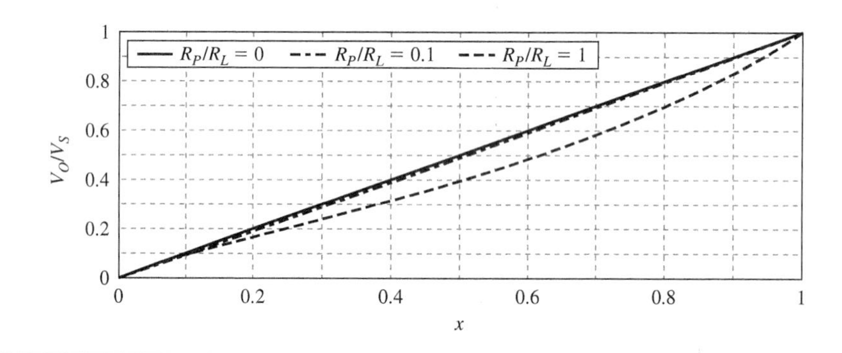

Example 8.2 Potentiometer

A single-turn rotary potentiometer with a 330° measurement range is used to provide angular-position feedback information for a positioning application. A 5 V DC voltage is applied across the potentiometer leads, and the potentiometer output is connected to a 12-bit A/D converter with a +/−5 V range. The potentiometer resistance is 50 Ω. Determine:

(a) The effective resolution of this sensor
(b) The power loss by the potentiometer, assuming a half-motion displacement

Solution:

(a) Assuming that the potentiometer uses a film as the resistance element, then the resolution is determined by the A/D resolution. The 330° motion range is mapped into the 5 V output, and the A/D total voltage range of 10 V is mapped into 2^{12} positions. Thus, the angular resolution is

$$10/4096 \times 330°/5 = 0.161°$$

(b) If the meter impedance is infinite, then all the current is passed through the potentiometer. In this case, the current is

$$5\text{ V}/50\ \Omega = 0.1\text{ A}$$

and the power loss is i^2R or 0.5 W.

8.3.2 LVDT

The Linear Variable Differential Transformer (LVDT) is an electromechanical transducer designed for accurately measuring mechanical displacement. Structurally, the LVDT (▶ available—V8.3) comprises a movable iron core positioned within a trio of transformer coils (see Figure 8.9) where a primary coil is centered between two symmetrically wound secondary coils. An external AC excitation voltage is applied to the central primary coil. The output is derived from the two secondary coils, which are wound in such a way that their induced voltages oppose when the core is perfectly centered. Displacement of the core from this centered position alters the magnetic coupling between the primary and each of the secondary coils. As a result, the output AC voltage from the secondary coils varies in amplitude and/or phase in proportion to the core's displacement. Typically, the excitation frequency for the LVDT ranges from 60 Hz up to 20 kHz, with several kHz being common. The output AC signal retains the same frequency as the excitation, but its amplitude and/or phase change conveys the core's position information. To derive a usable measure of displacement, the AC output signal is processed. This involves rectifying the signal to produce a DC voltage, filtering out high-frequency components, and amplifying to achieve a suitable output voltage level. For filtering, a low-pass filter may be employed; its cut-off frequency is selected based on application-specific requirements and the characteristics of the signal processing chain. Once processed, the resulting DC signal proportionally represents the displacement, making it suitable for digitization by an A/D converter or display on an appropriate device.

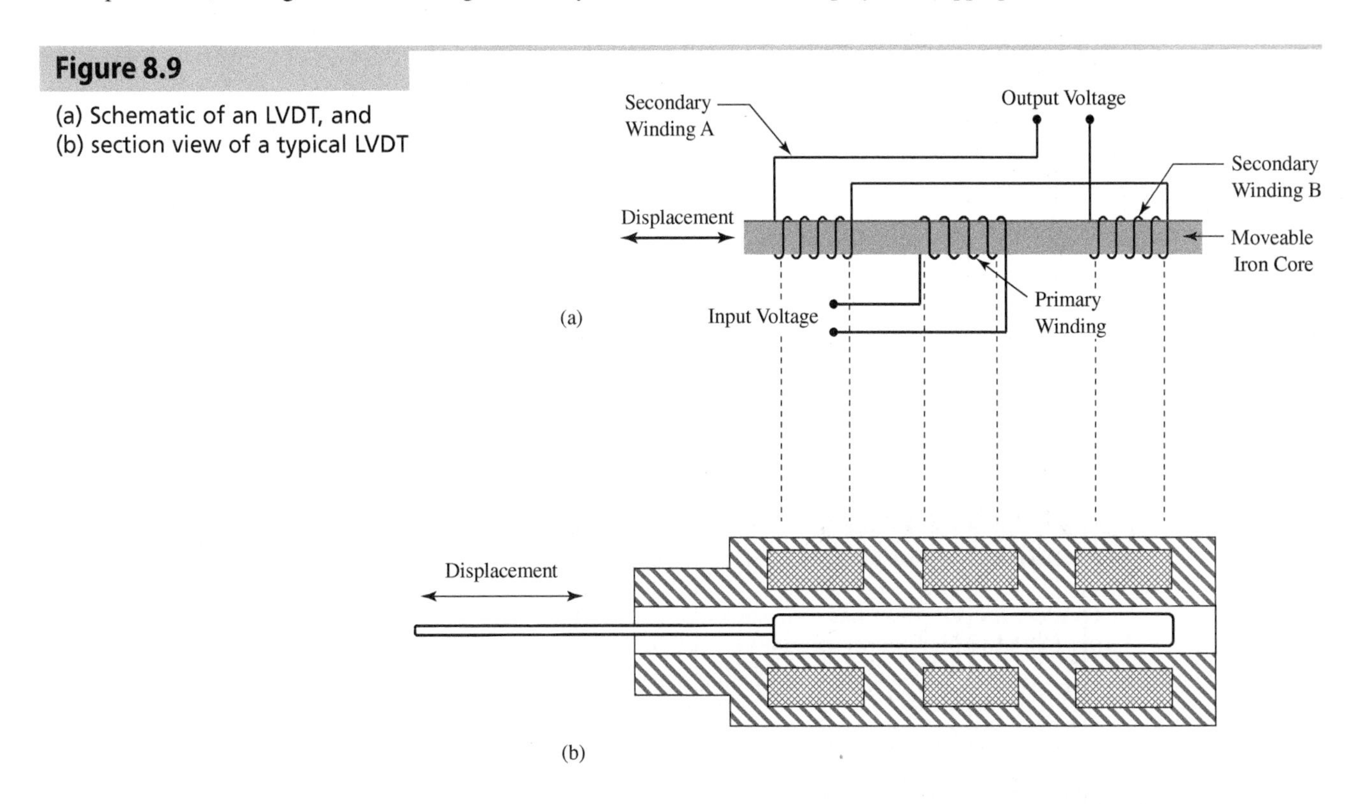

Figure 8.9
(a) Schematic of an LVDT, and
(b) section view of a typical LVDT

A typical output voltage versus core displacement of an LVDT is shown in Figure 8.10. When the core is at the null position or the center of the primary coil, the output voltage of an LVDT is zero. At this position, the induced voltages in the two secondary coils of the LVDT cancel each other out, resulting in a net output voltage of zero. The output voltage is very linear for displacements close to the null position, and the linear range is influenced by several factors, including the length of the secondary coils. As the core is displaced beyond the limits of its range, the output voltage becomes nonlinear.

An LVDT has the advantage that it can be constructed to measure displacement ranging from a few millimeters to hundreds of millimeters with almost infinite resolution. There is also no damage from overloading, as overloading simply separates the core from the device, and it is relatively insensitive to temperature changes. However, being a contact displacement sensor, the LVDT has a limited frequency response. Also, signal conditioning is required to process the output signal. LVDTs are available as DC or AC power operated. The DC configuration offers ease of installation and the ability to use battery power in measurement situations where AC power is not available, while the AC version results in a smaller body size and more accurate signal. Example 8.3 illustrates the computation of the voltage outputs in an LVDT, and Figure 8.11 shows a picture of commercial LVDTs.

Figure 8.10

Output voltage versus core displacement for an LVDT

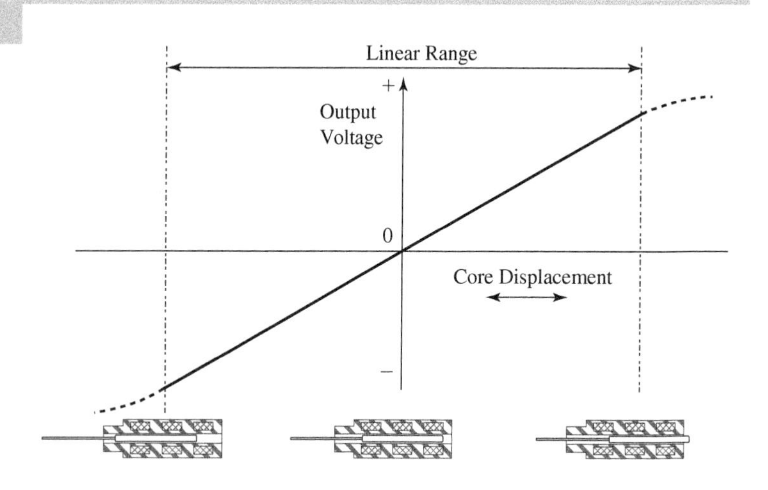

Figure 8.11

Examples of commercial LVDTs

(Courtesy of TRANS-TEK, Inc., Ellington, CT)

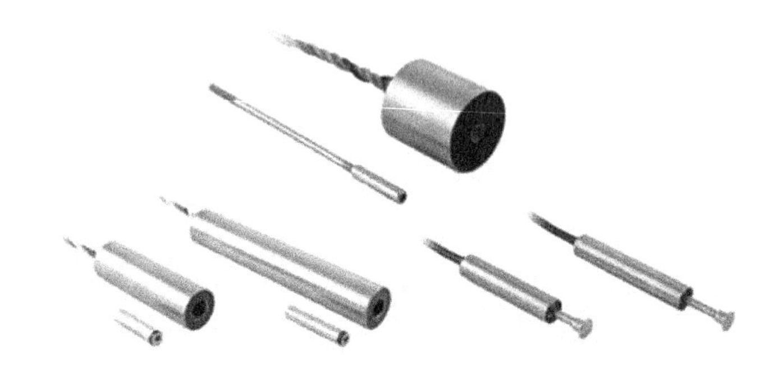

Example 8.3 Computation of the Voltage Outputs in a LVDT

Consider an LVDT that has a stroke length of ±100 mm and an output voltage sensitivity of 40 mV per mm of displacement. Determine:

(a) The LVDT's maximum output voltage
(b) The core position from the center when the output voltage is 3.2 volts
(c) The change in output voltage when the core is moved from +60 to −60 mm displacement

Solution:

(a) The LVDT's maximum output voltage is 4 V (100 mm × 40 mV/mm) in either direction from the null position, giving a range of ±4 V.
(b) The core position is obtained by dividing the voltage output by the voltage sensitivity, or (3.2 V/0.04 V/mm) which gives a core position of 80 mm.
(c) The output voltage will change from 2.4 to −2.4 V over this displacement range, or a change of 4.8 V from the initial reading.

8.3.3 Optical Incremental Encoder

Encoders (▶ available—V8.4) are digital devices used to measure position or changes in position. While there are several types of encoders available—including optical, magnetic, and inductive—optical encoders are particularly widespread. Although potentiometers can also measure linear and rotary positions, they are analog devices, whereas encoders provide digital outputs. This discussion will focus on rotary optical encoders, which are popular in motion-control applications. Rotary optical encoders can be categorized as either incremental or absolute. **Incremental encoders** measure changes in rotation relative to a reference position, while **absolute encoders** determine the actual angular position.

In its basic form, an optical incremental encoder is constructed from two light sources that shine light through a disk that has an alternating pattern of transparent and opaque regions. The light is sensed by two photodetector sensors that are located on the other side of the disk. To better understand the operation of an incremental encoder, first assume that we have only one light source and one sensor similar to that shown in Figure 8.12.

Figure 8.12

A simplified setup of an encoder that uses only one light source and one sensor

(Courtesy of Anaheim Automation, Anaheim, CA)

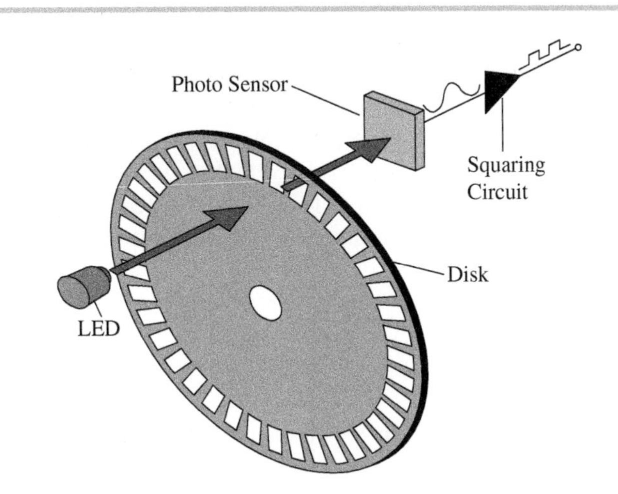

When the disk rotates, the light signal (as measured by the photodetector sensor) looks like that shown in Figure 8.13. The transparent regions on the disk allow light to pass through and reach the photodetector. This usually results in a 'low' state in the output pulse train. The opaque regions on the disk block the light. When the photodetector does not receive light, it typically produces a 'high' state in the output pulse train. To get angular displacement information, we simply count the number of pulses generated as the disk rotates from one position to another. Unfortunately, this simple scheme cannot provide us with direction information.

Figure 8.13

Output from a single light/sensor combination

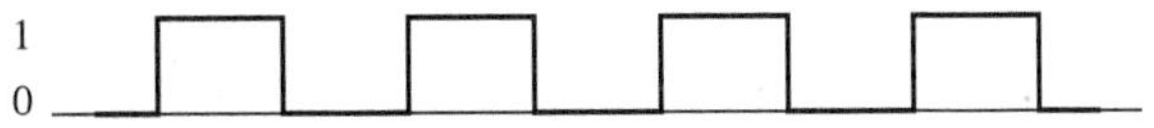

1
0

To get direction information, incremental encoders have another light source and sensor (called channel B). The channel B sensor is located with an offset from the first sensor and photodetector (channel A). There are two ways to implement this **offset** in practice.

1. There is only one track of lines: one light source-sensor combination is located to line up with the edge of one of the slots, while the other light source-sensor combination is located to have an offset of one half-slot with respect to the edge of one of the slots. This arrangement is shown in Figure 8.14(a), where for simplicity, the clear and the opaque slots on the disk are shown straight and not curved. This configuration is the most widely used.
2. Two concentric tracks of lines are used: the slots in one track have an offset of one-half slot with respect to the slots in the other track, but the sensors have no offset between them. This arrangement is shown in Figure 8.14(b), where for simplicity again, the clear and the opaque slots on the disk are shown straight and not curved.

Figure 8.14

Two configurations of the two light sensors on an incremental encoder disk: (a) single track and (b) two concentric tracks

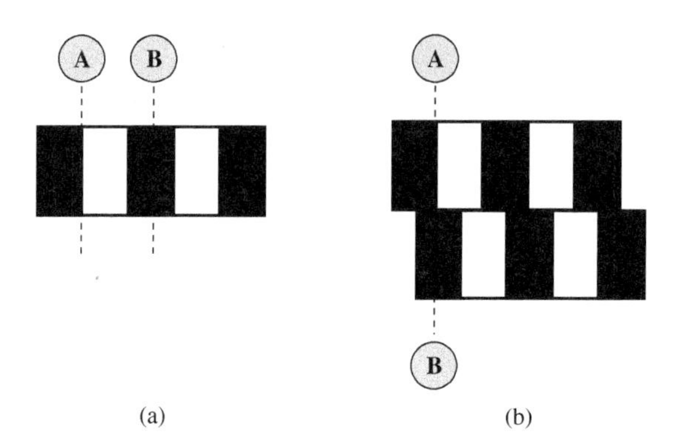

The pattern of the two sensor signals generated for clockwise (CW) and counterclockwise (CCW) rotations of a disk under constant angular speed is shown in Figure 8.15. Notice how the *B* signal leads the channel *A* signal by a quarter of a cycle for CW rotation, and it lags behind the channel *A* signal by the same amount for CCW rotation. The cycle that we are referring to is the chord distance made up of one transparent and one opaque strip. This lead/lag between the two channel outputs enables one to determine the direction of the rotation of the shaft that is attached to the encoder disk.

Figure 8.15

Output of an incremental encoder

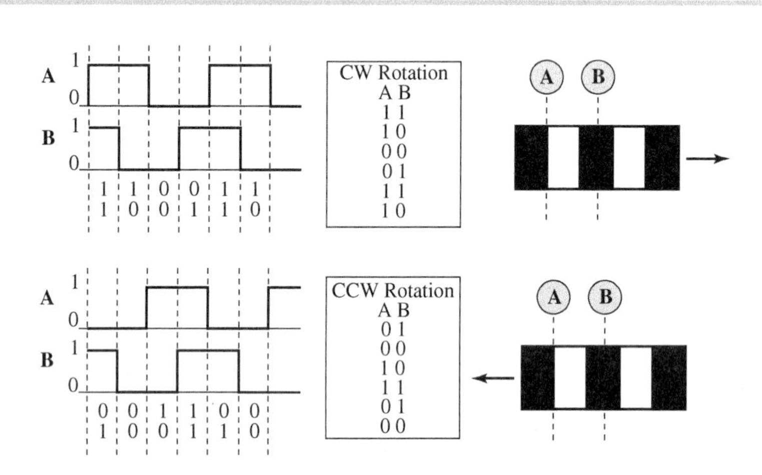

To understand this further, examine the *A* and *B* channel signal patterns for CW and CCW rotations that are shown in Figure 8.15. Notice that the output switches between one of four possible states for either rotation direction, and the order of these states is different for each direction. For example, if the photodetector sensor output is 00, then the next state will be 01 for CW rotation and 10 for CCW rotation. The different transitions for CW and CCW rotations enable one to write state-transition logic to determine the direction of rotation. Figure 8.16 shows an example of a state-transition diagram (see Section 6.5) that can do this job. Notice that, regardless of what state the output of the two sensors starts at, the diagram can determine the direction of rotation by examining the transitions from any one of the four possible states.

Figure 8.16

State-transition diagram for an incremental encoder

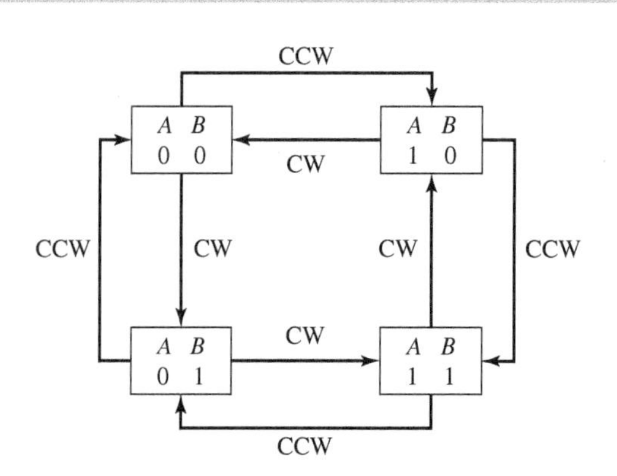

As noted with the use of two sensors, we get four distinct states for each strip on the disk. Thus, if an encoder has 1000 strips (or lines), we will get 4000 distinct states per one revolution of the encoder disk. Thus, the use of two sensors improves the resolution of the encoder by a factor of 4 since we can count 4000 leading and trailing edges per revolution compared to counting 1000 pulses per revolution for a single sensor. An encoder that gives four times the number of lines is operating in **quadrature** mode. In practice, the number of counts per second can get very large. For example, if the 1000-line encoder was rotating at 1000 rpm, then we would get 66.6e3 counts per second if the encoder were operating in quadrature mode. Most PCs or microcontrollers cannot keep up with counting at this rate if the output from the *A* and *B* channels is directly connected to the digital input port of the PC/microcontroller. In practice, hardware counters are used to process the *A* and *B* signals instead of using a software counting solution. These dedicated counters implement logic very similar to the method shown in Figure 8.16. The counter value is incremented by 1 on each state transition if the motion happens to be in one direction and is decremented by 1 for a motion in the opposite direction. To get the current position information, the PC or the micro-controller simply reads the output of the hardware counter. Thus, the processor does not have to worry about accuracy problems resulting from failing to count a particular transition. Example 8.4 illustrates the application of encoders.

Example 8.4 Incremental Encoder

A DC motor equipped with an incremental optical encoder is used to drive a lead-screw positioning table, as shown in Figure 8.17. The screw has a lead of 0.1 in./rev., the encoder disk has 1000 lines, and the encoder is operated in quadrature mode. Determine the measurement resolution of this encoder for the following.

(a) The setup shown in Figure 8.17
(b) The motor replaced with a geared one with a 5:1 gear ratio

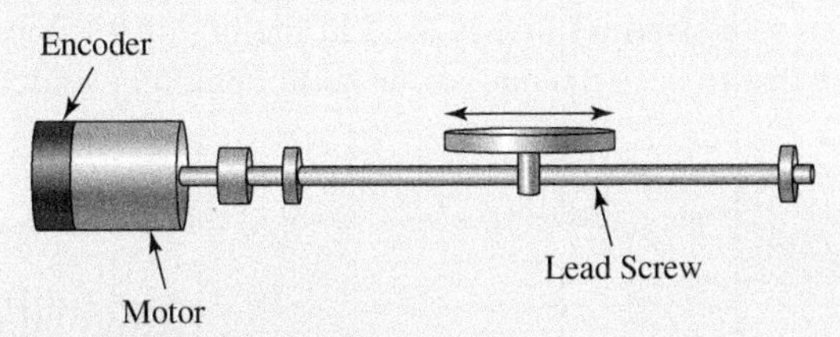

Figure 8.17 Positioning table system

Solution:

(a) The table travels a distance of 0.1 in. or 2.54 mm per one revolution of the motor. During this interval, the encoder will output 4000 counts (1000×4). Thus, the measurement resolution of this encoder setup is

$$2.54 \text{ mm}/4000 \text{ counts} = 0.635\ \mu\text{m per count}$$

(b) The encoder is mounted on the input side of the motor. Thus, if the motor rotates one revolution, the lead screw will rotate 0.2 revolution due to the use of a 5:1 gear on the output shaft of the motor. Hence, the encoder will generate 4000 counts for 0.508 mm ($2.54 \text{ mm} \times 0.2$) displacement of the table, and the measurement resolution in this case is

$$0.508 \text{ mm}/4000 \text{ counts} = 0.127\ \mu\text{m per count}$$

Note that while the measurement resolution is high, the part (b) configuration is not normally used for high-precision applications due to backlash in the gears and lead screw. A linear encoder mounted directly on the table is used instead.

Figure 8.18 shows an example of a **commercial counter IC**. The LS7166 IC is a 24-bit counter that can count in different modes, including up/down and quadrature. From Example 8.4 using 1000 rpm rotation speed and a 1000-line encoder operating in quadrature mode, this counter can count an interval exceeding 250 s before overflow. The incremental encoder *A* and *B* lines are connected to the *A* and *B* pins on this IC. This counter uses an 8-bit (pins D0 through D7) three-state I/O bus to communicate with external circuits. An 8-bit bus is used instead of a 24-bit to reduce the number of wires needed but also to be compatible with most micro-controllers/external devices, which have a limited number of I/O lines.

An external device can read the counter value by simply sending a preset control value over the I/O bus. This causes the 24-bit counter value to be transferred to the output port of the counter or the output latch. The three-byte contents of the output latch are then transferred by performing three successive read operations of the output latch where (after each byte is read) the address pointer for the next byte is automatically incremented.

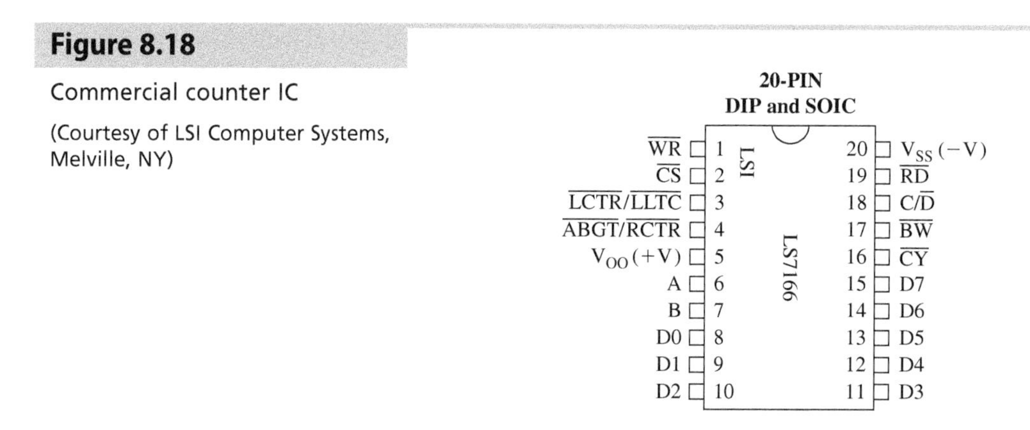

Figure 8.18

Commercial counter IC

(Courtesy of LSI Computer Systems, Melville, NY)

While incremental encoders do not have the limitations of potentiometers in terms of limited motion range and friction due to contact, they need to be 'homed' to establish reference information before they can be used in a motion-control application. In *homing*, the motor is rotated in one direction until a reference signal changes state. The output of the counter is recorded at this location, and displacements are measured with respect to this reference counter or home position. Most commercial incremental encoders have a third output called the **marker** or **z-channel** that is used in the homing sequence. Figure 8.19 shows a typical disk and the marker. Note that to have the marker, an additional light source and detector are needed.

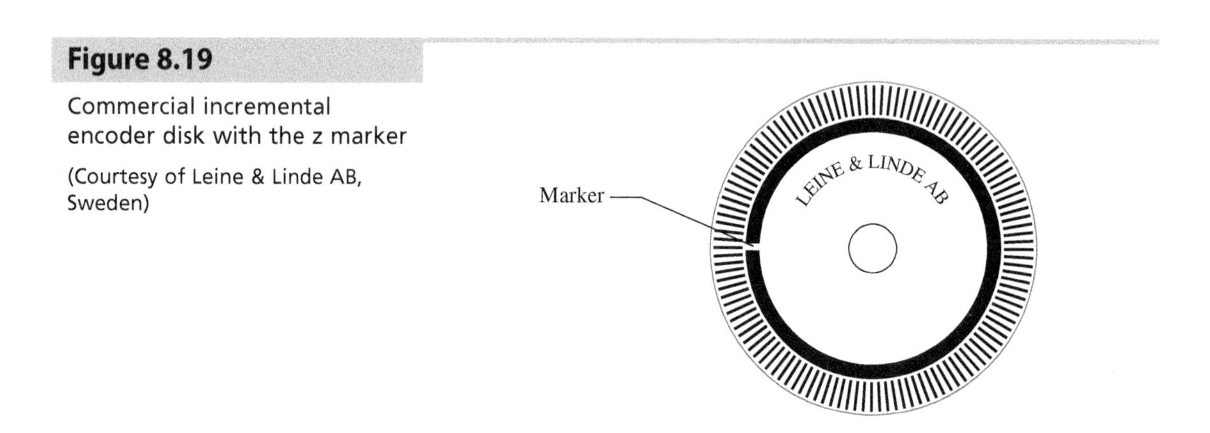

Figure 8.19

Commercial incremental encoder disk with the z marker

(Courtesy of Leine & Linde AB, Sweden)

A limitation of incremental encoders is that homing may not be safe to always perform. An example would be a robot arm that uses incremental encoders and is used for operations inside a vehicle frame or in regions with obstacles. If the robot happens to lose power while it is inside the vehicle, the robot will lose its current position information after the power is turned back on. In this case, the robot should not be homed automatically because of the possibility of the robot hitting the vehicle or one of the obstacles. It would be better in this case to use a position sensor that does not need to be homed. Such sensors are called absolute encoders and are discussed in the following section.

8.3.4 Optical Absolute Encoder

An optical absolute encoder (▶ available—V8.5) has different track information for different angular positions of the encoder disk. Figure 8.20 shows the layout of a commercial absolute encoder disk. There is no need for homing with an absolute encoder since each angular position of the disk has a unique output.

Figure 8.20

8-bit commercial absolute encoder disk

(Courtesy of BEI Sensors, Goleta, CA)

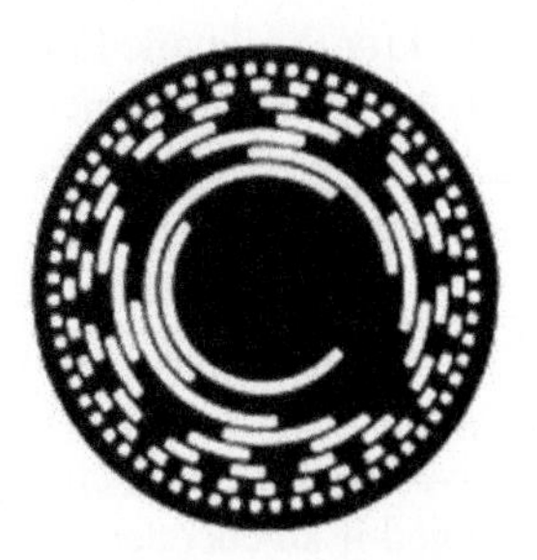

Absolute encoders are available with two different types of output: natural binary and gray code. In **natural binary**, the output of the encoder as the disk rotates changes in the normal way that binary numbers increase (i.e., 00, 01, 10, 11, . . .). In **gray code**, the output only changes one bit at a time as the disk rotates (i.e., 00, 01, 11, . . .). This makes gray code useful to reduce errors when reading the encoder if all the bits have not changed at the same time. Figure 8.21 shows the disk pattern (shown as linear for ease of display) and the corresponding output of the encoder for a 3-bit natural binary and gray code absolute encoder. Table 8.2 gives the binary output patterns for both types of encoders for the eight positions of the encoder disk.

Figure 8.21

Disk pattern and output from each track of an absolute encoder

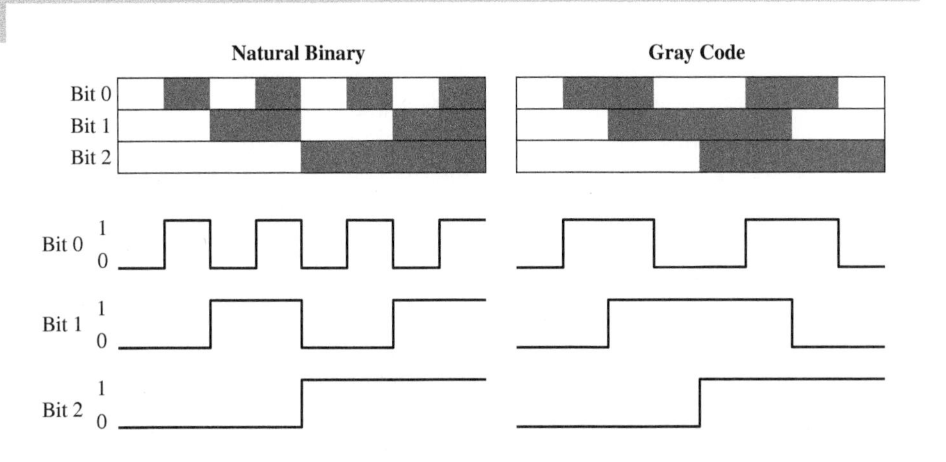

Table 8.2

Encoder output for natural binary and gray code

Position	Angular Segment (degrees)	Encoder Output	
		Natural Binary	Gray Code
1	0–45	000	000
2	45–90	001	001
3	90–135	010	011
4	135–180	011	010
5	180–225	100	110
6	225–270	101	111
7	270–315	110	101
8	315–360	111	100

Notice that this 3-bit absolute encoder (which has three tracks) can measure eight distinct absolute positions, each 45 degrees in size (i.e., 0–45, 45–90, . . . etc.). Commercial absolute encoders typically have 10-bit (or 10 tracks) or higher resolution, which gives them an angular resolution of 360/1024 degrees or less. With the use of a geared motor, the resolution of the angular measurement of the output shaft increases by the gear ratio factor.

To use an absolute encoder for **multi-revolution measurement**, multiple disks need to be used. A high-resolution disk is used for the detailed position information and one or more disks are used for counting turns. For example, to use an absolute encoder having a measurement range of 16 revolutions, two disks are used. The second disk will have four tracks (to indicate 16 different turns) and should be coupled to the high-resolution disk through a 16:1 gear ratio. If the primary disk has a 10-bit resolution, then this two-disk encoder will measure 16×1024 or 16384 discrete positions.

Commercial absolute encoders are available with different types of output. These include parallel digital output, which uses a single line for each bit. For a multi-turn encoder with a 14-bit disk, this results in an interface cable that has over 30 wires, which increases the cost of the encoder. A smaller-sized cable (and hence lower cost) is obtained if an encoder with SPI output is used. The SPI interface (see Chapter 5) uses only three wires for transmitting the data. Other output formats include DeviceNet™, Profibus, and Interbus.

8.3.5 Resolver

A resolver (▶ available—V8.6) is an absolute angular-displacement measuring device, similar to an absolute encoder, but giving analog voltages as an output. Resolvers were originally developed for military applications, and they normally are used in rugged, harsh environments where optical encoders may not be suitable.

There are several types of resolvers. The most common is the **rotary brushless resolver control transmitter**. A schematic of the construction of such a resolver is shown in Figure 8.22. It has two parts: a rotor and a stator. The rotor has a winding, called the reference winding, which gets energized by an AC voltage signal in a non-contact fashion using a rotary transformer. The stator has two windings, called the SIN and COS winding, which are offset from each other by 90°. The rotation of the rotor induces voltages in the SIN and COS windings. These voltages are a function of the angular position of the rotary shaft. The revolver gives two analog output signals: one from the SIN winding and the other from the COS winding. The ratio of the SIN and the COS outputs is the tangent of the shaft angle.

Figure 8.22

Schematic of brushless resolver control transmitter

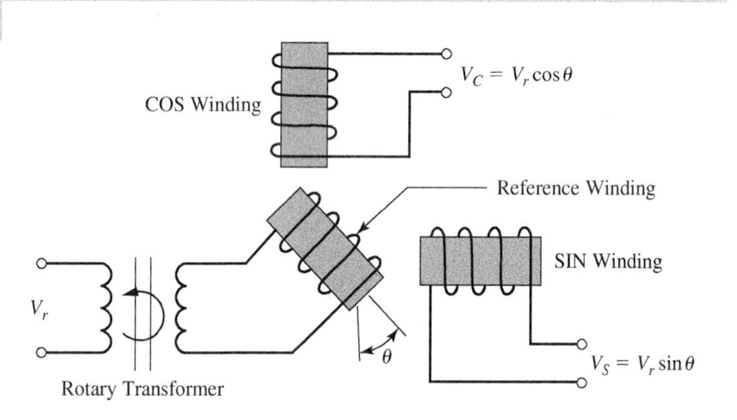

Much like absolute encoders, resolvers are available in single- or multi-turn configurations. The multi-turn configuration uses two resolvers that operate similarly to a vernier. Figure 8.23 shows several commercially available resolvers.

Figure 8.23

Commercially available resolvers

(Courtesy of TE Connectivity, Berwyn, PA)

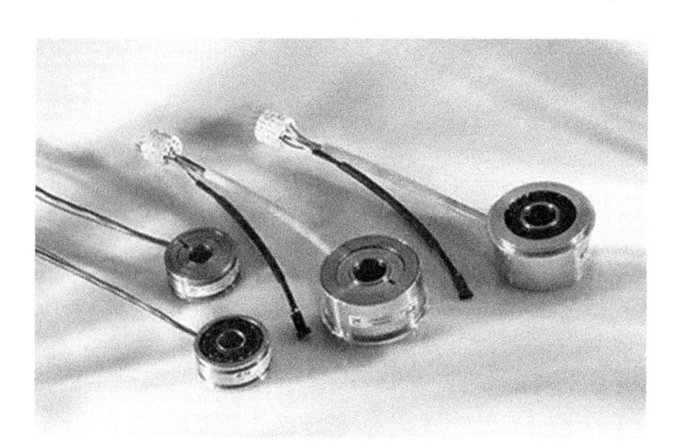

8.4 Proximity Measurement

A proximity sensor measures the presence or absence of an object. Proximity sensors are widely used in products in various industries, including automotive, appliance, and manufacturing. Examples include sensors to detect seatbelt on/off status in vehicles, door and lid open/close detection in appliances, obstacle presence in closing powered doors, rotor angle position in brushless DC motors, and end-of-travel detection in pneumatic actuators. There are several types of proximity sensors, including Hall-effect, inductive, capacitive, photoelectric, ultrasonic, and switch-type contact. Some of these sensor types will be discussed here.

8.4.1 Hall-Effect Sensors

A Hall-effect sensor (▶ available—V8.7) is a non-contact type sensor that is based on the Hall effect, which was discovered by Hall in 1877. The **Hall effect** states that a voltage difference is developed in a current-carrying conductor when subjected to a magnetic field. This voltage is perpendicular to both the current and the magnetic field.

The Hall effect is illustrated in Figure 8.24. Figure 8.24(a) shows a thin conducting plate in which a current is flowing. In the absence of an external magnetic field, there is no voltage difference across the sides of the plate. However, when a magnetic field is applied perpendicular to the direction of the current, as shown in Figure 8.24(b), it exerts a force on the charge carriers in the conductor. This force causes a redistribution of these charge carriers, leading to an accumulation on one side and a deficiency on the other, which results in a voltage difference known as the **Hall voltage**. This voltage can be described by

$$\vec{V}_H = \frac{\vec{I} \times \vec{B}}{n \times q \times d} \tag{8.2}$$

where $\boldsymbol{I}$ is the current vector, $\boldsymbol{B}$ is the magnetic flux vector, n is the charge carrier density, q is the elementary charge, and d is the thickness of the conductor. Note that the voltage difference is perpendicular to both the current flow and the magnetic flux direction. The amount of voltage that is generated is typically small and a differential amplifier is used to amplify this voltage signal.

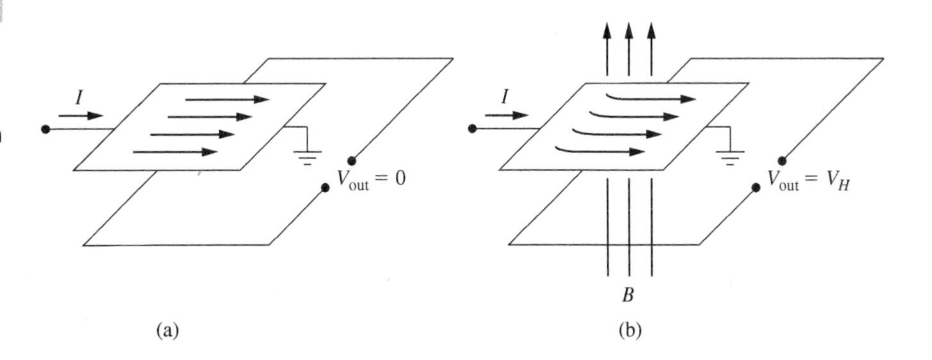

Figure 8.24

Illustration of the Hall effect: (a) current in a conductor with no magnetic field applied and (b) current in a conductor with a magnetic field perpendicular to the current flow

Hall-effect sensors are solid-state sensors that are constructed using semiconductor processing techniques. A Hall-effect proximity sensor consists of two pieces: a stationary sensor package and a magnet that is attached to the object whose presence needs to be detected, as seen in Figure 8.25. The magnet and the sensor package are separated by an air gap. There are two variations of Hall-effect sensors: unipolar and bipolar. In the **unipolar** design, when a south pole magnet approaches the designated package surface within a specified distance, the sensor turns ON. When the magnet is removed, the sensor turns OFF. In the **bipolar** design, the removal of the south pole does not cause the sensor to turn OFF; a north pole needs to approach the sensor to cause the sensor to switch OFF. A typical circuit for a Hall-effect digital proximity switch is shown in Figure 8.26.

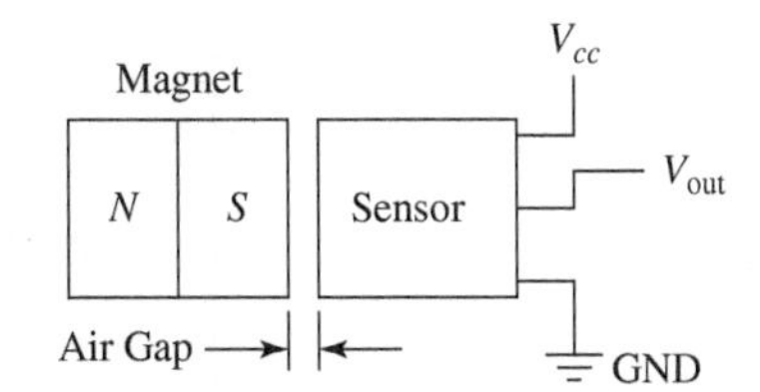

Figure 8.25

Hall-effect proximity sensor

Figure 8.26

Hall-effect proximity switch wiring

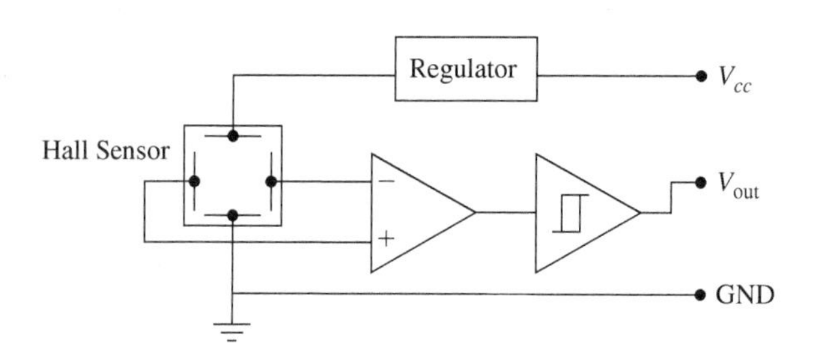

In this circuit, the supply voltage is connected to the Hall sensor through a voltage regulator. The Hall-effect voltage is processed through a differential op-amp (see Section 2.9.4) to amplify the voltage generated by the Hall-effect sensor. The output of the differential op-amp is connected to a **Schmitt trigger**. The Schmitt trigger (see Section 2.9.3) compares the output voltage from the differential op-amp to a preset voltage level. If the output voltage exceeds the preset voltage, the switch output will be set high. When the differential op-amp output falls below a threshold level, the switch output will be set low. The hysteresis of the Schmitt trigger is used to reduce the sensitivity of the sensor to noise and false triggering. The Schmitt trigger output can also be connected to a switching transistor.

An example of a commercially available Hall-effect sensor is shown in Figure 8.27.

Figure 8.27

Commercially available Hall-effect sensor

(Courtesy of OPTEK Technology, Carrollton, TX)

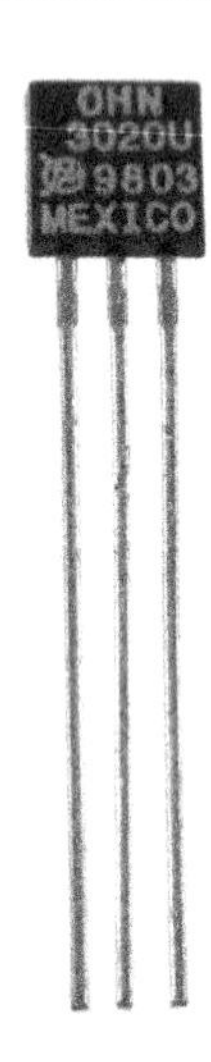

8.4.2 Inductive Proximity Sensors

Inductive proximity sensors (▶ available—V8.8) utilize the eddy current generated when a metallic element is placed within the proximity of an electromagnetic coil. The principle of operation of the sensor is shown in Figure 8.28. The sensor has an oscillator circuit that creates a magnetic field in front of the sensor through the coil inductance. When a metal target enters this magnetic field, it changes the magnetic field in the oscillator. This causes a swirling current, called an **eddy current**, to be generated in the coil. The change in current in the coil is detected by a circuit that is connected to a switching amplifier. The oscillator, the current detection, and the switching-amplifier circuits are all normally housed within the resin of the sensor.

Figure 8.28

Inductive proximity sensor

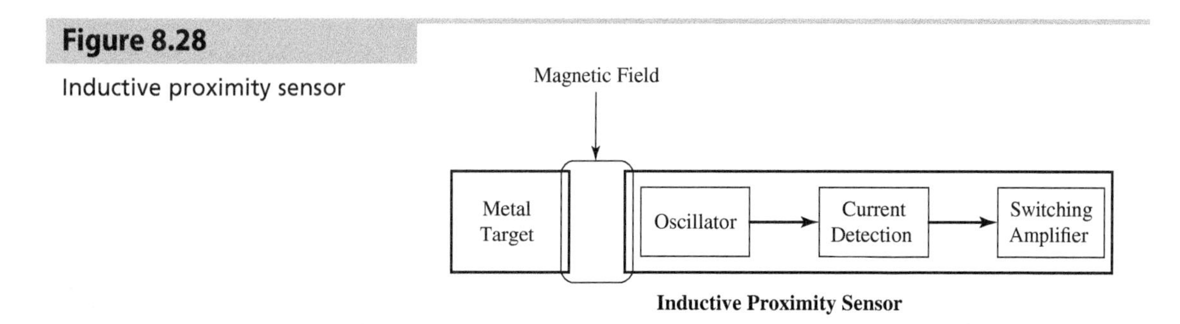

Figure 8.29(a) shows commercially available inductive proximity sensors. While most inductive sensors are cylindrical, rectangular-shaped sensors are also available. Cylindrical-shaped sensors are available with threaded or flat surfaces. Some units have an LED built into the sensor head to indicate object detection. The sensor electronics can be built into the sensor head or located separately from the head. Unlike Hall-effect sensors in which the target material is magnetic, inductive proximity sensors detect all metal objects at distances ranging from 1 to 30 mm. The larger the size of the sensor, the longer the detection range is. Standard inductive proximity sensors have a reduced detection range for nonferrous metals (such as copper, aluminum, and brass) than for ferrous metals (such as steel and iron, see Figure 8.29(b)). For non-metal objects, capacitive-type sensors can be used instead.

Figure 8.29

(a) Commercially available inductive proximity sensors and (b) detection range for a typical sensor for different metals

(Courtesy of Omron Corporation)

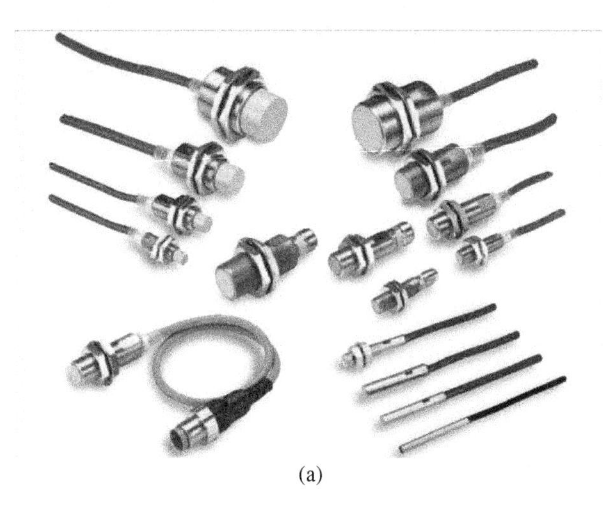

(a)

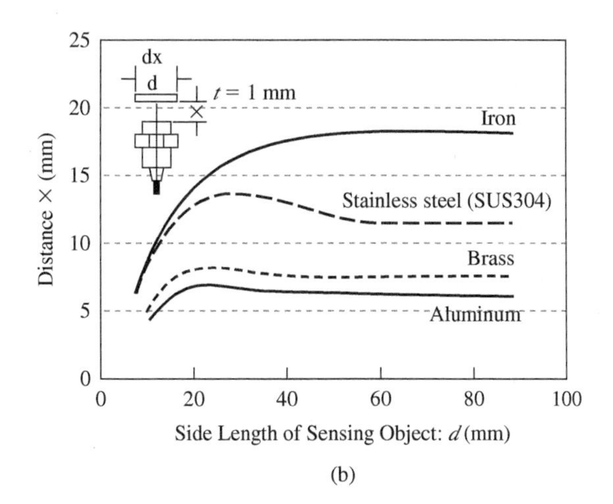

(b)

Inductive proximity sensors, as all on/off sensors or switches, are available in either NO or NC switch configuration. Furthermore, wiring to these sensors is available in either a two- or three-wire configuration. In the three-wire configuration, the output is available with either NPN or PNP transistor configuration. Example 8.5 illustrates the wiring circuit for a two-wire NO inductive proximity sensor used as a switch in a relay circuit.

Inductive proximity sensors are also used to detect vehicle presence at **intelligent traffic lights**. These traffic lights are commonly used in rural or country roads where the traffic volume is variable. The sensor takes the form of a wire loop that is placed in a groove that is cut into the asphalt surface. When a vehicle passes over the loop, the inductance of the loop is affected by the presence of the metallic body of the car. The electronics sense the vehicle's presence and use this information to adjust the traffic light timing.

Example 8.5 Two-Wire Inductive Proximity Sensor

Draw a wiring circuit for a two-wire NO inductive proximity sensor used as a switch in a relay circuit. The output circuit of the sensor is shown on the left side of Figure 8.30.

Solution:

The circuit is included in Figure 8.30. The supply voltage (typically 24 VDC) is connected to one end of the relay coil through a resistor *R*. The value of *R* is chosen to meet the current limit through the coil. The resistor R and the relay coil resistance will act as the load resistor on the sensor output circuit. The other end of the relay coil is connected to the load input on the sensor circuit. The other wire of the sensor circuit is connected to ground. Since this is a normally open sensor switch, the relay coil will not energize unless an object came within the detection range of the sensor. The detection of an object will thus cause the relay switch to close and to transmit power to the load connected to the relay.

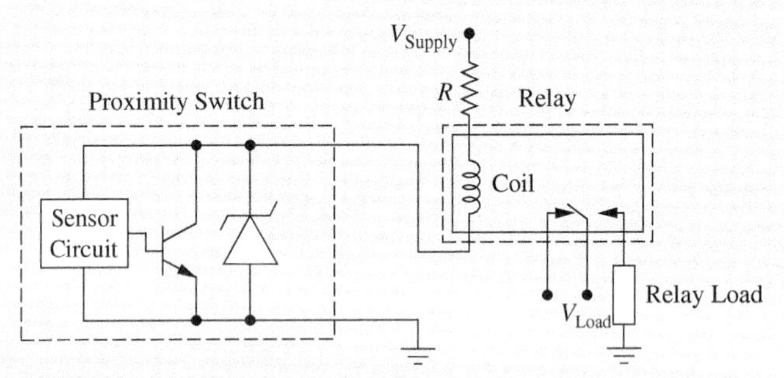

Figure 8.30 Wiring for a two-wire inductive proximity sensor

8.4.3 Capacitive Proximity Sensors

Capacitive proximity sensors (▶ available—V8.9) and inductive proximity sensors may appear similar in appearance, but they operate based on different principles. Capacitive proximity sensors operate based on the principle of measuring changes in capacitance. The capacitive sensor contains a conductive plate or electrode that generates an electrostatic field around it, unlike inductive proximity sensors which generate a magnetic field. When a target object comes near the sensor, it alters the electrostatic field, leading to a change in capacitance between the sensing electrode and the target object. The capacitance increases as the object gets closer to the sensor. The sensor is equipped with an oscillator circuit that generates an oscillating electrical signal. The change in capacitance caused by the target object alters the frequency or amplitude of the oscillator's output signal. The sensor's electronics process the changes in the oscillator's output signal and convert them into a usable digital switching output signal that indicates the presence or absence of the target object. In contrast to inductive proximity sensors, capacitive sensors can detect a wide range of materials, including both ferrous and non-ferrous metals as well as non-metal objects such as liquids and plastics. They also offer a larger detection range compared to inductive sensors. Additionally, capacitive sensors provide the flexibility to adjust their sensitivity for detecting specific targets, which is not possible with inductive sensors.

Figure 8.31 shows two examples of capacitive proximity sensors. Figure 8.31(a) shows a capacitive sensor with a detection range of up to 25 mm and with adjustable sensitivity. Figure 8.31(b) shows a sensor specifically designed for detecting the liquid level in a pipe. This sensor is unaffected by the color of the pipe or liquid.

Figure 8.31

(a) Long-distance capacitive sensor with adjustable sensitivity (Model E2K-C) and (b) liquid-level capacitive sensor (Model E2K-L)

(Courtesy of Omron Corporation)

(a)

(b)

Comparable to inductive proximity sensors, all on/off sensors or switches, are available in either NO or NC switch configuration. Furthermore, wiring to these sensors is available in either a two- or three-wire configuration. In the three-wire configuration, the output is available with either NPN or PNP transistor configuration.

8.4.4 Ultrasonic Sensors

Ultrasonic sensors (▶ available—V8.10) operate by emitting sound waves and detecting their reflections to discern the presence, absence, or distance of objects. Among the common types are **through-beam ultrasonic sensors**, which utilize separate transmitter and receiver units. The receiver identifies any disruption in the continuous sound wave sent by the transmitter, making it especially sensitive and reliable for object detection. **Proximity (or diffuse) ultrasonic sensors** combine the transmitter and receiver into a single unit. These sensors ascertain the presence or distance of an object by gauging the travel time of a high-frequency sound wave that reflects off an intervening object. The sensor's transducer periodically emits a burst of sound at high frequencies (30 kHz or higher) for a brief interval. Post-transmission of this burst signal, the sensor transitions to receiving mode, capturing the time of the echoed signal's arrival. With the speed of sound known in the transmission medium, such as air, the interval between the source signal's transmission and the reflected signal's receipt is used to deduce the object's position. **Retro-reflective ultrasonic sensors** are designed such that, in the absence of any intervening object, sound waves return to the sensor from a designated reflector. The appearance of an object between the sensor and the reflector interrupts this reflection, activating the sensor.

In comparison to inductive proximity sensors, ultrasonic sensors boast a significantly larger detection range (measured in meters versus millimeters). However, some ultrasonic sensors might struggle to detect very close objects (just a few centimeters away) due to potential overlap between transmission and reflection times. The detectability of objects in relation to the frequency of the sound signal is such that higher frequencies are better suited for smaller objects, while lower frequencies can detect larger ones. Ultrasonic proximity sensors can offer an analog output voltage related to an object's distance from the sensor or provide a binary output indicating the object's presence or absence within a predefined zone. Common applications include liquid-level measurements and production-line object detection. A noteworthy characteristic of ultrasonic sensors is their indifference to the color, transparency, or lighting conditions of the detected object. However, they may falter when tasked with detecting materials that absorb high-frequency sound, such as cotton or sponge.

A commercial ultrasonic sensor for use in industrial applications is shown in Figure 8.32. This through-beam sensor has a detection range of 500 mm and is suitable for the detection of transparent films, transparent bottles, and other similar workpieces. A low-cost proximity ultrasonic sensor (HC-SR04) that is widely used in educational mobile robots to detect obstacles and walls is shown in Figure 8.33. This sensor has a detection range of up to 4 m, operates at a frequency of 40 kHz, and is composed of two ultrasonic transducers, one to transmit ultrasonic sound pulses and the other to receive the reflected waves. There are several libraries, such as *NewPing,* available for the Arduino platform to process the signals from this sensor which simplifies the use of this sensor.

Figure 8.32

Compact ultrasonic sensor (Model E4E2)

(Courtesy of Omron Corporation)

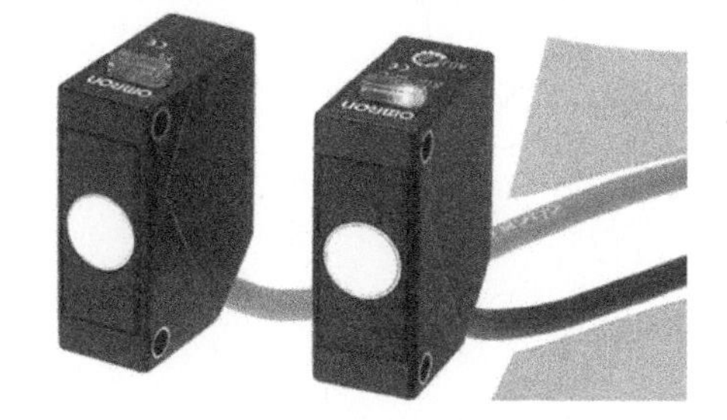

Figure 8.33

HC-SR04 ultrasonic sensor

(Jouaneh, University of Rhode Island)

8.4.5 Contact-Type Proximity Sensors

Contact mechanical switches known as '**limit switches**' are used in robotic and machine tool applications to detect the end of travel for a moving axis. They are also used in conveyor systems and transfer machines to detect objects and packages, as well as in elevators, scissor lifts, and safety guarding applications. These sensors are available with different 'operator' types that provide the interface between the contact object and the switch mechanism. These types include a roller plunger, a dome plunger, a roller lever, a telescoping arm, and a short lever. Figure 8.34 illustrates a few of these operator types. These sensors are rugged and can be used in harsh situations.

Figure 8.34

Operator types for limit switches

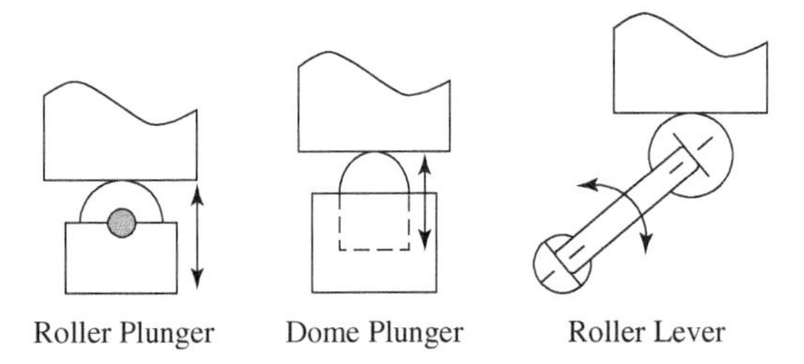

8.5 Speed Measurement

8.5.1 Tachometer

While speed information can be obtained by differentiating the position data, this approach is not very desirable as it amplifies the noise if the position signal is noisy. A better method is to use a sensor that can directly provide the speed information. A **tachometer** (see Figure 8.35) (▶ available—V8.11) is a speed-measuring device that provides an analog output voltage that is proportional to the speed. A tachometer is constructed similarly to a brush DC motor (see next chapter), but it is designed to operate in reverse. When the tachometer shaft rotates, the tachometer gives a DC output voltage. A characteristic of a tachometer is its sensitivity, which refers to the output voltage of the tachometer for a given speed. It is normally reported as several volts per 1000 rpm, but other speed units can be used. Another characteristic of a tachometer is its ripple. **Ripple** refers to the AC component of the output signal. Due to the use of a commutator in the construction of the tachometer, the output signal of the tachometer exhibits fluctuation which can be as high as 3 to 4% of the nominal output voltage.

Figure 8.35

A tachometer

(Jouaneh, University of Rhode Island)

The ripple affects the operation of a closed-loop speed control system since the control system responds to variation in the tachometer output voltage regardless of whether the variation is caused by a speed change in the motor or is due to the ripple effect. One way to eliminate ripple is to place a low-pass filter (see Section 7.6) constructed using an RC circuit on the output leads of the tachometer (see Figure 8.36). The R and C values should be chosen such that the cut-off frequency of the RC filter is below the ripple frequency. However, the use of an RC filter changes the dynamics of the feedback system.

Figure 8.36

RC filter on tachometer leads

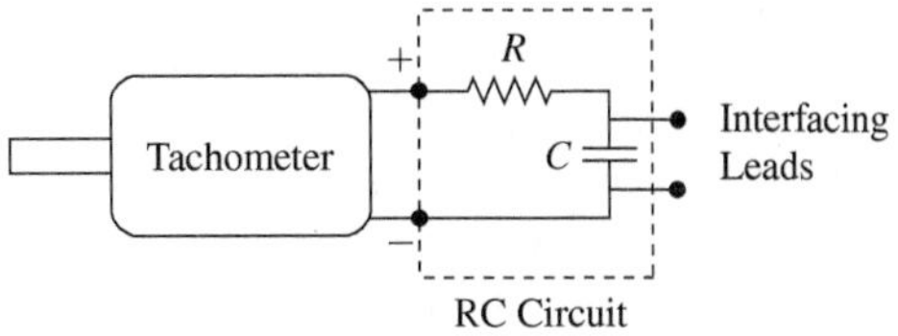

Figure 8.37 shows the output speed of a DC motor tachometer with and without using an RC filter to eliminate ripple.

Figure 8.37

Output speed of DC motor tachometer with and without an RC filter

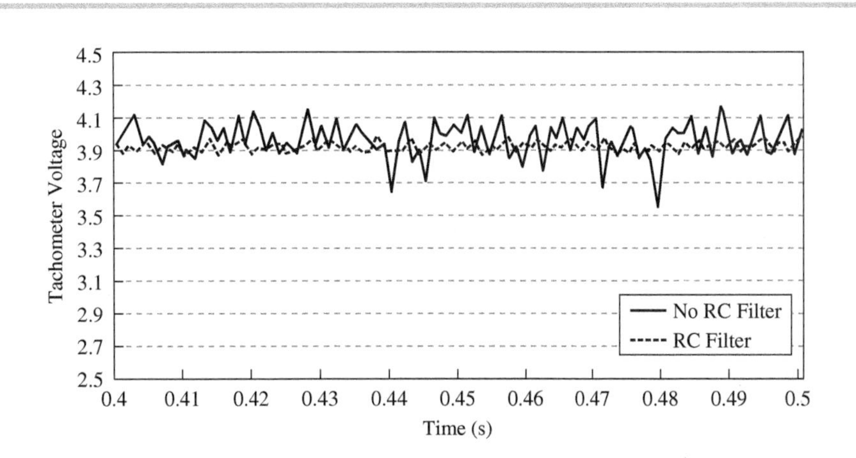

8.5.2 Encoder

Incremental encoders are also used to measure speed. Two techniques are normally used. These are the pulse-counting method and the pulse-timing method. In the **pulse-counting** method, the encoder count values are read at a certain fixed sampling frequency. The speed is obtained by dividing the difference between two successive encoder counter readings by the sampling time interval. If the encoder disk has l lines, the sampling interval is T in seconds, and the count difference between two successive readings is N, then the angular speed in rad/s is given by

$$\omega = \frac{2\pi N}{lT} \tag{8.3}$$

If quadrature is used, the previous expression needs to be multiplied by 1⁄4. The resolution of this technique increases with an increase in the speed because more counts are generated in the given sampling interval as the speed increases.

In the **pulse-timing** method, a high-frequency clock is used to record the time interval for the motion travel between two adjacent lines on the encoder disk. Assuming a clock frequency of f cycles/s, an encoder disk with l lines, and m clock cycles recorded, the angular speed ω is given by

$$\omega = \frac{2\pi/l}{m/f} = \frac{2\pi f}{ml} \tag{8.4}$$

This technique is particularly suitable for low-speed measurements. Note that as the speed increases, the resolution of this pulse-timing method decreases, since fewer clock cycles are used to record the motion travel between two adjacent lines on the encoder disk.

8.6 Strain Measurement

Strain is a basic quantity in solid mechanics. When a force (torque) acts on a member, it leads to a deformation of the member. The deformation is expressed in terms of strain. For elastic loading, the resulting stress (σ) and strain (ϵ) are linearly related through the modulus of elasticity of the material, E, or

(8.5)
$$\sigma = \epsilon E$$

Strain is quantified using a **strain gauge** (▶ available—V8.12), a device whose resistance changes proportionally to the mechanical deformation (or strain) it undergoes. The metal-foil strain gauge, illustrated in Figure 8.38, is the most prevalent type in modern applications. It represents an advancement over the earlier wire-resistance strain gauges that were developed over 80 years ago. There are also semiconductor strain gauges which, while offering sensitivities more than 100 times that of metallic gauges, are more susceptible to temperature effects.

Figure 8.38

Metal-foil strain gauge

(David J. Green - technology/Alamy Stock Photo)

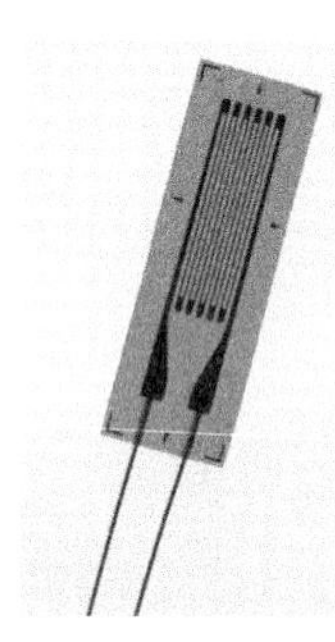

The metal-foil strain gauge consists of a metal alloy foil, typically constantan, in the form of a grid placed on a flexible polyimide backing. The backing material serves as an electrical insulator from the metal part to which the gauge is attached. The gauge is bonded using adhesive to the surface of the part whose strain needs to be measured. Some gauges are made with the lead wires already attached to the gauge, but other gauges provide an area where one can solder the lead wires. Standard rectangular gauges are made with grid gauge lengths varying from 1.5 to 25 mm and grid gauge widths varying from 1.2 to 8 mm.

Strain is defined as

(8.6)
$$\epsilon = \frac{\Delta l}{l}$$

where Δl is the change in length of a part of length l. Using a strain gauge, the measured strain is obtained using the relationship

(8.7)
$$\epsilon = \frac{1}{F}\frac{\Delta R}{R}$$

where F is called the **gauge factor** and its value is provided by the strain gauge manufacturer. We will show next how Equation (8.7) was obtained. For this, consider a bar with a cross-sectional area A and length L. The resistance of the bar is given by

(8.8)
$$R = \frac{\rho L}{A}$$

where ρ is the resistivity of the material. The area can be expressed as $A = CD^2$, where C is a constant and D is the section dimension. For a square section, $C = 1$; for a circular section, $C = \pi/4$; etc. Equation (8.8) can be written as

(8.9)
$$R = \frac{\rho L}{CD^2} = f(\rho, L, D)$$

Differentiating Equation (8.9), we get

(8.10)
$$dR = \frac{\partial f}{\partial \rho}d\rho + \frac{\partial f}{\partial L}dL + \frac{\partial f}{\partial D}dD = \frac{L}{CD^2}d\rho + \frac{\rho}{CD^2}dL - 2\frac{\rho L}{CD^3}dD$$

Dividing Equation (8.10) by Equation (8.9), we get

$$\frac{dR}{R} = \frac{d\rho}{\rho} + \frac{dL}{L} - 2\frac{dD}{D} \quad \textbf{(8.11)}$$

but

$$\epsilon_a = \frac{dL}{L}, \epsilon_l = \frac{dD}{D} \quad \text{and} \quad \nu = -\frac{\epsilon_l}{\epsilon_a} \quad \textbf{(8.12)}$$

where ϵ_a is the axial or longitudinal strain, ϵ_l is the lateral strain, and ν is the Poisson's ratio. Replacing the terms in Equation (8.11) by the equivalent terms in Equation (8.12), we get

$$\frac{dR}{R} = \frac{d\rho}{\rho} + \epsilon_a + 2\nu\epsilon_a \quad \textbf{(8.13)}$$

Dividing Equation (8.13) by ϵ_a, we get

$$\frac{dR/R}{\epsilon_a} = \frac{d\rho/\rho}{dL/L} + 1 + 2\nu \quad \textbf{(8.14)}$$

The right-hand side of Equation (8.14) is termed the gauge factor F. The value of F depends on the Poisson's ratio ν of the strain gauge material, as well as on how its resistivity changes with strain. For Constantan, F is 2.0. From Equation (8.14), if we replace dR with ΔR we thus get

$$\epsilon = \frac{1}{F}\frac{\Delta R}{R} \quad \textbf{(8.15)}$$

In most cases, the strain is a small quantity, and the term **microstrain** is used where the strain is multiplied by one million. Strain gauges are manufactured with standard resistances ranging from 120 to 1000 Ω, with both 120 Ω and 350 Ω being common, especially the 350 Ω in applications like load cells. When these gauges are subjected to mechanical deformation, the change in their resistance is typically small, often amounting to less than a fraction of one percent of the gauge's nominal resistance. Example 8.6 illustrates this point. To improve sensitivity and compensate for factors like temperature-induced resistance changes, strain gauges are typically incorporated into a Wheatstone bridge circuit (see Section 8.11.3). This configuration can measure resistance changes more precisely than a standard ohmmeter, especially when detecting the small resistance changes associated with strain

Strain gauges are used in a variety of applications. In addition to their use in directly measuring the strain and the resulting stresses on members subjected to loading, they are also used in the construction of force and torque sensors (see next section), some types of pressure sensors, and temperature sensors, since they can measure the elongation due to a temperature change. Due to its finite size, a strain gauge measures only the average strain over an area and not the exact strain. This approximation is acceptable in cases where the strain is uniform, but it can lead to errors in cases where the strain changes considerably, such as in stress concentration areas.

Example 8.6 Strain Under Axial Loading

A 2-cm diameter steel ($E = 200 \times 10^9$ N/m^2) rod is subjected to a tensile axial force of 2500 N. Assume a strain gauge with a resistance of 120 Ω and a gauge factor F of 2 is used to measure the strain due to this loading. Determine the change in resistance of the gauge under this loading.

Solution:

The stress due to this loading is given by

$$\sigma = \frac{F}{A} = \frac{2500}{\pi(0.01^2)} = 7.96 \text{ MPa}$$

The strain is obtained from Equation (8.5), which gives

$$\epsilon = \frac{\sigma}{E} = \frac{7.96 \times 10^6}{200 \times 10^9} = 39.8 \text{ microstrain}$$

The change in resistance of the strain gauge is then given by

$$\Delta R = \epsilon F R = 39.8 \times 10^{-6} \times 2 \times 120 = 0.00955\ \Omega$$

Note that the change in resistance is very small (0.008%) and cannot be precisely read from an ordinary ohmmeter, which does not have such a sensitivity.

While the single, linear-pattern strain gauge (Figure 8.38) is common, strain gauges are also made with many other configurations. These include the dual-grid gauge (Figure 8.39(a)) that is typically used to measure bending strain, the biaxial strain gauge (Figure 8.39(b)) to measure axial strain where the principal strain directions are generally known such as in pressure vessels, and the three-element rosette (Figure 8.39(c)) to measure strain in cases where the principal strain directions are not known in advance. The biaxial and the three-element rosette gauges are available with the grids stacked as in Figure 8.39(b) or in planar form (Figure 8.39(c)). The stacked configuration is more compact, but it is stiffer and less conformable than its planar counterpart.

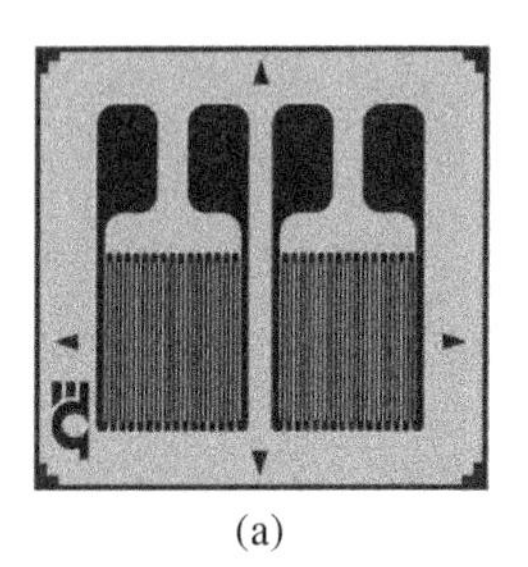

(a)

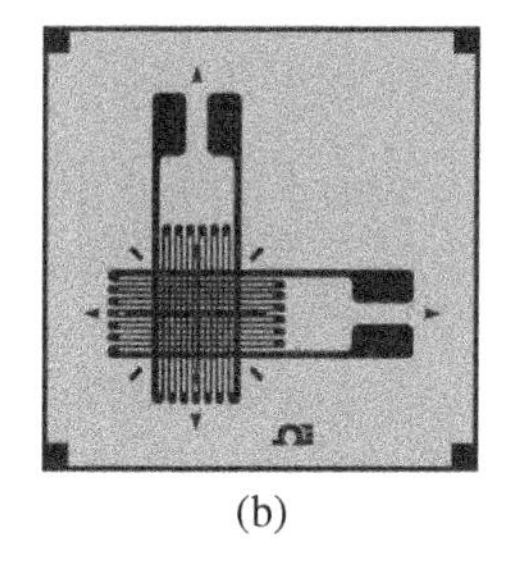

(b)

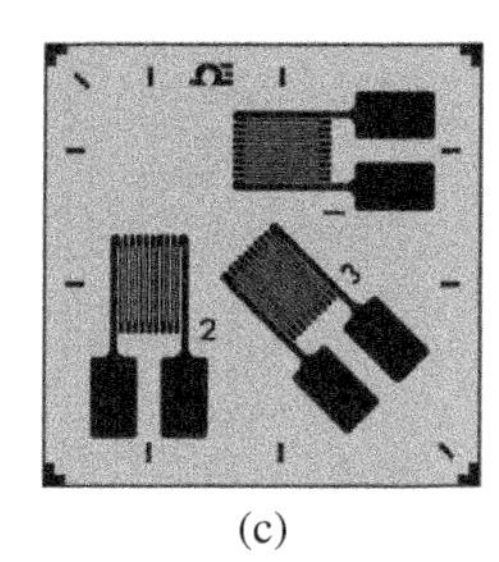

(c)

Figure 8.39

Other configurations of strain gauges: (a) dual-grid gauge, (b) biaxial, and (c) three-element rosette

(Reproduced with permission of Micro-Flexitronics Ltd. (MFL), courtesy of Omega Engineering, Inc., Stamford, CT 06907 USA www.omega.com)

8.7 Force and Torque Measurement

There are two methods to measure forces and torques. One is the direct comparison method, which is based on the use of some form of beam balance with known weights. The other is the indirect comparison method, which is based on the use of calibrated transducers. This textbook will focus on the second method since the output of the transducers can be easily interfaced with a PC or a microcontroller.

8.7.1 Force Sensors

Transducer-type force sensors or load cells (▶ available—V8.13) can be hydraulic, pneumatic, or strain-gauge-based. Hydraulic load cells measure the weight by sensing the pressure change in the fluid system, while pneumatic load cells measure changes in air pressure. Strain-gauge types are one of the most common types. They are based on the use of an elastic element combined with one or more strain gauges. The resistances of the gauges are processed by a Wheatstone bridge circuit (see Section 8.11.3). **Strain gauge load cells** are available in different configurations:

- Compression type
- Tension/compression type
- S-beam load cells
- Universal mounts
- Rectangular beam cells

A schematic of these configurations is shown in Figure 8.40. The compression-type cell is designed to handle compressive loads. It has a low profile and a small size, and it can be made to handle high loads that must be centered. The tension/compression type can handle both compressive and tensile center loads. The S-beam load cell can also handle both compressive and tensile loads, but it is better suited for harsh environments and offers good resistance to side loads. The universal mount is similar to the compression type, but it can handle off-center loads. The rectangular beam is a low-cost sensor that can handle compressive eccentric loading. This design is also known as the single-point load cell. Load cells typically use more than one strain gauge to increase the sensitivity of the sensor. In many cases, four strain gauges are used as seen in Figure 8.41, and they are laid out so that the change in resistance of the strain gauges under the applied loading adds to improve the output of the Wheatstone bridge (see Section 8.11.3). Strain gauges with a 350 Ω resistance are commonly used in load cells.

Figure 8.40

Different configurations of load cells

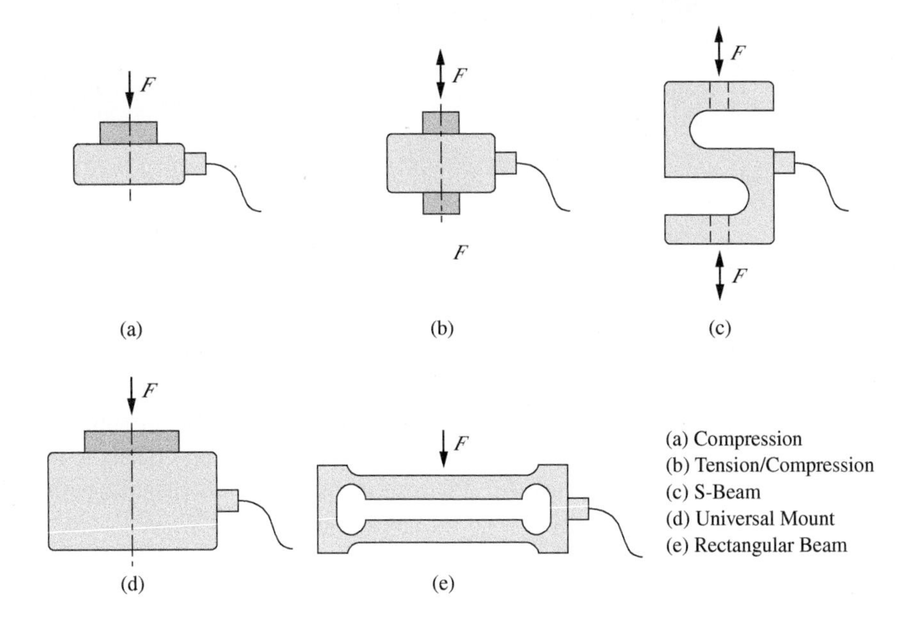

Figure 8.41

Four strain gauges used in a load sensor

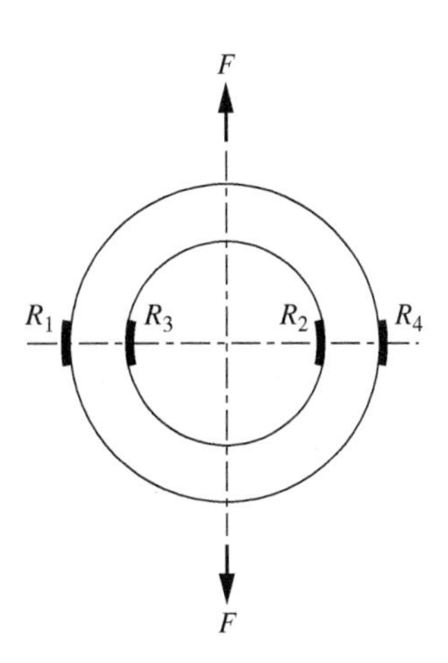

Manufacturers of load cells list the output of the load cells in mV/V (such as 2 mV/V). Due to the use of Wheatstone bridge circuitry, the output is directly related to the excitation input. For example, if the supply voltage is 10 V, then the full-scale output of the load cell is 20 mV for a load cell with a 2 mV/V output rating. To make this output more usable, especially with display devices or microcontrollers, an external amplifier can be employed to amplify the signal. Load cells are calibrated to ensure their outputs correspond to specific units of measurement, such as pounds (lbs) or Newtons (N). Like all sensors, load cells have inherent inaccuracies. They are particularly sensitive to thermal errors, often resulting from the thermal expansion or contraction of the materials used in the load cell's design.

8.7.2 Force-Sensing Resistor

A force-sensing resistor (FSR) (▶ available—V8.14) is a sensor that measures applied force through changes in its electrical resistance, and Figure 8.42 shows a photo of an FSR with a round active area. The sensor is constructed using a conductive polymer layer situated between two substrate layers, with the outer layers often having conductive materials serving as electrodes. When force is applied to the FSR, the conductive particles within the polymer come closer together, which causes a decrease in its resistance. However, it is important to note that this relationship is nonlinear; the resistance does not decrease proportionally with increased force.

Figure 8.42

Force-sensing resistor

(Courtesy of Interlink Electronics, Inc., Camarillo, CA)

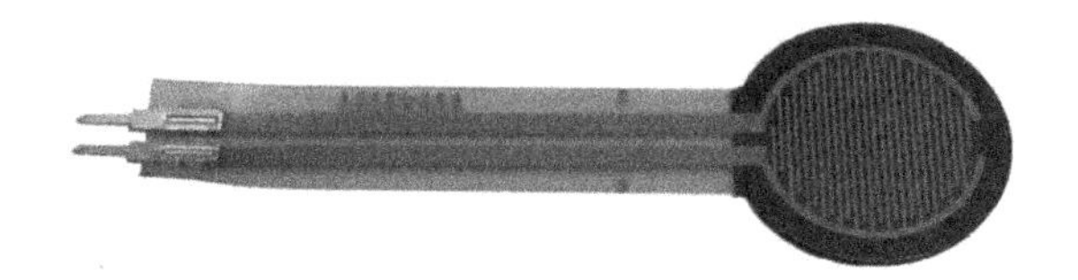

One of the primary applications of FSRs is in scenarios where the relative change or simple detection of force is more vital than the precise force measurement, such as in touch controls. Despite its low cost, thin profile, flexibility, and simplicity—highlighted by its two-lead design—the FSR comes with inherent disadvantages. It is known for its low accuracy, and its high resistance, especially under slight pressures, can lead to significant loading effects in circuits with lower input impedances. To address this, when integrating an FSR into circuits, it is often placed in a voltage divider configuration with a fixed resistor. The chosen value of this fixed resistor plays a crucial role in optimizing the sensor's sensitivity and range for specific applications. Moreover, to mitigate loading effects, a voltage follower (or buffer) circuit can be employed. Example 8.7 discusses the use of an FSR and its wiring to minimize loading effects. It is also worth noting that besides the low accuracy and potential loading effects, FSRs can exhibit hysteresis and might face durability concerns in high-cycle applications.

Example 8.7 Wiring of an FSR Sensor

Illustrate how a typical FSR sensor can be wired to minimize loading effects.

Solution:

A typical FSR wiring to minimize loading effects is depicted in Figure 8.43. An FSR is typically connected with one end to a voltage supply, V_{in}, and the other end to a pull-down resistor, R, grounded to create a voltage-dividing circuit (refer to Section 2.4). The voltage output taken from the point between the fixed pull-down resistor and the FSR serves as the FSR's output. The value of R usually lies between 10 and 100 kΩ. It is adjusted to fine-tune the sensitivity and range of the force sensor readings. Increasing the reference resistance R makes the sensor more sensitive (i.e., small changes in force result in larger voltage changes), but the downside is that the sensor can become saturated or maxed out at lower forces, hence reducing the overall force range over which the sensor can distinguish different force levels. The output from the voltage dividing circuit is subsequently passed to a voltage follower or buffer (refer to Section 2.9.3). This buffer offers a high input impedance to the FSR and a low output impedance to the following circuitry, ensuring precise voltage reading values.

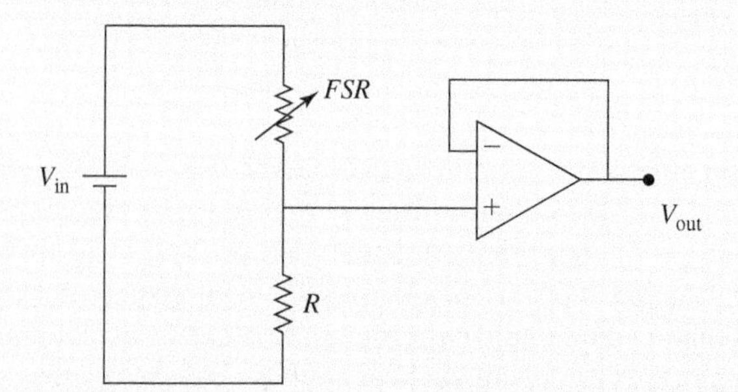

Figure 8.43 FSR wiring to minimize loading effects

8.7.3 Torque Sensors

Measurement of torque is done using two different configurations of sensors. These are the reaction torque sensors and the rotating torque sensors. Both configurations are based on the use of strain gauges that are mounted on elastic members. The elastic element in both configurations of torque sensors could be a solid or hollow circular shaft, a solid or hollow cruciform, or a solid square shaft (see Figure 8.44). Hollow cruciform is typically used for low-torque measurement applications, while solid circular and square shafts are used for high-torque applications.

Figure 8.44

Schematic of different elastic elements used in torque sensors

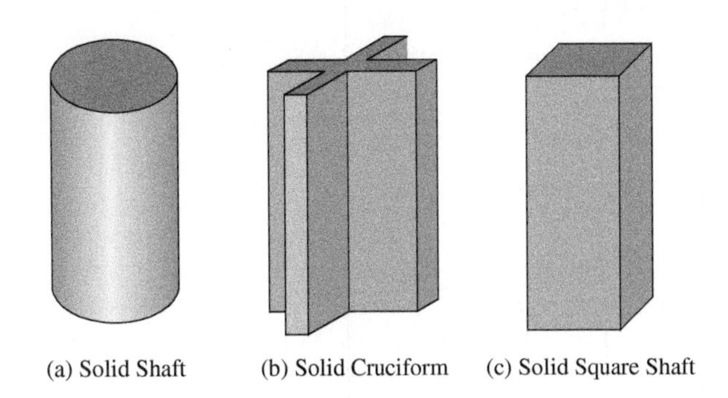

The **reaction torque sensor** (▶ available—V8.15) is used to measure torque in non-rotating applications. In this configuration, the sensor is stationary, and the shaft of the part of which the torque needs to be measured is connected through a coupling to the sensor. Reaction torque sensors are used, for example, to measure the motor torque output at zero speed or the starting torque. Other applications include bearing friction measurement and automotive brake torque sensing. The **rotary torque sensor** (▶ available—V8.16) on the other hand is used to measure torque between rotating devices. The sensor is typically mounted in-line between the torque source and the load. Typical applications for rotary torque sensors include engine dynamometer testing, fan and blower testing, and clutch testing. Figure 8.45 illustrates the use of these two sensors. Similar to load cells, torque sensors give an output voltage that is proportional to the applied torque.

Figure 8.45

Illustration of reaction and rotary torque sensors

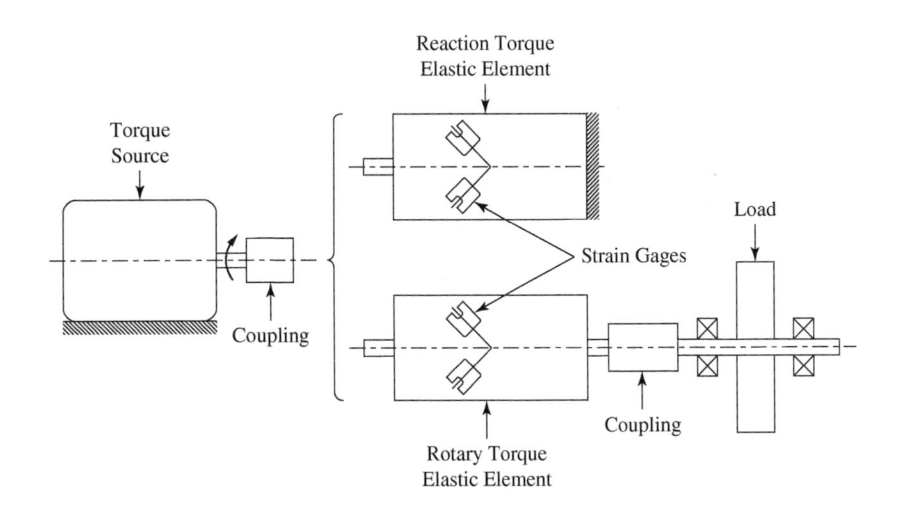

Since the sensing element is rotating in a rotary torque sensor, inertia effects are important. This is especially important during the power-up and power-down phases of the rotating member when the rotational speed is not constant. Thus, torque sensors with low inertia are desirable. Also, means must be provided to transmit the sensor signals from the rotating strain gauge transducer to the stationary electronics. Common methods for transmitting the signals include the use of slip rings and rotary transformers. Slip rings are similar to a commutator in a brush DC motor (Chapter 9) and are suitable for low–rotation speed applications. At speeds above 5000 rpm, the noise induced by brush friction makes them not very suitable. A rotary transformer is a non-contact device and is like a regular transformer, but the secondary coil is rotating relative to the primary coil. Two rotary transformers are used: one for transmitting the supply voltage to the Wheatstone bridge circuit and the other for transmitting the output from the bridge circuit (see Figure 8.46). Some rotary torque sensors also output the rotation angle of the sensor, which is obtained from an encoder that is built into the sensor. Reaction torque sensors have a higher torque-measurement capability than rotary torque sensors, and some units are made to measure torque values up to a few million-pound inches. End connections to both configurations include the use of a keyed shaft, a flange, or a spline. Figure 8.47 shows commercial reaction and rotary torque sensors with keyed shaft coupling.

Figure 8.46

Wheatstone bridge with rotary transformers

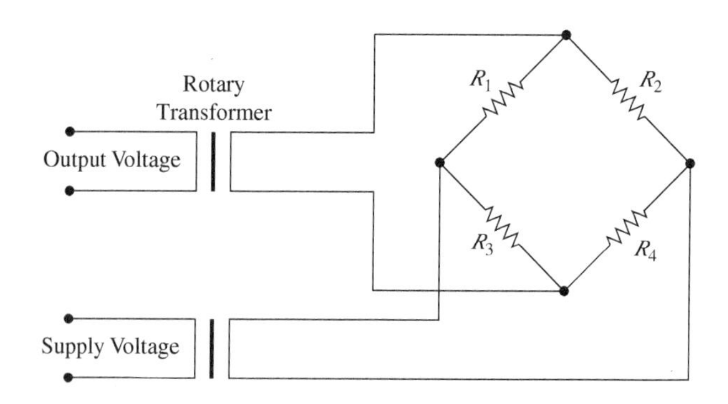

Figure 8.47

Commercial (a) reaction (Model CS1120) and (b) rotary torque (Model CD1095) sensors with keyed shaft coupling

(Courtesy of TE Connectivity, Berwyn, PA)

8.7.4 Multi-Axis Force/Torque Sensors

Many applications require the measurement of force and/or torque on more than one axis. These include robotic applications in which the robotic arm interacts with the environment such as assembly where force and torque sensing is used to detect misalignments for quality control and for ensuring proper assembly; material testing and research where force and torque measurements in multiple axes are used for structural analysis, material characterization, and studying mechanical properties under different loading conditions; and rehabilitation and prosthetic devices where force and torque sensors capture forces and torques applied during movements, enabling precise adjustments and customization of prosthetics for individual users.

A widely used multi-axes force/torque sensor is the six-axes force and torque sensor, also known as a six-axis load cell, which measures forces and torques along three orthogonal axes (X, Y, Z) and their corresponding moments or torques. The sensor typically consists of a rigid mechanical structure that can withstand and distribute applied forces and torques. Inside the sensor, strain gauges are bonded to the sensor's structure and strategically placed in specific locations to measure the strain caused by applied forces and torques. A picture of the inner details of a commercially available 6-axis sensor made by ATI Industrial Automation is shown in Figure 8.48. An inner hub to which the external forces and forces are applied is connected to the frame or outer wheel of the sensor through three beams or spokes that are spaced 120 degrees apart. Each beam has four strain gauges, with one gauge placed at each of the four surfaces of the beam. These gauges function in pairs, with each pair forming a half-bridge of the Wheatstone bridge (see Section 8.11.3) used to process the signals from these gauges.

Figure 8.48

Inner details of a commercial 6-axis force torque sensor

(Courtesy of ATI Industrial Automation, Apex, NC)

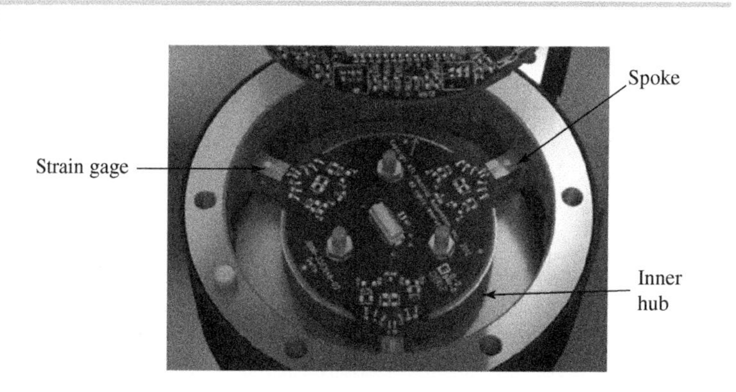

8.8 Temperature Measurement

Temperature is a basic quantity in process control systems, and there are several types of sensors available to measure temperature. These include thermistors, thermocouples, RTD, and IC sensors. These diverse types will be discussed in this chapter. The IC sensor is discussed in Section 8.10.1. Table 8.3 lists and compares several properties of these sensors.

Table 8.3

Comparison of different temperature sensors

Property	Thermistor	Thermocouple	RTD	IC
Resolution	Very high	Average	High	High
Temperature range	Small	Very broad	Broad	Limited
Output	Highly non-linear	Non-linear	Almost linear	Linear
Accuracy	Very High	Limited	High	Limited
Ruggedness	Fragile	Very rugged	Rugged	Fragile

8.8.1 Thermistors

A thermistor (▶ available—V8.17) is a resistance-based temperature measurement sensor made from semiconductor materials, giving it high sensitivity to temperature changes. Depending on the type of thermistor, its resistance can either decrease or increase with rising temperature. The most common type, the **negative temperature coefficient (NTC) thermistor**, sees its resistance decrease with increasing temperature, while the less common **positive temperature coefficient (PTC) thermistor** behaves oppositely. The output of a thermistor is notably nonlinear, requiring curve-fitting techniques for precise measurements. Typically, thermistors are used over a limited temperature range, often less than 300°C. A typical resistance versus temperature plot for a thermistor is shown in Figure 8.49. Note the highly nonlinear relationship between the temperature and resistance. An approximation of the resistance–temperature relationship for a thermistor is expressed by the exponential function

$$R = R_0 e^{\beta\left(\frac{1}{T} - \frac{1}{T_0}\right)} \tag{8.16}$$

where β is a constant that depends on the thermistor material used and R_0 is the resistance at the reference temperature T_0. In Equation (8.16), both T and T_0 are absolute temperatures that are measured in degrees Kelvin. Equation (8.16) can be written in the form shown in Equation (8.17) which allows the temperature measured by the thermistor to be determined from the measured resistance. A thermistor is sensitive to temperature changes, and the relationship between its resistance and temperature is consistent. With careful manufacturing and calibration, some thermistors can achieve a precision of 0.05°C or even better.

$$\frac{1}{T} = \frac{1}{T_0} + \frac{1}{\beta}\ln\left(\frac{R}{R_0}\right) \tag{8.17}$$

Figure 8.49

Typical resistance versus temperature plot for a thermistor

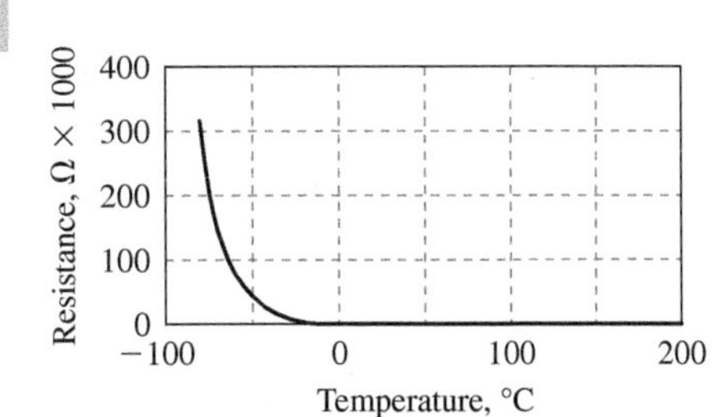

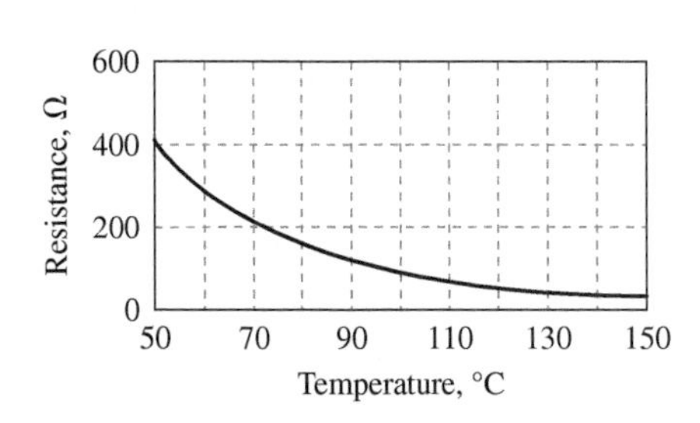

Thermistors are available in several forms including two-lead, surface mount, and leadless chip form. A typical two-lead thermistor is shown in Figure 8.50. For the two-lead thermistor, the thermistor can be epoxy coated or glass encapsulated.

Figure 8.50

A leaded thermistor

(sciencephotos/Alamy Stock Photo)

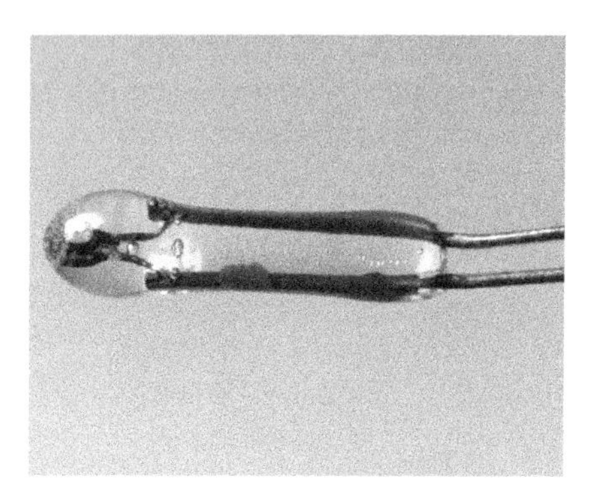

8.8.2 Thermocouples

Thermocouples (▶ available—V8.18) are one of the most widely used temperature sensors. A thermocouple is a **thermoelectric** type sensor and operates on the principle that an electromotive force (EMF) is created when two junctions of different metals are operated at different temperatures. This characteristic behavior was discovered by Seebeck in 1821. Figure 8.51 illustrates this situation and shows two dissimilar metals *A* and *B* connected at two different points. If one of the junctions is at a known reference temperature, then the voltage between the nodes is a function of the difference between the temperatures of the two junctions. This fact is used to indicate the temperature of the other node.

Figure 8.51

Thermocouple junctions

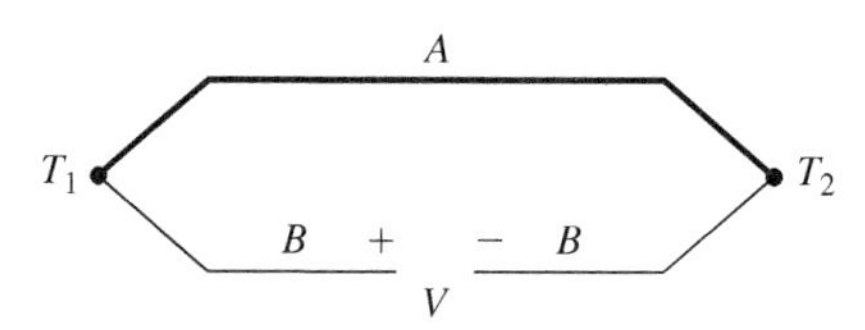

Several laws or properties apply to thermoelectric circuits.

Law of Intermediate Metals This law states that a third metal introduced into a thermocouple circuit will not affect the EMF output of the circuit provided that the two junctions introduced by the third metal are at the same temperature. This situation is illustrated in Figure 8.52 where a third metal *C* is introduced into the circuit. Provided that the two junctions of metal *C* with metal *A* are at the same temperature or $T_3 = T_4$, the EMF output of the circuit is not affected. Note that this law still applies if the third metal *C* was introduced at either junction of metals *A* and *B*.

Figure 8.52

Illustration of the law of intermediate metals

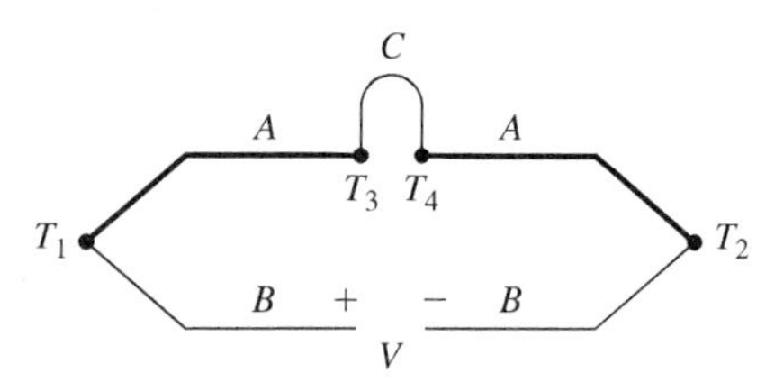

The application of this law allows for the insertion of a measuring device into the thermocouple circuit or the brazing or welding of the junction without altering the accuracy of the temperature measurement.

Law of Homogeneous Circuits This law states that if the thermocouple conductors are homogeneous, then they are not affected by intermediate temperatures of the conductors away from the junctions.

This situation is illustrated in Figure 8.53, where the lead wires away from the junctions have a temperature that is different from T_1 and T_2, but the difference in temperature does not affect the output voltage of the circuit. Application of this law permits the use of thermocouple-grade extension wires and implies that shielding of lead wires is not needed in thermocouple circuits.

Figure 8.53

Illustration of the law of homogeneous circuits

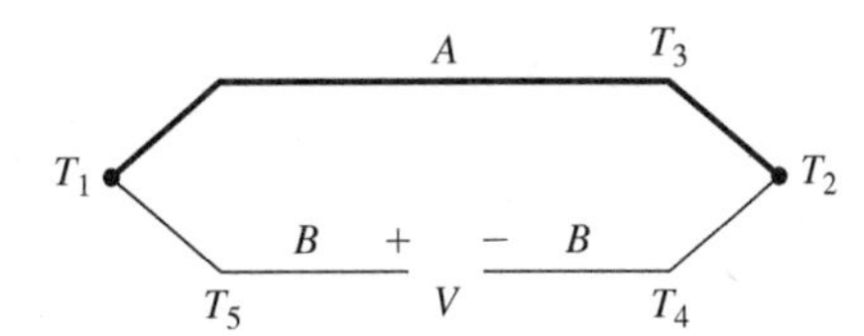

Law of Intermediate Temperatures This law dictates that if a thermocouple generates an EMF E_1 when one junction is at temperature T_1 and the other is at an intermediate temperature T_2, and it produces an EMF E_2 when the junctions are at temperatures T_2 and T_3, then the thermocouple will produce an EMF of E_1 and E_2 when its junctions are at temperatures T_1 and T_3. This law is illustrated in Figure 8.54.

Figure 8.54 Illustration of the law of intermediate temperatures

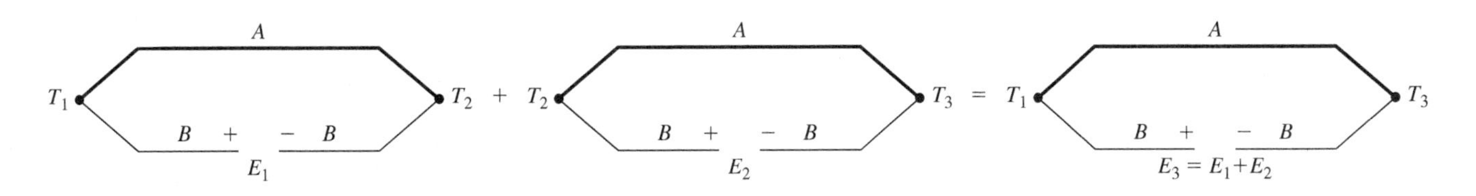

Essentially, the thermoelectric voltage between two temperatures can be determined as the algebraic sum of the voltages over the intermediate temperature range. This law facilitates the use of a thermocouple calibrated for one reference temperature to operate effectively with a different reference temperature.

There are several types of commercially available thermocouples. These include the 'J' thermocouple (iron-constantan), the 'K' thermocouple (chromel-alumel, where chromel is a nickel-chromium alloy and alumel is a nickel-aluminum or nickel-silicon alloy), and the 'T' thermocouple (copper-constantan). Type K is the most common thermocouple. Constantan is an alloy primarily composed of copper and nickel. The temperature measurement range for these types of thermocouples is shown in Table 8.4. Type J is typically preferred for reducing environments, whereas Type K is often chosen for oxidizing environments due to its nickel-chromium alloy's resistance to oxidation at high temperatures. Type T is suitable for ambient and sub-freezing environments.

Table 8.4

Temperature range of J, K, and T thermocouples

	Type J	Type K	Type T
Temperature range	−210 to 1200°C −346 to 2192°F	−270 to 1372°C −454 to 2502°F	−270 to 400°C −454 to 752°F

The measuring junction of a thermocouple can be enclosed in a probe cover or can be exposed directly to the measuring atmosphere. An exposed junction has a faster response time than an enclosed one, but it is not suitable for use in a corrosive environment. Historically, when using a thermocouple to measure temperature, the configuration shown in Figure 8.55 was typically used. One of the junctions was kept in an ice-water mixture to establish a reference junction with a known temperature of 0°C. The other junction measures the temperature of interest. The leads from the thermocouple are connected to a voltage-measuring device, which interprets the voltage difference, usually in millivolts, between the two junctions. Due to the thermocouple wires' potential to act as antennas, they can pick up electromagnetic interference. Therefore, some filtering is often needed to reduce this noise. To enhance the resolution, the output may be amplified before being interpreted by the voltage-measuring device. Modern thermocouple reading instruments have evolved significantly from their older counterparts. Instead of relying on an ice-water mixture for a reference junction, contemporary instruments use precision solid-state sensors or thermistors to measure and compensate for the reference junction's temperature. Additionally, many modern readers feature digital displays, user-friendly interfaces, and data-logging capabilities. They may also include built-in signal processing to filter out noise and electromagnetic interference, and some are equipped with wireless capabilities for remote monitoring and data transmission. Figure 8.56 shows a commercial Type K bead wire thermocouple temperature probe plugged into a thermocouple temperature meter.

Figure 8.55

Historical thermocouple measurement configuration

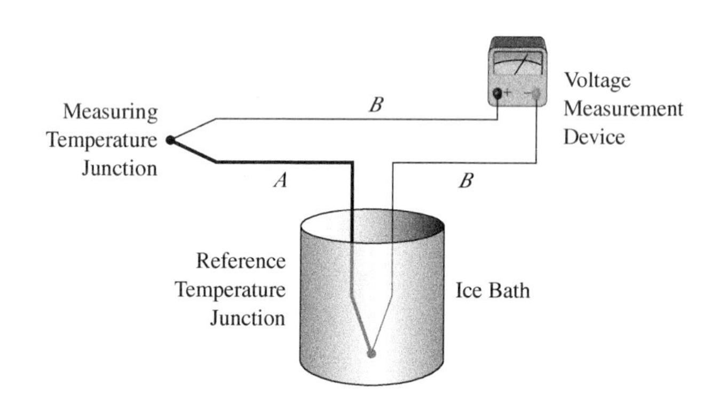

Figure 8.56

Commercial Type K bead wire thermocouple temperature probe plugged into a thermocouple temperature meter

(David J. Green/Alamy Stock Photo)

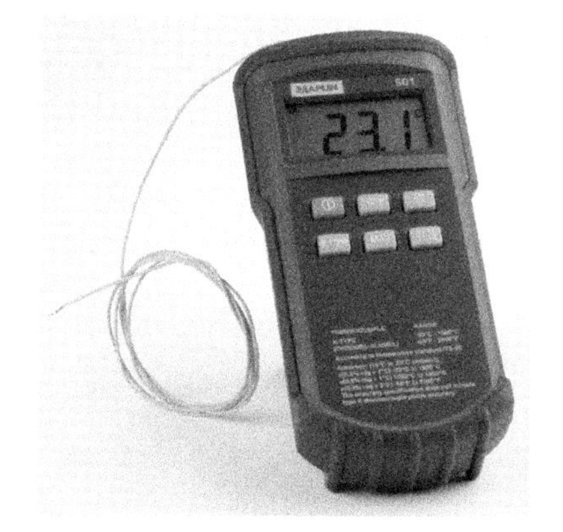

The output voltage of a thermocouple does not vary linearly with temperature except for K-type thermocouples, where for average accuracy one can assume a linear relationship between output voltage and temperature over the range of 0 to 1000°C. Table 8.5 shows the output in mV for Types T, J, and K thermocouples from −100°C to 1300°C, and Table 8.6 shows the detailed output of Type K thermocouples over the temperature range of 0 to 100°C. This data is obtained from the NIST ITS-90 online database [31], which has additional data for other types of thermocouples and for different temperature ranges.

Table 8.5

Thermoelectric voltage in mV as a function of temperature for −100 to 1300°C temperature range with a reference junction at 0°C for Types T, J, and K thermocouples (Data from NIST ITS-90 online database [31])

	Output in mV		
Temperature, °C	Type T	Type J	Type K
−100	−3.379	−4.633	−3.554
−50	−1.819	−2.431	−1.889
−25	−0.940	−1.239	−0.968
0	0.000	0.000	0.000
25	0.992	1.277	1.000
50	2.036	2.585	2.023
100	4.279	5.269	4.096
200	9.288	10.779	8.138
300	14.862	16.327	12.209
400	20.872	21.848	16.397
500		27.393	20.644
600		33.102	24.905
700		39.132	29.129
800		45.494	33.275
900		51.877	37.326
1000		57.953	41.276
1100		63.792	45.119
1200		69.553	48.838
1300			52.41

Table 8.6

Thermoelectric voltage in mV as a function of temperature for 0 to 100°C temperature range with a reference junction at 0°C for Type K thermocouple (Data from NIST ITS-90 online database [31])

°C	0	1	2	3	4	5	6	7	8	9	10
0	0.000	0.039	0.079	0.119	0.158	0.198	0.238	0.277	0.317	0.357	0.397
10	0.397	0.437	0.477	0.517	0.557	0.597	0.637	0.677	0.718	0.758	0.798
20	0.798	0.838	0.879	0.919	0.960	1.000	1.041	1.081	1.122	1.163	1.203
30	1.203	1.244	1.285	1.326	1.366	1.407	1.448	1.489	1.530	1.571	1.612
40	1.612	1.653	1.694	1.735	1.776	1.817	1.858	1.899	1.941	1.982	2.023
50	2.023	2.064	2.106	2.147	2.188	2.230	2.271	2.312	2.354	2.395	2.436
60	2.436	2.478	2.519	2.561	2.602	2.644	2.685	2.727	2.768	2.810	2.851
70	2.851	2.893	2.934	2.976	3.017	3.059	3.100	3.142	3.184	3.225	3.267
80	3.267	3.308	3.350	3.391	3.433	3.474	3.516	3.557	3.599	3.640	3.682
90	3.682	3.723	3.765	3.806	3.848	3.889	3.931	3.972	4.013	4.055	4.096

For other thermocouples or better accuracy, an inverse polynomial that provides the temperature in degrees °C from a given EMF voltage can be used. The **inverse polynomial** takes the form

$$T = d_o + d_1E + d_2E^2 + d_3E^3 + \cdots + d_nE^n \tag{8.18}$$

where T is the temperature in °C, E is the EMF voltage of the thermocouple in mV, and d_0, d_1,..., d_n are the polynomial coefficients, which are a function of the particular thermocouple material used. NIST publishes tables of these inverse coefficients for different types of thermocouples. The order n depends on the type of thermocouple and the temperature range and varies from 4 to 10. Using these coefficients for voltage-to-temperature conversion, the temperature accuracy is typically about +/−0.05°C.

Thermocouples come in two distinct grades of accuracy: **Standard Grade** and **Special Grade.** The level of accuracy depends not only on the thermocouple type—such as Type K, J, or T—but also on the temperature range in which it is used. For Type K thermocouples, Standard Grade offers an accuracy of either +/−2.2°C or 0.75% of the temperature reading above 0°C, whichever value is larger. On the other hand, Special Grade Type K thermocouples have a tighter accuracy specification, being accurate to either +/−1.1°C or 0.4% of the temperature reading, again choosing whichever value is greater.

Example 8.8 illustrates the use of thermocouples in temperature measurements.

Example 8.8 Temperature Measurement Using a Thermocouple

A Type K thermocouple produced a voltage output of 2 mV when its leads were connected to a voltmeter whose terminals are at a temperature of 20°C. Determine the temperature of the object measured by this thermocouple.

Solution:

Since the measurement in this setup is not referenced to 0°C, we need to use the Law of Intermediate Temperatures to determine the temperature of the object. This Law applied to this problem can be written as

$$EMF\_TRef0 = EMF\_TRef20 + EMF\_20Ref0$$

where *EMF_TRef0* is the EMF output of the thermocouple sensor at temperature *T* referenced to 0°C, *EMF_TRef20* is the thermocouple output at temperature *T* referenced to 20°C, and *EMF_20Ref0* is the thermocouple output at temperature 20°C referenced to 0°C. From Table 8.6, *EMF_20Ref0* = 0.798 mV. Adding this voltage to 2 mV, we get 2.798 mV for *EMF_TRef0*. From Table 8.6, and interpolating between the output voltages at 68°C and 69°C, this voltage output corresponds to a temperature of 68.7°C.

8.8.3 RTD

The resistance temperature detector (RTD) (▶ available—V8.19) is a resistance-based temperature sensor. It operates on the principle that the resistance of certain metals, such as platinum, copper, and nickel, increases predictably with rising temperature. Among these materials, platinum is the most commonly used in RTDs due to its broad temperature measuring range (−200 to 850°C), high-temperature stability, and superior accuracy.

An RTD does not have the high sensitivity of a thermistor, but its temperature resistance relationship is not highly nonlinear. Figure 8.57 shows a typical temperature resistance relationship for platinum RTD. Note that RTDs are made with different nominal resistances at 0°. A common value is 100 ohms, but RTDs with 500 or 1000 ohms are also made. RTDs are made with a specific **temperature coefficient (TC)** or alpha factor. Two commonly used TC factors for platinum RTDs are the European Standard and the American Standard. The European Standard has a TC value of 0.00385 Ω/Ω/°C over the range of 0 to 100°C, while the American Standard has a TC value of 0.00392 Ω/Ω/°C. Using the temperature coefficient, the resistance of an RTD can be approximated as

$$R = R_0(1 + \alpha(T - T_0)) \quad (8.19)$$

where R_0 is the nominal resistance at the nominal temperature T_0. Note that the exact resistance at any temperature can be obtained from RTD manufacturers who publish tables with exact values of resistance at different temperatures for a given TC and nominal 0° resistance. RTDs are also available with different tolerance levels.

Figure 8.57

Resistance versus temperature relationship for platinum RTDs

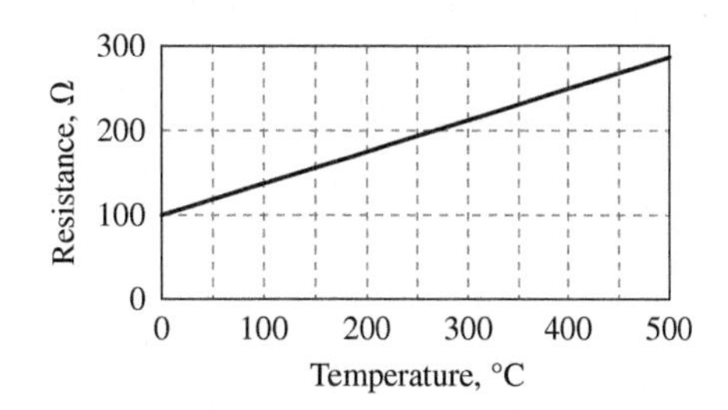

Although RTDs do not have the same large temperature measuring range as thermocouples, they are more linear than thermocouples and inherently more stable, and there is no need for a reference junction. However, an RTD has a slower response than a thermocouple. Example 8.9 illustrates the computation of the temperature output and sensitivity of an RTD.

Example 8.9 Temperature Output and Sensitivity of an RTD

Consider a platinum RTD sensor that has a nominal resistance of 100 Ω at 0°C and a TC factor of 0.00392 Ω/Ω/°C. Determine:

(a) The temperature read by the sensor if the RTD resistance is 300 Ω
(b) The resistance sensitivity of this sensor

Solution:

(a) We will use Equation (8.19) which relates the resistance and temperature of an RTD. Substituting the values for R and R_0, we get

$$300 = 100\,(1 + \alpha(T - T_0))$$

Dividing each term by 100, and then moving 1 to the left we get

$$3 - 1 = 2 = \alpha(T - T_0)$$

Substituting $T_0 = 0$, and solving for T, we get $T = 2/\alpha = 2/0.00392 = 510°C$.

(b) The resistance sensitivity of this sensor is obtained by differentiating Equation (8.19) with respect to temperature, or

$$\text{Sensitivity} = \frac{dR}{dT} = R_0\alpha$$

where the nominal resistance R_0 is a function of both the length and size of the sensor wire.

RTDs are available in different forms. These include thin film and wire-wound. In the **thin-film form**, a small layer of platinum is deposited on a substrate, and then wires are attached to the substrate. The substrate is then coated in epoxy. In the **wire-wound form**, a wire coil is either inserted inside a cylindrical glass or ceramic tube or wound around a glass or ceramic core and then covered with glass or ceramic material. RTDs are available with two-, three-, or four-wire configurations. The two-wire configuration is most sensitive to errors resulting from the additional resistance introduced by the lead wires. With the three- or four-wire configuration, a compensating bridge-type circuit (discussed in Section 8.11.3) can be constructed to compensate for the lead wire resistance. Figure 8.58 shows a commercial, platinum-type, four-wire RTD temperature sensor.

Figure 8.58

Commercial, platinum-type, four-wire RTD temperature sensor (Model Extech 850187)

(Courtesy of Teledyne FLIR)

8.9 Vibration Measurement

Vibratory motion (for further reading, see [32]) commonly occurs in machinery and flexible structures. Measurement of vibration is important for machine health monitoring of motors, pumps, fans, gearboxes, machine tool spindles, blowers, and chillers. Vibration measurement is also important in safety devices such as automotive airbags. Vibration is measured by either accelerometers or vibrometers. The two devices have a similar operating principle but differ in their natural frequency and damping. They are based on the measurement of the motion of a small, spring and damper-supported mass that is placed in a housing. The mass is commonly referred to as a **seismic mass**, and the housing is attached to the structure whose vibration motion needs to be measured. The motion of the structure is transferred to the seismic mass through the support spring and damper. The motion of the seismic mass can be obtained using different transducers. These include resistance-type strain gauges and piezoelectric and piezoresistive crystals. The following section illustrates the theory behind the operation of such a device.

8.9.1 Seismic Mass Operating Principle

With reference to Figure 8.59, let us assume that the support base has a displacement of y_b, and the seismic mass has a displacement of y_m. From a free-body diagram of the seismic mass, we can write the following equation of motion for the seismic mass:

(8.20)
$$-k(y_m - y_b) - c(\dot{y}_m - \dot{y}_b) = m\ddot{y}_m$$

Figure 8.59

Schematic of a seismic mass

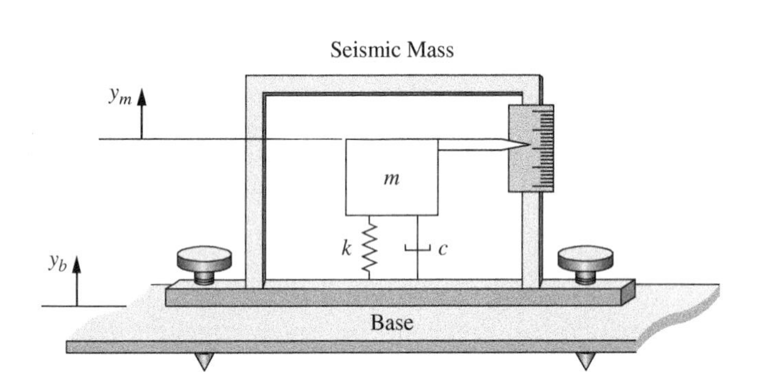

Let us define z **as the relative displacement** between the supported mass and the base or

(8.21)
$$z = y_m - y_b$$

Equation (8.20) can be written as

(8.22)
$$m\ddot{z} + c\dot{z} + kz = -m\ddot{y}_b$$

If we assume the base displacement to be a sinusoidal with amplitude Y and frequency ω as $y_b = Y\sin(\omega t)$, then Equation (8.22) can be written as

(8.23)
$$m\ddot{z} + c\dot{z} + kz = m\omega^2 Y\sin(\omega t)$$

Equation (8.23) is a second-order linear differential equation with a forced input. The steady-state solution of this equation is given by

(8.24)
$$z(t) = Z\sin(\omega t - \varphi_1)$$

where φ_1 is the phase shift angle and Z is the amplitude given by the expression

(8.25)
$$Z = \frac{m\omega^2 Y}{[(k - m\omega^2)^2 + (c\omega)^2]^{1/2}} = \frac{r^2 Y}{[(1 - r^2)^2 + (2\zeta r)^2]^{1/2}}$$

where r is the frequency ratio ω/ω_n and ω_n is the natural frequency given by $\sqrt{k/m}$. The phase angle φ_1 is given by the expression

$$\varphi_1 = \tan^{-1}\left(\frac{c\omega}{k - m\omega^2}\right) = \tan^{-1}\left(\frac{2\xi r}{1 - r^2}\right) \tag{8.26}$$

A plot of the ratio of Z/Y and the phase angle as a function of the frequency ratio r is given in Figure 8.60. The plot shows that the ratio Z/Y is very dependent on the damping ratio ζ, and Z/Y approaches 1 for r greater than 3 regardless of the damping ratio ζ. The ratio Z/Y equals 1 means that the relative displacement of the seismic mass is equal to the displacement of the base. Thus, if we can measure the relative displacement, then we have a measurement of the motion of the base. This is the principle behind the operation of **vibrometers**, which provide an output that is proportional to the displacement of the base. To make r greater than 3, the natural frequency of the vibrometer has to be small. This is achieved by using a large mass m and a support spring with a small stiffness k. This results in a bulky instrument.

Figure 8.60

Plots of (a) Equation (8.25) and (b) Equation (8.26)

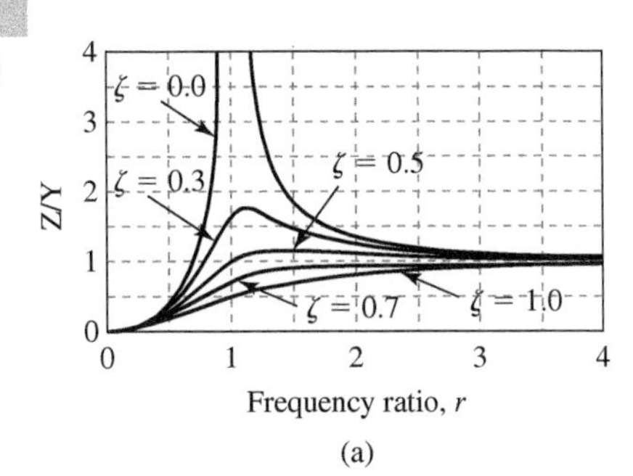

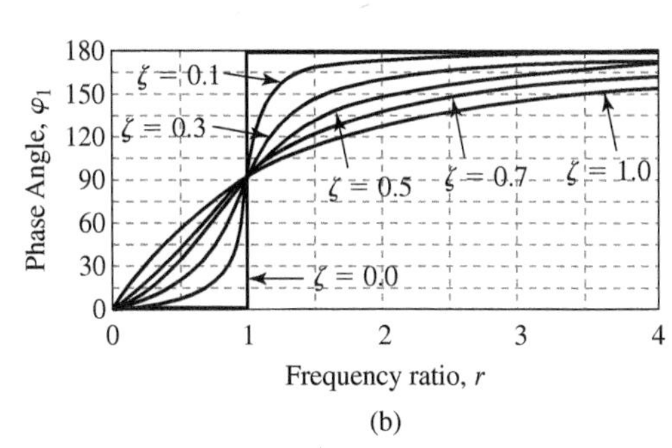

Notice the phase shift between the relative displacement and the base displacement in Figure 8.60(b). For ζ equal to 0, the phase shift is 180° for any r greater than 1. For ζ other than 0, the phase shift is dependent on the frequency ratio r. A phase shift means that relative displacement lags the base displacement by a time amount equal to φ_1/ω.

Since $r = \omega/\omega_n$, Equation (8.24) can be written as

$$-\omega_n^2 z(t) = \frac{-\omega^2 Y \sin(\omega t - \varphi_1)}{[(1 - r^2)^2 + (2\zeta r)^2]^{1/2}} \tag{8.27}$$

But the acceleration of the base is given by

$$\ddot{y}_b(t) = -\omega^2 Y \sin(\omega t) \tag{8.28}$$

This means that the term $-\omega_n^2 z(t)$ is proportional to the acceleration of the base. To make

$$-\omega_n^2 z(t) \cong -\omega^2 Y \sin(\omega t - \varphi_1) \tag{8.29}$$

the denominator of the right-hand side of Equation (8.27) must be equal to 1, or

$$\frac{1}{[(1 - r^2)^2 + (2\zeta r)^2]^{1/2}} = 1 \tag{8.30}$$

Figure 8.61 shows a plot of the left-hand side of Equation (8.30). The plot shows that the term is equal to 1 for all r less than 0.05, regardless of the value of the damping ratio. For ζ equal to 0.7, the term has a value of 1 for r less than 0.2. This means that if the seismic mass is made to have a very high natural frequency, then the setup shown in Figure 8.59 can be used to directly measure the acceleration of the base through measurement of the relative displacement between the seismic mass and the base for cases where r is very small. This is the principle of operation of **accelerometers**, which give an output that is proportional to the acceleration. Note that the term ω_n^2 in Equation (8.27) is fixed for a given device and does not change with the applied frequency. The phase lag expression given by Equation (8.26) also applies to this case. To achieve a low r ratio in practice, accelerometers are made with a small mass m and with a spring that has a large stiffness to result in a system with high natural frequency. Due to their compact size, accelerometers are more widely used than vibrometers, and the next section discusses piezoelectric accelerometers, one of the most used types of accelerometers. Other types of accelerometers include piezo-resistive, strain-gauge-based, and IC accelerometers. Example 8.10 illustrates the effect of damping and frequency ratio on the output of an accelerometer.

Figure 8.61

A plot of the ratio $1/[(1 - r^2)^2 + (2\zeta r)^2]^{1/2}$

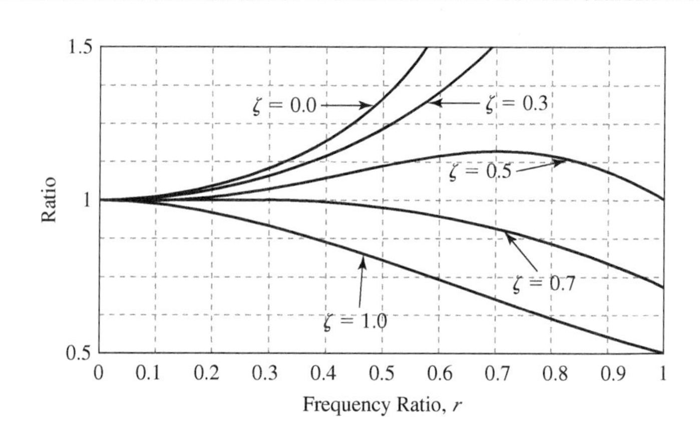

Example 8.10 Accelerometer Error

An accelerometer with a natural frequency of 1 kHz and a damping ratio of 0.6 is used to measure the vibration of a motor rotating at 3600 rpm. The accelerometer gave a reading of 1 g. What is the actual acceleration of the motor?

Solution:

The rotation frequency of the motor is

$$3600 \times \frac{2\pi}{60} = 377 \text{ rad/s}$$

The frequency ratio $r = \omega/\omega_n$ is therefore $377/(1000 \times 2\pi) = 0.06$

From Equation (8.27), the actual acceleration is equal to

$$\text{Measured Acceleration} \times [(1 - r^2)^2 + (2\zeta r)^2]^{1/2} = 1 \text{ g} \times [(1 - 0.06^2)^2 + (2 \times 0.6 \times 0.06)^2]^{1/2}$$

Or 1 g × 0.999 = 0.999 g. The error is therefore 0.1%, which is small.

8.9.2 Piezoelectric Accelerometers

Certain natural materials, such as quartz and Rochelle salt, as well as manufactured ceramic materials like lead zirconate titanate (PZT), exhibit a property known as the **direct piezoelectric effect** when subjected to force or pressure. These materials produce an electric charge proportional to the applied force. Conversely, when an electric field is applied in the direction of their polarization, they deform—a phenomenon known as the converse piezoelectric effect. The Curie brothers, Pierre and Jacques, discovered the piezoelectric effect in 1880. The direct piezoelectric effect is harnessed in the design of accelerometers and force sensors, while the converse effect is exploited in piezo actuators used for precision positioning and machining. Notably, the term 'piezo' originates from the Greek word for 'pressure'.

There are different designs of piezoelectric accelerometers (▶ available—V8.20). These include compression and shear type. A section view of a typical **compression-type piezoelectric accelerometer** is shown in Figure 8.62. The supported mass is sandwiched between a piezoelectric element and a compression spring. The compression spring could take the form of a bolt and a washer compressing the mass and the piezoelectric element. The accelerometer base is typically bolted or bonded to the structure whose acceleration needs to be measured, but accelerometers with magnetic bases are also available. The motion of the structure results in a motion of the supported mass, and the inertial force of the supported mass pushes against the piezoelectric element producing a charge signal. Since the supported mass is constant, the force exerted on the piezoelectric element is thus proportional to acceleration. In a **shear-type accelerometer**, the supported mass is designed to exert a shear force rather than a compressive force on the piezoelectric element. Shear-type accelerometers are used in flexible structure applications or where thermal gradients can cause distortion of the base. All the elements of the accelerometer are housed in a metal housing, typically made of stainless steel, which is sealed to prevent the entrance of dust, water, or dirt.

Figure 8.62

A section view of a compression-type accelerometer

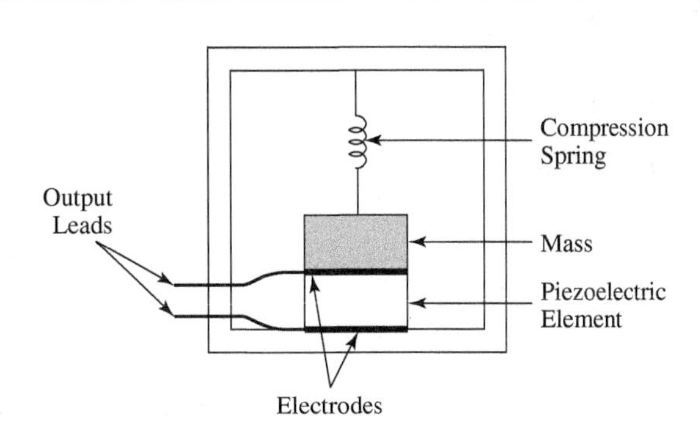

Piezoelectric accelerometers have the advantage of compact size, high sensitivity, high-frequency measurement range, and rugged construction. Figure 8.63 shows several commercially available piezoelectric accelerometers. Accelerometers are available with a top or side connector location; single axis, biaxial, or triaxial measurement capability; frequency measurement range up to 25 kHz; acceleration measurement range up to 500 g; and sensitivity that ranges from 10 mV/g to 10 V/g. Example 8.11 illustrates the relationship between accelerometer sensitivity and range. Note that piezoelectric accelerometers are not very suitable for measuring low-frequency (a few Hertz) oscillations, and they give a zero output at zero frequency.

Figure 8.63

Commercially available piezoelectric accelerometers

(Courtesy of Wilcoxon Research, Inc., Germantown, MD)

Piezoelectric accelerometers are available with two distinct types of output. These include **high-impedance charge output** and **low-impedance voltage output**. In the high-impedance version, the accelerometer gives a charge signal proportional to acceleration. The sensitivity of the accelerometer in this case is specified in pico (10^{-12}) Coulomb per g or pC/g. The charge signal cannot be read by a low-impedance device (such as a voltmeter) due to the large loading errors (see Section 2.6) caused by the high source

Example 8.11 Accelerometer Selection

An accelerometer is needed to measure acceleration with a range up to ±20 g. If the analog output of the accelerometer is read by an A/D converter with a ±10 volts input range, recommend a suitable sensitivity for the accelerometer.

Solution:

The voltage output of the accelerometer at the maximum desired acceleration affects the choice of the sensitivity. In this application, the accelerometer voltage output at 20 g should be equal or less than 10 V. This gives a maximum sensitivity of 10 V/20 g or 500 mV/g. Thus, any accelerometer with a sensitivity of 500 mV/g or less would work in this application. Note, however, that high sensitivity is preferred since it results in a better signal-to-noise ratio.

impedance and due to charge leakage through the load. The charge output is processed instead by a special amplifier called a **charge amplifier**, which takes the charge output from the accelerometer and produces a low-impedance analog output voltage. When using a charge amplifier, the capacitance of the cable that connects the accelerometer output to the charge amplifier input does not affect the output voltage of the amplifier. The output voltage is simply a function of the input charge and the feedback capacitance of the amplifier and is given by Equation (8.31):

(8.31)
$$V_{\text{out}} = \frac{q_{\text{in}}}{c_f}$$

where q_{in} is the charge produced by the accelerometer, and c_f is the feedback capacitance of the charge amplifier. This can be seen if we analyze the op-amp circuit shown in Figure 8.64. The input charge q_{in} is distributed as

(8.32)
$$q_{\text{in}} = q_c + q_{\text{inp}} + q_f$$

But the charge is related to voltage by

(8.33)
$$q = cV$$

Substituting Equation (8.33) into (8.32), we get

(8.34)
$$q_{\text{in}} = (c_c + c_{\text{inp}})V^- + c_f V_o$$

But $V^- = 0$ since the inverting input potential is equal to the noninverting input, and the noninverting input is grounded in this circuit. Thus, we get

(8.35)
$$q_{\text{in}} = c_f V_o$$

This is not the case if the charge output is directly connected to a high-impedance voltage amplifier. In that case, the capacitance of the sensor, the cable, and the amplifier must be considered.

Figure 8.64

Charge amplifier wiring

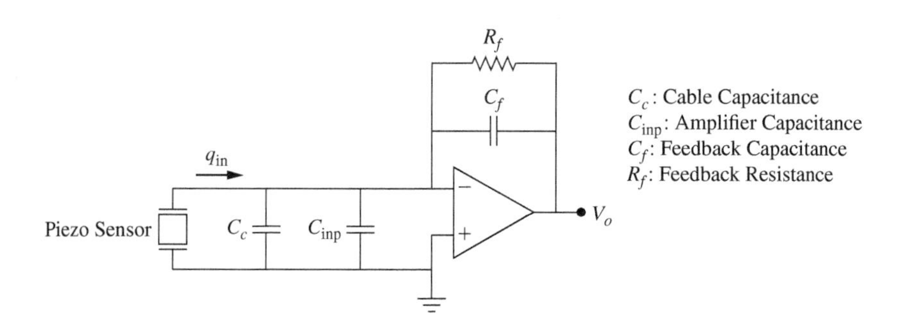

Voltage-output accelerometers include a small built-in circuit in the accelerometer housing that converts the charge output into a low-impedance output voltage. Charge output accelerometers are self-generating and require no external power; thus, they can be used in high-temperature or high-radiation applications without any damage. This is not the case with voltage-output accelerometers, where the electronics could be damaged under such conditions.

8.10 Integrated Circuit (IC) Sensors

Integrated circuit (IC) sensors, also known as sensor ICs or sensor integrated circuits, are semiconductor devices that combine sensor elements with signal-processing circuitry on a single chip. These sensors are designed to detect and measure various physical quantities such as temperature, pressure, or acceleration. By integrating the sensor and signal conditioning circuitry, IC sensors offer compact size, improved performance, and simplified system integration compared to discrete sensor components. This section discusses several of these sensors.

8.10.1 IC Temperature Sensors

Due to advancements in integrated circuit technology, IC temperature sensors are becoming widely available. These sensors are based on transistor technology, specifically the fact that the difference in forward voltage of a silicon pn junction is directly proportional to temperature. While IC temperature sensors have a smaller temperature measurement range than thermocouples or RTDs, they give an output that is linearly proportional to temperature, are inexpensive, and are fairly accurate. IC temperature sensors are available with either analog or digital output. The latter type includes an integrated A/D to convert the analog voltage or current signal into a digital signal that is transmitted using a PWM, an I^2C, or an SPI interface. An example of an analog IC temperature sensor is the **LM35C sensor** manufactured by National Semiconductor (now a division of Texas Instruments) and shown in Figure 8.65. Another is the AD590 sensor manufactured by Analog Devices. The LM35C sensor can measure temperature over the range of −40 to 110°C. An example of an IC temperature sensor with digital output is the TMP05 sensor manufactured by Analog Devices, which has an accuracy of +/−1°C and provides its output in PWM format.

The LM35C sensor is available in several packages (see Figure 8.65), including the hermetic TO-46 metal can package and the TO-92 plastic package. The sensor has three leads, one for 4 to 30 VDC input power, the other for ground, and the third for the analog voltage output from the sensor. The analog output of the sensor is proportional to temperature in degrees C with a sensitivity of 10.0 mV/°C. This sensor is very suitable for use with microcontrollers since it draws little current (less than 60 μA), and its output is directly calibrated in degrees Celsius, thus avoiding any conversion operations. Note that the LM35C is designed to measure temperature in ambient air and not to be immersed in a liquid. To use in liquid, the sensor must be encapsulated.

Figure 8.65

LM35C sensor (a) TO-46 metal can package and (b) TO-92 plastic package

(Courtesy Digi-Key Corporation)

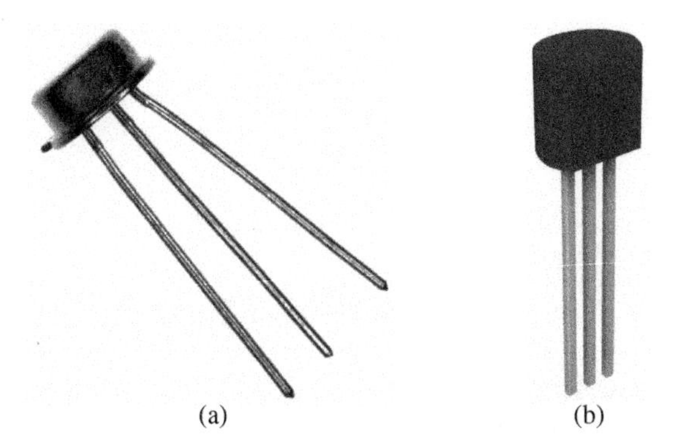

8.10.2 IC Accelerometers

IC or **MEMS** (Micro-Electro-Mechanical Systems) **accelerometers** (▶ available—V8.21), leveraging silicon capacitive micromachined technology, are cost-effective sensors widely employed in applications like airbag deployment, computer hard drive protection, and virtual reality input devices. Each accelerometer comprises a surface micromachined capacitive sensing cell (g-cell) and a CMOS signal conditioning circuit, both integrated within a single IC package. Manufactured using wafer processing techniques, the g-cell can be visualized, as depicted in Figure 8.66, as two stationary plates flanking a movable center plate. As the device experiences acceleration, the center plate deflects, altering the distance—and thus the capacitance—between itself and the stationary plates. The CMOS circuitry then translates these capacitance shifts into an interpretable output, representing the magnitude of acceleration.

Figure 8.66

Model of a silicon capacitive micromachined accelerometer

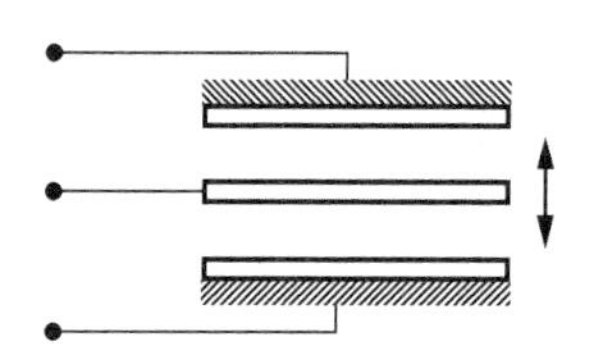

IC accelerometers are designed for mounting on circuit boards. They come in various sensitivities and acceleration measurement ranges. Figure 8.67 depicts an ADXL335 3-axis IC sensor, produced by Analog Devices and mounted on a circuit board, available from Seeed Technology. This sensor boasts a sensitivity ranging from 270 to 330 mV/g and a measurement range of ±3 g. Although the ADXL335 chip is rated for a 3 VDC input, when integrated with the Seeed board, users can provide an input voltage ranging between 3 and 5 VDC. A notable feature of this accelerometer is its ability to offer a separate analog output voltage for each x, y, and z acceleration it measures. These outputs can be easily read using an A/D converter. IC accelerometers' advantages include their affordability, low current consumption (less than 0.5 mA for this sensor), built-in signal conditioning, and linear output. Another benefit is that they do not necessitate a charge amplifier. However, their temperature operating range is narrower compared to charge output piezoelectric accelerometers, and they are not as robust. It is worth mentioning that many recent IC accelerometers favor digital outputs, like I^2C or SPI, over analog outputs.

Figure 8.67

ADXL335 sensor

(Jouaneh, University of Rhode Island)

8.10.3 IC IMU Sensors

An **Inertial Measurement Unit (IMU)** (▶ available—V8.22), also known as a motion sensor, is a highly versatile sensor used in a wide range of consumer and industrial applications to measure motion along multiple axes. The fundamental IMU sensor consists of an accelerometer for measuring linear acceleration and a gyroscope for measuring rotational speeds. More advanced IMU sensors may also incorporate a magnetometer to determine a compass heading by measuring the Earth's magnetic field. In automotive applications, IMUs play a vital role in several areas, including lateral stability and traction control, emergency maneuver assistance, and adaptive suspension systems. It is worth noting that IMUs are available in both integrated circuit (IC) form and non-IC form, but this section will primarily focus on the IC form. Integrated Case Study IV.D discusses an example of an IMU.

Most IMUs available today come with digital output rather than analog output. This digital output offers several advantages, including higher sensitivity compared to analog output. Many digital IMUs have a sensor output resolution of 16 or 32 bits, surpassing the typical 10-bit resolution of an analog-to-digital converter (ADC) used by microcontrollers to read analog IMU output. Digital IMUs transmit data in a digital format through a serial interface such as SPI or I^2C (as discussed in Chapter 5). Unlike analog IMUs, these sensors can be configured digitally. Most digital IMUs support features such as sensitivity adjustment, sampling rate modification, and even the ability to enable or disable specific functionalities like tap detection or freefall detection.

Integrated Case Study IV.D: IMU Sensor

For Integrated Case Study IV (Mobile Robot—see Section 11.5), the robot control board includes an LSM6DS33 inertial sensor that connects to the microcontroller using the I^2C bus (see Section 5.8). The LSM6DS33 integrates a digital, MEMS-type 3-axis accelerometer and a 3-axis gyroscope in a single package. It also has an embedded temperature sensor. The sensor provides different sensitivity levels for its six independent acceleration and rotation rate readings in the range of ± 2 g to ± 16 g for the accelerometer and ± 125°/s to ± 2000°/s for the gyroscope. While the ROMI robot uses the I^2C bus for interfacing with this sensor, the sensor can also be interfaced using the SPI protocol. The IC chip for this sensor has a small footprint ($3 \times 3 \times 0.86$ mm) so it is widely in consumer applications including smartphones, and the sensor is designed to be fully compliant with the Android operating system. It has built-in capabilities to detect events associated with smartphone usage such as free-fall, wakeup, and double tap-sensing. The pin connections and the measurement axes for this sensor are shown in Figure 8.68.

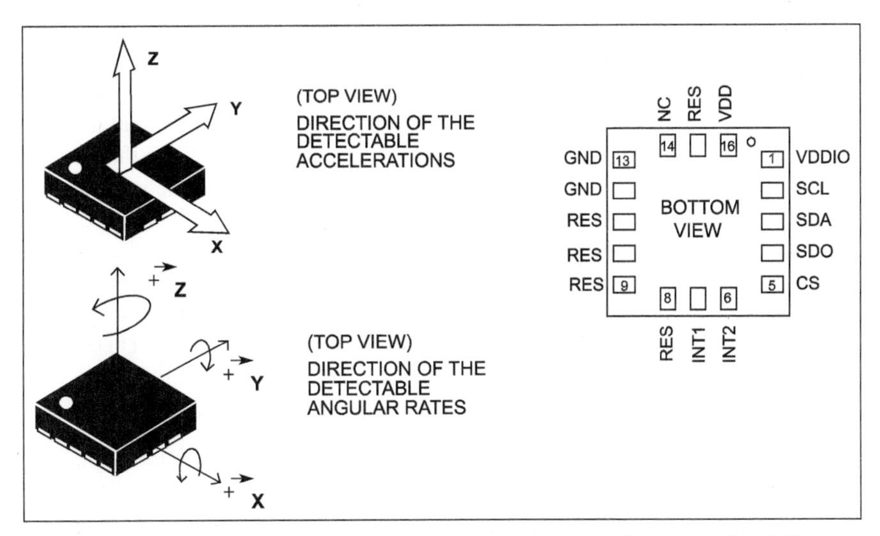

Figure 8.68 Pin connections and the measurement axes for the LSM6DS33

(Courtesy of STMicroelectronics NV)

One of the nice features of this sensor is that it can operate in different power modes to optimize power consumption. The other feature is that it can store up to 8K of data on the chip in first input first output fashion (FIFO). This feature provides power saving for the host system since the host system does not need to continuously check for data availability from the sensor but can get the data in burst mode out from the sensor when the host only needs to access the data.

This chip requires an analog supply voltage in the range of 1.7 to 3.6 V and thus cannot be directly powered by a 5 VDC system. To address this requirement, the ROMI control board which operates at 5 VDC includes level shifters to translate the logic signals from one level to another. **Level shifters** are electronic components that translate logic levels and can be unidirectional or bidirectional. A unidirectional level shifter translates logic level in one direction (e.g., from 5 to 3.3 VDC), while a bidirectional one can do the translation in both directions. The ROMI control board has bidirectional level shifters that are dedicated for interfacing with the IMU sensor as well as with an optional Raspberry PI that can be plugged into the board. It also has several non-dedicated unidirectional level shifters. Pololu makes a separate carrier board that includes this IMU sensor and allows operation using a supply voltage that ranges from 2.5 to 5.5 VDC. The carrier board can be mounted on a breadboard since it has the standard pin spacing for solderless breadboards.

8.11 Signal Conditioning

In many situations, the raw output of the sensor may not be in a form suitable to be interfaced with a measurement device or an A/D converter, and some form of signal conditioning needs to be applied to the sensor output. These signal conditioning operations include filtering, amplification, or using a bridge circuit, and are discussed next.

8.11.1 Filtering

In dynamic measurements, the output of a sensor or transducer comprises multiple frequency components, which can be accompanied by noise. This noise may arise from external factors like heat, vibration, imperfections in sensor materials, or interference from external signals. The presence of noise diminishes the sensor resolution and exacerbates repeatability errors. Noise can be introduced during signal acquisition, transmission, or processing. Filtering is the process of attenuating unwanted components or noise from the sensor output while allowing desired components to pass through. Section 7.6 discussed several types of filters including low-pass, high-pass, notch, and bandpass filters. These filters are named based on their desired frequency characteristics. Filtering techniques play a critical role in enhancing signal quality and reliability by reducing noise levels.

8.11.2 Amplification

In many cases, a sensor output voltage is too small to provide a meaningful resolution, and some form of amplification is needed to amplify the output voltage. When selecting an amplifier, consider the impedances of the sensor and the amplifier for accuracy. In general, a sensor should have a small output impedance, and the amplifier should have a large input impedance. However, amplifiers are not ideal devices, as they have finite impedances that can affect the accuracy of the amplification process. When the output impedance of the sensor is not small compared with the input impedance of the amplifier, the input voltage seen by the amplifier is different than the open circuit voltage of the sensor. This condition is called 'loading' and was discussed in detail in Chapter 2.

Amplifiers are available as single IC circuits or can be built using a combination of discrete elements such as transistors, diodes, and resistors. They can also be built using op-amps. Op-amp amplifiers have high input impedances and thus are very suitable for use in amplification circuits. Chapter 2 discussed different types of op-amp circuits. These included simple amplification, amplification with either integration or differentiation, voltage following, and differential amplification. The next chapter discusses amplifiers that are used for motor control applications.

8.11.3 Bridge Circuits

A bridge circuit is used to improve the accuracy and sensitivity of the output of certain sensors. It is commonly used to process signals from resistive, capacitive, and inductive type sensors. Consider the bridge circuit shown in Figure 8.69 made of resistive elements with a constant DC voltage applied to it. This four-arm bridge circuit is commonly known as the **Wheatstone bridge** (▶ available—V8.23) and is commonly used to process the signal output from strain gauges and resistance-based thermal sensors. The strain gauge resistance or the temperature sensor resistance forms one arm of the bridge. External power from a battery or a power supply is applied to two opposite vertices of the bridge, while the output of this circuit is read from the remaining two opposite vertices. Normally, the resistances of the arms of the bridge are chosen so that the bridge is initially balanced (output voltage is zero). A bridge circuit can then be used in one of two ways. Firstly (the **null method**), when one of the resistances changes (such as R_1) due to a change in the variable that is being measured, another resistance is changed (such as R_2) to bring the output of the bridge back to zero. This assumes that means are available to adjust the resistance R_2. This technique is only used with slowly varying resistance changes. In the other way (**deflection method**), the three resistances of the bridge are fixed, and only one is changed. When one of the resistances of the bridge changes, the change in the output of the bridge is used as a measure of the resistance change. The second way is suitable for dynamic measurements. The bridge is known as a deflection bridge in this mode of operation.

Figure 8.69

Wheatstone Bridge circuit

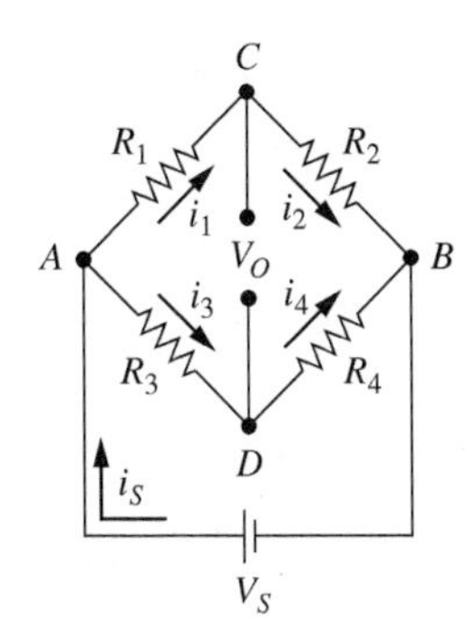

As a first step, determine the conditions that make the output of the bridge balanced. If the bridge is balanced, then the voltage V_O is zero. The voltage at point C is given by

$$V_C = \frac{R_2}{R_1 + R_2} V_S \tag{8.36}$$

because the resistances R_1 and R_2 act as a voltage-dividing circuit. Similarly, the voltage at point D is given by

$$V_D = \frac{R_4}{R_3 + R_4} V_S \tag{8.37}$$

V_O is then given as $V_C - V_D$, or

$$V_O = \left(\frac{R_2}{R_1 + R_2} - \frac{R_4}{R_3 + R_4} \right) V_S = \left(\frac{R_2 R_3 - R_1 R_4}{(R_1 + R_2)(R_3 + R_4)} \right) V_S \tag{8.38}$$

For V_O to be zero, the numerator in Equation (8.38) must be zero or

$$\frac{R_1}{R_2} = \frac{R_3}{R_4} \tag{8.39}$$

Thus, if the resistances of the bridge follow the ratios given by Equation (8.39), then the bridge will be balanced, and the output voltage V_O will be zero. Notice that the balance condition is independent of the magnitude of the supply source.

Now assume that the resistance R_2 of the bridge has changed by a small amount δR_2. We want to determine the resulting change in the bridge output due to this change. From Equation (8.38), the output of the bridge will be

$$V_O + \delta V_O = \left(\frac{R_2 + \delta R_2}{R_1 + R_2 + \delta R_2} - \frac{R_4}{R_3 + R_4} \right) V_S \tag{8.40}$$

The change in output is then given by subtracting Equation (8.38) from (8.40) or

$$(V_O + \delta V_O) - V_O = \left(\frac{R_2 + \delta R_2}{R_1 + R_2 + \delta R_2} - \frac{R_2}{R_1 + R_2} \right) V_S \tag{8.41}$$

If we let $R_1 = R_2 = R_3 = R_4 = R$ and $\delta R_2 = \delta R$, Equation (8.41) becomes

$$\delta V_O = \frac{\delta R}{4R + 2\delta R} V_S \tag{8.42}$$

or

$$\frac{\delta V_O}{V_S} = \frac{\delta R / R}{4 + 2\, \delta R / R} \tag{8.43}$$

Equation (8.43) is plotted in Figure 8.70, and it shows the output change of the bridge is nonlinear, especially for negative $\delta R / R$ values.

Figure 8.70

Plot of Equation (8.43)

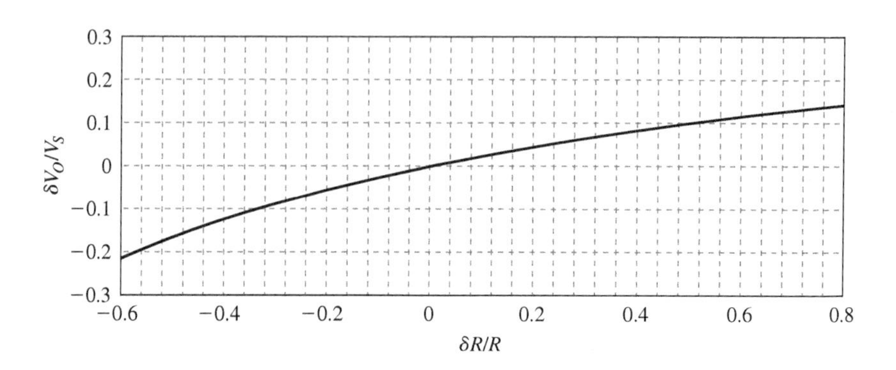

However, if we assume that $\delta R/R$ is small, then Equation (8.43) can be approximated as

(8.44)
$$\frac{\delta V_O}{V_S} = \frac{\delta R}{4R}$$

The factor 1⁄4 in Equation (8.44) is the sensitivity of the voltage output change due to resistance change. Equation (8.44) can be solved to get the resistance change as a function of the voltage output change of the bridge, or

(8.45)
$$\frac{\delta R}{R} = \frac{4\,\delta V_O}{V_S}$$

In some cases, more than one resistance in the Wheatstone Bridge can be active. From Equation (8.38), we can derive an expression for δV_O using differentiation rules from calculus. This gives

(8.46)
$$\frac{\delta V_O}{V_s} = \frac{(R_1\delta R_2 - R_2\delta R_1)}{(R_1 + R_2)^2} - \frac{(R_3\delta R_4 - R_4\delta R_3)}{(R_3 + R_4)^2}$$

For example, we can mount two strain gauges on a beam subjected to bending to increase the measurement sensitivity as shown in Figure 8.71. When the beam end deflects downward under the applied loading, the resistance of the upper gauge will increase, and the resistance of the lower gauge will decrease by the same amount. For the case where $R_1 = R_2 = R$, we have here $\delta R_1 = -\delta R$ and $\delta R_2 = \delta R$. Substituting into Equation (8.46), we get

(8.47)
$$\frac{\delta V_O}{V_s} = \frac{R\delta R + R\delta R}{(2R)^2} = 2\frac{\delta R}{4R}$$

Notice here how the bridge output is now double the case if we had just used a single strain gauge (Equation 8.44). We can generalize this case and rewrite Equation (8.47) in the form:

(8.48)
$$\frac{\delta V_O}{V_s} = K\frac{\delta R}{4R}$$

where K is the **bridge constant** and is defined as

(8.49)
$$K = \frac{\text{Output of bridge}}{\text{Output of bridge if only one strain gauge is active}}$$

Figure 8.71

Beam with two strain gauges

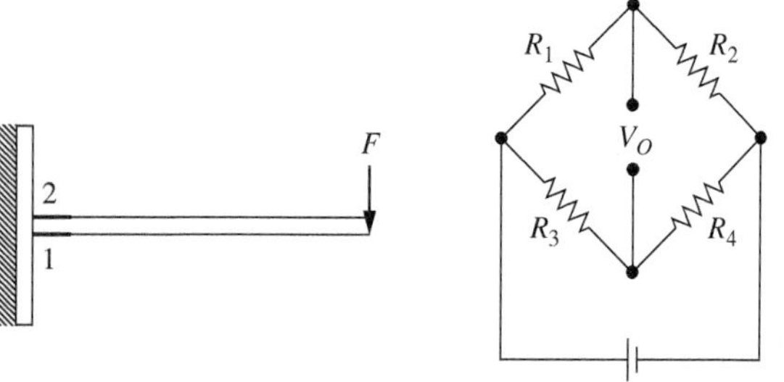

Examples 8.12 and 8.13 illustrate the use of bridge circuits.

Example 8.12 Strain Gauge Output

If the strain gauge considered in Example 8.6 was connected to a Wheatstone bridge with $V_s = 10\text{V}$, determine the output change of the bridge for the loading given in that example.

Solution:

For the 120 Ω gauge, the change in resistance was found to be 0.00955 Ω. Using Equation (8.44) (since δR/R is very small), the change in the bridge voltage output is given as

$$\delta V_O = \frac{\delta R}{4R} V_S = \frac{0.00955}{4 \times 120} 10 = 0.199 \text{ mV}$$

If the bridge was used in the deflection mode, this output can be amplified 1000 times or more before being read by an A/D due to the absence of a large, fixed voltage component. This is one of the main reasons why a Wheatstone bridge is used to process the output of a strain gauge. This is not the case if the gauge was one of the resistors in a two-resistor voltage-dividing circuit, as shown in Figure 8.72. The presence of the initial voltage drops across the gauge resistor makes it difficult to measure the minute change in voltage drop due to the applied loading.

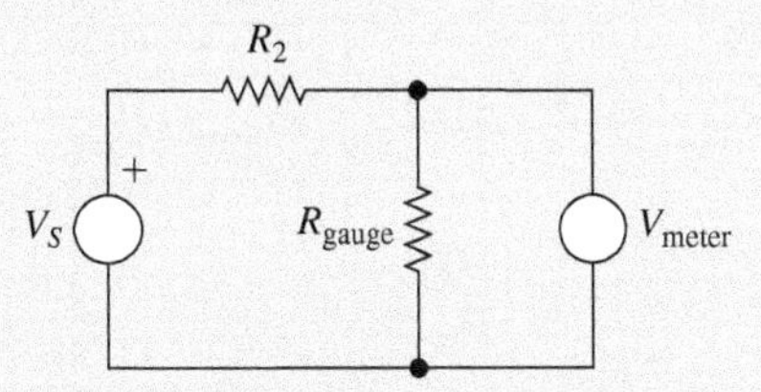

Figure 8.72 Two-resistor voltage-dividing circuit

Example 8.13 Bridge Circuit with an RTD Sensor

A Wheatstone bridge similar to that shown in Figure 8.69 is used to process the output from a two-lead platinum RTD. The RTD has an alpha of 0.00385 Ω/Ω/°C and a nominal resistance of 100 Ω at 0°C. The bridge was initially balanced when the temperature is 20°C. What will be the output voltage of the bridge if the sensor temperature was increased by 10°C? Assume that the bridge supply voltage is 10 VDC.

Solution:

The resistance of an RTD sensor is given by Equation (8.19). At $T = 20°\text{C}$, the resistance is $R_1 = R_0(1 + \alpha(T - T_0)) = 100(1 + 0.00385(20 - 0)) = 107.7\,\Omega$. Since the bridge is balanced at this temperature, we will assume the other three arms to have the same resistance.

For a 10° temperature increase, the change in resistance is equal to $(R_0\ \alpha\ \Delta T)$ or $\delta R = 100 \times 0.00385 \times 10 = 3.85\,\Omega$. Using the linear relationship first (Equation (8.44)), the output of the bridge circuit will be

$$\delta V_O = \frac{\delta R}{4R} V_S = (3.85 \times 10/(4 \times 107.7)) = 89.4 \text{ mV}$$

If we have used the exact relationship (Equation (8.42)), the output will be

$$\delta V_O = \frac{\delta R}{4R + 2\delta R} V_S = \frac{3.85}{4 \times 107.7 + 2 \times 3.85} 10 = 87.8 \text{ mV}$$

The error due to the linear approximation is 1.8%.

There are several varieties of bridge circuits. The bridge circuit that was previously analyzed is known as a **constant-voltage resistance bridge circuit** due to the use of a constant voltage source as the supply source. If the supply source is a constant-current source (by using a current-regulated DC power source), then the bridge circuit is known as a **constant-current bridge circuit**. The constant-current bridge has better linearity than the constant-voltage bridge, and the output voltage change for the case where $R_1 = R_2 = R_3 = R_4 = R$, and $\delta R_2 = \delta R$, (see Problem P8.57) is given by the relationship

$$\delta V_O = \frac{\delta R\, R}{4R + \delta R} i_S \tag{8.50}$$

where i_S is the supply current. Similarly, an **AC bridge circuit** is one in which the supply voltage is alternating. With an AC supply, the bridge can be used to measure inductance, capacitance, or resistance. A bridge circuit can be used to compensate for lead wire resistance. This is needed in situations where the sensor is located away from the signal conditioning equipment and long leads are used to connect the sensor. The RTD temperature sensor which was discussed in Section 8.8 is available with different leads configuration. In the **two-lead configuration** (see Figure 8.73(a)) the bridge output is sensitive to the variation of the resistance of the lead wire due to environmental influences. However, in the **three-lead configuration** (Figure 8.73(b)), any environmental effect on the resistance of the two lead wires will equally affect both 'legs' of the bridge assuming the same wire type, wire thickness, and wire length is used in both leads; thus, the sensor is insensitive to lead wire effects.

Figure 8.73

(a) Two-lead and (b) three-lead connections to a bridge

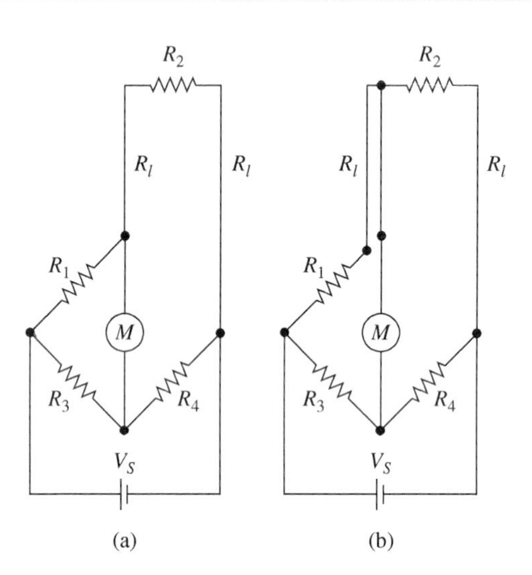

A bridge circuit can also be used to **compensate for temperature changes** in strain gauges. This is done by using a dummy strain gauge located next to where the actual measurement is made and placed on material similar to the test material. The connection diagram is shown in Figure 8.74. Any temperature change will change the R_1 and R_2 resistances by the same amount, making the circuit just sensitive to variation in the R_2 resistance due to loading.

Figure 8.74

Dummy gauge for temperature compensation

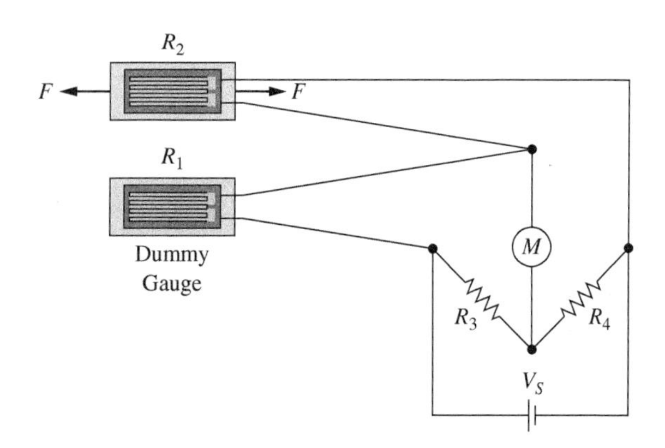

In the examples highlighted in this section, it is evident that the output voltage from a Wheatstone bridge is usually in the mV range. This small voltage range, although indicative of the parameters being measured, might not be sufficient for direct interpretation by most analog-to-digital (A/D) converters. A/D converters require a certain voltage level to produce a reading with a satisfactory resolution. To address this challenge, amplification of the bridge's output becomes necessary. In the realm of electronics, operational amplifier (op-amp) circuits, especially the differential amplifier and the instrumentation amplifier, serve this purpose efficiently. Both these circuits are designed to amplify the difference between two input signals. Among these, the instrumentation amplifier is the preferred choice as it provides superior input impedance characteristics compared to the differential amplifier. Figure 8.75 offers a visual representation, showing how a Wheatstone bridge can be interfaced with an instrumentation amplifier (refer to Section 2.9.5 for details).

Figure 8.75

Connections between a Wheatstone bridge and an instrumentation amplifier

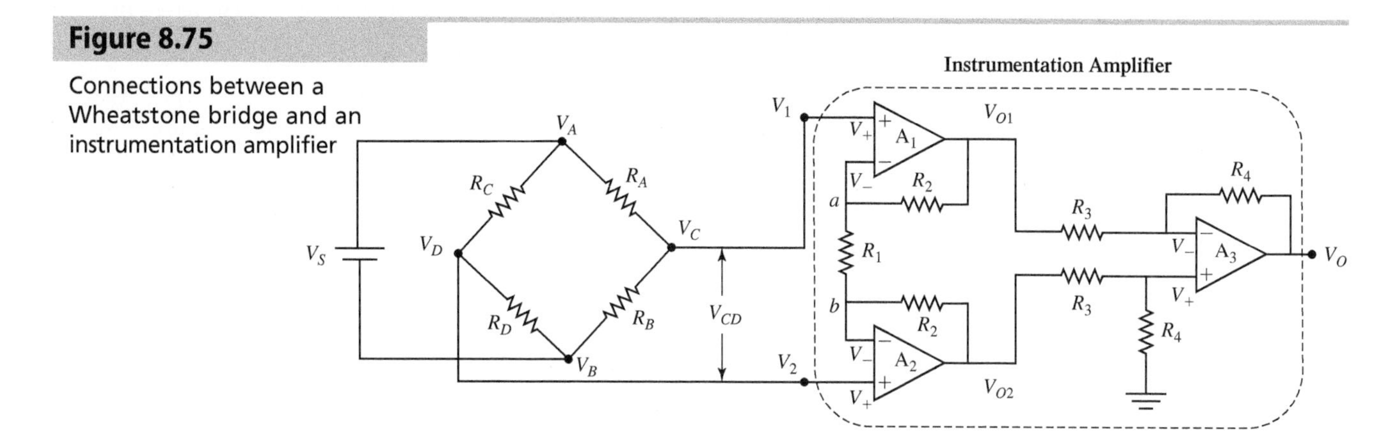

8.12 Understanding 4-20 mA Sensor Output Signals

Sensors can be broadly categorized based on their output signals: some produce analog outputs, while others provide digital outputs. Among those with analog outputs, a common standard is the **4-20 mA current signal**. This section will discuss this particular output method. In systems using the 4-20 mA standard (▶ available—V8.24), the sensor typically interfaces with a current transmitter. This transmitter produces a current signal that is proportional to a specific physical quantity, whether it is pressure, temperature, flow, or another measurable parameter. The 4-20 mA range represents the sensor's scaled range, with 4 mA typically signifying the lower range value (LRV) and 20 mA signifying the upper range value (URV). It is essential to note that these values do not always correspond to the absolute minimum and maximum measurable values of the sensor, but rather to the set operational range.

Current transmitters come in various configurations, with the **two-wire transmitter** being a prevalent choice. Figure 8.76 presents a schematic of a wiring diagram incorporating a two-wire transmitter. An external DC voltage powers the transmitter via a lead wire, exhibiting a resistance of R_L. The output current from the transmitter travels through another lead wire, also with a resistance of R_L, before passing through a receiver resistor characterized by a resistance of R_C. In this configuration, the transmitted current flows through every component in the loop. Given that voltage measurements are typically more straightforward than direct current measurements, the voltage drop across the precision receiver resistor is monitored. This voltage serves as a proxy for the sensor's output. For instance, when employing a 250 ohm receiver resistor, the voltage across it can range from 1 V (at 4 mA) to 5 V (at 20 mA) due to Ohm's law. Example 8.14 considers this output method.

Figure 8.76

Wiring for a two-wire current transmitter

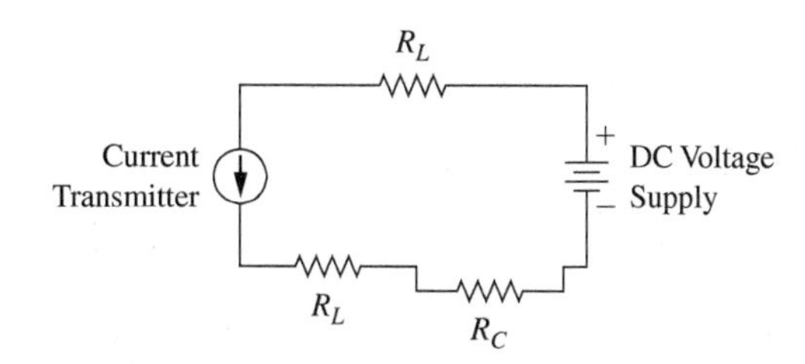

Example 8.14 Current Output Method

Consider a temperature sensor that outputs a current proportional to the temperature it measures. The sensor specification states that 4 mA corresponds to 0°C and 20 mA corresponds to 100°C. Suppose we want to connect this sensor to an A/D converter that accepts voltage inputs ranging from 0 V to 10 V. Determine:

(a) The size of the resistor to convert the 4-20 mA current to a corresponding voltage
(b) The temperature and the voltage output if the sensor output is an 8 mA signal

Solution:

(a) We will use Ohm's Law to determine the resistor value. We will consider that at the maximum output current of 20 mA, we should get a maximum voltage of 10 V at the A/D converter, or

$$10\text{ V} = 20\text{ mA} \times R$$

This gives R to be 500 Ω.

(b) From the specifications, we know that 16 mA (20 − 4) corresponds to a 100°C temperature range or a temperature sensitivity of 100/16 = 6.25°C/mA. Thus, at 8 mA output signal, this corresponds to temperature of (8 mA − 4 mA) × 6.25°C/mA = 25°C. The voltage output will be obtained by multiplying the current output by the resistance, or 8 mA × 500 Ω = 4 V.

An advantage of the 4-20 mA current loop standard, prevalent in process control, is its insensitivity to supply-voltage variations and resistance to noise in the circuit. The high impedance of the current transmitter, often in the range of megaohms, ensures that any noise voltage in the circuit predominantly drops across the transmitter, leaving minimal voltage variation across the receiver resistor. This resistance to electrical interference enables accurate transmission of sensor readings over extended cable runs, underscoring its popularity in many industrial applications. Additionally, the 'live zero' starting at 4 mA provides an intuitive distinction between the lowest sensor reading and a potential break in the loop. Any deviations outside the standard 4-20 mA range signal disruptions like open or short circuits.

8.13 Chapter Summary

This chapter delves deeply into the intricate world of sensors, pivotal elements in both mechatronic and measurement systems. Sensors are the eyes and ears of automated systems, transforming physical phenomena into electric signals that are processed, analyzed, and acted upon. At the outset, the chapter introduces readers to the foundational aspects of sensor performance: static and dynamic characteristics. Static characteristics define how a sensor behaves under steady conditions, while dynamic characteristics describe its performance under varying conditions. Together, they form the blueprint for assessing the quality, reliability, and suitability of a sensor for specific applications.

Diving into the vast array of sensors, the chapter methodically explains the operational principles behind several types:

Displacement Measurement: This involves detecting changes in position, and the key sensors in this category are potentiometers, LVDTs (Linear Variable Differential Transformers), encoders, and resolvers.

Proximity Measurement: Essential for determining the closeness of objects with or without physical contact, this category includes hall-effect sensors, inductive-type sensors, capacitive-type sensors, ultrasonic sensors, and contact-type proximity sensors.

Speed Measurement: For quantifying how fast something is moving, we use devices like tachometers and encoder-type sensors.

Strain, Force, and Torque Measurement: This segment introduces strain-gauge sensors, which are versatile in measuring strain, force, and torque, and force-sensitive resistors that are specific to force measurement.

Temperature Measurement: Temperature, being a vital parameter in many processes, is gauged using thermistors, thermocouples, RTDs (Resistance Temperature Detectors), and IC temperature sensors.

Vibration Measurement: Monitoring vibrations is crucial in various applications, and the chapter covers piezoelectric sensors and integrated circuit sensors for this purpose.

Inertial Measurement (IMU): The fundamental IMU sensor consists of an accelerometer for measuring linear acceleration and a gyroscope for measuring rotational speeds.

A section of the chapter is dedicated to signal-conditioning operations. Here, readers learn about vital processes like filtering (to remove unwanted noise), amplification (to boost the sensor's output to usable levels), and bridge circuits (to maximize precision). These techniques enhance the sensor output, making it compatible and ready to be interfaced with measurement devices or analog-to-digital converters. Concluding the chapter, there is an in-depth discussion on the 4-20 mA sensor output method. This current-based signaling method is a staple in the industry due to its robustness against interference, allowing accurate data transmission over long distances.

Questions

8.1 Name three static performance characteristics of sensors.

8.2 What is the difference between sensor accuracy and repeatability?

8.3 Explain why repeatability error cannot be improved by calibration.

8.4 What is the linearity error?

8.5 Define what is meant by sensor bandwidth.

8.6 List several sensors that are covered in this chapter that produce an analog output.

8.7 List several sensors that are covered in this chapter that produce a digital output.

8.8 List some disadvantages of potentiometers.

8.9 List some advantages of LVDTs.

8.10 What are the differences between resolvers and encoders?

8.11 Name one limitation of incremental encoders.

8.12 What is the Hall effect?

8.13 Explain the operation of ultrasonic proximity sensors.

8.14 Which type of non-contact proximity sensors can detect only metal objects?

8.15 What is 'ripple' in tachometers?

8.16 Name two methods to obtain speed measurement using an encoder.

8.17 What is the principle of operation of a strain gauge?

8.18 Explain the difference between reaction and rotary torque sensors.

8.19 What is a thermistor?

8.20 Which thermocouple type has almost linear output characteristics?

8.21 What is the minimum number of junctions needed in a thermocouple measuring setup?

8.22 What is the difference between a thermocouple and an RTD sensor?

8.23 Name some advantages of IC sensors.

8.24 What is an IMU sensor?

8.25 What is the difference between a vibrometer and an accelerometer?

8.26 What piezoelectric effect is used in the design of piezoelectric accelerometers?

8.27 Why is a Wheatstone bridge circuit used in some signal conditioning circuits?

8.28 What condition causes a Wheatstone bridge's arms to be balanced?

8.29 Why are filters used?

8.30 What is the advantage of the 4-20 mA sensor output method?

Problems

Section 8.1 Introduction

P8.1 Research and identify the type of sensors used in the following applications.

a. Kitchen oven

b. RPM indicator in vehicles

c. Back-up sensor in certain vehicles

d. Trunk compartment closure

e. Refrigerator door closure

f. Laptop cooling system

g. Vehicle engine cooling system

h. Servo robot

Section 8.2 Sensor Performance Terminology

P8.2 An infrared temperature sensor has a range of −50°C to 1000°C and an accuracy of ±1% of the reading. What is the maximum error in the temperature reading at a temperature of 500°C?

P8.3 A linear displacement sensor has a sensing range of 0 to 100 mm and a linearity error of ±0.2% FS. What is the maximum linearity error in millimeters at a displacement of 60 mm?

P8.4 A digital temperature sensor has a resolution of 0.0625°C and an accuracy of ±0.5°C. What is the maximum error in the temperature reading?

P8.5 A pressure sensor with a measurement range of 0 to 100 psi has the following specifications:

- Resolution: 0.5 psi
- Accuracy: 1.5% of full-scale reading
- Linearity Error: 0.75% of full-scale reading

Compute the maximum possible error (in PSI) one could expect when taking a reading with this sensor.

P8.6 A temperature sensor has a time constant of 4 seconds. The sensor, which was originally at a temperature of 20°C, was rapidly immersed in a fluid that has a temperature of 60°C. Approximately how long will it take for the sensor output to reach a steady state?

P8.7 A displacement sensor has a first-order system characteristic, a bandwidth of 1000 Hz, and a low-frequency gain of 10 V/mm. What will be the gain of the sensor at the following frequencies: 100, 500, and 800 Hz?

Section 8.3 Displacement Sensors

P8.8 A rotary potentiometer with a resistance of 5 kΩ and a measuring range of 325° uses a 10 V supply. The potentiometer output was read by a measuring device with a resistance of 100 kΩ. Determine the angle measured by the potentiometer if the measuring device output is 6 V. What would the measured angle be if the measuring device resistance is 1 MΩ instead?

P8.9 A linear variable differential transformer has a stroke length of ±150 mm and a sensitivity of 50 mV/mm when moved. Determine:

a. The LVDT's maximum output voltage

b. The output voltage when the core is moved 120 mm from its null position

c. The core position from the center when the output voltage is 3.75 volts

P8.10 Select an appropriate sensor to check for the 'quality' of the completed assembly at station #3 in Figure 6.13. Assume that the quality of the assembly is done by measuring the height of the two-part assembly relative to the indexing table surface. Justify your selection and explain how the selected sensor will be interfaced with a PC or a microcontroller system.

P8.11 A DC motor has a built-in incremental encoder. The encoder disk has 1250 lines. Determine the angular resolution of this encoder, assuming that the encoder is operated in quadrature mode.

P8.12 A geared DC motor has a built-in incremental encoder that is connected to the motor side. The encoder disk has 1250 lines, and the gear ratio is 8:1. Determine the angular resolution of this encoder, assuming that the encoder is operated in quadrature mode.

P8.13 A DC motor equipped with an incremental optical encoder is used to drive a lead-screw positioning table, as shown in Figure 8.17. The screw is double-threaded and has a pitch of 2 mm, the encoder disk has 1250 lines, and the encoder is operated in quadrature mode. Determine the measurement resolution of this encoder for this setup.

P8.14 A 32-bit counterboard is used to read the signals from an encoder that is integrated into a motor. The encoder disk has 1250 lines per revolution and is operating in quadrature mode. Determine how long the motor can run at a speed of 1000 rpm before the counterboard overflows.

P8.15 A single-turn, 10-bit absolute encoder has a gray code output. Determine (a) the resolution of this encoder and (b) the encoder output for the first 16 angular positions read by the encoder.

P8.16 Research where magnetic encoders are used and compare them to optical encoders.

Section 8.4 Proximity Measurement

P8.17 Draw a circuit to interface the output of a two-wire NO proximity sensor (see Example 8.5) that uses a 24 VDC power supply to the digital input line of an MCU that operates with a V_{DD} of 5 V.

P8.18 Consider a backup obstacle warning system (similar to that found in some vehicles) that just uses one ultrasonic distance sensor. Assume that the sensor provides a 0 to 10 VDC analog voltage output that is proportional to the distance of the object from the sensor. The sensor output is zero when the object is at the near-limit setting of the sensor. Design a circuit that can process the output of this sensor to inform the user of the closeness of the object to the sensor. Use multiple LEDs as an indication of the output. Make any reasonable assumptions.

P8.19 For the assembly system shown in Figure 6.13, select an appropriate sensor to detect the part's presence on conveyer *A* at the location where the part is picked up by robot #1. Assume that the part is metal, made of steel, has dimensions of $2 \times 2 \times 1.5$ in., and it is detected using one of its side surfaces. Justify your selection and explain how the selected sensor will be interfaced with a PC or a microcontroller system.

P8.20 A capacitive-type displacement sensor has a sensing range of 0 to 5 mm and a sensitivity of 0.1 mV/μm. If the displacement is 3.5 mm, what is the output voltage of the sensor?

Section 8.5 Speed Measurement

P8.21 A tachometer is used to measure the speed of a motor that is operating at a constant speed. The tachometer has a voltage gradient of 1.9 volts per 1000 rpm and a ripple of $+/-3\%$. What is the speed of the motor if the tachometer's average output is 4 V?

P8.22 A rotary incremental encoder is equipped with a disk that has 1000 lines and is operating in quadrature mode. The encoder count values are sampled at a rate of 1000 Hz, and the encoder count difference is 3000 over an interval of 10 samples. What is the angular speed measured by this encoder?

P8.23 A rotary incremental encoder is equipped with a disk that has 1000 lines. A clock with a frequency of 10,000 Hz recorded a time interval of 5 clock cycles for the motion between two adjacent lines on the encoder disk. What is the angular speed measured by this encoder?

Section 8.6 Strain Measurement

P8.24 A strain gauge is bonded to a bar subjected to tensile loading. The strain gauge has a gauge factor of 3. When the bar is subjected to a tensile load, the strain gauge resistance increases by 0.01 Ω from 120 Ω. Determine the microstrain experienced by the bar.

P8.25 A strain gauge with a resistance of 120 Ω and a gauge factor of 2 has a resistance change of 0.005 Ω.

a. Determine the microstrain measured by the gauge.

b. If this microstrain is due to the axial elongation of a rectangular steel ($E = 200 \times 10^9$ N/m^2) bar with a cross-sectional area of 0.4×10^{-3} m^2, determine the axial force acting on the bar.

P8.26 A steel ($E = 200 \times 10^9$ N/m^2) bar with a cross-sectional area of 4 cm^2 is subjected to a tensile axial force of 4 kN. Assume a strain gauge with a resistance of 120 Ω and a gauge factor of 3 is used to measure the strain due to this loading. Determine the change in resistance of the gauge under this loading.

P8.27 A cylindrical steel ($E = 200 \times 10^9$ N/m^2) rod with a diameter of 25 mm and a length of 500 mm is subjected to an axial tensile load of 10 kN. A strain gauge is fixed along the axial direction of the rod. If the strain gauge has a gauge factor of 2.1 and its initial resistance is 350 ohms, calculate the change in resistance of the strain gauge.

Section 8.7 Force and Torque Measurement

P8.28 A load cell with a sensitivity of 0.6 mV/V, a supply voltage of 10 VDC, and a maximum measurement capacity of 1250 kg is used for monitoring weight overload on an elevator. The elevator's rated load capacity is 1000 kg. Determine the output voltage from the load cell when the elevator is loaded at its rated capacity.

P8.29 In a vehicle testing facility, a car's drivetrain has a rotary torque sensor with a sensitivity of 2 mV/N·m installed on the shaft connecting the transmission to the wheels. The sensor's data acquisition system records a peak voltage of 0.8 V during hard acceleration from rest. At this moment, the shaft's rotational speed is at 2000 rpm. Determine (a) the power being delivered by the shaft at this instant and (b) the resolution of the torque sensing measurement if the A/D has a measurement range of $+/-5$ volts and a 12-bit resolution.

P8.30 Figure P8.30 shows an experimentally determined plot (from [33]) for the voltage output versus force for a force-sensing resistor (FSR).

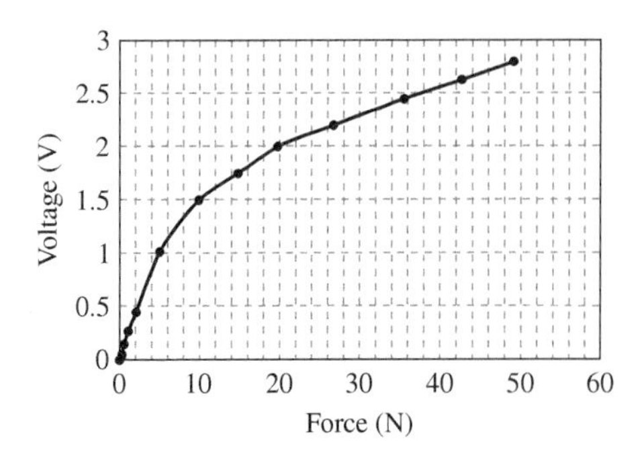

Figure P8.30

a. What is the approximate voltage sensitivity of this sensor for forces below 10 N and forces above 10 N?

b. If this plot was generated using a voltage dividing circuit where an 80 kΩ resistor was placed in series with the FSR and a 5 VDC voltage was applied to this circuit, determine the resistance of the FSR at 10 N and 20 N.

Section 8.8 Temperature Measurement

P8.31 An NTC thermistor with the following parameters is used to measure temperature: $T_0 = 298.15$ K, $\beta = 3950$, and $R_0 = 10$ kΩ. If the resistance of the thermistor is 15 kΩ, what is the temperature indicated by this thermistor?

P8.32 An NTC thermistor has a resistance of 10 kΩ at 25°C and a temperature coefficient of −5%/°C. What is the resistance of the thermistor at 75°C?

P8.33 You are provided with a 10 kΩ NTC thermistor and a 10 kΩ fixed resistor. Both are connected in series with a 9 V battery to form a voltage divider circuit. The midpoint between the thermistor and the resistor is connected to a small buzzer which activates when the voltage exceeds 5 V. For this problem:

a. Draw this circuit.

b. If the room temperature is such that the thermistor's resistance is 10 kΩ, what is the voltage at the midpoint, and is the buzzer active?

c. As the temperature rises and the thermistor's resistance drops to 5 kΩ, what happens to the voltage at the midpoint? and is the buzzer active?

P8.34 A thermocouple generates a voltage of 30 mV at 100°C and 20 mV at 50°C. What is the temperature of the thermocouple if the voltage output is 25 mV?

P8.35 A Type K thermocouple produced a voltage output of 1 mV when its leads were connected to a voltmeter whose terminals are at a temperature of 25°C. Determine the temperature of the object measured by this thermocouple.

P8.36 A Type K thermocouple produced a voltage output of 1.5 mV when its leads were connected to a voltmeter whose terminals are at a temperature of 22°C. Determine the temperature of the object measured by this thermocouple.

P8.37 A Type K thermocouple is used to measure temperature over the range of 0 to 1000°C. The voltage output of the thermocouple is read by an A/D converter with a voltage range of +/−1 volt and 16-bit resolution. Determine the temperature measurement resolution of this setup.

P8.38 A Type K thermocouple is used to measure temperature over the range of 0 to 200°C. The voltage output of the thermocouple is amplified by a gain factor K before being read by an Arduino Uno A/D converter that has a range of 0 to 5 volts and a 10-bit resolution. What should be the gain K so the temperature resolution of this setup is 1°C or better?

P8.39 A platinum RTD sensor has a nominal resistance of 100 Ω at 0°C and a TC factor of 0.00392 Ω/Ω/°C. What is the resistance of the sensor at 100°C?

P8.40 A platinum RTD sensor has a nominal resistance of 100 Ω at 0°C and a TC factor of 0.00392 Ω/Ω/°C. Determine the temperature read by the sensor if the RTD resistance is 200 Ω.

Section 8.9 Vibration Measurement

P8.41 An accelerometer generates an output voltage of 20 mV/g. If the acceleration of a machine is 5 g, what is the output voltage of the accelerometer?

P8.42 An accelerometer with a natural frequency of 1 kHz and a damping ratio of 0.5 is used to measure the vibration of a motor rotating at 5000 rpm. The accelerometer gave a reading of 1 g. What is the actual acceleration of the motor?

P8.43 In many realistic vibration measurement situations, frequencies that are multiples of the base frequency (called harmonics) can be measured. Consider a single-axis accelerometer that is used to measure the vibration of an industrial pump housing. The accelerometer registers consistent vibration amplitudes at 30 and 60 Hz frequencies. The pump is directly connected to a motor running at 1800 RPM. Are the measured frequencies related to the motor's operational speed, and what might have caused them?

Section 8.10 Integrated Circuits (IC) Sensors

P8.44 An IC digital temperature sensor has an output sensitivity of 10 mV/°C. The output of the sensor was read by a 10-bit A/D with a 5 V reference. Determine the actual temperature if the A/D reading is 84.

P8.45 An IC vibration sensor has a range of ±5 g and a sensitivity of 400 mV/g. The signal-conditioning circuit for this sensor gives an output of 2.65 V at 0 g. For positive acceleration, the output increases above 2.65 V, and for negative acceleration, the output decreases below 2.65 V. Determine the actual acceleration if the sensor output is (a) 2.0 V or (b) 3.8 V.

P8.46 An IMU is used in a drone flight stabilization system. The IMU's accelerometer has an adjustable sensitivity range, selectable between ±2 g, ±4 g, ±8 g, and ±16 g. During a calibration test, the drone's vertical motion during takeoff produces an actual maximum acceleration of 3 g for a short duration. Determine the following.

a. Which sensitivity setting would capture the drone's takeoff acceleration without clipping (i.e., maxing out)?

b. Given that the IMU's output data resolution improves with decreasing sensitivity settings (i.e., the output data will be more detailed at ±2 g compared to ±16 g), determine the optimal setting to capture the most detailed acceleration data of the drone's motion while still avoiding clipping.

Section 8.11 Signal Conditioning

P8.47 A Wheatstone bridge similar to that shown in Figure 8.69 has $R_1 = 150\ \Omega$, $R_2 = 180\ \Omega$, and $R_3 = 160\ \Omega$. What shall be the value of R_4 so the bridge is balanced?

P8.48 An electronics lab technician is trying to balance a Wheatstone bridge similar to that shown in Figure 8.69 for a specific application using commercially available standard resistors from the E12 series (see Appendix A). The E12 series consists of the following multiplier values: 1.0, 1.2, 1.5, 1.8, 2.2, 2.7, 3.3, 3.9, 4.7, 5.6, 6.8, and 8.2. For one leg of the bridge, the technician has selected the resistors $R_1 = 1.5 \times 10^2\ \Omega$ and $R_2 = 3.3 \times 10^2\ \Omega$. For the other leg, they selected $R_3 = 2.2 \times 10^2\ \Omega$. The technician needs to choose an R_4 value from the E12 series such that the bridge is as close to balance as possible. However, due to the limited values available in the E12 series, a perfect balance might not be achievable.

a. Which resistor value from the E12 series should the technician select for R_4 to get the bridge as close to balance as possible?

b. Calculate the mismatch percentage, i.e., by how much does the actual R_3/R_4 ratio differ from the ideal R_1/R_2 ratio.

P8.49 A Wheatstone bridge like that shown in Figure 8.69 is used to determine the unknown resistance R_1 by adjusting the resistance R_3. The resistance R_3 was found to be 151 Ω when the bridge was first balanced. When the resistances R_2 and R_4 were interchanged and the bridge was balanced again, R_3 was found to be 182.5 Ω. What is the value of R_1?

P8.50 A strain gauge with a resistance of 120 Ω was connected to a Wheatstone bridge with a supply voltage of 12 V. Determine the output voltage change of the bridge if the resistance change in the gauge is 0.01 Ω.

P8.51 A strain gauge with a resistance of 350 Ω was connected to a Wheatstone bridge with a supply voltage of 12 V. Determine the output voltage change of the bridge if the resistance change in the gauge is 0.02 Ω.

P8.52 A Wheatstone bridge similar to that shown in Figure 8.69 is used to process the output from a two-lead platinum RTD. The RTD has an alpha of 0.00385 Ω/Ω/°C and a nominal resistance of 100 Ω at 0°C. The bridge was initially balanced when the temperature was 20°C. What will be the output voltage of the bridge if the sensor temperature is increased by 30°C? Assume that the bridge supply voltage is 12 VDC.

P8.53 For the strain gauge considered in Example 8.12, design an op-amp circuit that uses a differential amplifier to amplify the bridge output voltage signal by a factor of 1000.

P8.54 An instrumentation amplifier (IA128—see Section 2.9.5) is used to amplify the output of a Wheatstone Bridge that is connected to a strain gauge. If the differential output of the Wheatstone bridge is 0.2 mV under the load applied to the member where the strain gauge is attached, select the value of the appropriate resistor for the instrumentation amplifier such that the voltage output of the bridge is amplified to 1 volt.

P8.55 The output of a Wheatstone bridge is connected to an instrumentation amplifier that has a gain of 1000 and whose output is connected to an A/D converter that has a range of 0 to 5 volts and 10-bit resolution. The Wheatstone bridge gives an output of 0.2 mV under the maximum load of 1000 N that is applied to the bar. Determine the force resolution of the output of the A/D for this setup.

P8.56 Show how Equation (8.46) was obtained.

P8.57 Derive Equation (8.50) for a constant-current source bridge.

Section 8.12 Understanding 4-20 mA Sensor Output Signals

P8.58 Consider a pressure sensor that is designed to measure pressures ranging from 0 to 10 bar. Its output is in the standard 4-20 mA format, where 4 mA corresponds to 0 bars and 20 mA corresponds to 10 bars. We wish to connect this sensor to an A/D converter that accepts voltage inputs from 0 to 5 V. Determine:

a. The appropriate resistor value to convert the 4-20 mA current into the desired voltage range

b. The corresponding pressure and the voltage across the resistor if the sensor outputs a current of 12 mA

Laboratory/Programming Exercises

L/P8.1 Use a hall-effect proximity sensor and a buzzer to simulate the operation of a single-sensor security system. Wire up the proximity sensor so the sensor output will be high when a magnet is not present next to the sensor and low when a magnet is present. Use the proximity sensor output to turn ON the buzzer when the magnet is not present.

L/P8.2 For the FSR in Problem P8.30 whose voltage-force relationship is shown in Figure P8.30, develop a program in Excel or MATLAB that can provide the force measured by this sensor as a function of the measured voltage.

L/P8.3 Develop code in Excel or MATLAB that can provide the temperature output as a function of the resistance obtained from an NTC thermistor with the following parameters: $T_0 = 298.15$ K, $\beta = 3950$, and $R_0 = 10$ kΩ.

L/P8.4 Using the Arduino Uno or similar board, develop an Arduino program to display the room temperature to the user every 1 second using the Serial Monitor. Use an NTC thermistor as the temperature sensor and build a voltage-dividing circuit to get the voltage output from the thermistor. Read the voltage output from the thermistor using the A/D converter on the board, convert the measured voltage to a temperature in °C, and display it to the user.

L/P8.5 Use the data from Table 8.6 to create a program in MATLAB or other software that can provide the temperature output in °C for a Type K thermocouple over the temperature range of 0 to 100°C if provided with the value of the output voltage read by a voltmeter to which the thermocouple leads were connected to, as well as the temperature of T of the leads where T is in the range of 15 to 30°C.

L/P8.6 Using an Arduino Uno board or a similar one, develop a program to read the voltage output from an analog output IC temperature sensor such as the LM35C temperature sensor. Convert the read voltage value to temperature in engineering units, and then display the temperature reading to the user every 1 second through the Serial Monitor.

L/P8.7 Build a bridge circuit that uses four resistors. Use three fixed-type resistors where each one has a different value for the R_1, R_3, and R_4 resistors, and a variable resistor for the R_2 resistor. Apply a DC voltage to the bridge such as 5 or 12 volts. Adjust the value of the R_2 resistor so the bridge is balanced. Measure and record the value of the four resistors when the bridge is balanced and compare the results obtained to the theory.

Chapter 9

Actuators

Chapter Objectives:

When you have finished this chapter, you should be able to:

- List the advantages of electrically powered actuators
- Explain the common configurations of brush-type DC motors
- Model the electro-mechanical behavior of a PM brush DC motor
- Explain the working principle of BLDC, AC, stepper, universal, and hobby servo motors
- Explain the function of a solenoid
- Explain the construction and drive techniques for stepper motors
- Interpret the torque-speed characteristics of an actuator
- Explain drive methods and amplifiers for different actuators
- Explain the operation of pneumatic and hydraulic actuators
- Explain the operation of piezoelectric and shape memory alloy actuators
- Select an appropriate actuator for a given mechatronic application

9.1 Introduction

Actuators are the key components of all mechanized equipment. An **actuator** is a device that converts energy to mechanical motion. There are many types of actuators available to drive machinery such as electric, internal combustion, pneumatic, and hydraulic. Electrically powered actuators are widely used in equipment (such as pumps, compressors, machine tools, and robots). Internal combustion actuators are normally used for mobile applications (such as in vehicles, boats, and power generation equipment). Pneumatic actuators, which use compressed air as the power source, are mostly part of machinery applications (such as pick-and-place robots and air-powered tools). Hydraulic actuators, which use pressurized oil to drive a piston, are used in applications for lifts and presses, among others.

This chapter focuses primarily on electrically powered actuators, which are commonly used in mechatronic systems. It also discusses pneumatic and hydraulic actuators, as well as piezoelectric and shape memory alloy actuators. Electrically powered actuators have the characteristics that they are clean (no leaking pressurized fluids), require no extra equipment (no need for air storage tanks, fuel tanks, or air filters), can operate indoors without the need for exhaust systems, and are easily controlled (especially DC-type electric motors). One disadvantage of electrically powered actuators is their low power-to-size ratio. In this chapter, the characteristics of several types of electric motors are discussed, along with information on how to select and size an actuator for a mechatronic system.

Electric motors are broadly classified into direct current (DC) and alternating current (AC), which refer to the type of electric power that drives the motor. Within each category, the motors are further classified depending on the physical construction of the motor and how the motor is wired. All electrically powered motors consist of two main components: a stator and a rotor, which are separated by an air gap. The **stator** is the fixed part of the actuator, while the **rotor** is the movable part of the actuator. In rotary-type electric motors, the stator takes the form of a hollow cylinder that is attached to the housing, while the rotor has a shaft that is supported by bearings. For further reading on the topics covered in this chapter, see [34-37].

9.2 DC-Motors

Direct-current (DC) motors are a type of motor that operates on steady, non-alternating voltages. They come in a wide range of power ratings, from very small units in milliwatts to large industrial motors that are rated in tens or hundreds of horsepower. There are two main categories of DC motors: brushed and brushless. Brushed DC motors have physical brushes that contact the commutator to supply current to the motor's armature. In contrast, brushless DC motors do not have brushes and instead use electronic control to manage the current through the motor windings. We will begin our discussion by focusing on brushed-type DC motors.

9.2.1 Brush DC

A brushed DC motor (▶ available—V9.1) is a widely used type of electric motor found in various applications, from small battery-operated devices like electric toys to larger industrial equipment such as conveyors, indexing tables, and material handling systems. As the name indicates, a brushed motor uses **brushes** to conduct electricity to the motor's rotating armature, ensuring a continuous flow of current in the proper direction as the rotor spins. The brushes, once made from solid copper or copper alloy materials, are now commonly constructed from spring-loaded graphite or carbon. These brushes press against the **commutator**, a ring made up of conducting copper segments separated by insulating material, that is affixed to the rotor. This design allows the commutator to reverse the current at precise intervals, maintaining consistent motor rotation. To illustrate the working principle of the brushed DC motor, imagine a simplified model with two brushes and a two-segment commutator shown in Figure 9.1.

Figure 9.1

Simplified construction of a brush DC motor

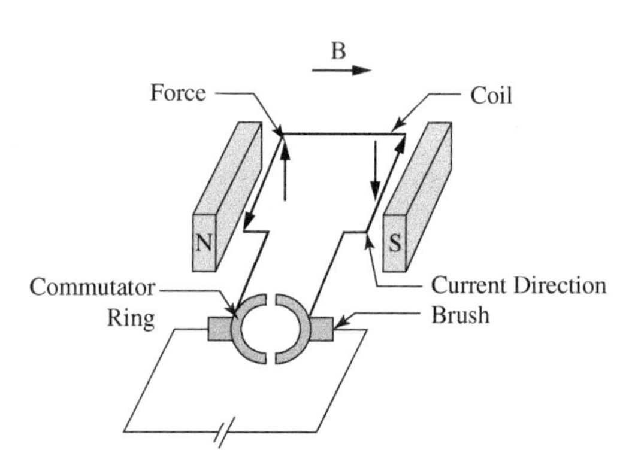

The rotor coil, which is placed within a magnetic field, has its ends attached to the commutator segments. When electric current flows through the rotor coils, it interacts with the magnetic field (supplied by either permanent magnets or electromagnets) to generate torque, which is the rotational force that turns the rotor. The torque is the result of the two equal but opposite forces acting on the sides of the coil and arises because of the Lorentz force, which acts on a current-carrying conductor in a magnetic field. According to **Lorentz's law**, the force ($\boldsymbol{F}$) experienced by a conductor carrying current ($\boldsymbol{I}$) in a magnetic field ($\boldsymbol{B}$) is described by the vector cross product equation

(9.1)
$$\boldsymbol{F} = \boldsymbol{I} \times \boldsymbol{B}$$

where the force is perpendicular to both the current and the magnetic field direction. For a simple two-piece commutator, as the rotor spins, the brushes maintain contact with the commutator segments to provide a continuous path for the current. These segments flip the current direction within the rotor coil every half turn to keep the motor running smoothly. Without this switching action by the commutator, the motor would stop every half turn as the forces would reverse and cancel out, creating no net movement. However, with just a two-segment commutator, the motor would experience a fluctuation in torque known as **torque ripple** (see Figure 9.2(a)), because the torque is not constant throughout the rotation—it is maximum when the coil is parallel to the magnetic field (where the moment arm is greatest) and drops to zero when the coil is in the vertical plane (where the moment arm is zero). In general, the torque is a function of the sine of the angle between the magnetic field direction and the vector normal to the plane of the coil, θ, where θ is 90° for the coil position shown in Figure 9.1. If we had used a six-piece commutator (and also three coils, one for each commutator pair) instead of a two-piece commutator, the torque would be smoother (Figure 9.2(b)) since the torque at any point in time is the sum of all the torques in all the coils. In commercial motors, commutators with many segments are used to produce a smoother torque. If the rotor has more coils, each connected to its own set of commutator segments, and the magnetic field is configured to be radial, the interaction between the rotor coils and the magnetic field is consistent, which further reduces torque ripple, allowing the motor to run more smoothly. In summary, the brushed DC motor relies on the clever use of electromagnetic forces to create rotation. The combination of a multi-segment commutator and multiple coils enables the production of smooth torque in practical motors.

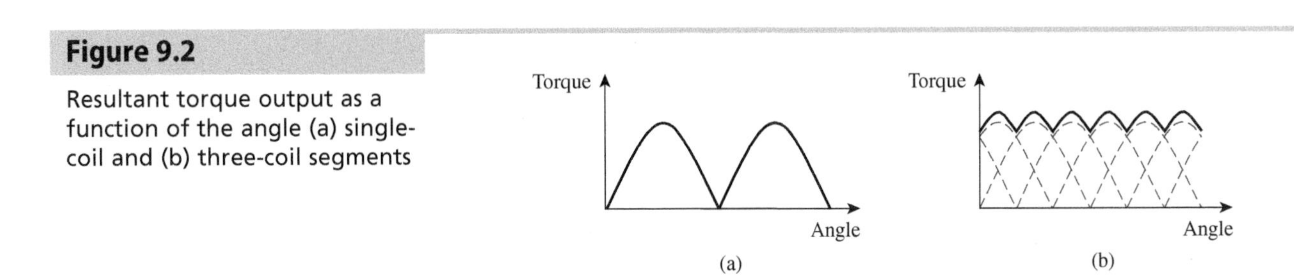

Figure 9.2

Resultant torque output as a function of the angle (a) single-coil and (b) three-coil segments

Figure 9.3 shows the brushes and commutator of a small DC motor. Note that the mechanical contact between the brushes and the commutator leads to wear of these components as well as the formation of arcs which require these components to be serviced periodically.

Figure 9.3

Commercial brush and commutator of a brush DC motor

(Jouaneh, University of Rhode Island)

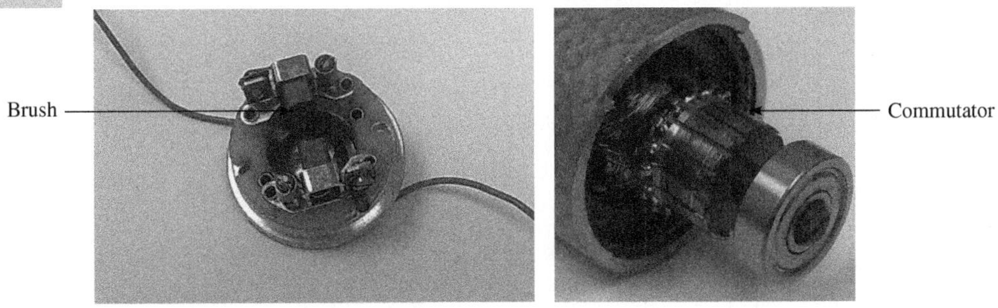

Brush

Commutator

One common configuration of brush DC motors is the **permanent magnet (PM)** DC motor where the stator is constructed of permanent magnets, and the rotor is made of wire coils (see Figure 9.4(a)). This configuration has the property of linear torque-speed characteristics. Other configurations are obtained by using electromagnets for the stator and by how the stator and rotor coils are wired. These configurations include series-wound motors, shunt-wound motors, and compound-wound motors (see Figure 9.4). In a **series-wound** motor, the stator and rotor coils are connected in series. In a **shunt-wound** motor, the stator coils and the rotor coils are connected in parallel, while in the **compound-wound** configuration, both series and parallel fields are used. Notice how the torque-speed curve is nonlinear for series, shunt, and compound-wound configurations. Also notice that in the **series-wound configuration**, the motor speed increases uncontrollably if the load acting on the motor is accidentally disconnected since the speed at zero torque is not limited. In small motors, the internal friction is usually sufficient to limit the breakdown speed, but in large motors, external safety devices need to be implemented. Series-wound motors are used where there is a need for a very large torque at low speed. An example would be moving a heavy load from rest such as an electric train, an elevator, or a hoist. **Shunt-wound motors** have a nearly constant speed under varying loads. This makes them attractive to drive machine tools and rotating equipment such as fans and blowers where it is desirable to have steady speeds. **Compound-wound motors** combine the characteristics of both series-wound and shunt-wound motors. They are used where there is a need for both a high starting torque and a constant speed operation such as in punch presses and shears. They are also safer to use in cases where the load might disconnect, such as in cranes since they have a controlled no-load speed.

Now develop a model of the dynamics of a PM brush DC motor. Such a motor is modeled as shown in Figure 9.5. On the electrical side, the rotor coil inductance L and its electrical resistance R are modeled as an inductor and a resistor in series. The **back electromotive force (EMF)** voltage due to the rotation of the conducting coil in the stator magnetic field is represented as a voltage source. On the mechanical side, the rotating coil is represented as a rotating mass with an inertia J with viscous friction acting on it that has a viscous coefficient of B. For the electrical part, applying KVL gives

(9.2) $$V_{in} - iR - L\frac{di}{dt} - V_{bemf} = 0$$

and for the mechanical part, applying Newton's Second Law gives

(9.3) $$T - B\omega = J\dot{\omega}$$

where T is the torque developed by the motor. This torque is proportional to the current through the motor coil, and is given by

(9.4) $$T = K_T\, i$$

where K_T is the torque constant. Since the back EMF voltage (V_{bemf}) is proportional to the rotating speed of the coil, we can write

(9.5) $$V_{bemf} = K_E\, \omega$$

where ω is the rotational speed of the motor and K_E is the back EMF constant. Equations (9.2)–(9.5) constitute the dynamic model of a PM brush DC motor which are solved (see Chapter 7) to obtain the current and speed of the motor as a function of time for a given input voltage and motor parameters.

Now examine the steady state torque-speed characteristics of a PM brush DC motor. At steady state, the voltage drop across the motor's inductor will be zero since the current is constant. From Equation (9.2), the current through the motor coil is then given by

(9.6) $$i = V/R = \frac{V_{in} - V_{bemf}}{R}$$

Combining Equations (9.4) through (9.6), we get

(9.7) $$T = K_T\frac{V_{in}}{R} - K_T\, K_E\frac{\omega}{R}$$

If we set $K_T\, V_{in}/R$ to be the starting torque T_s and $K_T\, K_E/R$ to be the constant α, then the torque-speed relationship is written as

(9.8) $$T = T_s - \alpha\, \omega$$

The relationship is plotted in Figure 9.6 for three different values (V_1, V_2, and V_3) of the supply voltage V_{in}, where $V_3 > V_2 > V_1$. The three plots have the same slope, but the starting torque increases (decreases) with an increase (decrease) in the supply voltage. For a given supply voltage, the torque-speed curve is linear and is defined by two parameters, the starting torque or the stalled torque, and the no-load speed. The **starting torque** is the maximum torque that is obtained from the motor. This maximum torque is obtained when the motor is stationary. As shown by Equation (9.7), as the motor starts rotating, the back EMF voltage generated due to the rotation of the conducting rotor through the magnetic field generated by the stator acts against the voltage applied to the motor leads. This causes the current through the motor coils to decrease, and thus the output torque, since the output torque is linearly related to the current through the coil. So, as the motor picks up speed, the torque further decreases and it will fall to zero at the

Figure 9.4

Common configurations of brush-type DC motors

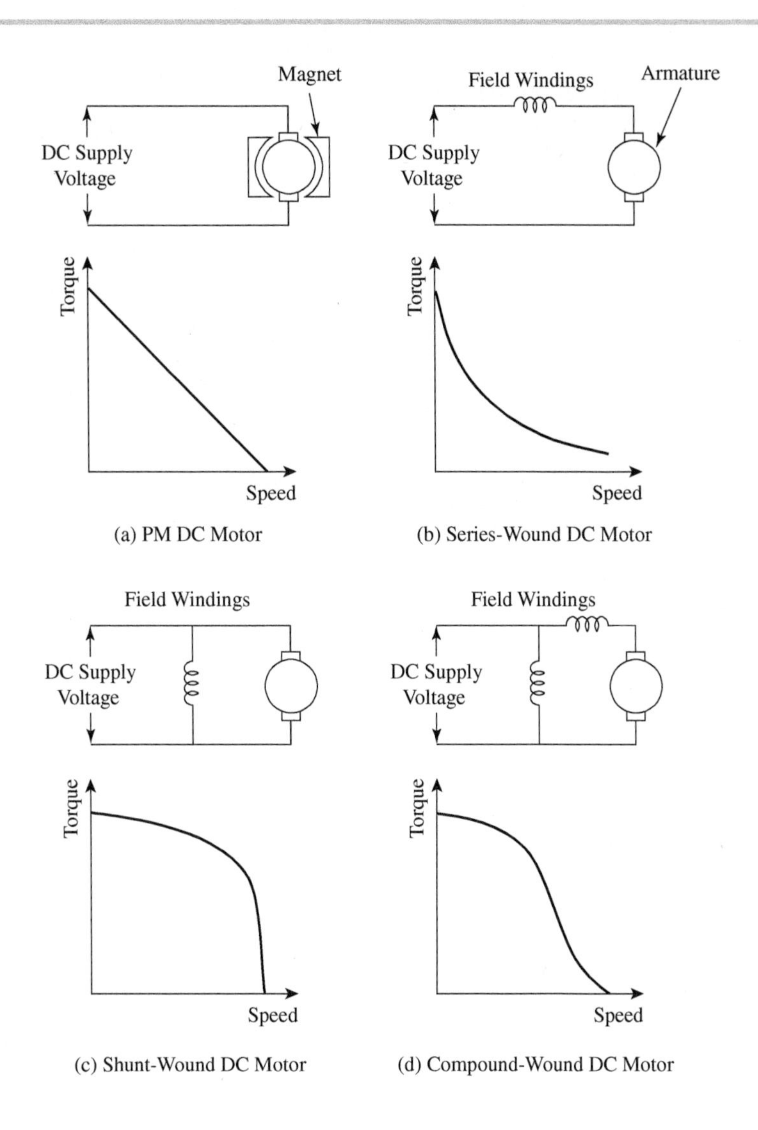

Figure 9.5

Electro-mechanical model of a PM brush DC motor

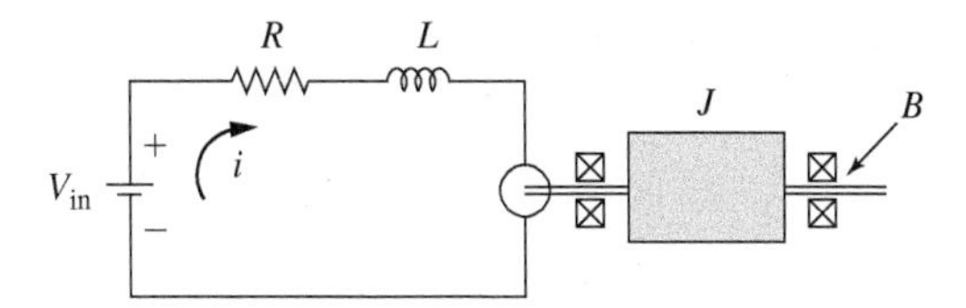

no-load speed, which is the maximum speed that is obtained from a DC motor. Figure 9.6 also shows the output power of the PM brush DC motor as a function of motor speed. Notice that the **maximum power** is obtained when the motor is operating at a speed equal to half the no-load speed. This is verified by writing an expression for the power, differentiating it, and setting it equal to zero to solve for the speed at which the power is maximum. When a motor drives a load, the operating speed of the motor will be less than the no-load speed, and it will be at the point where the load torque matches the motor torque as seen in Figure 9.6. Notice that if the load torque changes linearly with speed, the operating speed of the motor then linearly increases (decreases) with an increase (decrease) in the supply voltage. The easy adjustment of operating speed through voltage input control is one advantage of PM brush DC motors over other types of electric motors. Example 9.1 discusses the basic characteristics of a PM brush DC motor. Example 9.2 shows how to determine the operating conditions of a load driven by a PM brush DC motor, while Example 9.3 considers the dynamic modeling of a mechanical system driven by a PM brush DC motor.

Figure 9.6

Typical torque-speed characteristics of a PM brush DC motor

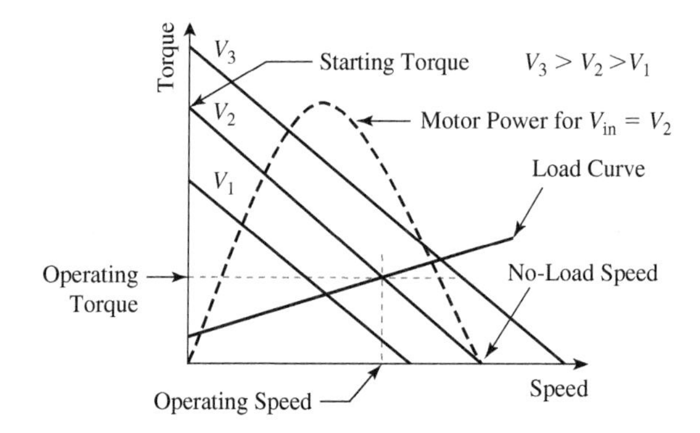

Example 9.1 Basic Characteristics of a PM Brush DC Motor

A PM brush DC motor has a nominal voltage rating of 24 V and a no-load speed of 5700 rpm. The armature resistance is 2.6 Ω, the motor's torque constant is 0.037 N · m/A, and the voltage constant is 0.037 V/(rad/s). Determine:

(a) The motor's speed and torque when the load current is 1.5 A
(b) The stalled current and torque of this motor

Solution:

(a) Using Equation (9.6) for the current in the motor gives

$$I = V/R = \frac{V_{in} - V_{bemf}}{R} = 1.5 = \frac{24 - V_{bemf}}{2.6} \quad (1)$$

Solving Equation (1) for V_{bemf} gives $V_{bemf} = 20.1$ V. But from Equation (9.5), $V_{bemf} = K_E\,\omega$. Thus $\omega = 20.1/0.037 = 540$ rad/s or 5200 rpm. The load torque is obtained from Equation (9.4), or

$$T = K_T I = 0.037 \times 1.5 = 0.056 \text{ N} \cdot \text{m}$$

(b) When the motor stalls, the motor speed is zero. Using Equation (1) and substituting $V_{bemf} = 0$ gives the stalled current as $I = V/R = 24/2.6 = 9.2$ A. Note how the stalled current is substantially larger than the load current. Using Equation (9.4), the stalled torque is $T = K_T I = 0.037 \times 9.2 = 0.34$ N · m.

Example 9.2 Operating Conditions of a Load Driven by a PM Brush DC Motor

A one-quarter hp PM brush DC geared motor is used in a lift mechanism to lift a load of 10 kg using a simple pulley arrangement, as shown in Figure 9.7. The no-load motor speed is 300 rpm, and the starting torque is 23.8 N · m. The frictional resistance in the pulley drive is 2 N · m. Neglecting the inertia of the rotor, the pulley, and the cable, determine:

(a) The initial acceleration of this load
(b) The steady-state lifting speed of the load
(c) Output horsepower of the motor at steady state

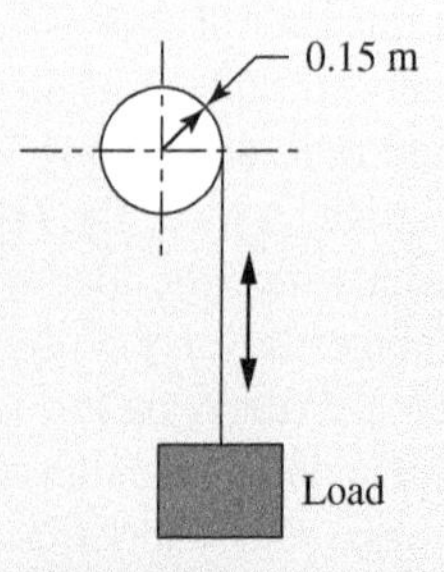

Figure 9.7 Lift mechanism

Solution:

(a) At startup, the available torque to accelerate the load is the starting torque minus the friction torque and the gravity torque. Or $23.8 - 2 - 10 \times 9.81 \times 0.15 = 7.1$ N · m. At a radius of 0.15 m, this corresponds to a starting force of $7.1/0.15 = 47.3$ N. Thus, the starting acceleration of the load is $47.3/10 = 4.73$ m/s^2. Note that as the motor speed starts increasing, the acceleration will decrease.

(b) At a steady state, the load is moving up at a constant speed. The torque that the motor needs to overcome will be the sum of the friction torque and gravity torque. This is given by:

$$T_{load} = 2 + 10 \times 9.81 \times 0.15 = 16.7 \text{ N} \cdot \text{m}$$

When the motor is used to lift the load, the steady-state speed is determined from

$$T_{motor} = 23.8 - (23.8/300)\,\omega = T_{load} = 16.7 \text{ N} \cdot \text{m}$$

$$=> \omega = 89.5 \text{ rpm}$$

$$V_{steady} = \omega\, r = (89.5 \times 2\pi/60) \times 0.15 = 1.41 \text{ m/s}$$

(c) At steady state, the output horsepower of the motor is equal to torque × speed = $16.7 \times 89.5 \times 2\pi/60 \times 1/746 = 0.21$ hp. Notice that the output hp is less than the rated hp for the motor since the steady-state speed is not half the no-load speed.

Example 9.3 Modeling of a Mechanical System Driven by a PM Brush DC Motor

For the drive system shown in Figure 9.8 with a gear ratio of N:1, assume that the motor is a PM brush DC motor. Develop a dynamic model that relates the input voltage applied to the motor to the motor speed as measured by a tachometer mounted on the motor shaft. The tachometer has a sensitivity of k_{tach} V/rpm. Let the viscous damping coefficient at the input shaft be b_1 and at the output shaft be b_2. Assume that the shafts are rigid, let I_1 represent the combined inertia of the motor shaft, input shaft, coupling, and the pinion, and let I_2 represent the combined inertia of the gear, the output shaft, and the load link.

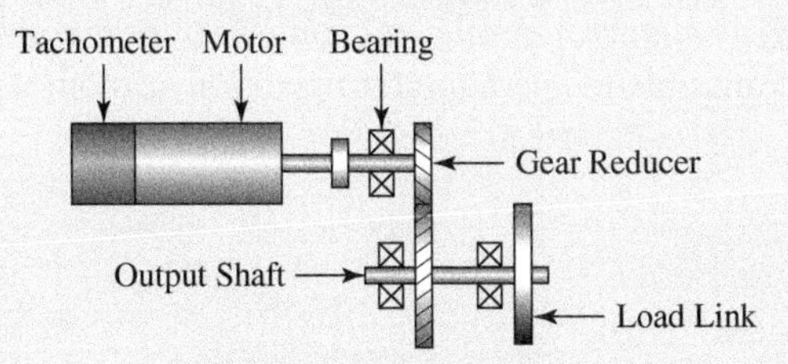

Figure 9.8 Drive system

Solution:

This system is similar to that considered in Example 7.3, but here a tachometer is added to the system.With the shafts assumed to be rigid and no external torques acting on the system, the torque (T_m) supplied by the motor is used to overcome the inertia and damping in the system. Thus, we can write

$$T_m = I_1\ddot{\theta}_1 + b_1\dot{\theta}_1 + \frac{(I_2\ddot{\theta}_2 + b_2\dot{\theta}_2)}{N} \quad (1)$$

where θ_1 is the angular velocity of the input shaft and θ_2 is the angular velocity of the output shaft. Due to gearing,

$$\dot{\theta}_2/\dot{\theta}_1 = 1/N \quad (2)$$

Using Equation (2) to express the velocity and acceleration of the output shaft in terms of the input shaft, we obtain

$$T_m = I_1\ddot{\theta}_1 + b_1\dot{\theta}_1 + \frac{(I_2\ddot{\theta}_1 + b_2\dot{\theta}_1)}{N^2} = I_{eff}\ddot{\theta}_1 + b_{eff}\dot{\theta}_1 \quad (3)$$

Where $I_{eff} = I_1 + \frac{I_2}{N^2}$, and $b_{eff} = b_1 + \frac{b_2}{N^2}$. Note that similar to inertia, the damping coefficient is reduced by the factor N^2 when reflected to the input shaft. Equation (3) is a model for this system that relates the input torque to the angular acceleration of the input shaft. Using the angular velocity of the motor shaft as the output variable, the model is

$$T_m = I_{eff}\,\dot{\omega}_1 + b_{eff}\,\omega_1 \tag{4}$$

Neglecting the motor inductance since it is usually small (see Example 9.4), and using Equation (9.7), the torque supplied by a PM brush DC motor is given as

$$T_m = K_T\frac{V_{in}}{R} - K_T K_E\frac{\omega_1}{R} \tag{5}$$

and the tachometer output voltage is related to the motor speed by

$$V_{tach} = k_{tach}\,k_2\omega_1 \tag{6}$$

where $k_2 = 60/2\pi$ is a conversion factor from rad/s to rev/min. Combining Equations (4) through (6), we obtain:

$$\dot{V}_{tach} = \frac{1}{RI_{eff}}\left(K_T V_{in} k_{tach}\,k_2 - (K_T\,K_E + b_{eff}\,R)\,V_{tach}\right) \tag{7}$$

Using Laplace transform, the transfer function between V_{in} and V_{tach} is

$$\frac{V_{tach(s)}}{V_{in(s)}} = \frac{K_T k_{tach} k_2}{RI_{eff}\,s + (K_T K_E + b_{eff}\,R)} = \frac{k_m}{\tau_m s + 1} \tag{8}$$

where k_m and τ_m are constants that can be obtained from Equation (8). As shown by Equation (7) or (8), the model is a first-order system.

In manufacturer data for a PM brush DC motor, the manufacturer typically lists the **nominal speed** (or rated or continuous speed) and **nominal torque** (or rated or continuous torque) values for the given motor in addition to the stall torque and no-load speed. The relationships between these parameters are shown in Figure 9.9. Typically, the nominal speed is 75 to 90% of the no-load speed, and the nominal torque is 10 to 25% of the stall torque. Note that PM brush DC motors are not suitable for continuous operation at their maximum torque level due to thermal effects. The nominal torque parameter defines the maximum torque that can be applied to the given motor for continuous duty operation with the temperature of the motor winding not exceeding the permissible temperature for operation at room temperature.

Figure 9.9

Nominal speed and torque for a PM brush DC motor

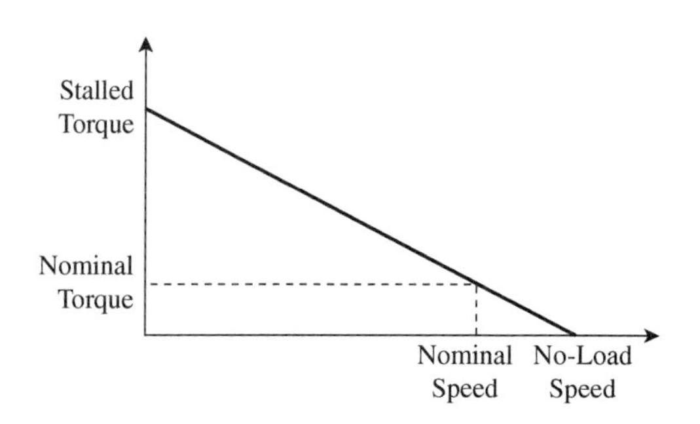

For intermittent duty operation where the motor is not running continuously but starts and stops with periods of rest in between, the applied torque can exceed the nominal torque. During these periods of rest, the motor can cool down, which allows it to handle higher loads for short durations without overheating. The duration of this over-torque can range from a few seconds to a few minutes. The exact duration would depend on factors such as the magnitude of the over-torque, the design of the motor, the ambient temperature, and how effectively the motor can shed the excess heat generated by the over-torque. Figure 9.10 shows manufacturer data for a small PM brush DC motor made by Pittman, and Example 9.4 explains some of this data.

Figure 9.10

Manufacturer data for a Pittman 9236 Series PM brush DC motor
(Courtesy of AMETEK, Kent, OH)

Brush Commutated DC Servo Motors
9236 Series

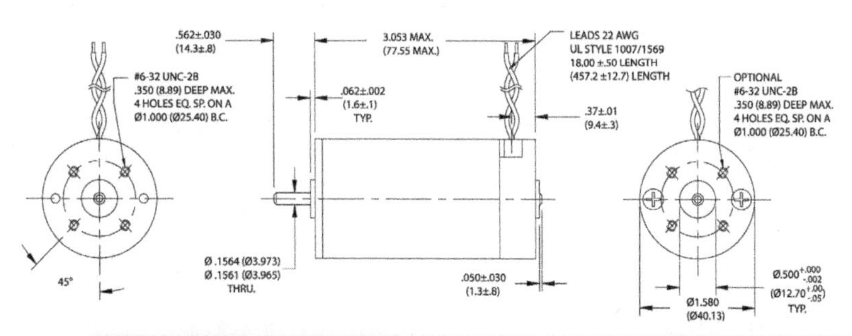

Specification	Units	Part/Model Number							
		9236 9.55 V	9236 12.0 V	9236 15.2 V	9236 19.1 V	9236 24.0 V	9236 30.3 V	9236 38.2 V	9236 48.0 V
Supply Voltage	VDC	9.55	12.0	15.2	19.1	24.0	30.3	38.2	48.0
Continuous Torque	oz-in	9.50	9.50	9.50	9.50	9.50	9.50	9.50	9.50
	Nm	0.0671	0.0671	0.0671	0.0671	0.0671	0.0671	0.0671	0.0671
Speed @ Cont. Torque	RPM	3530	3750	3850	3880	3980	3980	4010	3990
Current @ Cont. Torque	Amps (A)	4.52	3.65	2.88	2.26	1.82	1.44	1.14	0.90
Continuous Output Power	Watts (W)	25	26	27	27	28	28	28	28
Motor Constant	oz-in/sqrt W	3.7	3.9	4.0	4.1	4.1	4.2	4.2	4.2
	Nm/sqrt W	0.026	0.028	0.028	0.029	0.029	0.03	0.03	0.03
Torque Constant	oz-in/A	2.62	3.25	4.12	5.24	6.49	8.24	10.4	13.1
	Nm/A	0.019	0.023	0.029	0.037	0.046	0.058	0.073	0.093
Voltage Constant	V/krpm	1.94	2.40	3.05	3.87	4.80	6.09	7.66	9.69
	V/rad/s	0.019	0.023	0.029	0.037	0.046	0.058	0.073	0.093
Terminal Resistance	Ohms	0.50	0.71	1.07	1.64	2.49	3.91	6.14	9.72
Inductance	mH	0.43	0.66	1.06	1.72	2.63	4.24	6.70	10.7
No-Load Current	Amps (A)	0.40	0.33	0.26	0.20	0.16	0.13	0.10	0.080
No-Load Speed	RPM	4730	4800	4800	4750	4820	4790	4810	4770
Peak Current	Amps (A)	19.1	16.9	14.2	11.6	9.64	7.75	6.22	4.94
Peak Torque	oz-in	49.0	53.9	57.5	60.0	61.5	62.8	63.4	63.7
	Nm	0.3459	0.3805	0.406	0.4236	0.4342	0.4434	0.4476	0.4497
Coulomb Friction Torque	oz-in	0.80	0.80	0.80	0.80	0.80	0.80	0.80	0.80
	Nm	0.0056	0.0056	0.0056	0.0056	0.0056	0.0056	0.0056	0.0056
Viscous Damping Factor	oz-in/krpm	0.053	0.053	0.053	0.053	0.053	0.053	0.053	0.053
	Nm s/rad	3.56E-6	3.56E-6	3.56E-6	3.56E-6	3.56E-6	3.56E-6	3.56E-6	3.56E-6
Electrical Time Constant	ms	0.86	0.93	0.99	1.0	1.1	1.1	1.1	1.1
Mechanical Time Constant	ms	10	10	8.9	8.5	8.4	8.2	8.1	8.0
Thermal Time Constant	min	14	14	14	14	14	14	14	14
Thermal Resistance	Celsius/W	14	14	14	14	14	14	14	14
Max. Winding Temperature	Celsius	155	155	155	155	155	155	155	155
Rotor Inertia	oz-in-sec2	0.0010	0.0010	0.0010	0.0010	0.0010	0.0010	0.0010	0.0010
	kg-m2	7.06E-6	7.06E-6	7.06E-6	7.06E-6	7.06E-6	7.06E-6	7.06E-6	7.06E-6
Weight (Mass)	oz	13.8	13.8	13.8	13.8	13.8	13.8	13.8	13.8
	g	391.2	391.2	391.2	391.2	391.2	391.2	391.2	391.2

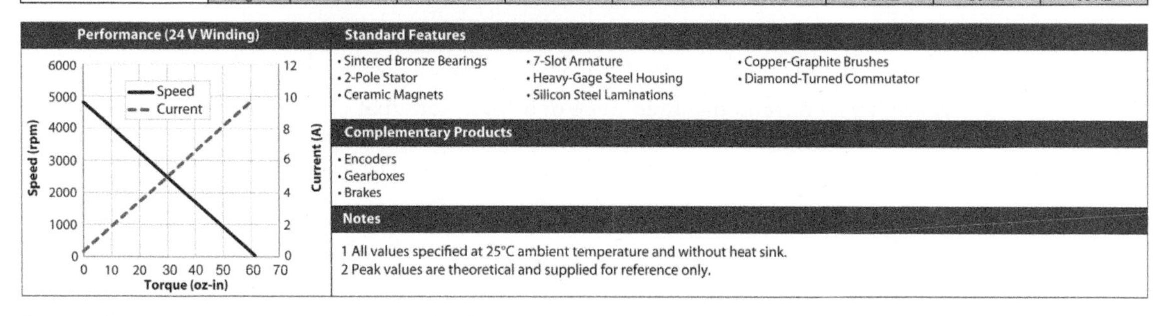

Standard Features

- Sintered Bronze Bearings
- 2-Pole Stator
- Ceramic Magnets
- 7-Slot Armature
- Heavy-Gage Steel Housing
- Silicon Steel Laminations
- Copper-Graphite Brushes
- Diamond-Turned Commutator

Complementary Products

- Encoders
- Gearboxes
- Brakes

Notes

1 All values specified at 25°C ambient temperature and without heat sink.
2 Peak values are theoretical and supplied for reference only.

AMETEK TECHNICAL & INDUSTRIAL PRODUCTS
343 Godshall Drive, Harleysville, PA 19438
USA: +1 215-256-6601 - Europe: +44 (0) 845 366 9664 - Asia: +86 21 5763 1258
www.ametektip.com

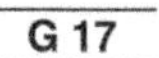

Example 9.4 Verification of the Characteristics of a PM Brush DC Motor

Using the motor data in Figure 9.10 for 24 V operation, do the following.
(a) Verify the listed peak torque and no-load speed values.
(b) Verify the listed electrical and mechanical time constants.

Solution:

(a) The torque-speed relationship for a PM brush DC motor is given by Equation (9.7). Using SI units, V_{in} is 24 V, K_T is 0.046 N · m/A, K_E is 0.046 V/rad/s, and R is 2.49 Ω. Substituting these values into Equation (9.7) and evaluating, we get

$$T = 0.443 - 8.50 \times 10^{-4}\,\omega \tag{1}$$

The peak (or stalled) torque is obtained at $\omega = 0$. This gives 0.443 N · m. The max no-load speed is obtained by solving Equation (1) for $T = 0$. This gives a rotational speed of 521 rad/s. These values are very close (0.443 versus 0.434 and 521 versus 505) to the listed values in Figure 9.10.

(b) The electrical time constant (τ_E) of the motor is a measure of the relationship between the input voltage to the motor and the output torque. If we do not neglect the armature inductance, then KVL applied to Figure 9.5 gives

$$V_{in} = L\frac{di}{dt} + iR + K_E\omega \tag{2}$$

Using Equation (9.4) and converting to the algebraic domain using the Laplace transform, we get the following expression for the motor torque

$$T(s) = \frac{K_T}{Ls + R}\left(V_{in}(s) - K_E\Omega(s)\right) \tag{3}$$

The transfer function between input voltage and output torque is then

$$\frac{T(s)}{V_{in}(s)} = \frac{K_T}{Ls + R} \tag{4}$$

Substituting the given values for K_T, L, and R, we get

$$\frac{4.6e^{-2}}{2.63e^{-3}s + 2.49} \text{ or } \frac{1.85e^{-2}}{1.06e^{-3}s + 1} \tag{5}$$

or an electrical time constant of 1.06 ms (or in general, τ_E is given as L/R), which matches the value listed in Figure 9.10. We note that if the electrical time constant is small (such as in this case), then the inductance can be neglected in dynamic modeling without appreciably changing the accuracy of the dynamic model as was done in Example 9.3.

The mechanical time constant can be found from Equation (8) in Example 9.3, or

$$\tau_m = \frac{RI_{eff}}{K_T K_E + b_{eff}R} \tag{6}$$

For the motor data, $b_{eff} = 3.56e^{-6}$ N-m-s/rad, and I_{eff} is $7.06e^{-6}$ kg m^2. Substituting these values and the values of K_T, K_E, and R into Equation (6), we get

$$\tau_m = \frac{2.49 \times 7.06e^{-6}}{4.6e^{-2} \times 4.6e^{-2} + 3.56e^{-6} \times 2.49} = 8.27 \text{ ms} \tag{7}$$

which is very close to the listed value of 8.4 ms. Note that in practice, the effective load inertia is added to the rotor inertia giving a larger time constant.

9.2.2 Brushless DC

Unlike a brushed DC motor, a brushless DC (BLDC) motor (▶ available—V9.2), as the name implies, does not contain brushes, resulting in reduced electrical and acoustic noise. It also eliminates the mechanical wear associated with brushes. This brushless design enhances the motor's reliability and extends its lifespan due to the absence of brushes that would otherwise require periodic replacement. In a BLDC motor, the rotor comprises permanent magnets, and the stator is equipped with coils, negating the need for electrical contact with the rotating part. The lighter rotor of a BLDC motor often allows for higher operational speeds. Additionally, BLDC motors generally offer improved efficiency over brushed motors, owing to the elimination of brush friction. They are also better equipped for thermal management because the heat-generating components, the coils, are on the stator, which is usually on the outside of the motor, facilitating more effective cooling. BLDC motors are favored in applications that demand rapid response, minimal heat generation, and long operational life. These characteristics make them suitable for sophisticated machine tools, robotic systems, aerospace technology, electric vehicles, and high-precision computer disk drives, as well as a variety of consumer electronic devices. The selection between a BLDC motor and a brushed motor depends on specific application requirements, which include not just performance characteristics but also cost constraints and system design complexities.

In a BLDC motor, commutation is electronically controlled rather than mechanically, as with a brushed motor. Electronic commutation shifts the stator magnetic fields to synchronize with the rotor's position. This rotor position is typically detected using non-contact, Hall-effect sensors (see Section 8.4.1) mounted on the stator, although encoder feedback may also be employed for greater precision. Some BLDC motor controllers operate without sensors by detecting the induced EMF voltage in the unpowered coils to determine the appropriate commutation moments. BLDC motors are available in single-phase, two-phase, and three-phase configurations, where the phase refers to the number of independent windings on the stator, with the three-phase being the most common for industrial use due to its efficient and smooth operation. The windings in a typical **three-phase motor** are connected in either a delta or a Y configuration (see Figure 9.11), with the Y configuration often preferred for its electrical efficiency. A three-phase BLDC motor cable usually has three wires for the motor phases, with additional wires if Hall effect sensors are present, generally comprising three for the sensor signals and two for the power supply and ground. The brushless driver typically energizes two out of the three motor phases sequentially over the electrical cycle, effectively driving the motor through each segment of rotation.

Figure 9.11

(a) Y wiring and (b) delta wiring of a three-phase BLDC motor

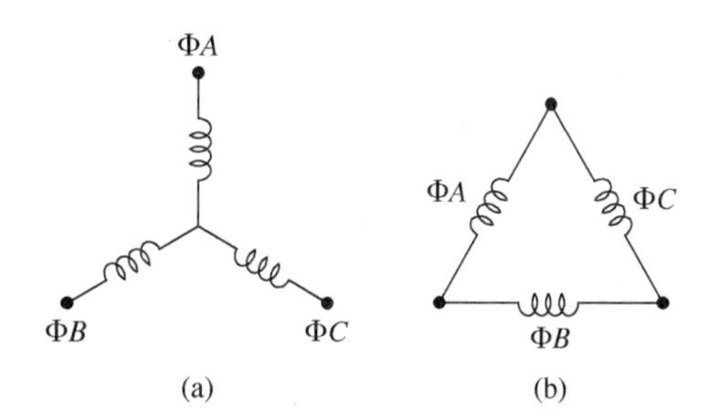

To illustrate the operation principle of a BLDC motor, consider a simplified three-phase BLDC motor with a two-pole rotor as shown in Figure 9.12. The stator coils are labeled phase A (ΦA), phase B (ΦB), and phase C (ΦC) and wired using a single circuit similar to that shown in Figure 9.11(a). Real motors have multiple circuits that are wired in parallel to each other, and a corresponding number of multi-pole rotors [38]. In the simplified schematic of Figure 9.12, one electrical revolution of the motor corresponds to one mechanical revolution. If two electrical circuits were used, then there would be two electrical revolutions per mechanical revolution. The figure also shows three Hall-effect sensors labeled *A*, *B*, and *C* that are placed on the stator. Each sensor outputs a high signal for 180° of electrical rotation and a low signal for the other 180°. The sensor output is high when the north pole of the rotor is pointing toward the sensor. There are six combinations of the sensors' output, with one combination or state for each 60° of electrical rotation. Each combination is indicated in Figure 9.12 using the notation [CBA], where the least significant bit corresponds to sensor *A* output, and the most significant bit corresponds to sensor *C* output. In practice, the particular labeling of each sensor (i.e., whether it is *A* or *B*) is not important. What is important is the association of the sensors' output states with the position of the rotor.

In a BLDC motor, the phases are **electronically commutated** as the rotor moves from one sensor state to another. For each combination of the sensors' output, two of the three phases are activated such as to produce an angle close to 90° between the stator and rotor flux vectors [39]. There are six possible **stator flux vectors** shown as arrows in Figure 9.12. A particular stator flux vector is generated for a given position of the rotor and a desired direction of rotation. For the rotor position shown in Figure 9.12, the stator flux vector should be horizontal and pointing toward the left for CCW rotation or horizontal and pointing to the right for CW rotation. Figure 9.13 shows the generation of this stator flux vector for CCW rotation

Figure 9.12

Schematic of a simplified three-phase BLDC motor

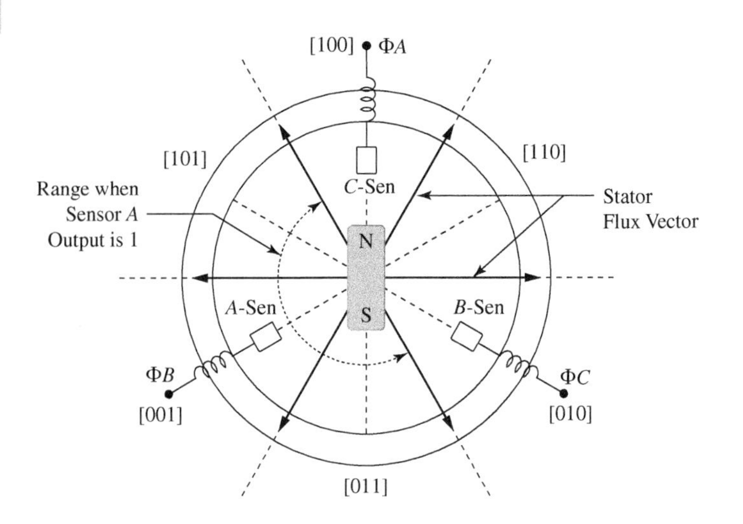

through activation of phases C and B and leaving phase A floating, with phase C connected to high voltage and phase B connected to low voltage. The commutation sequence for all possible sensor states is listed in Table 9.1 for both CW and CCW rotation. The reader can verify the commutation sequence for other positions. Using this commutation sequence, one can create a drive timing diagram for the CW rotation of the rotor as shown in Figure 9.14. The 0° electrical position corresponds to the 12 o'clock position.

Figure 9.13

Illustration of phase activation to produce a particular stator flux vector

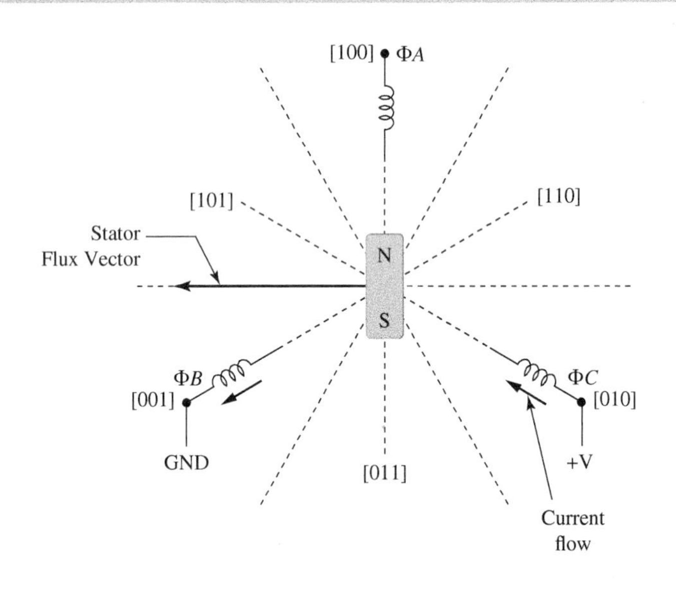

Table 9.1

Commutation sequence for CW and CCW rotation

Sensor Output			CW Rotation			CCW Rotation		
C	B	A	ΦA	ΦB	ΦC	ΦA	ΦB	ΦC
1	0	0	NC	Hi	Low	NC	Low	Hi
1	0	1	Low	Hi	NC	Hi	Low	NC
0	0	1	Low	NC	Hi	Hi	NC	Low
0	1	1	NC	Low	Hi	NC	Hi	Low
0	1	0	Hi	Low	NC	Low	Hi	NC
1	1	0	Hi	NC	Low	Low	NC	Hi

Figure 9.14

Drive timing diagram for CW rotation

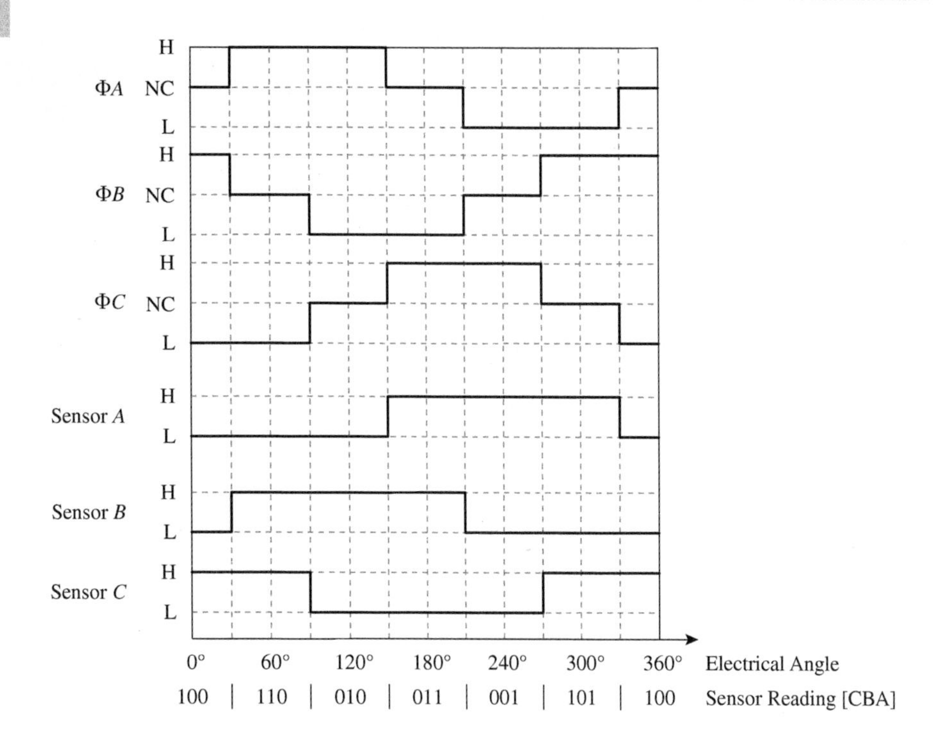

A brushless motor requires a special driver that can provide the proper excitation voltages to the stator coils. The driver is of the bipolar type and is commonly referred to as a **three-phase bridge driver** (see Figure 9.15). Such a driver consists of three parallel half H-bridge legs. As seen in Figure 9.15, closing switch *S1* on the first leg and switch *S4* on the second leg and keeping the remaining switches open causes power to be applied to phases *A* and *B* of the motor with Phase *C* floating. To reverse the current flow for phases A and B, close switch *S3* on the second leg and switch *S2* on the first leg while keeping the remaining switches open. The particular switches to close are determined from a **commutation table** (such as Table 9.1), which provides the commutation sequence for correctly driving the motor. The switches in Figure 9.15 are a representation of transistors in actual implementation.

Figure 9.15

Three-phase bridge driver for driving a BLDC motor

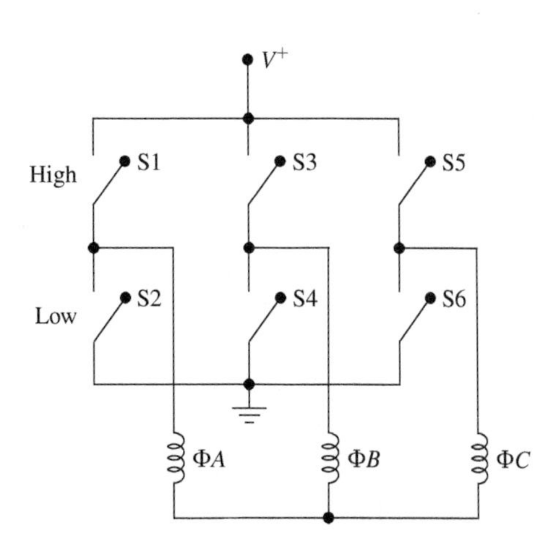

An example of a brushless DC motor is the motor that powers the small cooling fans in personal computers (see Figure 9.16). Since BLDC motors are very light and produce little electrical and acoustic noise, they are preferred for this application. These fans are typically constructed using a two-phase BLDC motor.

Figure 9.16

Brushless DC cooling fan
(Photos.com)

A layout of the components of a BLDC fan is shown in Figure 9.17. The rotor has surface-mounted permanent magnets, while the stator has two-phase coils. The fan uses a single Hall-effect sensor that is mounted on the stator circuit board. When the rotor axis is aligned with the sensor, the sensor sends out two complementary 50% duty cycle waves. These cycles are fed to the transistor gate input that controls the current flow through each of the two coils. Since the two square waves are complementary, only one coil will be active at a time. Increasing the voltage supplied to the motor causes the motor to increase its speed. Some BLDC fans come with an output that indicates the speed of the fan.

Figure 9.17

Components of a BLDC fan
(Jouaneh, University of Rhode Island)

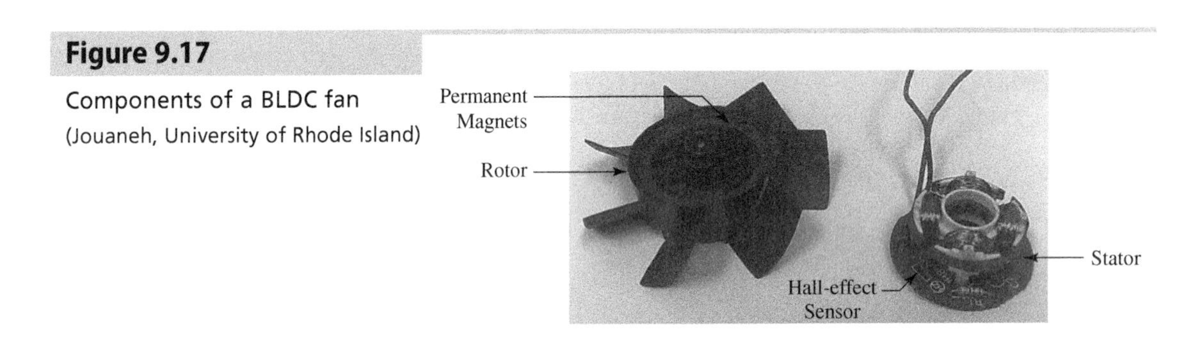

The torque-speed characteristics of a BLDC motor differ from those of a brush-type motor. Typical characteristics are shown in Figure 9.18. For a BLDC motor, the peak torque typically remains constant over a range of speeds up to a certain point before it begins to decrease. The nature of this decrease may approximate a linear relationship as the motor approaches its no-load speed, although the actual curve can vary depending on the specific motor design. The speed range over which the peak torque remains constant can differ widely and usually becomes larger as the design or operating voltage for the motor increases. Motors designed for lower voltages, like 12 or 18 volts, may exhibit a torque reduction behavior starting from a lower speed, which is more characteristic of PM brush DC motors. To ensure reliable operation without overheating, manufacturers specify an optimal operating zone, delineated by a continuous rated torque that lies below the peak torque curve, and a rated speed that is lower than the no-load speed of the motor. However, this is not universally followed, and some manufacturers extend the right end of the operating zone to meet the torque-speed curve. This operating zone ensures the motor operates within its thermal limits for continuous use. Note that the rated current for a BLDC motor is significantly smaller than the peak current. The rated current refers to the maximum current that the motor can handle continuously without overheating, while the peak current is the maximum current that the motor can handle for a short period without damage. A servo drive for a brushless motor should be selected to accommodate the peak current requirements to manage situations like start-up or sudden loads. Example 9.5 illustrates the basic characteristics of a BLDC motor.

Figure 9.18

Torque-speed characteristics of a BLDC motor

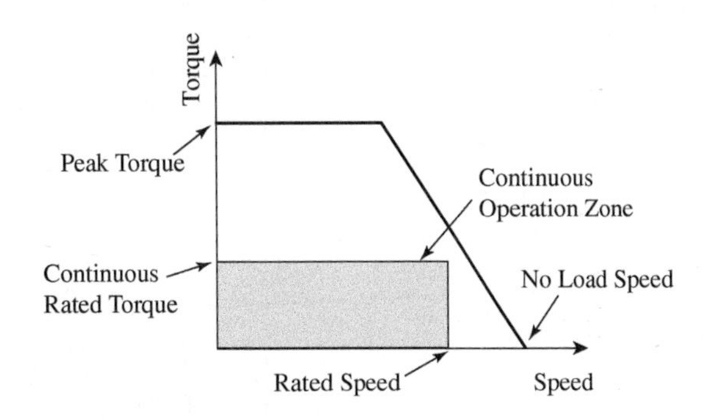

Example 9.5 BLDC Motor Characteristics

A BLDC with torque-speed characteristics similar to those shown in Figure 9.18 has the following specifications: peak torque of 0.35 N · m, no load speed of 5700 rpm, rated torque of 0.11 N · m, rated speed of 3700 rpm, and the knee point (where the peak torque starts decreasing from a constant value) in the torque-speed curve is at 1200 rpm. For this motor,

(a) Determine the peak torque at a speed of 3000 rpm.
(b) Determine the rated power of this motor.
(c) Determine the peak and rated current for this motor if the torque constant K_T is 0.04 N · m/A.

Solution:

(a) For an operating speed of 3000 rpm, the peak torque will lie on the sloping line in the torque-speed curve. The slope of that line is $(0.35 - 0)/(1200 - 5700) = -7.77 \times 10^{-5}$ N · m/rpm. The peak torque is then equal to $0.35 - 7.77 \times 10^{-5}(3000 - 1200) = 0.21$ N · m.
(b) The rated power will be determined at the rated torque and speed or

$$\text{Power} = \text{torque} \times \text{speed} = 0.11 \text{ N} \cdot \text{m} \times 3700 \text{ rev/min} \times 2\pi/60 \text{ (rad/s)/rev} = 43 \text{ W}.$$

(c) The current will be determined from the relationship $T = K_T I$. The peak current is $0.35/0.04 = 8.8$ A, and the rated current is $0.11/0.04 = 2.8$ A. Note that the peak current is much larger than the rated current.

9.2.3 Servo Drives

For motion control applications, servo drives or amplifiers (▶ available—V9.3) are utilized to amplify the reference signal sent to the actuator, consolidating all the necessary amplification components into one convenient package. Drives are broadly categorized as either digital or analog. **Digital drives** accept various input signal forms, such as analog, digital (step and direction or PWM and direction), RS-232, or Ethernet. Conversely, **analog drives** typically receive analog signals (e.g., +/−10 VDC from a D/A converter) or PWM signals with direction. Configuration of digital drives is performed through software on a PC, offering a user-friendly and versatile setup, while analog drives are configured manually using switches and potentiometers. An example of an analog servo drive is the AB15A100 made by Advanced Motion Controls and shown in Figure 9.19. This drive is suitable for driving small motors, takes +/−10 VDC as input, and requires a single unregulated DC power supply that can supply a DC voltage in the range of 20 to 80 V. This amplifier allows the user to adjust the following parameters using 14-turn potentiometers: loop gain, current limit, reference gain, and offset.

Loop Gain: The loop gain factor when the amplifier is operated in the voltage or velocity mode (see below).

Current Limit: The peak and continuous current that is supplied by the amplifier. When the current limit is adjusted, the ratio of the continuous to peak current is maintained.

Reference Gain: The ratio between the output variables (voltage, current, or velocity) and the input signal

Offset: A signal that is used to adjust any imbalance in the input signal or the amplifier output.

Figure 9.19

AB15A100 analog servo drive

(Courtesy of ADVANCED Motion Controls, Camarillo, CA)

This amplifier can operate in various modes including voltage, velocity, and current (torque) modes. In **voltage mode**, the amplifier produces an output voltage signal that is proportional to the input voltage signal regardless of variation in the supply power. The output amplification or loop gain is variable and set by adjusting the *Loop Gain* potentiometer. In **velocity mode**, the amplifier supplies a voltage to maintain the motor at a specific speed. For this mode, the amplifier uses the actual motor speed, as measured by the tachometer that is attached to the motor shaft, as a feedback signal. In **current mode**, the amplifier produces an output current signal to the motor. Because the output torque of DC motors is proportional to the input current, the current operation mode is sometimes referred to as torque mode.

This amplifier, like many other commercial servo amplifiers, gives a high frequency (33 kHz) pulse width modulated output (or PWM output). Rather than changing the amplitude of the output signal, a PWM amplifier varies the duty cycle of a fixed-amplitude, fixed-frequency signal as a means of modulating the output power. This results in a very efficient way to deliver power to the load since it reduces the power lost in the output stage of the drive. Integrated Case Study II.C discusses the use of a previous version of this servo drive (BE12A6) to drive a brushless DC motor.

Another drive that is suitable for driving small brushed and brushless servomotors, as well as stepper motors (discussed in Section 9.5), is depicted in Figure 9.20. This digital drive is configured to accept a variety of external command signals, including ± 10 V analog, PWM with direction, and 5 V digital step with direction, making it well-suited for interfacing with microcontrollers. Additionally, many modern servo drives, including this one, incorporate networking protocols such as RS-485/232 or Modbus RTU for communication.

Figure 9.20

DZXRALTE-008L080 digital servo drive

(Courtesy of ADVANCED Motion Controls, Camarillo, CA)

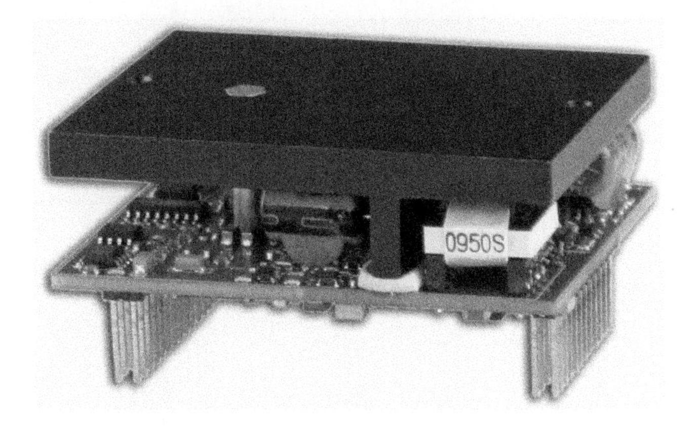

A low-cost, commercially available driver that uses the L298 chip that was discussed in Section 3.10 is shown in Figure 9.21. This motor driver module has two full H-bridges that are independently controlled and can supply a maximum of 2 A current for each bridge. One feature of this module is that it has flyback diodes, which the L298 chip does not include.

Figure 9.21

L298N H-bridge driver module

(Jouaneh, University of Rhode Island)

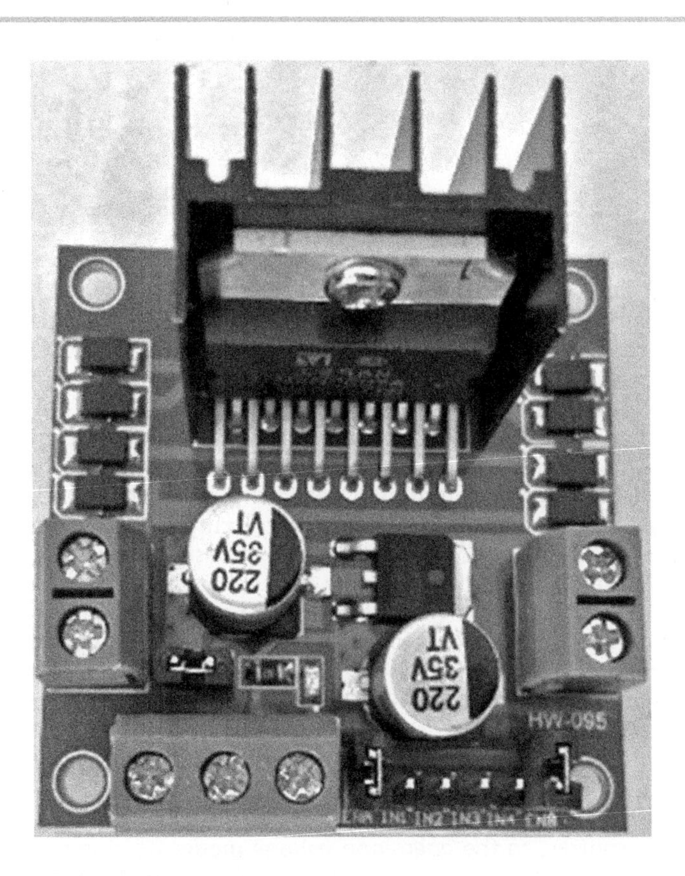

Integrated Case Study II.C: Brushless DC Motor and Servo Drive

For Integrated Case Study II (DC Motor-Driven Linear Motion Slide—see Section 11.3), the slide is driven by a NEMA 23 brushless DC motor. A servo drive from Applied Motion Control (Model BE12A6—see Figure 9.22) was used as the drive for the motor, which is very similar to the AB15A100 drive shown in Figure 9.19, which replaced this discontinued drive.

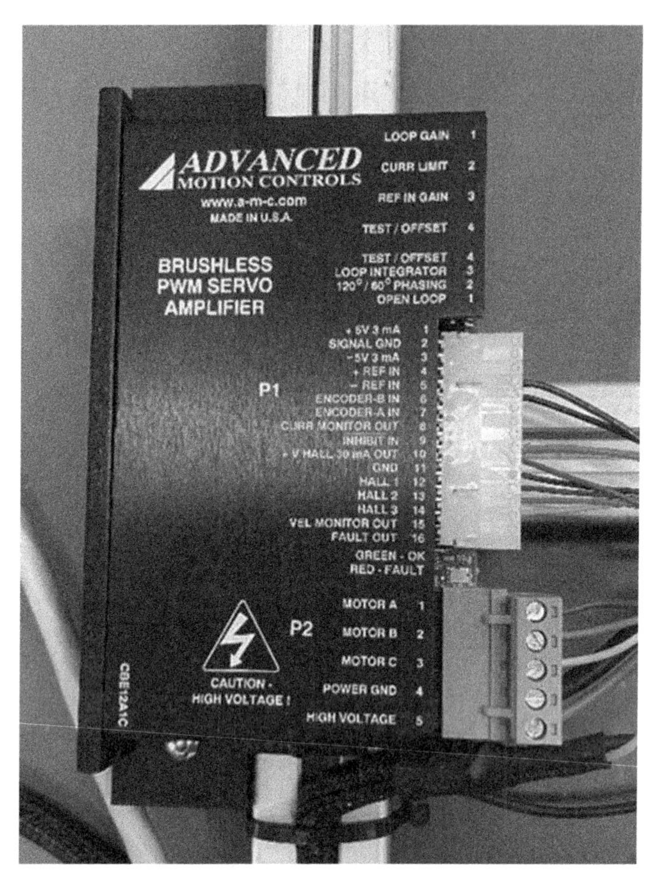

Figure 9.22 BE12A6 servo drive
(Jouaneh, University of Rhode Island)

This drive can power both brush and brushless DC motors. A DC brushless motor was chosen for this application since DC brushless motors offer high efficiency, smooth motion, and high torque output. Furthermore, there is a very limited choice of DC brush motors with an integrated encoder, which is not the case for brushless motors. This drive uses analog +/−10 volts as the input signal, which is provided by the D/A converter that is part of the Measurement Computing data acquisition board that is used for the control of this setup. This drive accepts a supply voltage in the range of 20-60 VDC and is powered by a 24 VDC signal from a switching power supply (see case Study II.A). This drive supplies a maximum peak output current of 12 A and a maximum continuous output current of 6 A, well above the 3.2 A rated current for the motor. A brushless motor has eight wires for interfacing, one for each of the three phases, one wire for each of the three Hall Effects sensors, plus 5 VDC and ground signals for the sensors.

The drive is versatile, offering several operational modes including current mode, open-loop voltage mode, and encoder velocity mode. For the given application, the current mode was selected. In this mode, the drive utilizes the +/−10 V analog input signal as a reference for controlling the motor's current, which directly correlates to the torque output. It operates in a closed-loop control system, typically employing a Proportional-Integral (PI) control loop (see Chapter 10). While the current loop parameters are factory-set to ensure optimal performance, the driver allows for user adjustment of the reference gain. This setting dictates the relationship between the input voltage and the target current.

In contrast to the open-loop voltage mode, which merely amplifies the input voltage, the current mode offers enhanced stability by improving the phase margin (see Chapter 10) in the servo loop and reducing sensitivity to motor inductance variations. Additionally, this drive has comprehensive protection features that guard against common faults such as over-voltage, under-voltage, over-current, over-heating, and short circuits in motor, ground, and power leads. It also features an output for real-time monitoring of the motor current, facilitating performance tracking and system diagnostics.

9.2.4 PWM Control of DC Motors

There are two methods of using PWM signals in motor control. The first is called the sign-magnitude method and the other is called the locked anti-phase method. In the **sign-magnitude method**, the controller (such as an AVR MCU) provides two signals: a PWM signal to control the magnitude of the voltage supplied to the load and a digital two-level signal to control the direction or sign. In the sign-magnitude method, the higher the duty cycle of the PWM signal, the higher the average voltage (or magnitude) delivered to the load. At 0% duty cycle, no voltage is supplied to the load, while at 100% duty cycle, the full voltage signal is supplied to the load. For example, the sign-magnitude method can be used to drive the digital servo drive shown in Figure 9.20 or an H-bridge driver such as the L298 shown in Figure 3.62 (or Figure 9.21). In using the L298 bridge driver, additional circuitry is needed to interface the sign and magnitude signals to the *In1*, *In2*, and *EnA* leads of Figure 3.62 (see Figure 9.23(a)).

In the **locked anti-phase method** (▶ available—V9.4), sometimes referred to as 50/50 PWM method, a single PWM signal is used to control the voltage delivered to the load and the direction of rotation. In this method, opposite pairs of switches in an H-bridge circuit (such as S1 and S4 or S2 and S3 in Figure 3.60) alternate between being ON and OFF according to the duty cycle. If the duty cycle is 50%, the opposite pairs of switches in the H-bridge circuit will close for the same duration in each PWM cycle, resulting in the net voltage applied to the motor being zero and hence no rotation of the motor. If the duty cycle is 60% for example, then a net 20% (60–40%) of the available voltage is applied to the motor leads in each PWM cycle. The direction of rotation is set by which motor lead has a more positive net voltage in each cycle. Thus, when the duty cycle is less than 50%, the motor will rotate in one direction, and when the duty cycle is above 50%, the motor will rotate in the opposite direction. At both 0 and 100% duty cycles, the maximum voltage is applied to the motor in each case but in opposite directions.

The locked anti-phase method, utilizing a single PWM signal for motor control, stands out for its efficient dynamic braking and smooth, rapid direction changes. Dynamic braking is achieved effectively in this method; when the PWM duty cycle is set to 50%, the motor experiences opposing voltages that quickly bring it to a stop rather than allowing it to coast. This feature is particularly advantageous in applications where precise control of motor position or abrupt stops are necessary. Additionally, the method facilitates smooth and rapid changes in motor direction. By adjusting the duty cycle above or below 50%, the motor can seamlessly transition from forward to reverse rotation and vice versa, enabling agile responses ideal for applications like robotics. However, these benefits are balanced against challenges such as increased ripple currents, reduced low-speed efficiency, potential for greater mechanical wear, and higher heat generation. This method's choice depends on the specific needs of the application, weighing its streamlined control against these considerations.

In servo drives (such as the one shown in Figure 9.20) or H-bridge drivers (such as LMD18200) with dedicated direction and PWM input lines, the PWM input line is wired to high logic, and the PWM control signal is applied to the direction line in using the locked anti-phase drive method. For the left H-bridge on the L298 IC shown in Figure 3.62, it is wired as shown in Figure 9.23(b) using the locked anti-phase drive method.

Figure 9.23

Wiring of left H-bridge on L298 IC for (a) sign-magnitude drive method and (b) locked anti-phase drive method

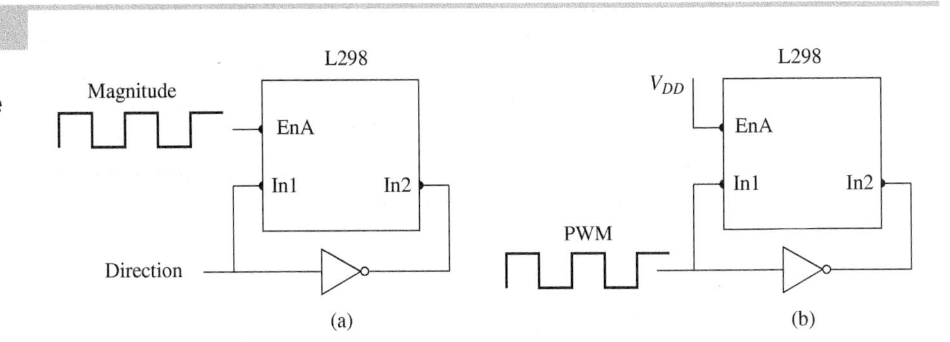

9.3 Solenoids

A solenoid (▶ available—V9.5) is an actuator that is commonly powered by a DC signal (although it can also be designed to operate with AC power) that converts electrical energy into mechanical motion. It is commonly used for various on-off applications, including electromechanical relays (see Section 2.3.2), door locks, ratcheting devices, electrical valves (see Section 9.8), and gate diverters. It serves as an example of an inductive element widely utilized in electrical systems.

The solenoid operates as an electrically actuated mechanical device that can exist in two states: retracted and extended. Its construction typically involves a movable armature core that moves inside a stationary iron core, as illustrated in Figure 9.24. When the armature coil is energized with current, it extends by closing the air gap between the cores, thereby increasing the flux linkage. The movable core is spring-loaded and will retract when the current is switched off. This motion is used to perform mechanical work, such as opening or closing a valve, actuating a switch, or moving a mechanical component.

Figure 9.24

Schematic of a solenoid

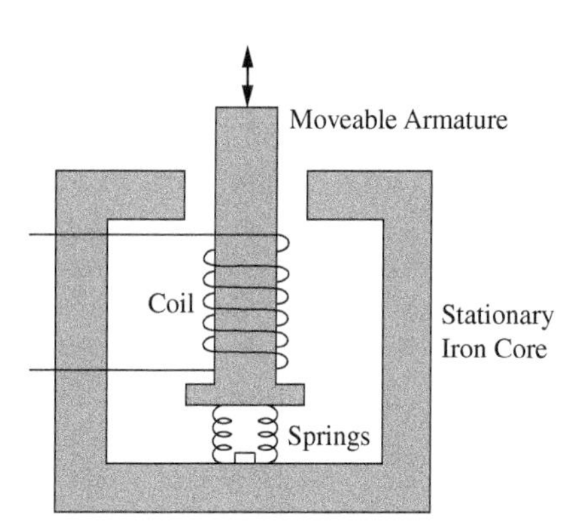

Solenoids are commercially available in two main forms: linear and rotary. The linear solenoid is further classified as either pull-type or push-type. In a **pull-type solenoid**, the force is directed toward the solenoid body when the solenoid is energized. Conversely, in a **push-type solenoid**, the force is directed away from the solenoid body upon energization. Linear solenoids typically have a stroke length of less than 25 mm. On the other hand, rotary solenoids exhibit a stroke of approximately 45°, allowing for rotational motion. These solenoid variations provide versatility in their applications. For example, linear pull-type solenoids are often employed in mechanisms where a pulling force is needed, such as door locks, while push-type solenoids find utility in applications requiring a pushing force, such as certain types of switches or latches. Rotary solenoids, with their ability to generate rotational motion, are suitable for tasks like indexing, valve actuation, and other rotary mechanical systems. Figure 9.25 shows a picture of a linear push-type solenoid.

Figure 9.25

Push-type solenoid

(Jouaneh, University of Rhode Island)

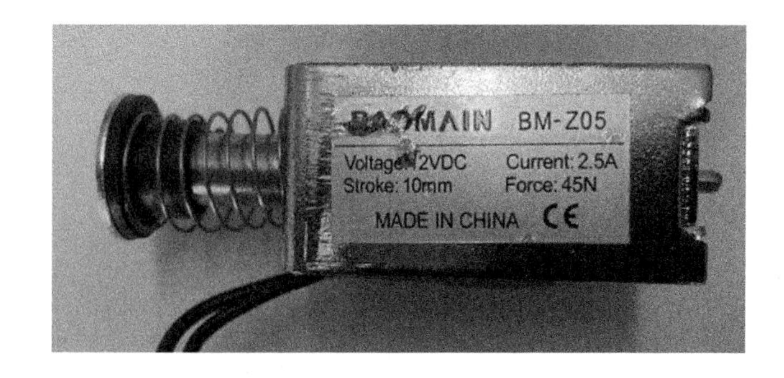

9.4 AC Motors

AC motors (▶ available—V9.6) are broadly classified into single-phase and multi-phase (typically three-phase) types. **Single-phase AC motors** are commonly used with the electrical power available in homes and light industrial environments. These motors are ideal for lower power applications such as household appliances and small machinery. On the other hand, **three-phase AC motors** utilize three voltage signals that are 120° out of phase with each other. This configuration enables more consistent power delivery, making three-phase motors suitable for industrial and high-power applications.

AC motors are either asynchronous (commonly known as **induction motors**) or synchronous. Asynchronous AC motors, particularly single-phase ones, are among the most widely used, found in applications like fans, pumps, and compressors. The rotor in an induction motor, often referred to as a squirrel cage (see Figure 9.26), consists of a stack of steel laminations. These laminations contain holes for copper or aluminum bars, which are looped around to form continuous coils or windings. This design contributes to the efficiency and robustness of the motor. AC motors are known for their ruggedness and versatility, with power ratings that can reach several thousand horsepower. This range demonstrates their suitability for a wide spectrum of applications, from small domestic devices to large-scale industrial machinery.

Figure 9.26

Rotor (squirrel cage) of an AC induction motor

(Jouaneh, University of Rhode Island)

The stator of an induction motor is made up of coils to which the electrical leads are connected. The lack of any wiring to the rotor leads to a simple construction that is rugged and requires very little maintenance. When the time-varying AC voltage signal is applied to the stator windings, voltages are induced in the rotor. The induced voltage causes the rotor to rotate, but the rotor rotates at a speed that is slightly lower than the speed of rotating stator fields or the synchronous speed. This speed reduction is called **slip** and is about 3 to 5% for most motors. Because the motor rotates at a speed that is lower than the excitation speed, these motors are known as asynchronous motors. The synchronous speed of an AC motor is a function of the excitation frequency (50 or 60 Hz) and the number of poles in the stator. For four-pole motors at 60 Hz frequency, the synchronous speed is 1800 rpm, and the operating speed is about 1725 to 1750 rpm at full load due to slip. In general, the **synchronous speed** (in rev/min) is given as

$$N_S = 120\frac{f}{p} \tag{9.9}$$

where f is the frequency of the AC power supply in Hz and p is the number of poles. The operating speed is

$$N = N_S\left(\frac{100 - \%\,Slip}{100}\right) \tag{9.10}$$

In a **synchronous motor**, alternating voltage is supplied to the stator windings, while the rotor is excited with direct current (DC) or uses permanent magnets. This configuration causes the rotor to rotate in synchronization with the stator's magnetic field, hence the name 'synchronous motor'. The rotor's DC excitation is delivered via **slip rings** and brushes, a setup somewhat similar to the commutator in brush DC motors but with a different function and operation principle. The key feature of synchronous motors is their ability to maintain a constant speed that matches the frequency of the supply voltage. This characteristic makes them ideal for applications where precise speed control and synchronization are essential, such as in timers, instruments, and synchronized operations of multiple conveyor belts.

One of the key challenges with single-phase AC induction motors is their inability to start on their own. To overcome this, an additional mechanism, often involving an auxiliary coil or a start capacitor, is introduced. This setup creates a phase shift between currents in different windings, generating an initial rotating magnetic field. This field applies a starting torque to the rotor, enabling it to begin rotating. Once the motor reaches a certain speed, this auxiliary system is usually disengaged. Another characteristic of AC induction motors is their speed, which is generally fixed and determined by the frequency of the supplied AC voltage and the motor's number of poles. To facilitate variable speed operation, Adjustable Frequency Controls or **Variable Frequency Drives** (VFDs) are employed. These devices alter the frequency of the supply voltage, thereby allowing control over the motor's rotational speed. Regarding power ratings, single-phase AC induction motors are typically available up to about 3 horsepower (hp). However, this can vary based on the design and manufacturer. For applications requiring higher power, three-phase motors are preferred due to their greater efficiency and capacity for handling larger loads.

Several varieties of AC single-phase motors affect how the motor starts.

Shaded Pole: A short circuit is used to make one side of the field magnetize before the other side

Split Phase: Uses two windings, one with higher resistance than the other

Capacitor Start: Uses two windings and a capacitor on one of the windings to create a leading phase

The torque-speed characteristics of an AC induction motor depend on the design of the motor. A typical characteristic is shown in Figure 9.27. On the vertical axis, the torque developed by the motor as a percentage of full load (or rated load) is shown. On the horizontal axis, the rotational speed as a percentage of synchronous speed is shown. Similar to a DC motor, the starting torque is larger than the rated torque. After an initial dip, the motor torque increases with speed until it reaches the breakdown point, at which point the torque starts decreasing. The steeper the torque-speed curve after the breakdown point, the more the motor has almost constant speed operation as the load varies.

Figure 9.27

Typical torque-speed characteristics of a single-phase AC induction motor

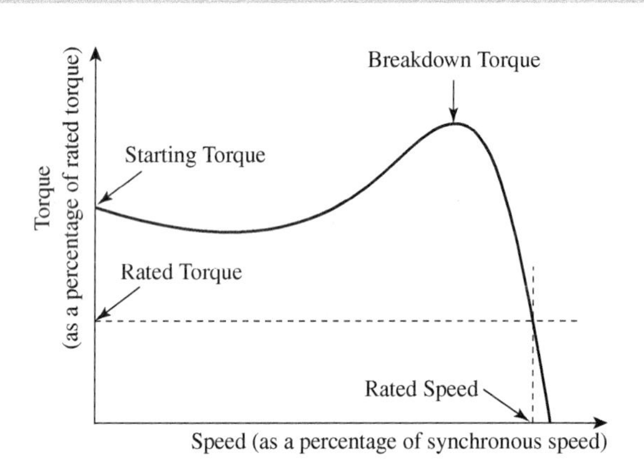

Typical torque-speed data for a 1 hp induction motor made by LEESON Electric are shown in Figure 9.28. Notice how the speed of this motor changes as the load changes. At the rated load (4.1 N · m (or 3 lb-ft) of torque) and 230 V operation, the rated speed is 1747 rpm. If the load reduces by 50%, the speed increases to only 1774 rpm. Similarly, if the load increases by 50%, the speed decreases to just 1708 rpm. The motor speed changes by less than +/−2.5% from its rated speed for a load change of +/−50% from the rated load. This 'near-constant' speed operation under load variation is typical of several types of AC motors. Example 9.6 illustrates the use of an AC motor's torque-speed characteristics in determining the operating conditions, while Example 9.7 explains some of the characteristics shown in Figure 9.28.

Example 9.6 Operating Conditions of a Load Driven by an AC Induction Motor

Assume that the PM brush DC-motor in Example 9.2 was replaced by a geared one-quarter hp AC single-phase motor with a 20:1 gear ratio. If the motor has the torque-speed data given in Table 9.2 (with no gearing), determine the steady-state lifting speed of the load.

Table 9.2*

Torque-speed data of a single-phase AC motor

Rated Load	1.02 N · m					
% of Rated load	25	50	75	100	125	150
Speed (rpm)	1786	1772	1757	1748	1735	1719

*Data is for a General Purpose, AC single-phase motor under 230V/60Hz excitation

Solution:

From Example 9.2, the steady-state load torque at the gear output side is 16.7 N · m.

Because of the 20:1 gear ratio, the load torque as seen by the motor is

$$16.7/20 = 0.835 \text{ N} \cdot \text{m}$$

For the above motor, this load represents

$$0.835/1.02 \times 100\% = 82\%$$

of the rated load.

Interpolating from the above table gives the motor speed as

$$1757 + (82 - 75) \times (1757 - 1748)/(75 - 100) = 1754 \text{ rpm}$$

and the lifting speed of the load is

$$V_{\text{steady}} = \omega\, r = (1754/20 \times 2\pi/60) \times 0.15 = 1.38 \text{ m/s}$$

Example 9.7 Characteristics of an AC Induction Motor

Explain the performance data for the AC motor of Figure 9.28 under 230 V, 60 Hz operation.

Solution:

We will explain the performance data using three load conditions: 50%, 100%, and 150% of rated load. Table 9.3 shows the output mechanical horsepower, the input mechanical horsepower, and the real electrical power delivered to the motor for these three load conditions.

The output mechanical horsepower is given by the formula:

$$\text{Output mechanical power} = T\,n/7125$$

where T is in N · m and n is in rpm. For a 50% rated load, the output mechanical horsepower is then ½ × 3 × 1.3558 × 1774/7125 = 0.506 hp.

The input mechanical horsepower is the output mechanical horsepower divided by the motor efficiency. For the 50% rated load case, the input mechanical horsepower is

$$0.506/0.697 = 0.726 \text{ hp}$$

The real electrical power delivered to the motor should be equal to the input mechanical horsepower. From Equation (2.36), the real power is given as

$$\text{P (in Watts)} = V_{\text{rms}}\, I_{\text{rms}}\, \text{Power\_factor}$$

For the 50% rated load case, the real electrical power is

$$230 \times 5.17 \times 0.456 = 542 \text{ W} = 0.727 \text{ hp}$$

which is the same as the input mechanical horsepower. The data for the 100% and 150% load given in Table 9.3 show a similar agreement.

Table 9.3

Power data for motor

	50% Rated Load	100% Rated Load	150% Rated Load
Output mechanical horsepower	0.506	0.997	1.46
Input mechanical horsepower	0.726	1.33	2.08
Real electrical horsepower	0.727	1.33	2.08

Figure 9.28

Torque-speed data for a 1 hp, single-phase AC induction motor made by LEESON Electric
(Courtesy of Leeson Electric Corporation)

Catalog No	110167.00	Model	M6C17DB7J
Product type	AC MOTOR	Stock	Stock
Description	1HP..1725RPM.56.DP./V.1PH.60HZ.CONT.MANUAL.40C.1.15SF.RIGID.GENERAL PURPOSE.M6C17DB7J		

Information shown is for current motor's design — View Outline | View Connection

Engineering Data

Volts	115	Volts	208-230	Volts	
F.L. Amps	12.8	F.L. Amps	6.4	F.L. Amps	
S. F Amps	13.6	S. F Amps	6.8	S. F Amps	
RPM	1800	Hertz	60		
HP	1	Duty	CONTINUOUS	TYPE	CD
KW	.75				
Frame	E56	Serv. Factor	1.15	Phase	1
Max Amb	40	Design	N	Code	K
Insul Class	B	Protection	MANUAL	Therm.Prot.	CEJ50CA
Eff 100%	75	Eff	75%	PF	68
UL	Yes	CSA	Yes	Bearing OPE	0
CC Number		CE	No	Bearing PE	0
Load Type		Inverter Type	NONE	Speed Range	NONE

Performance

Torque UOM	LB-FT	Inertia (WK^2)	.1 LB-FT^2	
Torque	3(Full Load)	6.9(Break Down)	6.7(Pull Up)	9(Locked Rotor)
CURRENT (amps)	6.4(Full Load)	0(Break Down)	0(Pull Up)	33(Locked Rotor)
Efficiency (%)	0(Full Load)	72.8(75% Load)	69.7(50% Load)	56(25% Load)
PowerFactor	0(Full Load)	58(75% Load)	45.6(50% Load)	31.1(25% Load)

Load Curve Data @60 Hz, 230 Volts, 1 Horsepower

Load	Amps	KW	RPM	Torque	EFF	PF	Rise By Resis	Frame Rise
0.0	4.6	0.148	1798	0.0	0.0	14.0	0.0	-
0.25	4.76	0.34	1786	0.75	56.0	31.1	0.0	-
0.5	5.17	0.542	1774	1.5	69.7	45.6	0.0	-
0.75	5.8	0.774	1762	2.25	72.8	58.0	0.0	-
1.0	6.42	0.993	1747	3.0	75.0	67.2	52.0	40.0
1.15	6.85	1.135	1737	3.45	75.0	72.0	56.0	44.0
1.25	7.56	1.271	1728	3.75	72.4	73.1	0.0	-
1.5	8.82	1.552	1708	4.5	70.3	76.5	0.0	-

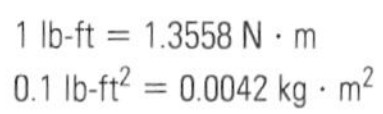
1 lb-ft = 1.3558 N · m
0.1 lb-ft^2 = 0.0042 kg · m^2

9.5 Stepper Motors

A stepper motor (▶ available—V9.7) is classified as a DC motor since it is driven by non-alternating voltages, but its construction and operation are distinct from a DC motor. Stepper motors, as the name suggests, can move in small angular increments, or steps, ranging from 0.9° per step to 90° per step, depending on the construction of the motor and on how it is driven.

One feature of stepping motors is that they can be used in position control applications without the need for a position sensor. As long as the motor is operated within its specified limits, the nominal position of the stepper motor is controlled by the number of steps that were sent to the motor. Another feature is that they are easily controlled with digital circuits, since the stepper motor driver, which generates the appropriate signals to drive the poles of the motor, requires two digital input signals: a pulse signal and a direction signal. A third feature is that there are no wires connected to the rotor, which eliminates the need for brushes and a commutator. A fourth feature is that they can generate a large torque at low speeds, which eliminates the need for gears.

There are three types of stepper motors. These are permanent magnet (PM), variable reluctance (VR), and hybrid. Figure 9.29 shows photos of a PM and a hybrid stepper motor. The configurations differ primarily in the construction of the rotor. In a **PM stepper motor**, the rotor is a permanent magnet and has no teeth, while in a **VR motor**, the rotor is constructed from non-magnetized soft iron material and has teeth. A VR motor has the advantage of a faster dynamic response. PM motors are widely used in nonindustrial applications such as computer printers and typewriters. A **hybrid motor**, as the name suggests, combines features from both PM and VR motors. Its rotor is a permanent magnet but also has teeth. Furthermore, the magnet is magnetized along the axis of the rotor, with the upper half of the rotor having one polarity while the lower half has another polarity. The hybrid configuration is the most widely used in industrial applications. In all configurations, the stator is constructed from pairs of electromagnets commonly referred to as poles. When a stepper motor is not powered, it typically offers a small resistance to rotation, known as **detent torque**. This torque is a characteristic feature of both PM and hybrid stepper motors. In PM stepper motors, the detent torque arises due to the interaction between the permanent magnets in the rotor and the stator's teeth, even in the absence of electrical power. Hybrid stepper motors, which combine features of both PM and VR motors, also exhibit detent torque due to their incorporation of permanent magnets. In contrast, VR stepper motors do not have permanent magnets. Their torque is exclusively generated by the magnetic attraction between the energized stator and the rotor's soft iron. Consequently, when a VR stepper motor is not energized, it generally shows very little or no detent torque, as there are no permanent magnets to create a residual holding force.

Figure 9.29

(a) PM and (b) hybrid stepper motors

(Courtesy of Anaheim Automation, Anaheim, CA)

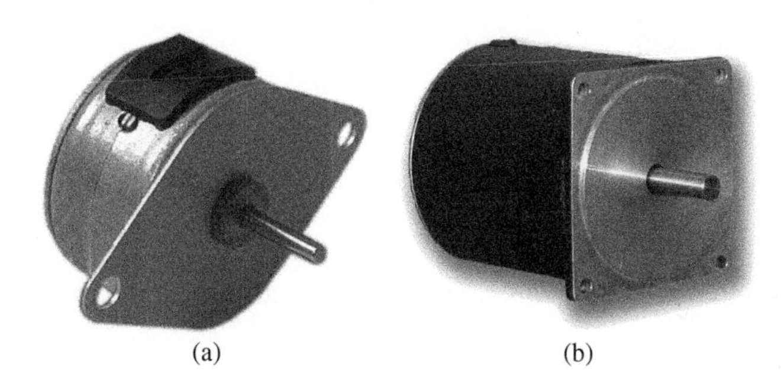

9.5.1 Drive Methods

To understand how stepper motors work, consider the two-phase PM stepper motor shown in Figure 9.30. A **phase** refers to a coil winding; thus, a two-phase motor has two separately activated coil windings, and the coil windings are placed perpendicular to the rotor. This motor has four **poles** with two poles for each phase. The motion of the motor depends on how the stator coils or phases are actuated. There are four possible ways of actuating the phases:

- Wave Drive
- Full Stepping
- Half Stepping
- Microstepping

Figure 9.30

A schematic of a two-phase PM stepper motor

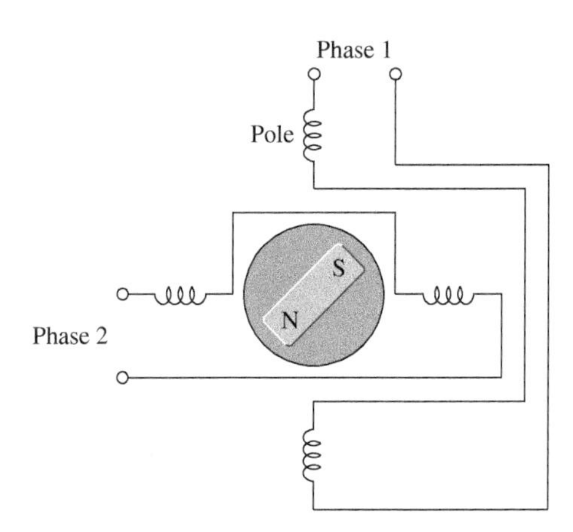

We will illustrate these drive techniques for the two-phase PM motor. Figure 9.31 shows the actuation steps for a **wave drive**. The two phases are labeled *A* and *B*. The rotor goes through the positions 1, 3, 5, and 7 in 90° steps in this drive method. To do a complete rotation, the motor has to go through the four steps shown. If the phases are activated in the reverse fashion, the motor will rotate in the opposite direction. Notice that in each of these steps, the rotor as shown is in equilibrium, and moves only if the polarity of the stator coils has changed. Also notice that in a wave drive, only one phase is active or on at a time. This means that only 50% of the available coils are active, which limits the torque applied to the rotor.

Figure 9.31

Wave drive actuation steps

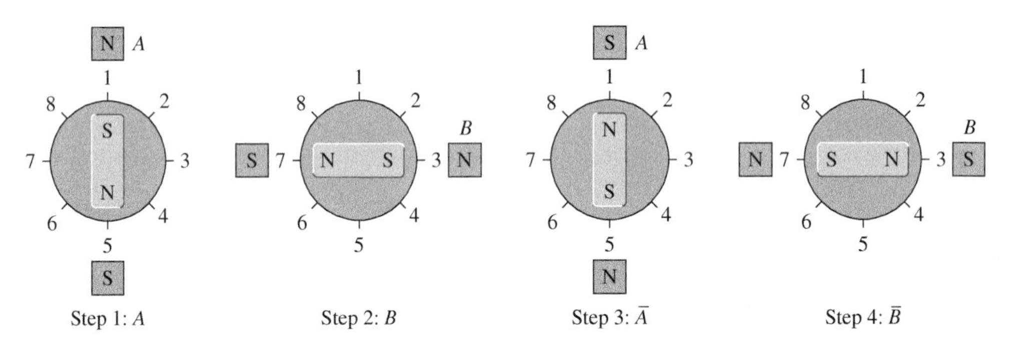

The **full-stepping actuation** sequence is shown in Figure 9.32. Here both phases of the stator are active at any point. The resulting motion is similar to wave drive actuation (90° between steps) but the rotor moves through positions 2, 4, 6, and 8 in this case. Due to both phases being on, the torque applied to the rotor is higher in this case than in the wave drive, and at any point in time, current is flowing in both windings.

Figure 9.32

Full-stepping actuation

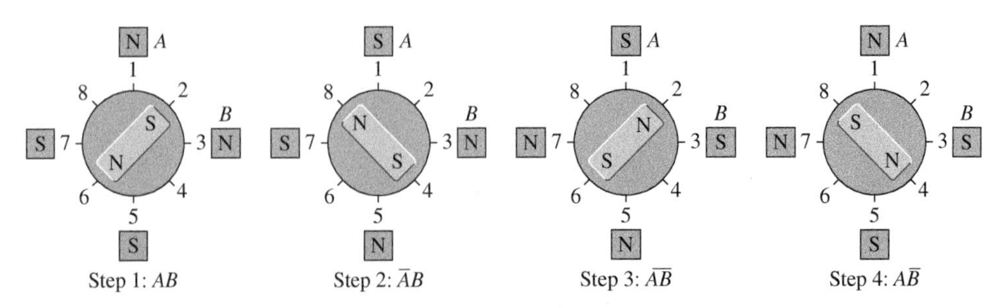

The **half-stepping actuation** sequence is shown in Figure 9.33. This actuation method alternates between activating one phase and two phases at a time, and it takes eight steps to complete one rotation. The rotor in this case travels in 45° steps from position 1 to position 8. Similar to the previous two actuation methods, the direction of rotation is reversed by simply reversing the sequence of actuation steps.

Figure 9.33

Half-stepping actuation

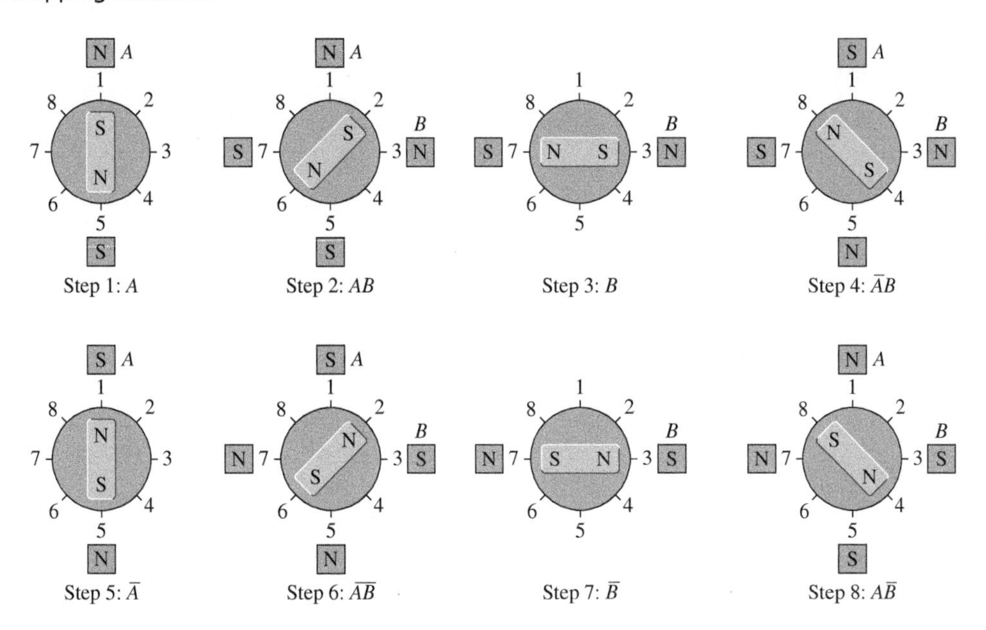

Notice that in the above three drive techniques, the voltage polarity to the stator coil is reversed in some of the steps. This is called **bipolar excitation** and requires supplying a voltage of opposite polarity to the coil. If we are only interested in a unipolar voltage excitation, then we need to use a four-phase motor instead. A **four-phase motor** has four coils, each of which are activated separately. In practice, a four-phase motor is constructed such that there are two coils for each set of stator poles. These two coils are wound oppositely, and only one of them is energized at a time. This is called **bifilar winding** as opposed to **unifilar winding**, which is the winding used with the two-phase motor. Using a four-phase motor, the excitation steps for full stepping are shown in Figure 9.34. The four phases are labeled *A*, *B*, *C*, and *D*. Notice here how the *A* and *C* phases activate the same set of stator poles but in an opposite fashion. Only one-half of the coils available are excited at any point in time compared to all the coils in the bipolar two-phase motor case. For example, in step 1 coils *A* and *B* are ON while *C* and *D* are OFF.

Figure 9.34

Full-stepping excitation for a four-phase unipolar PM rotor

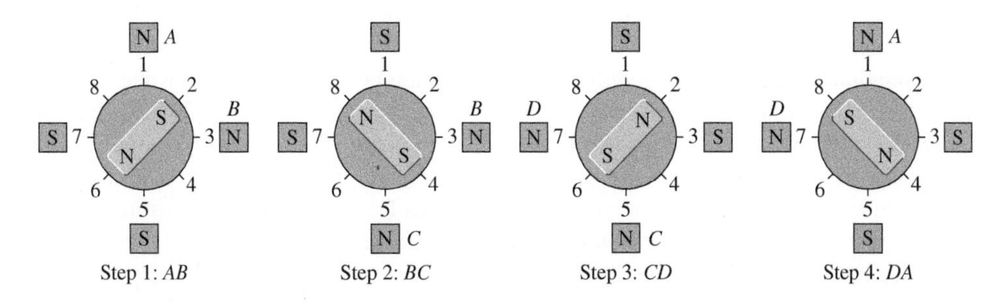

Rather than having the phases fully ON or OFF as illustrated above, in **microstepping drive**, the currents to both phases are varied in small steps as shown in Figure 9.35. This allows the motor to have a smaller resolution than that of full or half-stepping since it increases the number of equilibrium positions for the rotor. The resolution can be increased by a factor of up to 250 or more in microstepping actuation. Microstepping actuation results in a smoother motion of the motor with less vibration, but the applied torque to the rotor is reduced by 30 to 40% compared to full-stepping actuation.

Figure 9.35

Microstepping excitation

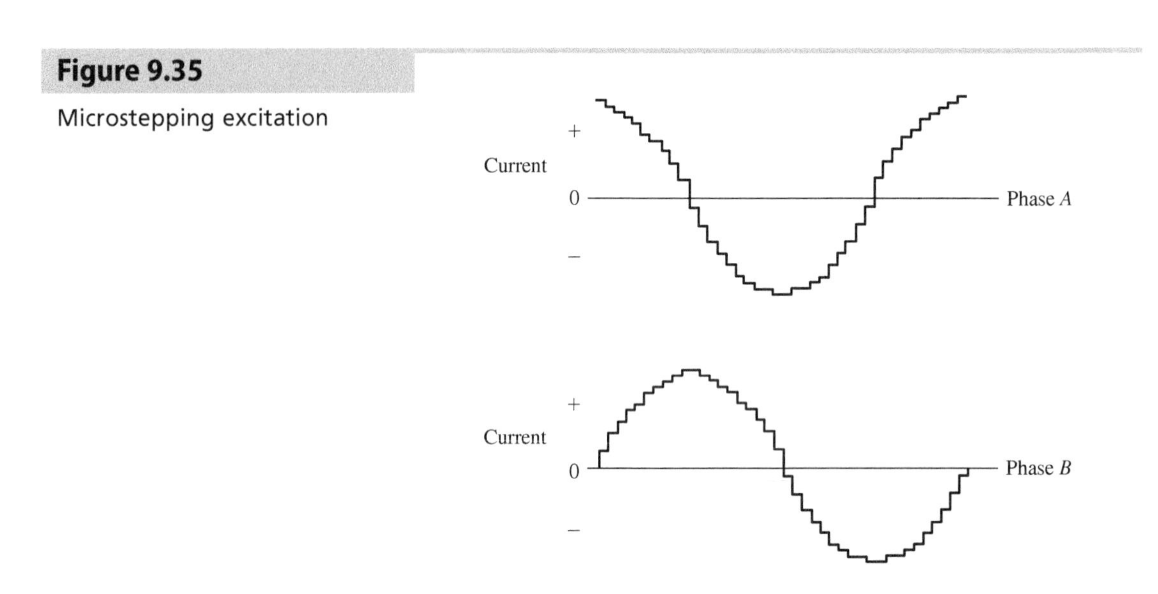

Notice that in the two-phase PM motor discussed above the step angle is 90° for full stepping. In general, the **full-stepping step angle** of a PM stepper motor is given by the formula

(9.11)

$$\Delta\theta = \frac{360}{2PS}$$

where

P is the number of rotor pole pairs

S is the number of stator pole pairs

In the two-phase PM motor discussed before, the motor has one rotor pole pair and two stator pole pairs. Thus, its angular resolution is 90° according to Equation (9.11).

For industrial applications, the **hybrid motor** is widely used. A cut-out view of a hybrid motor is shown in Figure 9.36(a). The rotor has two toothed cups, each of which has a separate polarity (N or S). Each cup has 50 teeth that are equally spaced, and the teeth in one cup are offset from those in the adjacent cup by a half a tooth pitch or 3.6°. The stator also has teeth. A typical cross-section of a two-phase hybrid motor is shown in Figure 9.36(b). This motor has four poles per phase, with the poles 180° from each other having the same polarity, and those 90° from each other having opposite polarity. In this configuration, the motor advances 1.8° per step in wave or full-stepping mode. The motion of this motor for the different actuation methods is very similar to that of a PM stepping motor shown in Figures 9.31 through 9.33 but replacing the 90° step with a 1.8° step.

From the previous discussion, we see that PM stepper motors and BLDC motors have similar structures, with both featuring rotors made of permanent magnets and stators with multiple coils. However, they differ in control and feedback mechanisms. Stepper motors operate in an open-loop fashion, moving in fixed angular steps without feedback sensors, ideal for predictable, precise control. Alternatively, BLDC motors typically use Hall-effect sensors in closed-loop systems for finer speed and position control. Operating BLDC motors in an open loop can result in coarse, less stable control, unlike the precise, predictable movement of stepper motors.

Figure 9.36

Hybrid stepper motor (a) major components and (b) cross-section
(Images Courtesy of Oriental Motor Corp USA)

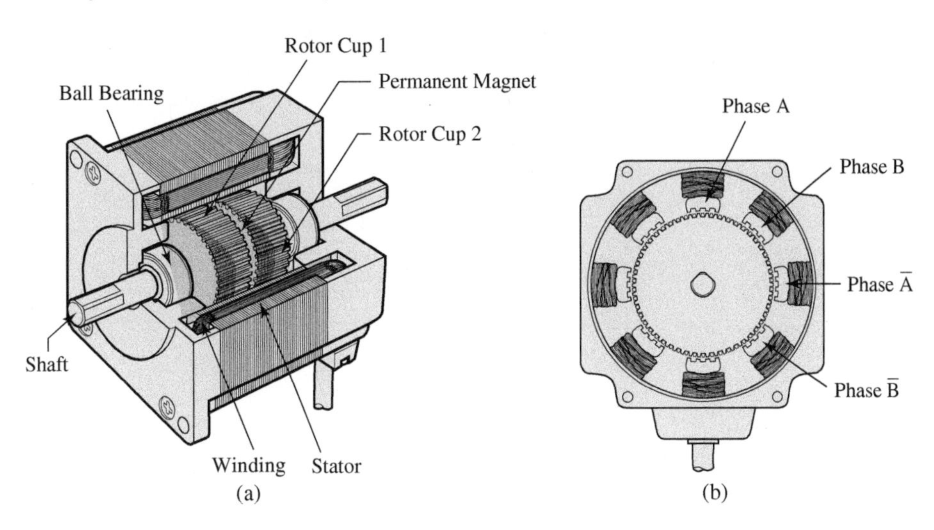

9.5.2 Wiring and Amplifiers

Stepper motors are available with different lead configurations (see Figure 9.37). The four-lead configuration is not a bifilar wound type and is only used with bipolar excitation. The six-lead configuration is very commonly used for four-phase unipolar motors, but can also be used with bipolar excitation. Note that in this configuration, one lead serves as a common connection for each pair of bifilar wound coils. The five-lead configuration is not very common. In this configuration, the common connection to all the coils is brought out as one lead. In the eight-lead configuration, the leads of each bifilar wound coil are brought out separately. This configuration gives the greatest flexibility in the wiring options for the motor.

Figure 9.37

Lead wires for stepper motors

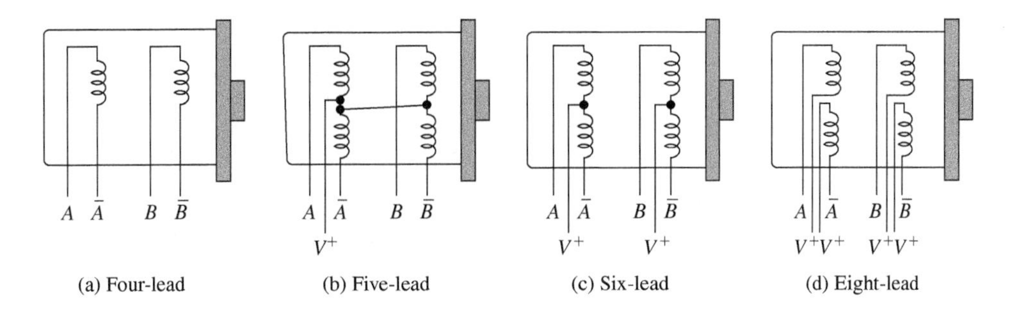

The driver for a stepper motor is constructed in different ways. On a basic level, one can use a transistor to drive each phase or coil winding for unipolar drives, or an H-bridge for bipolar drives (see Figure 9.38).

Alternatively, if the output current per phase is less than 500 mA, then one can use a transistor array IC such as ULN2003A. Figure 9.39 shows a commercially available board that has this IC. This board is a practical and efficient solution for driving small unipolar stepper motors in various electronic projects. The ULN2003A IC features seven Darlington transistor pairs capable of handling higher currents, making it ideal for controlling inductive loads like stepper motors. This setup is particularly favored in DIY projects due to its ease of interfacing with microcontrollers like Arduino.

Figure 9.38

Unipolar and bipolar drive wiring: (a) bipolar wiring for phase *A* and (b) unipolar wiring for phases *A* and $\bar{A}$

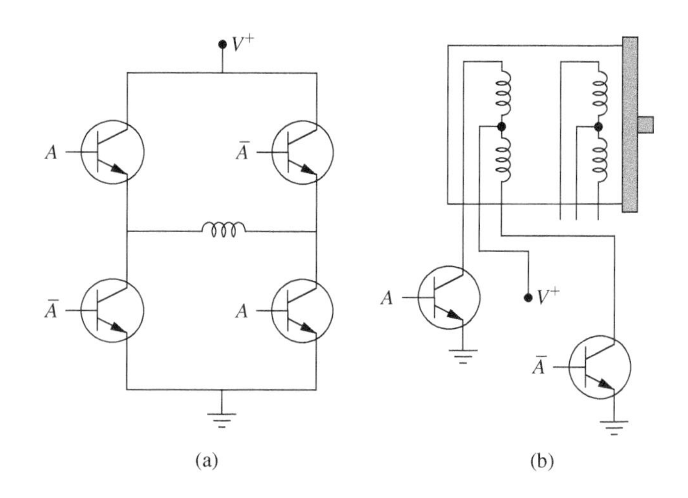

Figure 9.39

ULN2003 interface board

(Jouaneh, University of Rhode Island)

In industrial applications, a stepper motor drive system is set up as shown in Figure 9.40. It consists of a computer or a programmable logic controller (PLC) that interfaces with an indexer. The indexer in turn interfaces to a driver (▶ available—V9.8), which is connected to the stepper motor.

Figure 9.40

Typical stepper motor drive system

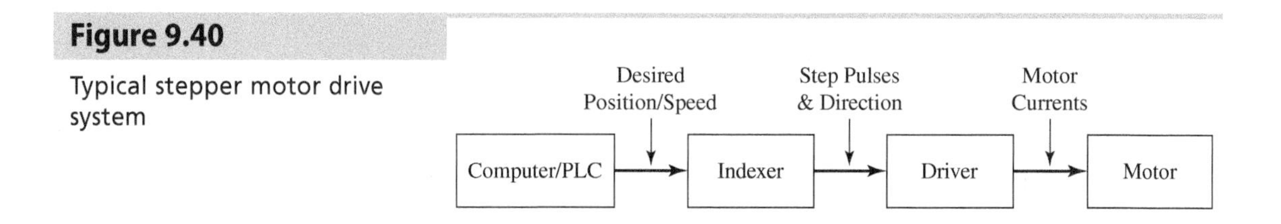

The computer or PLC sends the desired position and speed information to the indexer. The **indexer** converts this information into a sequence of pulses at the appropriate frequency and a direction signal. The driver interprets the step pulses and the direction signals to generate the voltages and currents that drive the different phases of the motor to obtain the desired motion. In certain applications, the indexer is eliminated, and the computer can directly send the step pulses and direction information to the driver. Figure 9.41 shows a **commercial stepper motor driver** that requires 12-48 VDC for operation. The driver accepts pulse and direction inputs to control the motor. The current per phase, as well as different stepping modes, are set with dip switches. This driver is used in Integrated Case Study I.C which is presented at the end of this section. The stepper motor leads are connected to the motor connector pins (A+, A−, B+, and B−) at the bottom right side of the driver. Four-lead stepper motors are simply connected to these pins. For a six-lead motor, the motor leads are connected in one of two ways: series connection and center tap connection. Figure 9.42 shows these two connections. In the series connection, the motor torque will be higher than the center-tap connection. For an eight-lead motor, many possibilities are available for connection, including parallel, series, and two of four windings (see Figure 9.43).

Figure 9.41

STR2 stepper motor driver
(Jouaneh, University of Rhode Island)

Figure 9.42

Connections for a six-lead stepper motor with a four-position amplifier: (a) series connection and (b) center tap connection

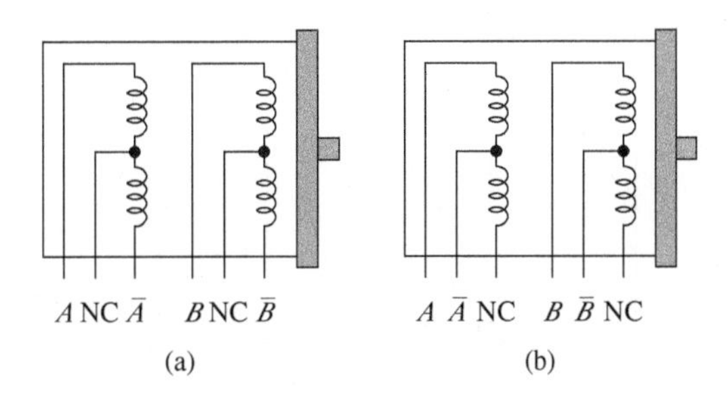

Figure 9.43

Connections for an eight-lead stepper motor with a four-position amplifier: (a) parallel connection, (b) series connection, and (c) two windings

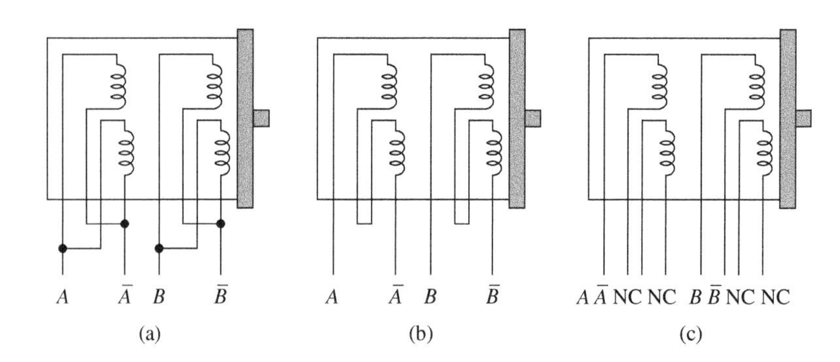

Typical torque-speed characteristics of a stepper motor are shown in Figure 9.44. The speed is typically given in terms of pulses per second (PPS) or Hz. The figure shows two regions of operation. These are the start/stop region or the locked step region, and the slewing region. In the **start/stop region**, the motor can start, stop, or reverse direction instantly without losing any steps. In the **slewing region**, on the other hand, the motor cannot be instantaneously started, stopped, or reversed. There must be a gradual acceleration of the motor to enter this region, and a gradual deceleration of the motor to leave this region. To operate in this region, the motor must start first in the start/stop region. The curve that defines the torque-speed limits of the start/stop region is called the **pull-in torque** curve, while the curve that defines the limits of the slew region is called the **pull-out torque** curve. For a stepper motor, the torque at zero speed is defined as the **holding torque**, which represents the maximum torque that can be applied to a powered (but not rotating) motor without moving it from its rest position and causing spindle rotation. It should be noted that the torque-speed characteristics are a function of both the motor and the driver that is used to actuate the motor phases. For a given motor, its torque-speed characteristics will change if the driver is changed or the driving mode is changed. Stepper motor manufacturers usually provide only the pull-out torque curve. Example 9.8 examines the performance data of a commercial stepper motor.

Figure 9.44

Torque-speed characteristics of a stepper motor

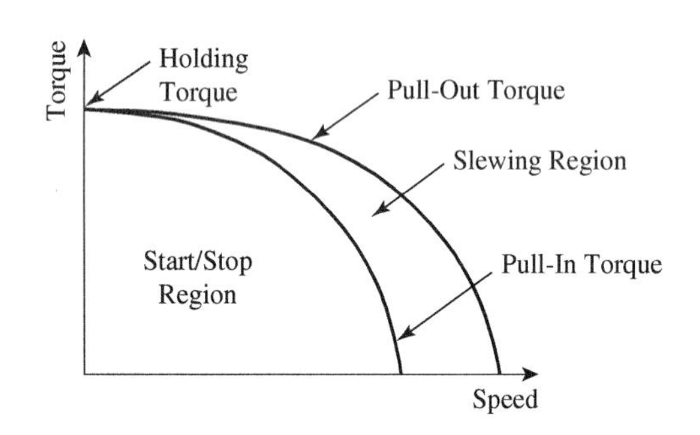

Example 9.8 Stepper Motor Characteristics

A two-phase hybrid stepper motor with six-lead wires has the specifications given in Table 9.4.

Table 9.4*

Stepper motor specifications

Connection Type	Holding Torque (N · m)	Rated Current A/phase	Voltage VDC	Resistance Ω/phase	Rotor Inertia (kg · m²)
Bipolar-series	0.43	0.85	5.6	6.6	68×10^{-7}
Unipolar	0.32	1.2	4	3.3	

*Data is for Oriental Motor PK245-01 motor

Explain the torque and current characteristics of this motor.

Solution:

Since this motor has six leads, it has bifilar windings. The wiring diagram for bipolar-series connection is as shown in Figure 9.42(a), while for unipolar connection it is as shown in Figure 9.37(c). Since both coils are active in bipolar wiring, a correspondingly higher holding torque is obtained as shown in the above table. The rated current per phase is simply the voltage divided by the phase resistance, or 5.6/6.6 = 0.85 A for bipolar series-connection. Note the resistance for unipolar connection is half that of bipolar-series connection due to the fact that only half of the available coils are activated in unipolar connection.

Integrated Case Study I.C: Stepper Motor and Driver for the System

For Integrated Case Study I (Stepper Motor-Driven Rotary Table—see Section 11.2), the stepper motor (Applied Motion Products HT23-594) used in this project is a NEMA 23, two-phase bipolar, 8-lead motor (see Figure 9.37(d)). The motor leads were connected to the A and B connections on the STR2 driver (see Figure 9.41) in a series fashion (see Figure 9.43(b)). To properly drive a stepper motor, the driver used should be able to meet the current demands of the motor, which for this motor, is 1.41 A/phase at an operating voltage of 3.95 V. Note for a bipolar stepper motor, the current per phase for full stepping mode is simply the voltage divided by the phase coil resistance (1.41 A = 3.95/2.8). Like most drivers for stepper motors, the STR2 driver has a set of dip switches that can be used to set the appropriate current to the motor. The 1.41 A/phase current specification is the RMS value, and the selected current setting on the driver should be equal or slightly higher than this value; supplying much more current than the specified value will lead to heating of the motor and the driver. The 2.04 A/phase current limit was selected for this setup along with the 70% maximum current setting to give an effective maximum current of 1.43 A/per phase ($2.04 \times 0.7 = 1.43$). A stepper motor can be operated with a stepper driver that supplies less than the current rating per phase, but the torque produced by the motor will be reduced. The optimal current setting is the one that allows the motor to move at the desired speed with no loss of steps but without excessive heating of the motor or the driver. The proper voltage to drive the motor is set automatically by the driver provided that the voltage supplied to the driver is larger than the rated voltage for the motor. This driver accepts 12-48 VDC as the input voltage and is operated with a 24 volt power supply.

The interface circuit for this system is shown in Figure 9.45. This driver accepts step and direction signals, which are sent to the driver from the Arduino Uno board. The step signal is sent as a PWM signal using pin #11 on the Arduino board, while the direction signal is sent as a digital output signal. The timer setting for the PWM pin (see Section 4.6.2) determines the operating speed of the motor. For this application, the default setting for *Timer2* (which controls pin #11) was used, which gives a PWM frequency of 490.2 Hz and a rotating speed of 2.45 rev/sec (490.2/200) at a stepping resolution of 200 steps/rev. The driver also has an optional *Enable (EN)* connection, but the default (no connection) setting was used here. The control signals are optically isolated and configured as open-collector signals.

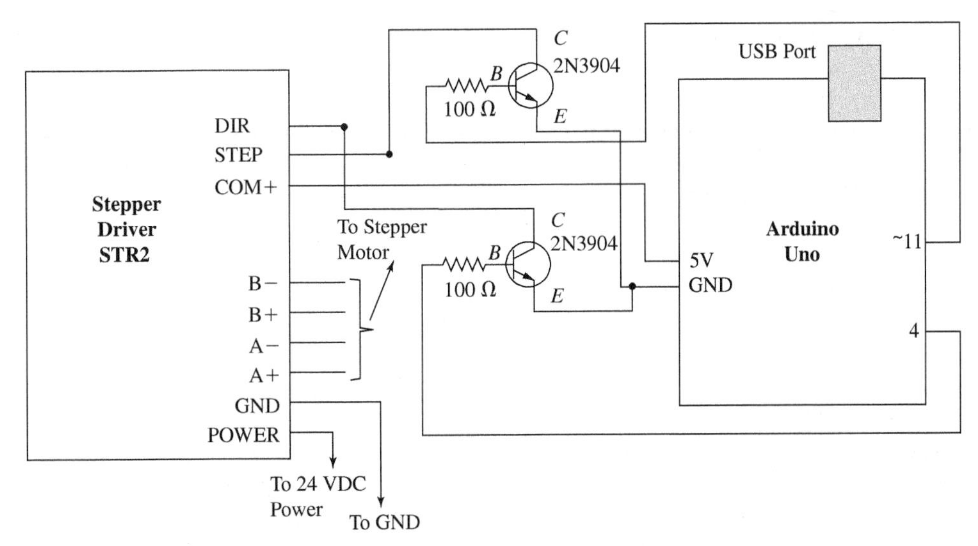

Figure 9.45 Interface circuit for the stepper motor-driven rotary table

As shown in the internal circuit diagram provided by the manufacturer (see Figure 9.46), sending a 5 V signal to the base of the 2N3904 transistor (see Section 3.4) whose collector is connected to the STEP pin causes the 2N3904 transistor to saturate and a current to flow through the internal optoisolator in the drive which will cause the step signal to be high. A similar wiring method is used for the direction signal. The power signals (5 volts and ground) for the control signals are provided by the Arduino board and are separate from the power connections that supply the current to the motor phases.

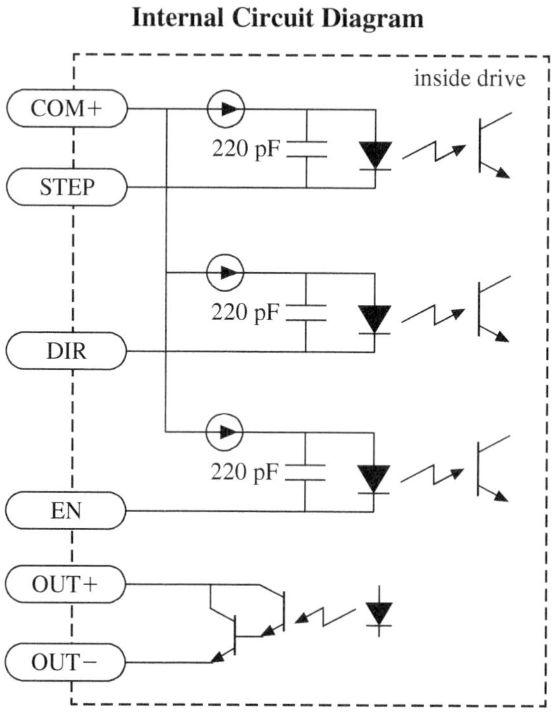

Internal Circuit Diagram

Figure 9.46 Connection diagram for control signals to the driver

(Courtesy of Applied Motion Products, Morgan Hill, CA)

This driver has multiple settings for the steps per revolution of the motor ranging from the standard 200 steps/revolution to 20,000 steps/revolution, which can be set through dip switches. Using microstepping gives a smaller positioning resolution of the motor.

9.6 Other Motor Types

Universal Motor: A **universal motor** (available—V9.9) is an electric motor that can be operated using both AC and DC voltage signals. It is mostly used in hand tools such as drills and appliances such as vacuum cleaners, mixers, and blenders. The motor uses brushes for commutation, and its construction is similar to that of a series-wound DC motor with a wound rotor and a wound stator. The currents in the rotor and the field coils change polarity at the same time, making the direction of the resultant force acting on the rotor constant. The universal motor is also known as an **AC series motor** or an **AC commutator motor**. One feature of the universal motor is that it allows variable speed control of the motor similar to a DC motor but using AC power, thus eliminating the need for an AC-to-DC transformer. The speed control can be implemented using phase control with SCR (see Chapter 3), a rheostat, or a chopper drive (uses a PWM duty cycle to control the effective voltage). Universal motors have a high power-to-size ratio, and they have a high no-load speed (20,000–40,000 rpm) which is much higher than the line frequency of either 50 or 60 Hz. However, a DC motor of the same size as a universal motor is more efficient than a universal motor.

Servo, Gear, and Brake Motors: Many of the motors that were discussed before can be referred to by different names depending on the application. For example, a **servomotor** (available—V9.10) is a motor that is equipped with either a position or velocity feedback device to be used in closed-loop control applications. The motor itself can be a DC or AC type. The control system for the servomotor adjusts the current supplied to the motor windings to accurately control the motor's output and achieve the desired motion characteristics. Servo motors offer high speed or position accuracy due to the feedback device and closed-loop control and are also known for their excellent dynamic response and ability to quickly change speed and direction. Another example is a **gear motor**, which is a motor that has a gear attached to it to reduce the speed of the motor and increase the output torque of the motor. A third example is the **brake motor**, which has an attached brake to prevent the shaft from rotating when the power is disconnected from the motor.

Integrated Servo Motor: An **integrated servo motor** (▶ available—V9.11) integrates into a single package a motor, a servo drive, an encoder, and a motion controller. This integration provides the user with a perfectly matched motor and drive and eliminates the need for separate drive enclosures and extensive cabling, resulting in a more compact and space-saving solution. Integrated servo motors also often come with user-friendly interfaces and software tools that simplify system configuration, setup, and programming. Figure 9.47 shows an example of an integrated servo motor that is made by Teknic. This particular integrated servo system combines a Brushless DC motor with an encoder along with a digital servo drive and controller. A single package that just combines a DC motor and a drive is called an **integrated motor drive**, although some of these also include an encoder. Integrated motor drives are also available for stepper motors.

Figure 9.47

Integrated servo motor
(Jouaneh, University of Rhode Island)

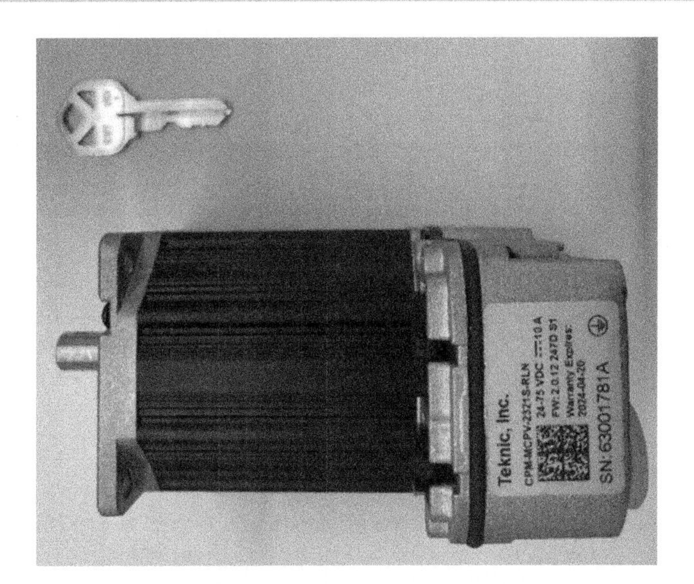

Hobby Motors: A special class of motors is called **hobby** or **RC servo motors** (▶ available—V9.12). These motors are widely used in radio-controlled cars, planes, and boats. A typical standard hobby servo motor is shown in Figure 9.48. Hobby servo motors are relatively inexpensive, are driven by low voltages (about 5 VDC), and are available in several sizes including standard, mini-micro, and quarter scale. The standard size is the most common and has the advantage that its physical size and mounting holes are the same regardless of the manufacturer. The mini-micro size is half the size or smaller than the standard servo. The rotational speed of the hobby servo is about 0.2 s for 60° of angular travel.

The hobby servo motor consists of four components that are packaged together. These are a small PM brush DC motor, a gear reducer, a potentiometer, and a control board. These servos have three leads labeled power, ground, and control signal. The power signal ranges from 4.8 to 7.2 V and can be conveniently obtained from battery power packs. The control signal is a PWM signal (5 V) at a frequency of 20 to 60 Hz. This signal can be conveniently generated from a microcontroller or a commercial control board.

Figure 9.48

A standard-size hobby servo motor

(Courtesy of Hitec RCD USA, Poway, CA)

There are two versions of hobby servo motors. One is the limited motion servo, and the other is the unlimited motion servo. In the **limited motion servo,** also known as a standard servo, the pulse width of the control signal, which ranges from about 0.7 to 2.3 ms for most servos, controls the position of the servo. At 0.7 ms, the servo is at one extreme of its motion range, while at 2.3 ms it is at the other extreme. At 1.5 ms pulse width, the servo is in the center or mid-position. Most servos have a motion range of ±90°. Figure 9.49 shows the typical relationship between the pulse width and the hobby servo position.

Figure 9.49

Limited motion hobby servo position as a function of pulse width

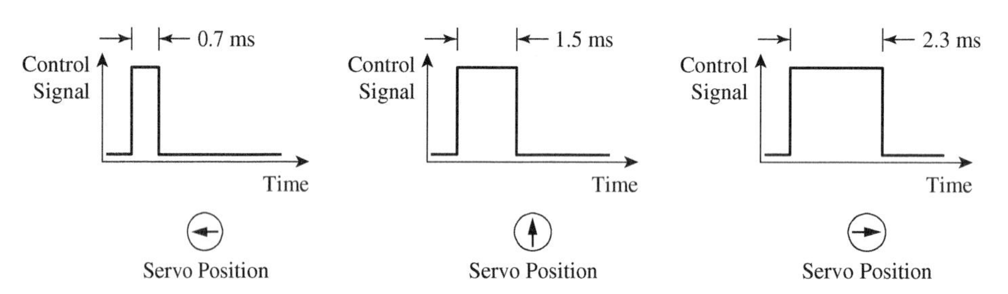

Limited motion hobby servos operate in a closed-loop position control fashion (see Chapter 10). On the control board, the duration of the PWM control signal is converted to a voltage signal. This voltage signal is compared with the voltage output from the potentiometer that is connected to the motor shaft. The difference between these two voltage signals is then used to drive the motor so the error between these two signals goes to zero.

The **unlimited motion servo**, often referred to as a continuous rotation servo, is an RC servo motor that has been modified to run in velocity mode instead of position mode. It does not operate in closed-loop position control. The PWM signal controls the speed of the motor. In continuous rotation servos, the PWM signal typically still ranges from about 0.7 to 2.3 ms. However, instead of controlling position, this range controls the speed and direction of rotation. Around 1.5 ms usually corresponds to a stop or zero speed, while values lower or higher than 1.5 ms control the speed and direction (forward or reverse) of the rotation.

9.7 Electric Motor Selection

When selecting an electric motor (▶ available—V9.13) for a mechatronic application, the selection should include the type of motor (DC, AC, Stepper, etc.), the power rating and speed, the operating voltage and frequency, the motor frame size, and mounting details.

To provide means for the interchangeability of motors, commercial motors are made in standard mounting sizes called **NEMA** (National Electrical Manufacturers Association) **frame sizes**. These frame sizes facilitate compatibility across motors from different manufacturers. The NEMA frame sizes (▶ available—V9.14) are categorized by numbers, with stepper motors often using sizes such as 17, 23, 34, and 42, and industrial AC and DC motors typically starting at frame size 42 and going upwards. The NEMA frame size determines key mounting dimensions, such as the distance between mounting holes and the height of the shaft centerline above the mounting surface, which is referred to as the 'D' dimension. This **'D' dimension** is the distance from the center of the motor shaft to the bottom of the base mount (see Figure 9.50). For a two-digit frame size, the 'D' dimension in inches is found by dividing the frame size by 16. For a three-digit frame size, the 'D' dimension in inches is calculated by dividing the first two digits by 4. For instance, a frame size of 34 implies a 'D' dimension of 2.125 inches or 53.98 mm (34 divided by 16), and a frame size of 405 translates to a 'D' dimension of 10 inches or 254 mm (the first two digits, 40, divided by 4). While the NEMA frame size primarily specifies the motor's mounting dimensions, it also indirectly relates to other aspects of the motor's physical characteristics, such as the shaft's length and diameter, but it is not a direct indicator of the motor body diameter. When selecting a motor, it is important to consider the entire set of dimensions and specifications as provided by NEMA standards to ensure the motor will suit the application.

Figure 9.50

Illustration of the NEMA *'D'* dimension

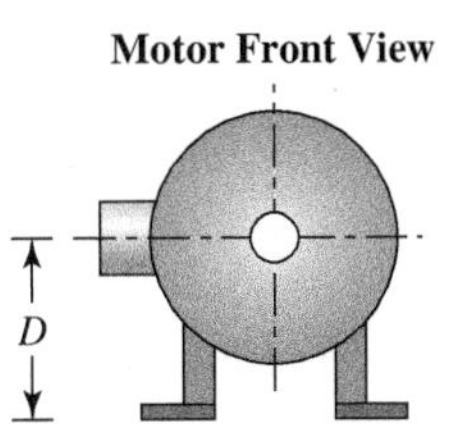

Motors can also differ in the details of their mounting and the type of enclosure they have. Common **mounting configurations** include foot mounted, cushion base, C-Face mounting, and vertical mounting. The common **enclosure types** are shown in Table 9.5. Many things should be considered when selecting an actuator. The most important of these would be that the torque-speed characteristics of the actuator should match those of the desired application. Secondary considerations include control method, cost, size, and ease of maintenance. Table 9.6 compares DC, AC, and stepper motors in several categories.

Table 9.5

Common enclosure type for electric motors

Enclosure Abbreviation	Description	Comments
ODP	Open Drip Proof	For normal applications where low cost is important
TEFC	Totally Enclosed Fan-cooled	Commonly used motor enclosure. The enclosure is dust-tight but cannot stand high water pressure. Uses a fan to cool the motor
TEAO	Totally Enclosed Air Over	Normally used with fans or blowers and utilizes the air drawn by the fan or blower for cooling
TEBC	Totally Enclosed Blower Cooled	Uses a blower for cooling. The enclosure is dust-tight but cannot stand high water pressure
TENV	Totally Enclosed Non-Ventilated	Use with small hp rated motors and utilize fins on the motor body for cooling
TEWC	Totally Enclosed Water Cooled	Expensive motors that utilize a double shell body through which water flows for cooling

Table 9.6

Comparison of brush DC, brushless DC, AC, and stepper motors

Characteristic	Brush DC	Brushless DC	AC	Stepper
Supply voltage	Needs a simple DC voltage power source	Uses DC voltage but requires an electronic speed controller to modulate the DC voltage and phase timing	Can be readily run from the AC line voltage	Uses DC voltage but requires a stepper motor driver to drive each phase
Direction change	Easily done by reversing the polarity to the motor leads	By activating the phases in a reverse fashion	By changing the wiring in the starting circuitry in single-phase motors (but not always reversible) and by swapping two of the three-phase wires in three-phase motors	By activating the phases in a reverse fashion or by changing the direction signal if a stepper motor driver is used
Speed change	Easily done by changing the value of the input voltage	By changing the rate of activating the phases	More difficult to do and requires voltage control or capacitance changes for single-phase, and Variable Frequency Drive (VFD) for three-phase	By changing the rate of activating the phases or the pulse rate if a stepper driver is used
Starting	Self-starting	Self-starting	Some single-phase AC motors may not be self-starting without auxiliary circuitry such as start and/or run capacitors. Three-phase motors are self-starting	Self-starting
Maintenance	Need to periodically replace the brushes and resurface the commutator	No need to replace brushes or resurface the commutator	Requires less maintenance, especially for AC induction motors	No wear problems due to the absence of brush contact but they may still require lubrication and other mechanical maintenance
Available sizes	Few Watts to a few hundred hp	Few Watts to several hundred hp	Single phase up to a few hp and multiphase up to several thousand hp	Stepper motors do not have a hp rating since they do not rotate continuously. The equivalent max hp rating is a fraction of 1 hp, but they offer a substantial torque at low speeds. They could have torque ratings up to a few thousand N-cm.

9.8 Pneumatic and Hydraulic Actuators

Pneumatic and hydraulic actuators are mechanical devices that convert the pressure of compressed gas or fluid into mechanical force and motion. They are part of systems where compressors or pumps are used to create gas or fluid pressure, and their operation is controlled by valves. However, there are distinct differences between the two types of actuators. **Pneumatic actuators** employ a compressible gas, typically air, to generate force and motion. **Hydraulic actuators** use relatively incompressible fluids, such as hydraulic or mineral oil, ethylene glycol, or water, to generate force and motion. These fluids provide greater resistance to compression, allowing hydraulic actuators to exert higher forces compared to pneumatic actuators. One notable advantage of hydraulic actuators is their high load-to-size ratio. They can exert substantial force even in relatively compact designs, making them suitable for applications where high force output is required within limited space constraints.

Both pneumatic and hydraulic actuators find widespread use in industrial and manufacturing applications for controlling and moving machinery and equipment. Pneumatic actuators are known for their low cost and simplicity and are commonly employed in pick-and-place automation equipment, air tools in the construction industry, and work-holding devices. On the other hand, hydraulic actuators are known for their high force and precision, making them ideal for heavy-duty applications such as construction equipment, injection molding machines, compactors, and presses. Unlike electrical actuators, which can operate independently, pneumatic and hydraulic actuators require additional components for their functioning. These components are:

Storage Tank or Reservoir: A storage tank is necessary for pneumatic systems to store compressed air, while hydraulic systems require a reservoir to hold hydraulic fluid. These containers ensure a constant and ready supply of the working medium.

Compressor or Pump: Pneumatic systems rely on a compressor to compress air and provide the required pressure, while hydraulic systems employ a pump to force the hydraulic fluid through the system.

Power Source: An electric motor or another power source is needed to drive the compressor in pneumatic systems or the pump in hydraulic systems. It supplies the energy required to operate the compression or pumping mechanism.

Valves: Valves play a critical role in controlling the direction, pressure, and flow rate of the compressed air or hydraulic fluid. They enable precise regulation and manipulation of the working medium, allowing for smooth and controlled actuation.

Actuator: The actuator itself is the component responsible for converting the energy of the compressed air or hydraulic fluid into mechanical force or torque to perform useful work. Actuators can take the form of cylinders, providing linear motion, or motors, delivering rotary motion.

Piping: Piping or tubing is necessary to transport the compressed air or hydraulic fluid throughout the system. It ensures the proper flow and distribution of the working medium to the actuator and other components.

Regulators: Regulators are utilized to set and control the air or hydraulic pressure within the system. They help maintain the desired pressure levels, ensuring consistent and reliable actuator performance.

Gauges: Gauges are employed to monitor and measure the air or hydraulic pressure. They provide feedback on the pressure levels within the system, enabling operators to ensure optimal functioning and detect any abnormalities.

These components work together to generate, control, and regulate air or fluid pressure, enabling the actuators to perform their intended tasks.

Pneumatic and hydraulic actuators (▶ available—V9.15) typically consist of a cylinder and piston assembly, with the piston attached to the equipment to be moved. When compressed air or hydraulic fluid is pumped into the cylinder, it exerts pressure on one side of the piston. This pressure imbalance causes the piston to move, generating force and initiating the desired motion. These actuators are available in different formats, including single-acting and double-acting (see Figure 9.51). A **single-acting cylinder** operates by using compressed air or fluid to exert force on the piston in one direction. When the supply of air or fluid is halted, the cylinder relies on external factors such as the load force, gravity, or a return spring to bring the piston back to its original position.

Figure 9.51

(a) Single-acting and (b) double-acting pneumatic/hydraulic actuators

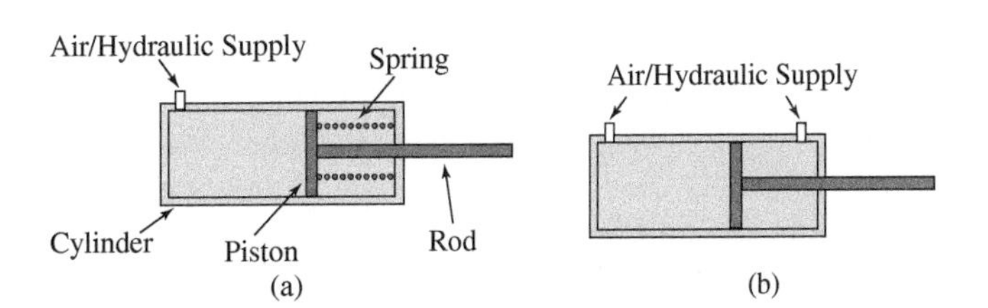

In contrast, a **double-acting cylinder** employs pressurized fluid or air to apply force on both sides of the piston. To extend the piston, the air or fluid is introduced into one port of the cylinder, while to retract the piston, the air or fluid is applied to the other port. Due to this bidirectional force application, there is no need for a spring to return the piston, as the air or fluid itself pulls it back. This design offers the advantage of improved control and flexibility in performing both extension and retraction actions. Figure 9.52 shows a picture of a commercially available double-acting pneumatic actuator.

Proximity sensors can be effectively attached to both pneumatic and hydraulic cylinders to indicate whether the piston is extended or retracted. This capability provides essential feedback for automated systems. The two most common types of sensors used for this purpose are Reed and Hall Effect sensors, both of which utilize a magnetic field to detect the piston's position. Therefore, it is common for cylinders used in conjunction with these sensors, whether pneumatic or hydraulic, to have a magnet located internally on the piston. **Reed sensors**, operated by magnetic fields, are known for their simplicity, reliability, and cost-effectiveness in detecting the presence of a magnetic piston. They consist of two ferromagnetic contacts in a hermetically sealed glass envelope, which close in the presence of a magnetic field, thereby completing an electrical circuit. This makes them ideal for straightforward, binary position indication. **Hall Effect sensors** (see Section 8.4.1) offer a solid-state solution, using a magnetic field to detect piston position and providing greater durability and precision. The selection of an appropriate sensor type depends on factors such as the operating environment, required accuracy, system compatibility, and the specific needs of the hydraulic or pneumatic system. Both types of sensors enhance the functionality and safety of actuator systems in various industrial applications.

Figure 9.52

Double-acting pneumatic actuator

(Surasak_Photo/Shutterstock.com)

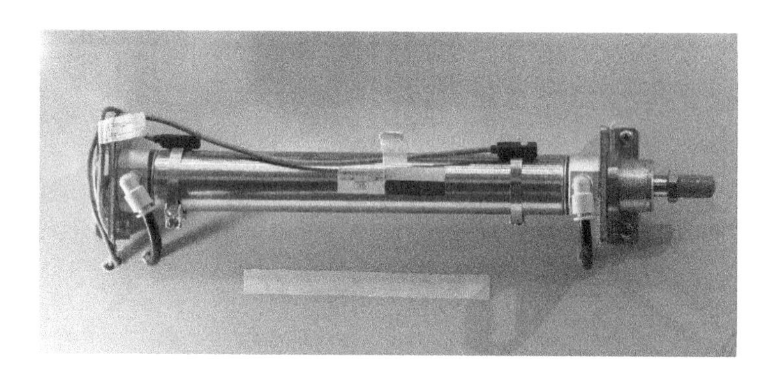

Directional control valves are essential components in pneumatic and hydraulic actuators used for controlling the direction of air or fluid flow within the system. Among these, **solenoid control valves** (▶ available—V9.16) are particularly common. These valves utilize a solenoid mechanism, as detailed in Section 9.3, to move a plunger or armature. This movement controls the opening and closing of the valve, thereby regulating the flow of air or fluid. Solenoid control valves offer quick response times, accurate positioning, and seamless integration with electronic control systems, making them highly efficient for various applications. Different types of solenoid valves, illustrated in Figure 9.53, cater to specific needs:

Two-Way Solenoid Valves (2/2 Valves): The simplest form of solenoid valves, these have two ports and two positions (open or closed) and thus their name (2/2 valves). They are designed to either allow or stop the flow of fluid or air. Each valve has an inlet and an outlet, functioning in an 'on' or 'off' mode.

Three-Way Solenoid Valves (3/2 Valves): These valves are equipped with three port connections and are used for controlling single-acting cylinders. When energized, the solenoid allows fluid or air to flow from the inlet to the cylinder. Conversely, when de-energized, the supply is cut off, and the fluid or air in the cylinder is released through an exhaust outlet.

Four-Way Solenoid Valves (4/2 Valves): More complex in design, these valves are suited for double-acting cylinders, featuring four port connections. They direct fluid or air from the supply port to one side of the cylinder while allowing the other side to exhaust. The energizing of the solenoid shifts the valve, routing the fluid or air to extend the cylinder. When de-energized, the valve reverses the flow, leading to cylinder retraction. This dual-directional control is crucial for applications requiring bidirectional motion.

Although not discussed here, 5/2 and 5/3 valves also exist, offering more complex control options. Solenoid valves are available in normally closed (NC) and open (NO) configurations. An **NC valve** remains closed in the absence of an electrical current and opens when the solenoid is energized. In contrast, an **NO valve** stays open when de-energized and closes upon solenoid activation. It is important to note that while solenoid valves are highly effective, they have limitations such as sensitivity to dirt, limited flow capacity, and reliance on a power source. Understanding these characteristics is crucial for selecting the right valve for a specific application.

Figure 9.53

Different types of NC solenoid valves: (a) 2-way valve, (b) 3-way valve, and (c) 4-way valve

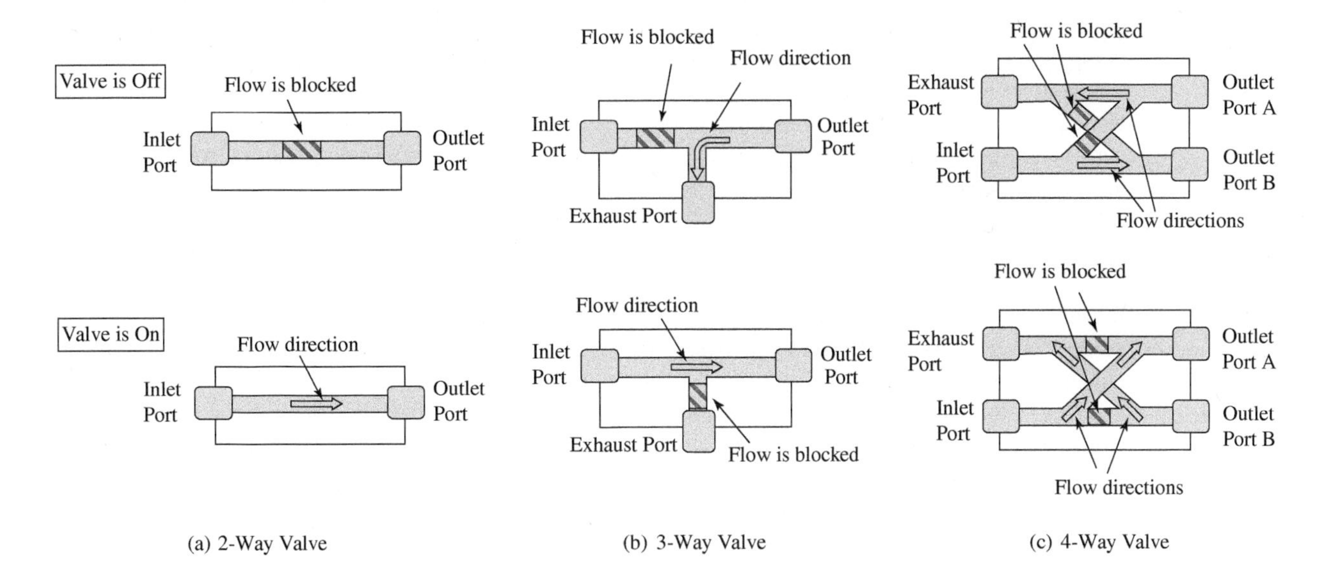

When showing pneumatic/hydraulic components and their connections on a schematic, engineers and technicians use basic symbols that are defined in a widely used international standard called **ISO 1219-1** [40]. This standard not only specifies individual symbols for components like valves, pumps, and cylinders, but also provides guidelines for assembling these symbols into complex system schematics. While universally recognized, it is important to note that certain industries or regions may have variations or additional standards. Overall, ISO 1219-1 serves as the foundational reference for accurately conveying designs and functions in fluid power systems. Example 9.9 explains the symbols for two solenoid valves.

Example 9.9 ISO 1219-1 Symbols for Solenoid Valves

Explain the ISO 1219-1 symbols for the solenoid valves shown in Figure 9.54(a) and (b).

Figure 9.54 ISO 1219-1 symbols: (a) 2/2 NC valve and (b) 3/2 NO valve

Solution:

The ISO 1219-1 symbols for the solenoid valves depicted in Figure 9.54 (a) and (b) illustrate a 2/2 NC valve and a 3/2 NO valve, respectively. Each symbol contains two squares, indicating the two positions of the valve: the default unactuated (Off) position on the right, and the actuated (On) position on the left. The number of T-intersections in each square denotes the number of ports: two in Figure 9.54(a) for a 2-way valve and three in Figure 9.54(b) for a 3-way valve. In Figure 9.54(a), the absence of a connection between the inlet and outlet ports in the right square indicates a Normally Closed (NC) valve, which blocks flow when not actuated. Conversely, in Figure 9.54(b), the presence of a path between the inlet and outlet ports, even when the solenoid is not energized, signifies a Normally Open (NO) valve. The actuating solenoid for each valve is represented on the left side of the symbol by a rectangle with a diagonal line and indicates that the valve is actuated electronically. The spring symbol opposite the solenoid denotes a spring return, ensuring the valve returns to its original position when the solenoid is de-energized.

9.9 Other Actuator Types

In addition to electric motors and pneumatic and hydraulic actuators, there are several other types of actuators available for different applications. These include piezoelectric actuators [41] which utilize the converse piezoelectric effect, shape memory alloy (SMA) actuators [42] which can change shape in response to temperature changes, thermal actuators [43] which rely on the thermal expansion or contraction of materials to produce motion, and electroactive polymer (EAP) actuators [44] which are made from materials that can change shape or size in response to electrical stimulation. This section discusses piezoelectric actuators and shape memory alloy actuators.

9.9.1 Piezoelectric Actuators

Piezoelectric actuators (▶ available—V9.17) generate mechanical deformation in response to an applied voltage. They operate based on the **converse piezoelectric effect**, which is the ability of certain materials, not just limited to ferroelectric ones, to deform or change shape when subjected to an electric field. When an electric potential or voltage is applied across the electrodes of a piezoelectric material, it induces expansion or contraction, leading to mechanical strain or displacement. This deformation is the operational principle behind piezoelectric actuators. These actuators can be made from a range of piezoelectric materials, encompassing piezoceramics, polymers, and composites. The most common configuration of piezoelectric actuators is the stack design. In this configuration, referred to as a **stacked piezoceramic actuator**, multiple layers of piezoelectric ceramic materials (such as Lead Zirconate Titanate, or PZT) are assembled with metal electrodes interleaved between them (see Figure 9.55(a)). Each layer is 'poled' so that the electric field applied aligns in the same direction across all layers, producing a cumulative effect. When voltage is applied to a stack actuator, all layers respond by expanding or contracting in unison, resulting in linear motion.

Figure 9.55

(a) Construction of a piezoceramic stack actuator and (b) commercial piezoceramic stacked actuators (Courtesy of Physik Instrumente (PI) GmbH & Co. KG, www.pi.ws)

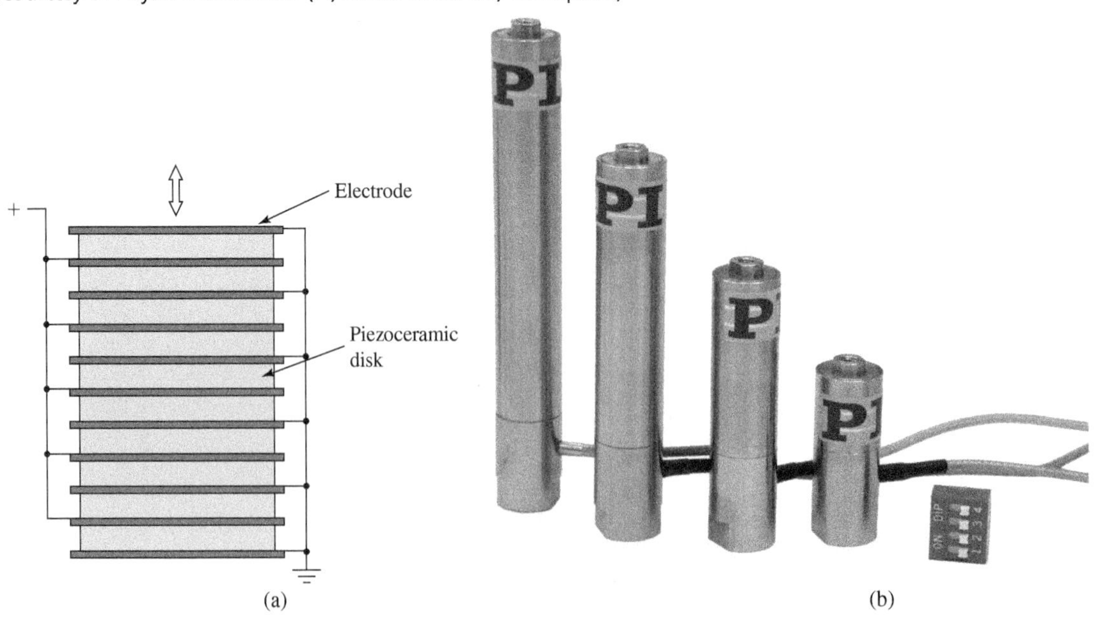

The expansion or contraction of a piezoelectric actuator is generally a smooth function of the applied voltage, though this relationship can be nonlinear, especially near the limits of the operating voltage. The characteristics of a piezoelectric actuator, such as high stiffness, infinite resolution (limited only by the precision with which the applied voltage can be controlled), extremely low power consumption in an energized state, operation over millions of cycles without wear or deterioration, rapid expansion (limited only by the inertia of the object being moved and the output capability of the control system), and compact configuration, make piezoelectric actuators excellent candidates for high precision motion applications where there is a need for fast response, high resolution, and precise motion. These actuators are commonly employed in applications requiring nano- and micro-positioning, such as electron microscopy and high-precision optics. However, one major limitation of stacked piezoceramic actuators is that their maximum displacement is only about 0.1% of the length of the actuator. For example, a 100 mm long actuator will have a maximum displacement of 100 μm. To increase the effective displacement obtained from piezoceramic actuators, displacement amplification mechanisms that utilize levers, flexure springs, and hinges are often used. Another limitation of piezoceramic actuators is their lack of inherent accuracy due to hysteresis and drift. Because piezoceramic materials are ferroelectric, they exhibit a fundamentally nonlinear response to an applied electric field, manifesting as a hysteresis effect between the displacement and the electric field. This effect becomes more pronounced with increased field strength and the piezoelectric sensitivity of the material. To address these issues, precision control applications typically employ closed-loop position control systems, and many stacked piezoceramic actuators are equipped with built-in displacement sensors for feedback control.

Commercially, piezoceramic actuators are available in low-voltage (<200 V) and high-voltage (>1000 V) forms. Special amplifiers are used to drive these actuators, tailored to their voltage requirements. During operation, piezoceramic actuators exhibit characteristics akin to an electric capacitor. This capacitive nature affects the speed of the actuator's expansion and contraction, as it determines how quickly the actuator can charge and discharge during operation. To protect piezoceramic actuators from bending loading, which they are generally not well-suited to withstand, these actuators are typically preloaded in a spring structure that includes flexure hinges. This preloading helps to maintain an optimal operating condition and prolongs the life of the actuator. As piezoceramic actuators are strain-type actuators, the force they apply is dependent on their boundary conditions. This means that the effective expansion of a piezoceramic actuator under preload is typically less than its free expansion. The relationship between the preloaded actuator expansion and its free expansion can be complex and depends on the details of the preloading mechanism. For simple axial preloading, the expression for the preloaded expansion Δ_p is given by

$$\Delta_p = \frac{k_p}{k_p + k_s}\Delta_0 \tag{9.12}$$

where k_p is the piezoceramic actuator stiffness, k_s is the spring structure stiffness, and Δ_0 is the free expansion of the actuator. Example 9.10 illustrates the effect of preloading on piezoceramic actuators.

Example 9.10 Expansion of a Piezoceramic Actuator

A piezoceramic actuator with a free expansion of 30 μm and an axial stiffness of 20 N/μm was preloaded into a spring structure as shown in Figure 9.56 which has axial stiffness of 5 N/μm. Determine (a) the expansion of the preloaded actuator and (b) the overall axial stiffness of the structure and the preloaded actuator.

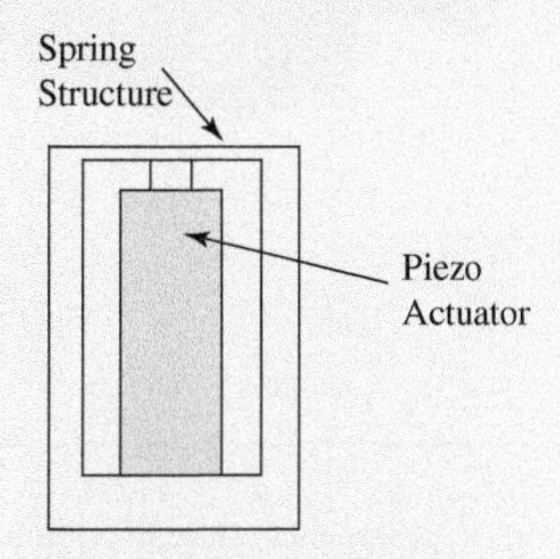

Figure 9.56 Preloaded piezoceramic actuator

Solution:

(a) Using Equation (9.12), the preloaded expansion of the actuator is given as

$$\Delta_p = \frac{k_p}{k_p + k_s}\Delta_0 = \frac{20}{20 + 5}30 = 24\ \mu m$$

(b) The actuator and the spring structure behave as two springs in parallel as experienced by the system when a force is applied to the top of the structure. Thus, the effective stiffness is the sum of the two stiffnesses, or

$$k_{overall} = k_p + k_s = 20 + 5 = 25\ \text{N}/\mu\text{m}$$

9.9.2 Shape Memory Alloy Actuators

Shape memory alloy (SMA) actuators (▶ available—V9.18) are based on shape memory alloys, with the most widely used one being Nitinol, an alloy of nickel and titanium. SMA actuators can change shape in response to temperature changes, typically reverting to their original form when heated above their specific transformation temperature, which varies based on the alloy's composition. This process is reversible, with the material returning to its deformed shape when cooled to ambient temperature. The most common form of an SMA actuator is wire, which contracts when heated and stretches when cooled. SMA actuators are lightweight, offer a high force-to-weight ratio, and are used in various applications, including robotics, aerospace, and medical devices.

Commercially available Shape Memory Alloy (SMA) wires are offered in a wide range of diameters, typically from about 0.1 to 5 mm, with some options extending even larger to suit various applications. A wire with a diameter of approximately 0.5 mm is capable of generating a pulling force of around 30 N. This force is contingent on factors such as the specific alloy composition and its heat treatment processes. To achieve higher force outputs, options include using wires with larger diameters or configuring multiple strands of SMA wire to operate in parallel. An example of one of the options is depicted in Figure 9.57, which illustrates a commercially available SMA actuator. This particular actuator employs several SMA wires arranged in parallel, optimizing its mechanical output. It boasts a motion range of up to 7 mm and is capable of exerting a force up to 650 N. SMA actuators require a power supply, which should be selected to effectively heat the SMA wires, thus facilitating the actuation process. The characteristics of the power supply, such as voltage and current output, are essential for determining the actuator's heating rate, actuation speed, and overall efficiency.

Figure 9.57

Commercially available actuator based on SMA technology

(Courtesy of Kinitics Automation, Vancouver, BC)

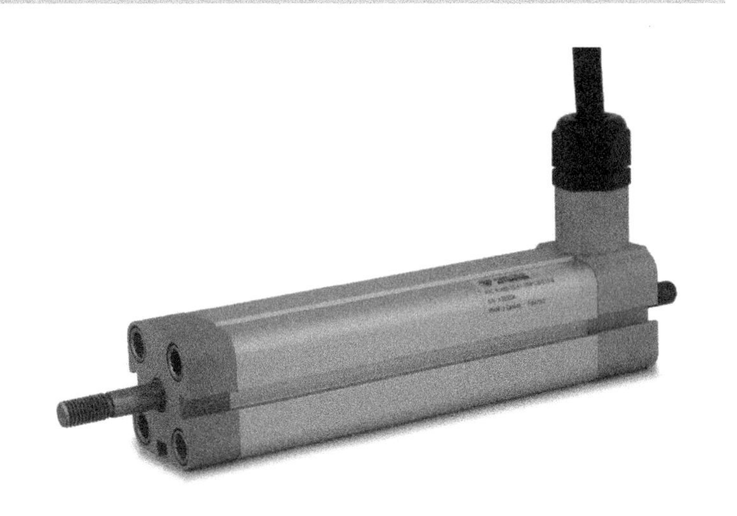

9.10 Chapter Summary

This chapter covered various types of actuators with a focus on electric actuators. The chapter covered brush and brushless DC motors, AC motors, stepper motors, universal motors, hobby servo motors, pneumatic and hydraulic actuators, and piezoelectric and shape memory alloy actuators. Brush DC motors are commonly used in many consumer and industrial applications, and they use brushes to transfer the electric current to the rotating coil to allow a continuous flow of current in the same direction as the rotor rotates. Brushless DC motors are more reliable than brush DC motors. In a brushless DC motor, the rotor is made of permanent magnets, and the stator is made of coils. There is no wiring to the rotor. Brushless DC motors use electronic commutation rather than mechanical commutation. AC motors are very rugged, are very widely used in industrial applications, and are available in power ratings up to several thousand hp. Stepper motors are typically used in open-loop position control applications because they can operate without the need for a position sensor. They move in small angular increments or steps. A universal motor is an electric motor that can be operated using both AC and DC voltage signals. Hobby servo motors are widely used in radio-controlled planes and boats, and they consist of four components that are packaged together. These are a small PM brush DC motor, a gear reducer, a potentiometer, and a control board. Pneumatic and hydraulic actuators are mechanical devices that convert fluid pressure into mechanical force and motion. Pneumatic actuators are known for their low cost and simplicity, while hydraulic actuators are known for their high force and precision. Unlike electrical actuators, which can operate independently, pneumatic and hydraulic actuators require additional components for their functioning. Piezoelectric actuators, which utilize the converse piezoelectric effect, are used for high-precision motion applications where there is a need for fast response, high resolution, and precise motion. Shape memory alloy actuators, which can change shape in response to temperature changes, are lightweight and offer high force-to-weight ratio actuation.

The operating principles of each type of actuator are given as well as the torque-speed characteristics for some of the actuators. Driving methods, amplifiers, and drive circuitry are also provided for some of the actuators. The information given in this chapter will enable the user to select the appropriate actuator for a given mechatronic application.

Questions

9.1 Name several advantages of electric-type actuators.

9.2 What type of DC motor has linear torque-speed characteristics?

9.3 What type of DC motor has a nearly constant speed over a large torque range?

9.4 Why does a PM brush DC motor require a large current at startup?

9.5 What is the no-load speed of a PM brush DC motor?

9.6 What property of a motor relates the motor torque output per current input?

9.7 Compare brush DC and brushless DC motors.

9.8 Explain what is meant by the electronic commutation of brushless DC motors.

9.9 What type of driver is used to drive brushless DC motors?

9.10 Name several operating modes of servo drives.

9.11 Explain the sign-magnitude method for PWM control of DC motors.

9.12 What is the difference between single and three-phase AC motors?

9.13 Explain what is meant by 'slip' in AC induction motors.

9.14 Name some limitations of the AC single-phase induction motor.

9.15 List several advantages of stepper motors.

9.16 Compare full-stepping and half-stepping actuation of stepper motors.

9.17 Explain what is meant by microstepping.

9.18 Name the two wiring methods used for stepping motors.

9.19 What is a universal motor?

9.20 What is an integrated motor?

9.21 What type of sensor does a standard hobby servo motor have?

9.22 What is the difference between the two versions of hobby servo motors?

9.23 What is meant by NEMA 34 size motor?

9.24 What is the difference between single-acting and double-acting pneumatic/hydraulic actuators?

9.25 Which solenoid valve is needed to operate double-acting pneumatic/hydraulic actuators?

9.26 Name the two types of proximity sensors that can be attached to pneumatic/hydraulic actuators.

9.27 Name some limitations of piezoceramic actuators.

9.28 Explain how SMA actuators work.

Problems

Section 9.2 DC Motors

P9.1 A PM brush DC motor has a nominal voltage rating of 12 V and a no-load speed of 4800 rpm. The armature resistance is 0.71 ohms, the motor's torque constant is 0.023 N · m/A, and the voltage constant is 0.023 V/(rad/s). Determine the starting torque and current for this motor.

P9.2 A PM brush DC motor has a nominal voltage rating of 24 V and a no-load speed of 4820 rpm. The armature resistance is 2.49 ohms, the motor's torque constant is 0.046 N · m/A, and the voltage constant is 0.046 V/(rad/s). Determine the motor's speed and torque when the load current is 1 A.

P9.3 A PM brush DC has a torque constant of 0.092 N · m/A and an armature resistance of 3.8 Ω. If a torque load of 0.2 N · m is applied to the motor, what is the current flowing through the motor?

P9.4 Determine the operating speed and power of a PM brush DC motor with the following operating characteristics: starting torque = 1.4 N · m, no-load speed = 4000 rpm, and torque load = 0.56 N · m.

P9.5 Determine the operating speed and power of a PM brush DC motor with the following operating characteristics: starting torque = 1.4 N · m, no-load speed = 4000 rpm, and torque load (in N · m) = 0.2 + 0.0015 ω, where ω is the motor speed in rad/s.

P9.6 A geared PM brush DC motor has a gear ratio of 50:1. The input speed is 4000 rpm at an input power of one-half horsepower. Determine the output torque and speed of this motor, assuming a 5% power loss in the gear drive.

P9.7 Show that the maximum power for a PM brush DC motor is obtained when the motor is operating at a speed equal to half the no-load speed.

P9.8 Determine the electrical and mechanical time constants for a drive system that uses a PM brush DC motor to drive a rotational mass as shown in Figure P9.8 that has an inertia of 1.0×10^{-4} kg · m^2 (which includes the inertia of the shaft and the coupling) and a viscous damping coefficient of 2.0×10^{-4} N · m · s/rad at the support bearings. The motor has the following specifications: armature resistance is 6.14 ohms, the armature inductance is 6.7 mH, the motor's torque constant is 0.073 N · m/A, the voltage constant is 0.073 V/(rad/s), the rotor inertia is 7.06×10^{-6} kg · m^2, and the rotor viscous damping coefficient is 3.56×10^{-6} N · m · s/rad.

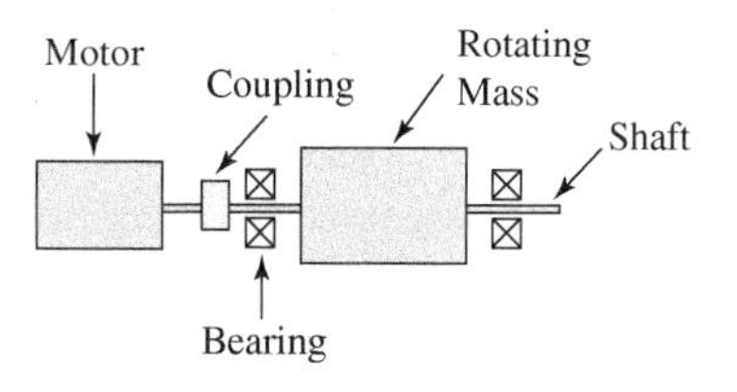

Figure P9.8

P9.9 For the setup and data in Problem P9.8, and neglecting the armature inductance, determine the transfer function between the input voltage and the motor speed in rad/s.

P9.10 For the setup and motor data in Problem P9.9, determine the maximum acceleration of the load using this motor if the inertia of the rotational mass is 3.0×10^{-3} kg · m^2 and the motor operates at a voltage of 20.

P9.11 A PM DC brush motor (data is for Pittman 14207/24 V motor) is used to drive a lead-screw table motion system similar to that shown in Example 8.4. The table has a mass of 5 kg, the combined inertia of the coupling and the lead screw is 2.0×10^{-4} kg m^2, the viscous damping coefficient due to the bearings is 0.001 N · m · s/rad, and the lead is 2 mm. The motor has the following specifications.

Continuous (nominal) torque =	0.353 N · m
Speed at continuous torque =	2810 rpm
Stalled (peak) torque =	2.85 N · m
No-load speed =	3160 rpm
Rotor inertia =	4.73×10^{-5} kg m^2

Determine the maximum acceleration of the table. Make any reasonable assumptions in solving this problem.

P9.12 The speed response of a PM brush DC motor when a 12 V step input is applied to it is shown in Figure P9.12. Use this response data to obtain the transfer function between the motor velocity and voltage input.

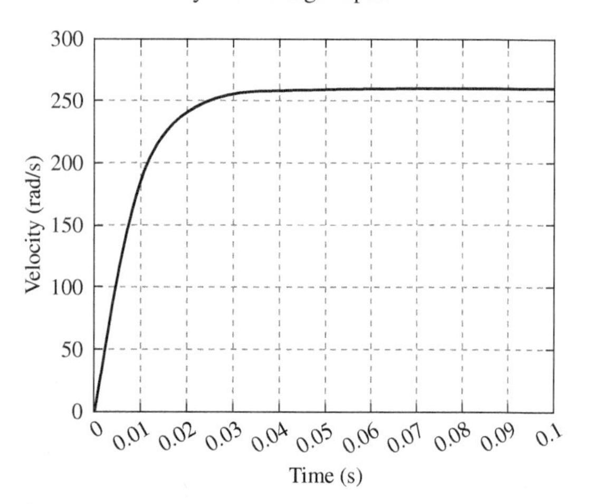

Figure P9.12

P9.13 A BLDC with torque-speed characteristics similar to those shown in Figure 9.18 has the following specifications: peak torque of 1.34 N · m, no-load speed of 4300 rpm, rated torque of 0.34 N · m, rated speed of 3600 rpm, and the knee point (where the peak torque starts decreasing from a constant value) in the torque-speed curve is at 1600 rpm. For this motor:

a. Determine the peak torque at a speed of 2500 rpm.

b. Determine the maximum angular acceleration if this motor was used to drive a rotating mass that has an inertia of 0.01 kg m^2.

P9.14 Draw a wiring diagram to operate a PM brush DC motor that uses an Arduino Uno or a similar board and an L298N motor driver module.

P9.15 Research and explain dynamic and regenerative braking in motor systems.

Section 9.3 Solenoids

P9.16 Research and identify three household, commercial, or automotive applications that use solenoids. For each application, specify the type of solenoid used.

P9.17 A solenoid has the following specifications: operating voltage = 12 VDC and coil resistance = 60 Ω. Suggest an interfacing method to drive this solenoid using an MCU such as Arduino Uno.

Section 9.4 AC Motors

P9.18 Research and explain how (a) the speed and (b) the direction of an AC induction motor can be changed.

P9.19 A 5 hp induction motor with 80% efficiency is operated from a 220 V line. If the power factor is 0.7, determine the amount of current drawn by the motor.

P9.20 The input electrical power to a 1 hp induction motor operating from a 115 V line is 995 W. What is the power factor if the current drawn by the motor is 13.7 A?

P9.21 An AC single-phase motor has a torque-speed specification as shown in Table P9.21. Determine the operating speed of the motor if it is subjected to a load of 1.16 N · m.

Table P9.21

Rated Load	1.36 N · m					
% of Rated Load	25	50	75	100	125	150
Speed (rpm)	1779	1763	1745	1725	1705	1680

Section 9.5 Stepper Motors

P9.22 A stepper motor is driven by a microstepping driver with a step resolution of 16 microsteps per step. If the motor has a step angle of 1.8 degrees per full step, what is the step angle in degrees per microstep? What should be the microstep pulse frequency (in pulses per second) to cause the motor to rotate at a speed of 200 rpm?

P9.23 A stepper motor has a specification of 400 full steps per revolution or a step angle of 0.9 degree. It is driven by a microstepping driver that can divide each full step into 64 microsteps. The desired operating speed of the motor is 900 rpm.

a. What is the pulse frequency (in pulses per second) required to achieve this speed for operation in full stepping mode?

b. What is the pulse frequency (in pulses per second) required to achieve this speed for operation in microstepping mode?

c. If the microstepping driver uses 48 kHz as the maximum step pulse frequency, what is the maximum motor speed that can be achieved when operating in microstepping mode?

P9.24 A conveyor belt system uses a stepper motor that has a specification of 200 steps/revolution. The conveyor requires the motor to rotate at a speed such that the belt moves at a linear speed of 12 cm per second. The diameter of the drive pulley attached to the motor shaft is 60 mm. For this setup, determine the pulse frequency (in pulses per second) needed to achieve the desired belt speed, assuming the motor driver is set to operate in half-stepping mode.

P9.25 A stepper motor has a specification of 200 steps/revolution. The motor shaft drives a linear positioning table using a lead screw with a lead of 2.5 mm/rev. Determine the linear speed of the table if the motor is operated in half-stepping mode at a rate of 500 pulses/s.

P9.26 A stepper motor with a step angle of 1.8 degrees is connected to a lead screw with a lead of 2.5 mm/rev. This setup is used as a linear actuator in a 3D printer for the z-axis movement. Calculate how many full steps the motor needs to take to move the print bed 10 mm up. What is the 3D printer motion resolution in the z-axis?

P9.27 Draw a wiring diagram for driving a small bipolar stepper motor (less than 500 mA per phase) with an Arduino Uno or a similar board, utilizing the ULN2003A board as an interface between the Arduino and the stepper motor.

P9.28 Draw a wiring diagram to operate a two-phase stepper motor that uses an Arduino Uno or a similar board and an L298N motor driver module.

P9.29 You have a stepper motor driver designed to drive a variety of stepper motors. The driver has four output leads, labeled A, B, $\overline{A}$, and $\overline{B}$. It can be configured to work with both bipolar and unipolar stepper motors. Given the following stepper motors, describe how you would wire each motor to the stepper motor driver.

a. Six-lead Unipolar Stepper Motor: This motor has leads labeled as 1A, 1B, 1C, 2A, 2B, and 2C. 1A and 1C are connected to one coil, while 2A and 2C are connected to another coil. 1B and 2B are the common leads.

b. Four-lead Bipolar Stepper Motor: This motor has four leads labeled A, B, C, and D. Leads A and C are connected to one coil, while B and D are connected to the other coil.

P9.30 A bipolar stepper motor with 200 steps/rev is rated for a maximum current of 2 A per phase and has a winding resistance of 1 Ω per phase. What is the maximum power dissipated in one phase of the motor's windings when the motor is stationary and when the rated current is applied directly?

P9.31 Consider a hybrid stepper motor with the following specifications: a detent torque of 0.15 N · m and a holding torque of 1.0 N · m.

a. If an external torque of 0.25 N · m is suddenly applied to the motor shaft when the motor is not energized, determine whether the shaft will rotate or remain stationary. Justify your answer based on the detent torque.

b. Calculate the minimum external torque required to cause the motor shaft to move from its position when the motor is energized but stationary.

Section 9.6 Other Motor Types

P9.32 A hobby servo motor, which has a motion range of 180 degrees, is controlled using a PWM signal generated from an Arduino Uno MCU. The PWM frequency is 61.035 Hz (refer to Table 4.7). The duration of the PWM pulse is 0.7 ms at 0 degrees and 2.3 ms at 180 degrees. Determine the motion resolution of the servo motor.

P9.33 Research and explain how a standard hobby servo motor uses the pulse width of the control signal to control the desired position of the motor.

Section 9.7 Electric Motor Selection

P9.34 Research and identify the type of actuator used in the following devices. Also list the approximate power rating and nominal voltage level for the actuator.

a. Kitchen sink garbage disposal

b. Powered car window

c. Powered car mirror

d. Food blender

e. Cordless electric drill

f. Camera focus system

g. Hybrid electric vehicle

h. Residential garage door opener

P9.35 Suggest brush DC, brushless DC, AC, or stepper motors for the following applications and explain your selection.

a. Low-maintenance, outdoor, and constant-speed operation

b. Low heat generation and controlled speed operation

c. Low-cost and controlled position operation

P9.36 Suggest the type of motor enclosure for the following operating conditions and explain your selection. Make your selection from the enclosure types listed in Table 9.5.

a. Hot and dusty environment

b. Wet environment

c. Dry and clean environment

Section 9.8 Pneumatic and Hydraulic Actuators

P9.37 A pneumatic system uses a double-acting cylinder with a bore diameter of 40 mm and a stroke length of 100 mm. What is the minimum air pressure required to generate a force of 200 N at the end of the stroke?

P9.38 A hydraulic actuator is needed to lift a load of 10 kN. A cylinder with a diameter of 50 mm and a stroke length of 200 mm is proposed to be used for this application. The pressure supply is 100 bar and the hydraulic fluid has a density of 850 kg/m^3. Determine if the cylinder can provide enough force to lift the load and the volume of fluid displaced during the stroke.

P9.39 The hydraulic actuator of Problem P9.38 is operated by a pump that has a flow rate of 20 L/min. What is the maximum speed at which the cylinder can extend?

P9.40 A single-acting pneumatic cylinder is used to push a product along a conveyor belt. Describe a practical setup using a solenoid valve to control this cylinder. Include a basic connection diagram and explain the operation of the system.

P9.41 Research the applications of single-acting and double-acting pneumatic actuators in automated manufacturing systems. Provide two specific examples where each type would be more advantageous than the other.

Section 9.9 Other Actuator Types

P9.42 A stacked piezoceramic actuator is constructed using 50 layers of piezoceramic disks. Each disk has an axial stiffness of 600 N/μm and undergoes an expansion of 1 μm when subjected to an applied voltage of 1000 V. Determine (a) the total expansion of the stack when it is subjected to 1000 V, and (b) the effective axial stiffness of the entire stack

P9.43 A piezoceramic actuator with a free expansion of 40 μm and an axial stiffness of 15 N/μm was preloaded into a spring structure as shown in Figure 9.56. If the preloaded expansion of the actuator is 30 μm, what is the stiffness of the spring structure?

P9.44 An SMA actuator is composed of a bundle of 10 wires, each of which is capable of generating a pulling force of 10 N. Additionally, assume each wire has a resistance of 15 ohms and requires heating to a specific temperature for actuation, which is achieved by passing an electric current through it. For this actuator, determine the following.

a. The total pulling force generated by the bundle when all wires are actuated

b. The total current required to heat the entire bundle simultaneously, assuming each wire requires a current of 1 ampere to reach the actuation temperature

c. The total power consumed by the bundle during actuation, based on the current calculated in part (b) and the given resistance of the wires

Laboratory/Programming Exercises

L/P 9.1 Model in MATLAB the mathematical model for a PM brush DC motor discussed in Section 9.2.1 using the data for 24 V operation listed in Figure 9.10 for the motor's parameters. Obtain the speed and current time response of the motor to a 24 V step input.

L/P9.2 Using an Arduino Uno or a similar board and a relay, build an interface circuit to drive a solenoid. Use a push-button switch to activate the solenoid.

L/P9.3 Using an Arduino Uno or a similar board and an L298N motor driver module, wire and drive a small PM brush DC motor. Issue commands using the Serial Monitor in Arduino to cause the motor to rotate right or left.

L/P9.4 Using an Arduino Uno or a similar board and an L298N motor driver module, wire and drive a two-phase stepper motor. Issue commands using the Serial Monitor in Arduino to cause the motor to rotate right or left. Use the PWM feature in the MCU to generate the pulses. Play with the timer frequencies associated with the PWM channel to allow the stepper motor to be driven at different frequencies.

L/P9.5 Use the Arduino Uno or a similar board and four transistors to build a system that drives a four-phase stepper motor. In this exercise, each phase of the motor is driven by a transistor using control signals that are sent by the MCU. The MCU should use one digital output line for each phase. The code inside the MCU should send out timed signals (use a short delay between actuation for each step) to drive the four phases of the motor in either full or half-stepping modes. The full/half stepping mode is set by a digital I/O line that the user sets high/low.

L/P9.6 Use the Arduino Uno or a similar board to drive a hobby servo motor such as the Hitec HS-311 by outputting a pulse signal to the hobby servo motor. Use the PWM feature in the MCU to generate the pulse. Use a rotary potentiometer to adjust the pulse width (or duty cycle) setting of the control signal to change the angle of the motor. When the rotary potentiometer is at one extreme of its travel, the pulse width setting should be at its minimum setting. Turning the potentiometer clockwise from that position should increase the pulse width setting. When the rotary potentiometer is at the other extreme of its motion, the pulse width setting should be at its maximum setting. Implement an infinite do-loop to read the desired pulse width and adjust the duty cycle of the PWM signal. Add a small delay (100 ms) in each run through the loop.

10 Chapter

Feedback Control

Chapter Objectives:

When you have finished this chapter, you should be able to:

- Explain the difference between open- and closed-loop control systems
- Derive the closed-loop transfer function of a control system
- Explain the operation and limitations of an on-off controller
- Obtain the steady-state error for first- and second-order systems under P-, PI-, PD-, or PID-control
- Explain the operation of a cascaded control loop
- Determine the stability of a control system
- Obtain the initial gains for a PID controller using the Ziegler-Nichols tuning method
- Explain control robustness and phase and gain margins
- Explain the digital implementation of a PID controller
- Simulate in Simulink a closed-loop control system
- Explain the effect of nonlinearities on control system behavior
- Explain the operation of a state feedback controller

10.1 Introduction

Feedback control systems are integral to the conveniences and luxuries of modern living. Systems ranging from thermostats regulating our indoor temperatures to advanced high-speed elevators rely heavily on feedback control. Essentially, a **feedback control system** strives to maintain a set relationship between its output and a reference input. It achieves this by continuously comparing the two and using the resulting difference to adjust the system's response [45]. A feedback control system is usually comprised of several components:

Controller: Responsible for the decision-making processes.

Plant or Process: The actual system or process under control.

Actuator: Executes the control actions dictated by the controller onto the plant or process.

Sensor or Measuring Element: Gathers data from the plant or process for comparison with the reference input.

In the previous chapters, we have covered the details, models, and working principles for many of these components. This chapter covers the basics of feedback control systems. Its objective is to familiarize the reader with the design, simulation, and implementation of these systems. We begin by comparing open-loop and closed-loop control methods, after which we review the fundamental feedback control topics. Subsequently, the on-off controller and its limitations are discussed. This is followed by an in-depth examination of the PID controller, one of the most used controllers in the industry. Using the velocity and position control of a simple rotational mass as an example, the chapter delves into the effects of various control actions—P, PI, PD, and PID—on the system's response. These control actions are illustrated through MATLAB simulations. The cascaded control loop structure, which is widely used in positioning control applications, is also discussed. Additionally, the chapter addresses control design considerations such as controller stability, bandwidth, tuning, robustness, and digital implementation. Nonlinear control effects and state feedback controllers are also discussed in this chapter.

10.2 Open- and Closed-Loop Control

Before we discuss in detail feedback control systems, which are also referred to as closed-loop control systems (▶ available—V10.1), let us contrast them with open-loop control systems. In an **open-loop system**, the actual output of the system does not influence the input to the system, while in a **closed-loop system,** the input to the system is a function of *both* the actual output and the reference input. The designations 'open-loop' and 'closed-loop' stem from the way each element in the control system is visually represented. Each element is depicted by a block, illustrating its input-output relationship. In an open-loop system (refer to Figure 10.1), there is not a loop linking these blocks, whereas in a closed-loop system (as shown in Figure 10.2), a distinct loop is present, indicating feedback.

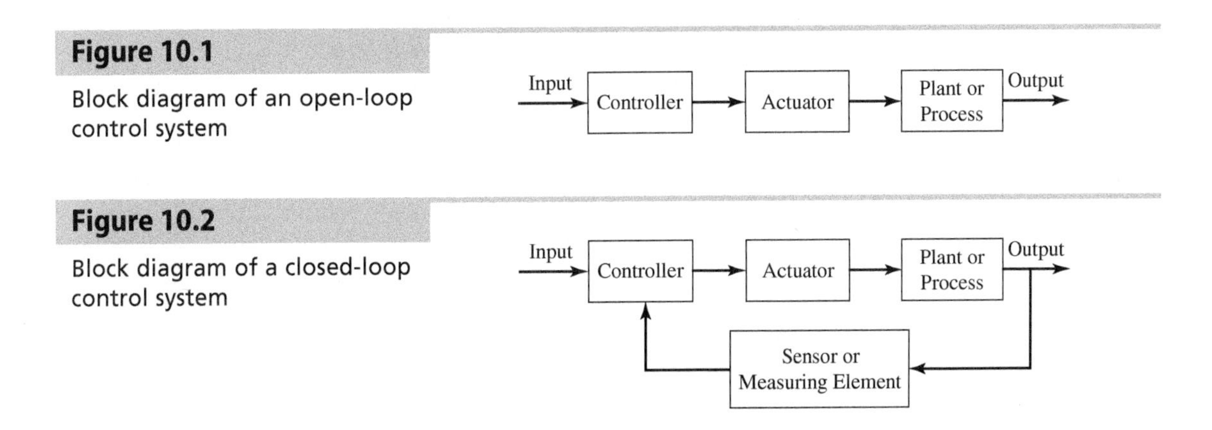

Figure 10.1

Block diagram of an open-loop control system

Figure 10.2

Block diagram of a closed-loop control system

Examples of open-loop systems include a stepper motor-driven stage operating without feedback sensors and a window air-conditioner functioning without thermostat control. Although more straightforward to implement (given that stability is not a primary concern as in closed-loop systems), open-loop systems exhibit certain limitations:

- Their accuracy relies heavily on consistent calibration between input and output. Any operational change, like load variations in motor systems, necessitates recalibration.
- Their performance can be compromised by internal or external disturbances, such as noise, wear, dirt, or temperature shifts.

Conversely, closed-loop systems, by continuously monitoring and adjusting their output, inherently counteract these limitations. Moreover, they offer adaptability, precision, and robustness against unforeseen disturbances, although ensuring their stability during design is crucial.

10.3 Design of Feedback Control Systems

Designing a feedback control system (▶ available—V10.2) involves choosing a specific control law and determining the controller's parameters to achieve desirable transient and steady-state performances. Transient performance refers to how the system behaves in response to changes or disturbances, while steady-state performance describes the system's behavior after it has settled over time. Key performance metrics include overshoot, rise time, and steady-state error. Due to the inherent risk of instability in closed-loop systems, which can result in unbounded outputs for a given bounded input, it is imperative to meticulously analyze and simulate the proposed feedback controller. Ensuring the controller behaves as intended before deploying it on the actual system is crucial. A misbehaving or unstable closed-loop control system can cause irreparable damage or malfunction to the physical system.

For accurate analysis and simulation, a mathematical dynamic model of the system in question is essential. These dynamic models can either stem from foundational physical laws, such as Newton's laws (see Appendix D), or be derived from experimental response data. However, it is important to note that even with a comprehensive model, the real system might not precisely mirror the simulated one. This discrepancy arises because capturing every nuance in a simple model, such as friction, nonlinearities, and other real-world effects, can be challenging. As such, real-life system behavior may deviate from model predictions.

While all mechanical plants or processes (such as motors, heaters, and mixing tanks) operate in continuous time, their associated controllers can be either analog or digital in nature. **Analog controllers** are built around the principles of analog electronics, utilizing key components such as op-amps (see Section 2.9). Op-amps are versatile devices that can amplify signals and execute various mathematical operations, depending on how they're configured. In contrast, **digital controllers** operate in the realm of binary signals and digital computations. They leverage computers and microcontrollers, as discussed in Chapters 4 and 6. Digital controllers offer a notable advantage: the controller gains can be conveniently adjusted in software by modifying parameter values within the control law, eliminating the need for hardware modifications or circuit rewiring. As a testament to their versatility, most modern control systems predominantly employ digital controllers. When utilizing a digital controller, appropriate interfacing is required to bridge the gap between the digital controller and the continuous-time plant, typically through A/D or D/A converters, as explored in Chapter 5. It is worth noting that while a digital controller operates on a discrete-time basis, updating its output only at specific sampling instances, previous control action remains effective between these instances. Thanks to innovative processors that support high sampling rates, digital controllers can, in many cases, be modeled as near-continuous systems with minimal accuracy loss. Given this backdrop, the ensuing discussions in this chapter will primarily focus on continuous-time models for both plants and control laws.

Once a mathematical model is available for a control system, analysis can be performed in either state space form (which represents the system as a set of differential equations) or in transfer function form. For systems that are linear or can be linearized by approximating the nonlinearities, the Laplace transform can be applied. The Laplace transform is an integral transformation that converts time-domain differential equations into algebraic equations in the complex variable s. For single-input, single-output (SISO) systems, analyzing the system in the algebraic domain using the Laplace-transformed functions is often more straightforward. Section 7.4 gives some background about the Laplace Transform, and Appendix C gives further details about it. All control textbooks have coverage of the Laplace Transform (see for example [25]), so its detailed principles will not be covered here.

10.4 Control Basics

This section covers some basic control systems material. We start by discussing transfer functions (▶ available—V10.3) and the poles and zeros of transfer functions. The transfer function (refer to Section 7.4) defines the input-output relationship for each system component or block in a control system. The transfer function is written as the ratio of two polynomials $B(s)$ and $A(s)$ in the parameter s, as shown in Equation (10.1).

(10.1)
$$G(s) = \frac{B(s)}{A(s)}$$

The roots of the numerator polynomial ($B(s)$) are called the **zeros** of the transfer function. They influence both the transient and steady-state behavior of the system response. The roots of the denominator polynomial ($A(s)$) are called the **poles** of the transfer function. The location of these poles is crucial: for a continuous-time system to be stable, all poles must possess a negative real part. If any pole lies in the right half of the s-plane (having a zero or a positive real part), the system is considered unstable since its transient response does not decay. A block diagram representation of a transfer function is shown in Figure 10.3, where $U(s)$ is the input and $Y(s)$ is the output.

Figure 10.3

Block representation of a transfer function

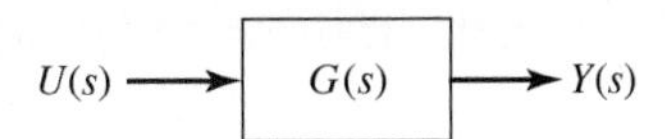

Consider three blocks in series (meaning the output of one system feeds directly into the input of the next), with one to represent a controller with a transfer function $G_c(s)$, another to represent the actuator with a transfer function $G_a(s)$, and a third to represent a plant with a transfer function $G_p(s)$, as seen in Figure 10.4. $R(s)$ is the reference or set value, and $Y(s)$ is the output or actual value of the system. The overall transfer function that represents the effect of the controller, actuator, and plant combined is given by the product of the three transfer functions $G_c(s)$, $G_a(s)$, and $G_p(s)$. This overall transfer function is called $G_{open}(s)$. The term '$G_{open}(s)$' implies an 'open-loop' transfer function, which means that there is not any feedback mechanism influencing the input.

Figure 10.4

Combined transfer function, $G_{open}(s)$

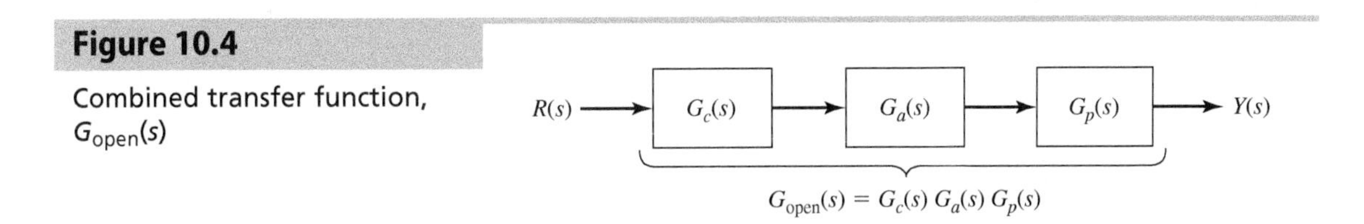

Now assume we added a sensor with a transfer function $H(s)$ to measure the output of the plant, and then we used the sensor signal to place the system under closed-loop feedback control as shown in Figure 10.5. Using block diagram rules, the output of the system, $Y(s)$, is given by

$$Y(s) = G_p(s)G_a(s)G_c(s)E(s) = G_p(s)G_a(s)G_c(s)(R(s) - H(s)Y(s)) \tag{10.2}$$

where $E(s)$ is the error and is defined as $R(s) - H(s)Y(s)$. The **overall closed-loop transfer function**, $G_{closed}(s)$, is defined as the ratio between the actual output of the system, $Y(s)$, and the reference input, $R(s)$. This ratio can be obtained from Equation (10.2), and is

$$G_{closed}(s) = \frac{Y(s)}{R(s)} = \frac{G_p(s)\,G_a(s)\,G_c(s)}{1 + H(s)\,G_p(s)\,G_a(s)\,G_c(s)} = \frac{G_{open}(s)}{1 + H(s)G_{open}(s)} \tag{10.3}$$

Note that $G_{closed}(s)$ is equivalent to the effect of all the components in the dashed block in Figure 10.5. Also note the form of the closed-loop transfer function. In the numerator, it has the product of all the blocks in the forward loop between $R(s)$ and $Y(s)$ or $G_{open}(s)$, while in the denominator, it has the form of 1 plus the product of all the blocks in the feedback loop between $R(s)$ and $Y(s)$, or $1 + H(s)G_{open}(s)$.

Figure 10.5

Overall closed-loop transfer function

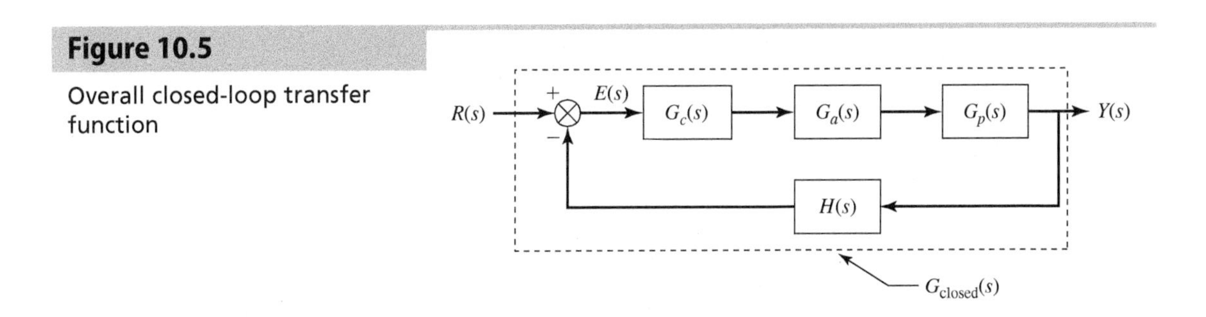

The overall closed-loop transfer function can be of first, second, or a higher order, depending on the plant dynamics model and the type of controller used. It should be noted that the order of a transfer function is the highest power of the variable s in the denominator. Examples 10.1 and 10.2 illustrate the determination of the closed-loop transfer functions.

Example 10.1 Closed-Loop Transfer Function

Determine the closed-loop transfer function between the output signal $Y(s)$ and the reference signal $R(s)$ for each of the systems shown in Figure 10.6.

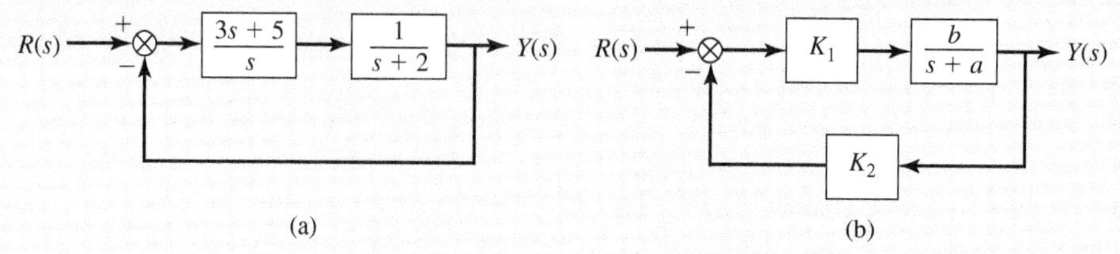

Figure 10.6 Block diagrams of the two systems

Solution:

We will use Equation (10.3) for the solution. For the system in Figure 10.6(a), $G_{open}(s)$ is

$$G_{open}(s) = \frac{3s+5}{s(s+2)} \tag{1}$$

and $H(s) = 1$. Substituting into Equation (10.3), we get

$$G_{closed}(s) = \frac{Y(s)}{R(s)} = \frac{\dfrac{3s+5}{s(s+2)}}{1 + \dfrac{3s+5}{s(s+2)}} = \frac{3s+5}{s(s+2)+3s+5} = \frac{3s+5}{s^2+5s+5} \tag{2}$$

For the system in Figure 10.6(b), $G_{open}(s)$ is

$$G_{open}(s) = \frac{K_1 b}{s+a} \tag{3}$$

and $H(s) = K_2$. Substituting into Equation (10.3), we get

$$G_{closed}(s) = \frac{Y(s)}{R(s)} = \frac{\dfrac{K_1 b}{s+a}}{1 + K_2\dfrac{K_1 b}{s+a}} = \frac{K_1 b}{s+a+K_2K_1 b} \tag{4}$$

For case (a), the closed-loop transfer function is a second-order, while for case (b), it is a first-order.

Example 10.2 Overall Closed-Loop Transfer Function

Determine the overall closed-loop transfer function for the control system represented with the block diagram shown in Figure 10.7.

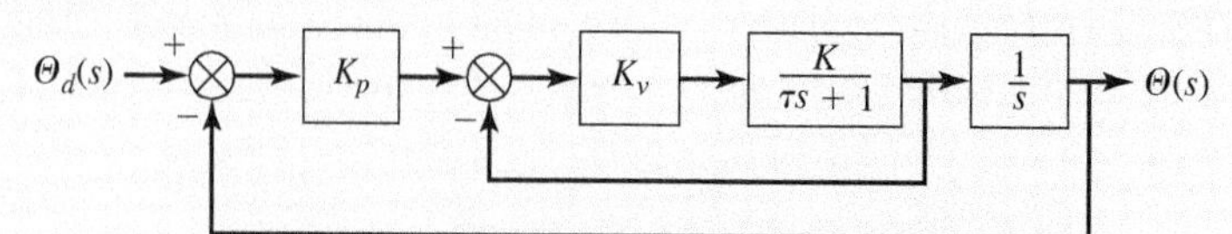

Figure 10.7 Block diagram of control system

Solution:

The given block diagram has two loops, an inner and an outer loop. From Equation (10.3), the closed-loop transfer function for the inner loop is given by

$$G_{Inner}(s) = \frac{K_v\dfrac{K}{\tau s+1}}{1 + K_v\dfrac{K}{\tau s+1}} = \frac{KK_v}{\tau s + KK_v + 1} \tag{1}$$

The block diagram can now be represented as shown in Figure 10.8 with only one loop.

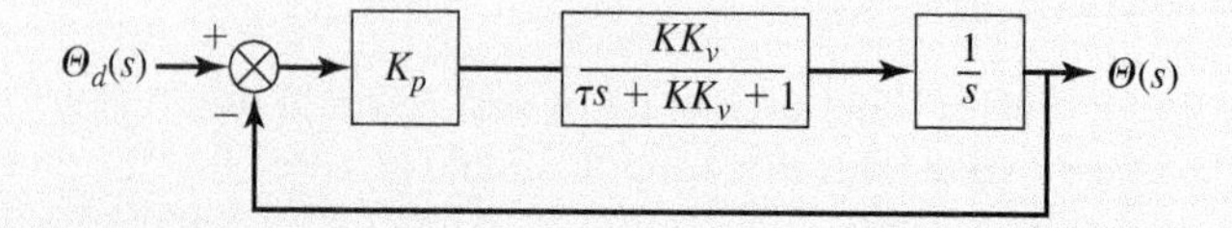

Figure 10.8 Simplified block diagram of control system

The overall transfer function is:

$$G_{Overall}(s) = \frac{\Theta(s)}{\Theta_d(s)} = \frac{K_p \frac{KK_v}{\tau s + KK_v + 1} \frac{1}{s}}{1 + K_p \frac{KK_v}{\tau s + KK_v + 1} \frac{1}{s}} = \frac{K_p K K_v}{\tau s^2 + (KK_v + 1)s + K_p K K_v} \quad (2)$$

The transfer function is a second order and represents the dynamics between the desired position $\Theta_d(s)$ and the actual position $\Theta(s)$.

10.5 On-Off Controller

The **on-off controller** (▶ available—V10.4) is a simple but nonlinear controller in which the controller is either on or off. A heating/cooling system controlled by a thermostat is an example of an on-off controller. Chapter 6 illustrates the simulation of such a controller in discussing the implementation of a digital thermostat state-transition diagram in software using different computing platforms. An on-off controller can be easily implemented in a PC or MCU using a combination of a digital output line and a relay or a transistor to turn on/off the control voltage to the actuator or heater. Let us define:

$e(t)$: the error signal at time t, which is the difference between the reference or setpoint and the actual or measured variable.

$u(t)$: the controller output at time t.

C_{on}: the output value when the controller is on.

C_{off}: the output value when the controller is off.

Δ: the half-width of the error band

The on-off controller can then be written in equation form as:

$$u(t) = \begin{cases} C_{\text{on}} \text{ if } e(t) > \Delta \\ C_{\text{off}} \text{ if } e(t) < -\Delta \\ \text{last } u(t) \text{ otherwise} \end{cases} \quad \textbf{(10.4)}$$

In Equation (10.4), a dead band is used where the controller retains its last value if the error is within the error band. This is done to reduce the frequent switching of the actuator. Because the on-off controller is nonlinear (the output is not proportional to the error but is fixed at either one of two values), traditional linear analysis, including the concept of a transfer function, is not applicable to such systems. However, we can simulate the behavior of the controller using MATLAB/Simulink. Figure 10.9 shows a simulation of a heated plate in which the controller output is either 0 or 12 V (the model of this system is discussed in Section 11.4.5) and an error band of 0.2 degrees is used. The on-off controller was implemented as a Simulink *relay block* with a sample time of 1 second. Figure 10.9(a) shows the temperature response of the plate for a 10° step change in temperature, as well as the output from the on-off controller. As seen in the plot, when the plate temperature was below the desired temperature, the controller was fully on. After the plate reaches the desired temperature at $t = 148$ s, the on-off controller alternates between being fully on and off. Since the controller cannot supply any cooling, the off interval is longer than the on interval, as seen in Figure 10.9(b).

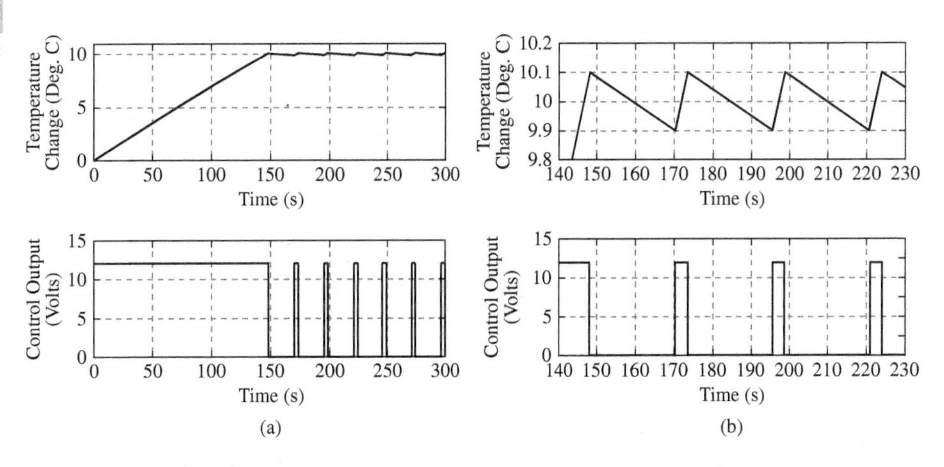

Figure 10.9

Simulink simulation of an on-off controller for the heater system considered in Section 11.4: (a) Response and controller output and (b) detailed view of response and controller output

Despite their simplicity and applicability in certain systems, on-off controllers come with several limitations. One of the most notable limitations of on-off controllers is that they can lead to cycling or oscillations around the setpoint. This happens because the controller will frequently switch the actuator on and off as the process variable crosses the setpoint. Furthermore, some on-off controllers incorporate a dead zone or hysteresis to prevent rapid switching, such as the one used in the simulation to generate Figure 10.9. While this can mitigate cycling, it introduces a zone where the controller does not take any action, leading to less accurate control. Another issue is that in systems with significant time delay or too aggressive actuation, an on-off controller can introduce instability. Despite these limitations, on-off controllers are still widely used due to their simplicity, low cost, and effectiveness in certain applications where the limitations are not critical, such as household thermostats and refrigerators. However, in industrial settings or applications requiring more precision, more advanced control strategies like proportional-integral-derivative control are often favored. The next section discusses the PID controller.

10.6 PID Controller Introduction

The proportional, integral, derivative (PID) controller (▶ available—V10.5) is one of the most widely used controllers. More than 90% of all industrial controllers are implemented using this popular controller [46]. In continuous time, the controller takes the form:

(10.5)
$$Y(t) = K_p e(t) + K_i \int e(t)dt + K_d \frac{d}{dt} e(t)$$

As seen in Equation (10.5), the output of the controller is proportional to the error signal (P-term), the integral of the error signal (I-term), and the derivative of the error signal (D-term) through the gains K_p, K_i, and K_d, respectively. The transfer function of the PID controller is given by Equation (10.6).

(10.6)
$$G_c(s) = \frac{Y(s)}{E(s)} = \left(K_p + \frac{K_i}{s} + K_d s\right)$$

In practice, variations of the above control law are also implemented, such as a **P-control**, which has only the P-term, a **PI-control**, which has only the P and I terms, or a **PD-control**, which has only the P and D terms. To illustrate the effect of different control actions, we will consider in the next two sections the control of first- and second-order systems using variations of this control law. This will include speed and position control of a rotational mass, as shown in Figure 10.10.

Figure 10.10

Simple rotating mass model

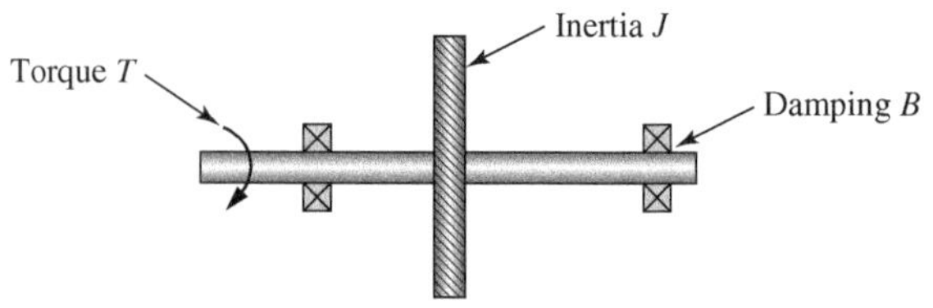

This model can represent the dynamics of a rotational mass driven by a PM DC motor which has a small inductance (see Example 10.4). In this case, the dynamics of the electrical parts of the motor can be neglected, and only the dynamics of the mechanical part need to be considered. The equation of motion of this system is given by

(10.7)
$$T = J\ddot{\theta} + B\dot{\theta}$$

where θ is the angular position of the mass, J is the inertia of the rotating parts, and B is the viscous damping coefficient. The transfer function between the torque input, $T(t)$, and the rotational speed, $\omega(t)$, is given by

(10.8)
$$\frac{\Omega(s)}{T(s)} = \frac{1}{Js + B} = \frac{1/B}{J/Bs + 1} = \frac{K}{\tau s + 1}$$

Note in the above form, τ is the time constant (see Section 7.2) for this first-order system and is given by the ratio of J to B. When this first-order system is subjected to a unit torque step input, the Laplace Transform of the output speed is given by

$$\Omega(s) = \frac{K}{\tau s + 1}\frac{1}{s} \tag{10.9}$$

In the time domain, the output is given by Equation (10.10) as

$$\omega(t) = K(1 - e^{-t/\tau}) \tag{10.10}$$

A plot of this equation is shown in Figure 10.11 for a system with $J = 0.7$ kg m^2 and $B = 0.5$ N · m s/rad. These same parameters are used for all the simulations shown in this chapter.

Figure 10.11

Open-loop unit step response of a rotational mass

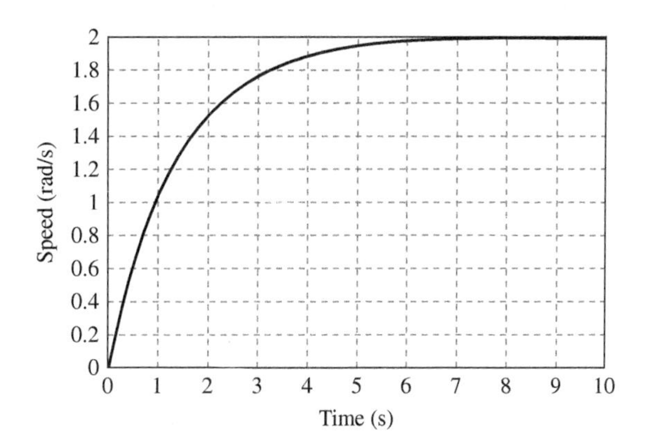

The final steady value of this system is a function of the parameter K, which is equal to K (2 in this case) for a unit step input. From differentiating the expression for the speed and evaluating it at time zero, we see that the slope of the speed response curve at time zero is the ratio of K to τ. This fact can be used to experimentally determine the parameters K and τ of a first-order system from a step response plot (see Integrated Case Study III.C in Section 7.2).

10.7 PID-Control of a First-Order System

This section will look at the application of P, PI, PD, and PID controllers (▶ available—V10.6) to the control of a first-order system. The system that we will consider is given by the transfer function in Equation (10.8).

10.7.1 P-Control of a First-Order System

The block diagram when the simple rotational mass system is placed under P-control is shown in Figure 10.12. The closed-loop transfer function for the overall system is $K_p/(Js + B + K_p)$ and is still a first-order system.

Figure 10.12

Block diagram of a first-order system with P-control

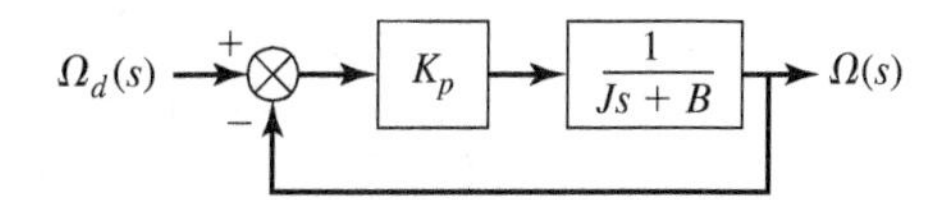

Under a unit step input and $K_p = 3$, the output of this closed-loop system is shown in Figure 10.13. As seen in the figure, the steady-state value does not reach the reference input of 1, and the system has a steady-state error of about 0.14 rad/s.

Figure 10.13

Speed response with P-control ($K_p = 3$)

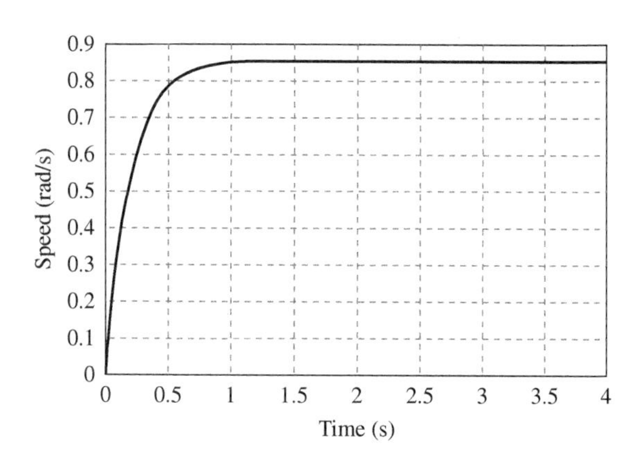

The value of this steady-state error can be obtained by first applying the **Final Value Theorem**, which is one of the properties of the Laplace Transform (see Appendix C) which allows us to directly obtain the steady-state or final value of a given transfer function in the time domain as time tends to infinity without obtaining the inverse Laplace transform. This gives

(10.11)
$$\omega(t)_{t\to\infty} = \lim_{s\to 0} s\Omega(s) = \lim_{s\to 0} s\frac{K_p}{Js + B + K_p}\frac{1}{s} = \frac{K_p}{B + K_p}$$

Subtracting this final output value from the unit reference input yields the steady-state error, as depicted in Equation (10.12). As indicated by Equation (10.12), the steady-state error approaches zero when the gain K_p becomes extremely large. However, attaining such elevated gain values might not be practical or feasible in real-world scenarios. While a P controller does not result in a zero steady-state error, it has the advantage that it does not produce oscillation when controlling a first-order system.

(10.12)
$$e(t)_{t\to\infty} = 1 - \frac{K_p}{B + K_p} > 0$$

Example 10.3 illustrates the selection of the K_p gain in a P-control, while Example 10.4 illustrates the modeling of a PM brush DC motor under P-control.

Example 10.3 Selection of the K_p Gain in a P-Control

For the first-order model given by Equation (10.8), with $J = 0.7$ and $B = 0.5$, select the gain K_p for each of the following cases:

(a) The steady-state error is less than 5% of the reference value.
(b) The time constant of the closed-loop system is 1/5 that of the open-loop system.

Solution:

(a) The expression for the steady-steady error is given by Equation (10.12). Setting the error e as 0.05 for a unit input, and using the given value of B, we get

$$1 - \frac{K_p}{0.5 + K_p} < 0.05 \tag{1}$$

Solving Equation (1) for K_p, we get that K_p needs to be greater than 9.5 to reduce the steady-state error to less than 5%.

(b) The closed-loop transfer function of the system under P-control is given by

$$\frac{K_p}{Js + B + K_p} \tag{2}$$

Equation (2) can be written as

$$\frac{K_p}{Js + B + K_p} = \frac{\frac{K_p}{B + K_p}}{\frac{J}{B + K_p}s + 1} = \frac{K_2}{\tau_c s + 1} \tag{3}$$

Thus, the time constant τ_c of the closed-loop system is $J/(B + K_p)$. From Equation (10.8), the time constant for the open-loop system is J/B. Setting $\tau_c = 1/5 J/B$, and solving for K_p, we get $K_p = 4B = 2$. Note that for any gain $K_p > 0$, the closed-loop system will have a faster response (or smaller time constant) than the open-loop system.

Example 10.4 P-Control of a PM Brush DC-Motor

Develop a block diagram model of a speed P-control system that uses a PM brush DC motor to drive the rotating mass system shown in Figure 10.10. An amplifier with a gain K_a volt/volt amplifies the voltage signal sent to the motor. The shaft speed is obtained from a tachometer with a sensitivity of k_{tach} volts/rpm.

Solution:

A block diagram of the components of this system is shown in Figure 10.14. The inductance of the PM brush DC motor is neglected here. The PM brush DC motor characteristics are given in Equations (9.2)-(9.5). The diagram has two loops: an inner loop to represent the PM DC motor characteristics; and an outer loop to represent the P-control action. The shaft speed measured by the tachometer is compared against the reference speed $\Omega_{desired}$, which should be in voltage units, to obtain the error voltage V_e. The factor $60/2\pi$ in the outer loop is a conversion factor from rad/s to rpm units. The error voltage is multiplied by the gain K_p (with units of volt/volt) to obtain the input voltage V_{in} that is sent to the amplifier. The amplifier output voltage V_m is sent to the motor. The back EMF voltage generated due to motor shaft rotation is subtracted from V_m, and the net voltage is divided by the motor resistance R to get the armature current I_a. The armature current is multiplied by the motor torque constant K_T to get the torque T_m sent to the rotating system.

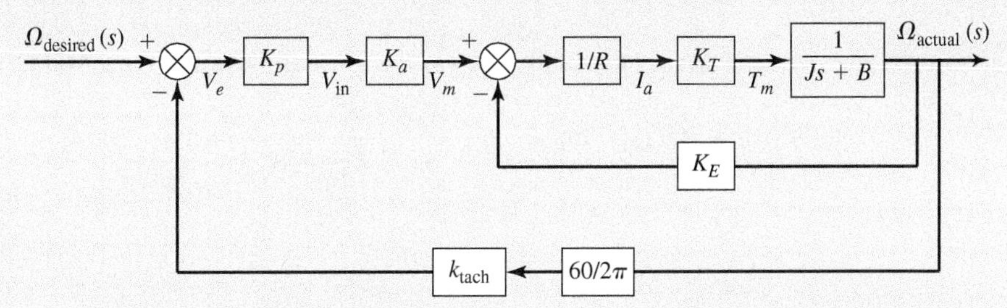

Figure 10.14 Block diagram of the components of the system

The inner loop in the above diagram can be represented by the following transfer function:

$$\frac{\Omega_{actual}(s)}{V_m(s)} = \frac{K_T}{RJs + RB + K_T K_E}$$

This transfer function has the same structure as that of the rotating mass transfer function given in Equation (10.8) but with different parameter values. Thus, one can see that using the model given by Equation (10.8) in the control analysis done in this chapter gives the same information as using a more detailed model such as the one above.

10.7.2 PI-Control of a First-Order System

From the previous section, we found that the P-control was not able to eliminate the steady-state error. In this section, we will examine the PI-control and see how it handles the steady-state error. A block diagram of the system with PI-control is shown in Figure 10.15.

Figure 10.15

Block diagram of a first-order system with PI-control

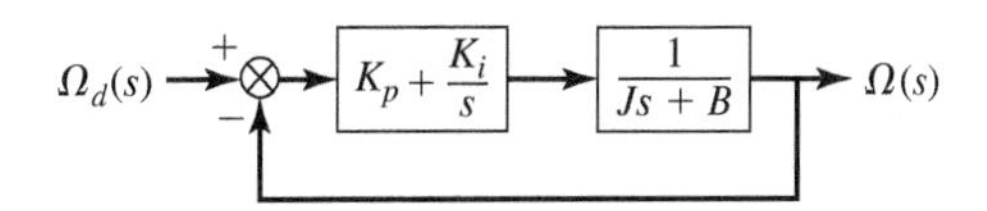

The closed-loop transfer function for the overall system with PI-control can be determined using Equation (10.3), or

$$G_{closed}(s) = \frac{\Omega(s)}{\Omega_d(s)} = \frac{\frac{K_p s + K_i}{s}\frac{1}{Js + B}}{1 + \frac{K_p s + K_i}{s}\frac{1}{Js + B}} = \frac{K_p s + K_i}{s(Js + B) + K_p s + K_i} = \frac{K_p s + K_i}{Js^2 + Bs + K_p s + K_i} \quad \textbf{(10.13)}$$

Notice how the system is now of second order. This means that the response could exhibit oscillation depending on the value of the parameters of the system (see Section 7.3). The speed response of the system for a unit step input is shown in Figure 10.16 for two values of K_i and shows no steady-state error.

Figure 10.16

Speed response with PI-control

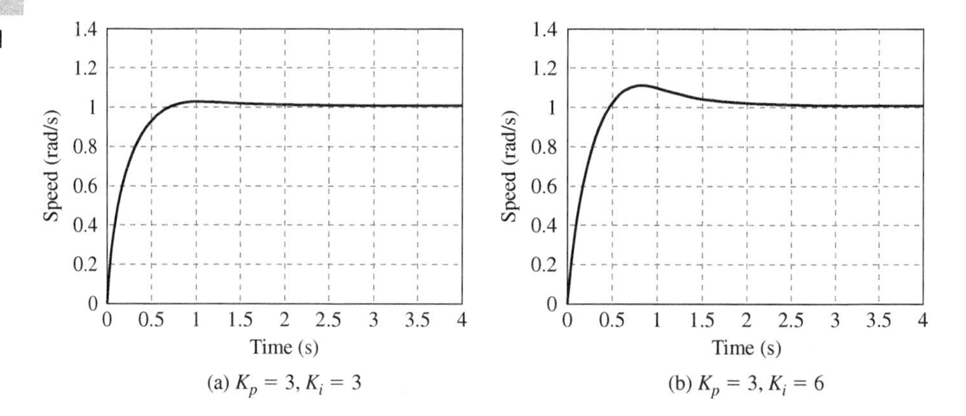

This is also confirmed by determining the final steady-state error as shown in Equation (10.14).

(10.14)
$$\omega(t)_{t\to\infty} = \lim_{s\to 0} s\Omega(s) = \lim_{s\to 0} s\frac{K_p s + K_i}{Js^2 + Bs + K_p s + K_i}\frac{1}{s} = 1$$

So

$$e(t)_{t\to\infty} = 1 - 1 = 0$$

Unlike the P-control, adding the I-term to the controller results in an elimination of the steady-state error, but at the same time increases the order of the given system by one. In PI-control, two control gains (K_p and K_i) need to be selected, and Example 10.5 illustrates the selection of these gains to satisfy a given criterion.

Example 10.5 Design of a PI Speed Controller

The model given by Equation (10.8) is placed under a PI speed controller. Determine the proportional and integral gains of the PI controller such that the controlled system has a critical damping ratio and a desired time constant of τ_d.

Solution:

The closed-loop transfer function of the model given by Equation (10.8) under a PI-control was given in Equation (10.13). The characteristic equation, which is second order, is

$$Js^2 + (K_p + B)s + K_i = 0 \quad (1)$$

The above characteristic equation can be equated to the characteristic equation of a second-order system that is given in Section 7.3 (see Equation (7.7)). This gives

$$s^2 + \frac{(K_p + B)}{J}s + \frac{K_i}{J} = s^2 + 2\zeta\omega_n s + \omega_n^2 \quad (2)$$

For critical damping, $\zeta = 1$. From Table 7.1, we know that the time constant and the natural frequency of a second-order system are related by

$$\tau = \frac{1}{\zeta\omega_n} \quad \text{for} \quad \zeta \le 1 \quad (3)$$

Using Equations (2) and (3), the gains K_p and K_i are then obtained as

$$K_p = \frac{2J}{\tau_d} - B = B\left(2\frac{\tau}{\tau_d} - 1\right) \quad (4)$$

and

$$K_i = \frac{J}{\tau_d^2} \quad (5)$$

where $\tau = J/B$. For $J = 0.7$, $B = 0.5$, and $\tau_d = \tau/4$, K_p is 3.5 and K_i is 5.7. Figure 10.17 is the step response of the system using these gains.

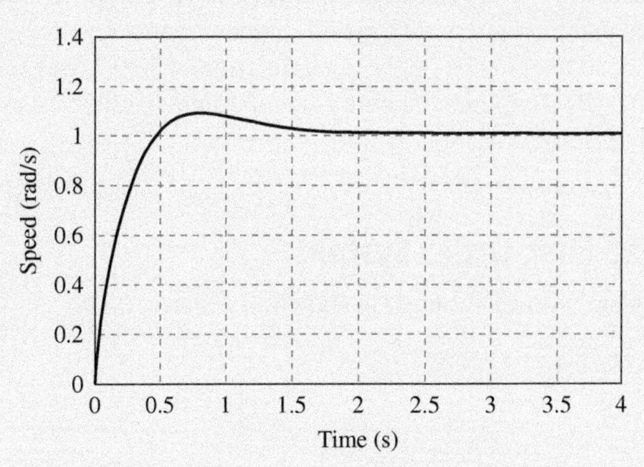

Figure 10.17 Step response of system using gains determined from Equations (4) and (5)

Note that a typical second-order system with no numerator dynamics such as that given by Equation (7.14) will have no overshoot if it is critically damped unlike the response shown in Figure 10.17. However, because our closed-loop transfer function has the term $K_p s + K_i$ in the numerator, the obtained response of our system is slightly different than that of the typical second-order system.

In real systems, **disturbances,** which are unwanted inputs that act on a control system and can affect its performance, are present. Examples of disturbances include unmodeled dynamics of the system, wind gusts acting on a vehicle, sudden external forces acting on a mechanical system, or fluctuations in the main voltage supply in an electrical system. In a closed-loop feedback system, a disturbance is modeled as an additional input that acts on the plant under control. Figure 10.18 shows the PI-control system of Figure 10.15 with an input disturbance added. To determine if a controller can eliminate disturbances acting on the system, we will determine the transfer function between the system's output and the disturbance and use that expression to obtain the steady-state output value due to the disturbance.

Figure 10.18

Block diagram of a first-order system with PI-control with an added disturbance input

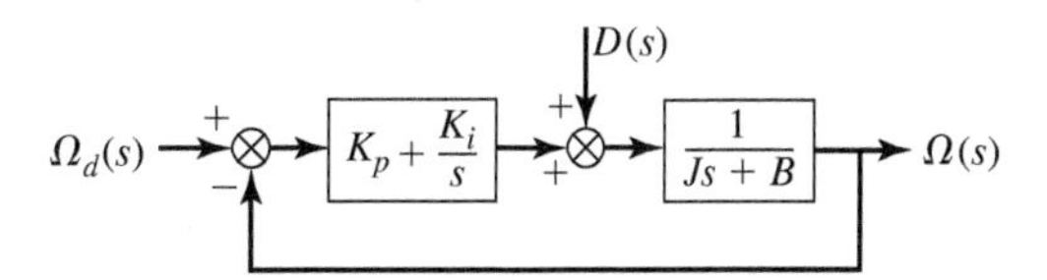

Using block diagram rules, the transfer function between the speed $\Omega(s)$ and the disturbance $D(s)$ is:

$$\frac{\Omega(s)}{D(s)} = \frac{\dfrac{1}{Js + B}}{1 + \dfrac{K_p s + K_i}{s}\dfrac{1}{Js + B}} = \frac{s}{s(Js + B) + K_p s + K_i} = \frac{s}{Js^2 + Bs + K_p s + K_i} \qquad \textbf{(10.15)}$$

Assuming a unit step disturbance input ($D(s) = 1/\mathrm{s}$), the steady-state output due to the disturbance is given by Equation (10.16) and is zero.

$$\omega(t) = \lim_{t\to\infty} s\Omega(s) = \lim_{s\to 0} s\frac{s}{Js^2 + Bs + K_p s + K_i}\frac{1}{s} = 0 \qquad \textbf{(10.16)}$$

This means that the PI-control is robust, and constant disturbances do not affect the output of the system at steady state. While the specific details are not presented here, it is important to note that the steady-state output of a first-order system under P-control is influenced by disturbances. Consequently, PI-control offers enhanced disturbance rejection capability compared to P-control, especially at steady state. It should be noted that using an integral (I-action) alone in control systems is not recommended, as it can lead to slow response due to its reliance on the accumulation of errors over time. Without the immediate correction of the proportional (P) component, a pure I-action might result in an overshoot, as it can continue adjusting even after the error is minimal.

10.7.3 PD-Control of a First-Order System

A block diagram of the system with PD-control is shown in Figure 10.19.

Figure 10.19

Block diagram of a first-order system with PD-control

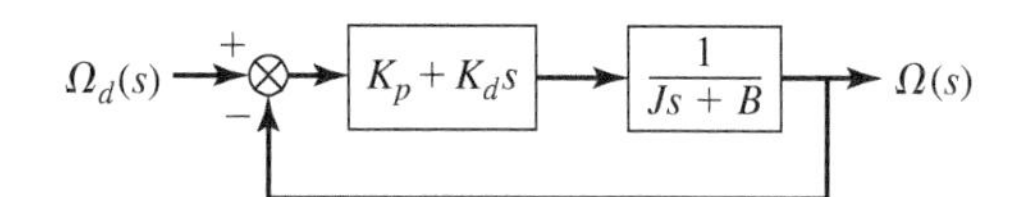

The closed-loop transfer function for the overall system with PD-control can be determined using Equation (10.3), or

$$G_{\text{closed}}(s) = \frac{\Omega(s)}{\Omega_d(s)} = \frac{(K_p + K_d s)\frac{1}{Js + B}}{1 + (K_p + K_d s)\frac{1}{Js + B}} = \frac{K_d s + K_p}{(J + K_d)s + B + K_p} \tag{10.17}$$

The order of the system has not changed with PD-control. As we have previously done for other controller types, we will use the Final Value Theorem to determine the final output under this controller as time tends to infinity. This gives

$$\omega(t) = \lim_{t \to \infty} s\Omega(s) = \lim_{s \to 0} s\frac{K_d s + K_p}{(J + K_d)s + B + K_p}\frac{1}{s} = \frac{K_p}{B + K_p} \tag{10.18}$$

This is the same expression that we obtained in Equation (10.11) for the final steady-state value with P-control, which means that adding the D-action does not affect the steady-state value. Similar to P-control, PD-control is also not robust to disturbances. Unlike a P-control, a PD-control includes a term that is proportional to the rate of change of the error. A D-action is normally added to reduce overshoot and improve the transient response. However, adding a derivative action might not be necessary for controlling a first-order system, since first-order systems already exhibit a naturally damped behavior with no overshoot. The derivative action is often more valuable for higher-order systems where oscillations and overshoots are more prevalent. This will be illustrated when we discuss the control of second-order systems in Section 10.8.3. It should also be noted that derivative action should not be used by itself because it does not reduce the error, only the rate of change of it.

10.7.4 PID-Control of a First-Order System

Figure 10.20 shows a block diagram of the first-order system that we have been discussing with PID-control.

Figure 10.20

Block diagram of a first-order system with PID-control

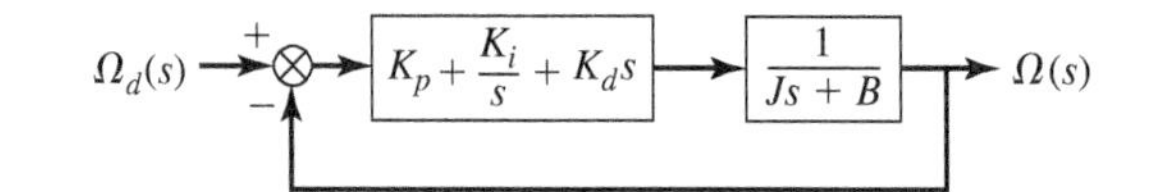

The closed-loop transfer function for the overall system with PID-control can be determined using Equation (10.3), or

$$G_{\text{closed}(s)} = \frac{\Omega(s)}{\Omega_d(s)} = \frac{\dfrac{K_p s + K_i + K_d s^2}{s}\dfrac{1}{Js + B}}{1 + \dfrac{K_p s + K_i + K_d s^2}{s}\dfrac{1}{Js + B}} = \frac{K_p s + K_i + K_d s^2}{s(Js + B) + K_p s + K_i + K_d s^2}$$

$$= \frac{K_d s^2 + K_p s + K_i}{(K_d + J)s^2 + (B + K_p)s + K_i} \qquad \textbf{(10. 19)}$$

Similar to PI-control, the closed-loop transfer function is second-order in this case. A PID-control combines the best aspects of the PI-control (zero steady error) and PD-control (reduced overshoot), but similar to PD-control, adding the D-action is not very beneficial for first-order systems, and if the gains are not chosen carefully, it may make the response worse. Figure 10.21 shows the step response of the system using the same gains that were used for the PI-control case (see Figure 10.16(a)), but with D-action added ($K_d = 0.25$). Note how the response is slower in this case, and the overshoot now lasts longer. Notice also that due to application of a step input which has a sudden change at time $t = 0$, the speed response shows a sudden jump at $t = 0$ due to the anticipatory behavior of the D-action with this first-order system. The derivative component of a PID controller reacts to the rate of change of the error. A step input, which is a sudden change, will cause a significant initial response from the derivative component.

Figure 10.21

Speed response with PID-control ($K_p = 3$, $K_i = 3$, $K_d = 0.25$)

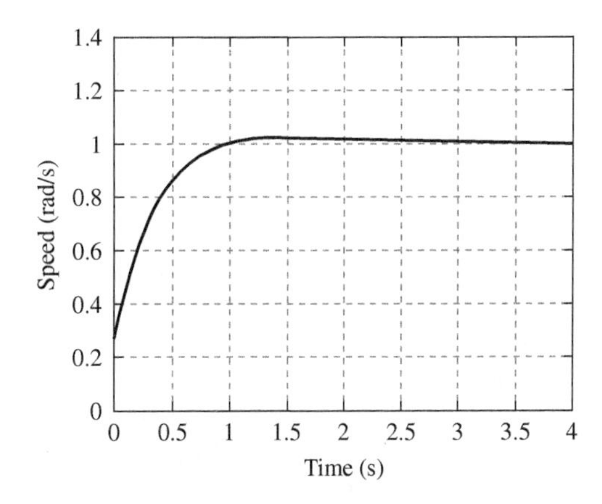

10.8 PID-Control of a Second-Order System

In this section, we will apply P, PI, PD, and PID controllers (available—V10.7) to the control of a second-order system given by Equation (10.20). This transfer function is similar in structure to the transfer function of the mass-spring-damper system considered in Section 7.4.

$$G(s) = \frac{1}{as^2 + bs + c} \qquad \textbf{(10.20)}$$

10.8.1 P-Control of a Second-Order System

A block diagram of the system with P-control is shown in Figure 10.22. The diagram shows an input disturbance that is also applied to the system.

Figure 10.22

Block diagram of a second-order system with P-control

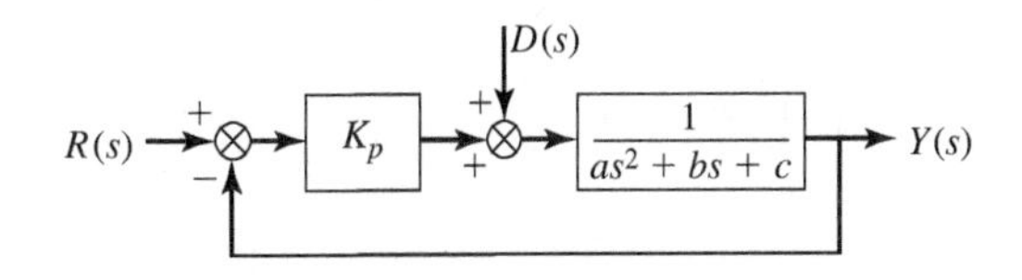

Using block diagram rules, we can determine an expression for the controlled output $Y(s)$ as a function of both the reference input $R(s)$ and an input disturbance $D(s)$. For linear systems, the response to multiple inputs can be determined by computing the response to each input separately and then summing these individual responses. For a step reference input $R(s) = R_d/s$ and a constant disturbance $D(s) = D/s$, the output is given by

$$Y(s) = \frac{K_p}{as^2 + bs + c + K_p} R(s) + \frac{1}{as^2 + bs + c + K_p} D(s) \tag{10.21}$$

So the steady-state output is

$$\underset{t \to \infty}{y(t)} = \lim_{s \to 0} sY(s) = \frac{K_p}{c + K_p} R_d + \frac{1}{c + K_p} D \tag{10.22}$$

and the steady-state error is

$$\underset{t \to \infty}{e(t)} = R_d - y(t) = \frac{c}{c + K_p} R_d - \frac{1}{c + K_p} D \tag{10.23}$$

The steady-state error here is non-zero with or without a disturbance input. However, this result does not apply to all second-order systems. If our second-order system has an integral action in its dynamics (i.e., has one root at the origin, or c in Equation (10.20) is zero), then we obtain zero steady-state error if no disturbance is applied, since the R_d term in Equation (10.23) will go to zero when $c = 0$. An example of a such a system is the rotating mass system we considered in Figure 10.10 if we control its position instead of its speed. The transfer function for this system in this case is

$$G(s) = \frac{\Theta(s)}{T(s)} = \frac{1}{s(Js + B)} \tag{10.24}$$

Since disturbances are always present in a real system, there will always be a steady error if only P-action control is used to control a second-order system of the form given in Equation (10.20) regardless of whether the parameter c is zero or not. Figure 10.23 shows the step response with P-control of the transfer function given in Equation (10.24) without and with constant disturbance ($K_p = 0.2$ and $D = 0.05$). Notice the large steady-state error when the disturbance is present.

Figure 10.23

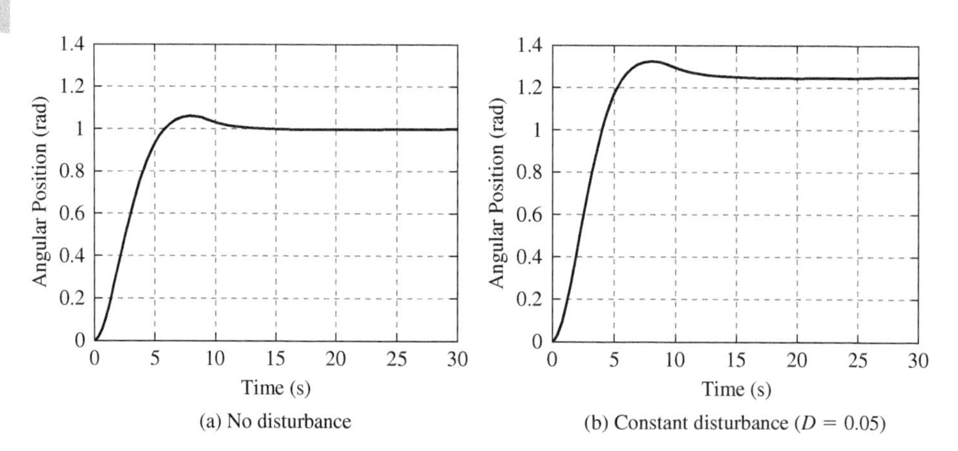

Position response of the system in Equation (10.24) under P-control without and with constant disturbance ($K_p = 0.2$)

10.8.2 PI-Control of a Second-Order System

A block diagram of the system with PI-control is shown in Figure 10.24. The diagram shows an input disturbance that is also applied to the system.

Figure 10.24

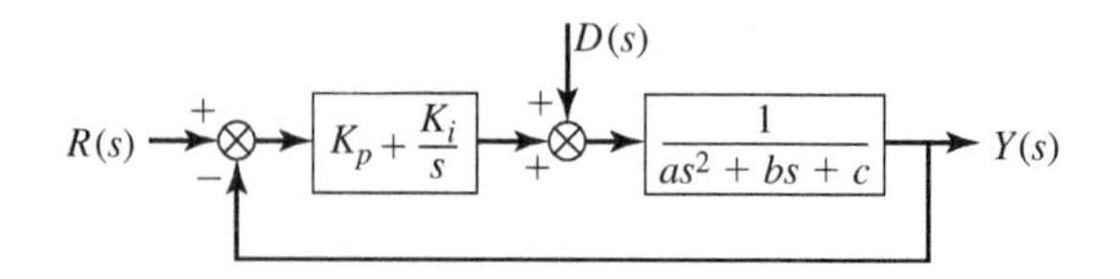

Block diagram of a second-order system with PI-control

Similar to what we did in P-control, we can determine an expression for the controlled output $Y(s)$ as a function of both the reference input $R(s)$ and an input disturbance $D(s)$. For a step reference input $R(s) = R_d/s$ and a constant disturbance $D(s) = D/s$, the output is given by

$$Y(s) = \frac{K_p s + K_i}{as^3 + bs^2 + (c + K_p)s + K_i} R(s) + \frac{s}{as^3 + bs^2 + (c + K_p)s + K_i} D(s) \quad \textbf{(10.25)}$$

So, the steady-state output is

$$\underset{t \to \infty}{y(t)} = \lim_{s \to 0} sY(s) = R_d + 0 = R_d \quad \textbf{(10.26)}$$

and the steady-state error is

$$\underset{t \to \infty}{e(t)} = R_d - y(t) = 0 \quad \textbf{(10.27)}$$

Unlike P-control, the addition of the I-action will produce a response with zero steady-state error even with the presence of disturbances and independent of whether the parameter c in Equation (10.20) is zero or not. Thus, for closed-loop control of a second-order system, a PI-control should give a response with no steady-state error. Figure 10.25 shows the step response of the system given by Equation (10.24) under closed-loop PI-control ($K_p = 0.2$, $K_i = 0.02$) without and with input disturbance present.

Figure 10.25

Position response of the system in Equation (10.24) with PI-control without and with disturbance ($K_p = 0.2$, $K_i = 0.02$, and $D = 0.05$).

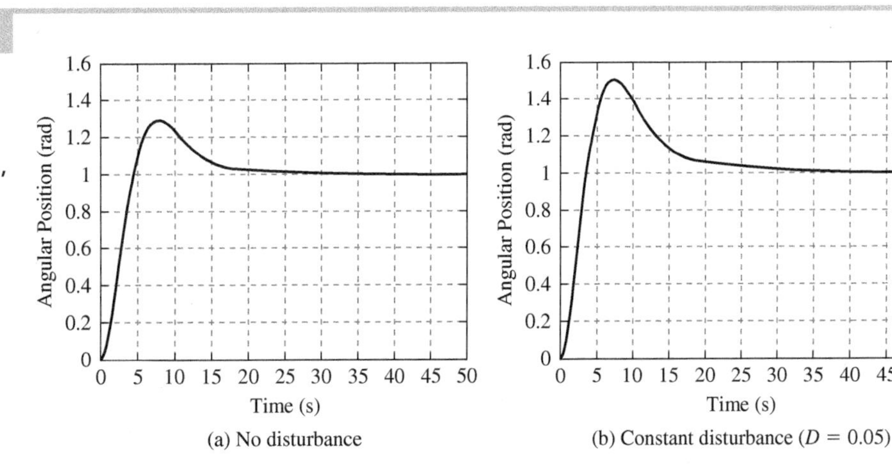

When employing PI-control with a second-order system, the resulting closed-loop transfer function is of third-order, as illustrated in Equation (10.25). Designing control gains for this third-order system can be more intricate because the direct formulas associated with second-order system characteristics (see Table 7.1), such as damping ratio or time constant, are not applicable. Analytical solutions exist but require more advanced techniques.

10.8.3 PD-Control of a Second Order System

A block diagram of the system with PD-control is shown in Figure 10.26. The diagram shows an input disturbance that is also applied to the system.

Figure 10.26

Block diagram of a second-order system with PD-control

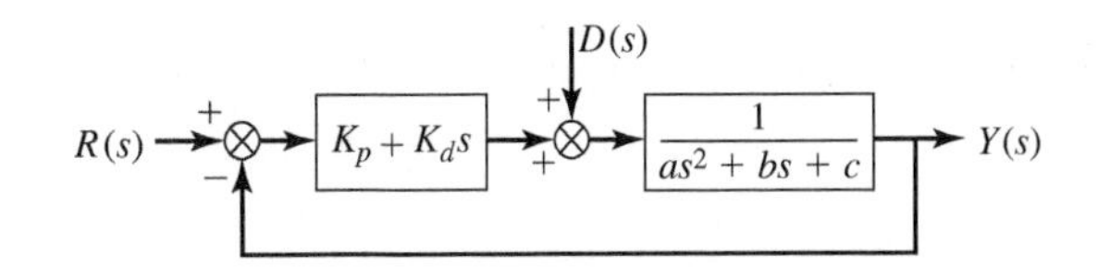

Similar to what we did in P- and PI-control, we can determine an expression for the controlled output $Y(s)$ as a function of both the reference input $R(s)$ and an input disturbance $D(s)$. For a step reference input $R(s) = R_d/s$ and a constant disturbance $D(s) = D/s$, the output is given by

$$Y(s) = \frac{K_d s + K_p}{as^2 + (b + K_d)s + c + K_p} R(s) + \frac{1}{as^2 + (b + K_d)s + c + K_p} D(s) \quad \textbf{(10.28)}$$

Similar to P-control, the order of the system under PD-control does not change. Note that the D-action gain, K_d, is now added to the coefficient b of the s-term in the denominator of the transfer function. This s-term dictates the system's damping, so adding the D-action effectively increases the system's damping. Similar to PD-control in a first-order system, the D-action in a second-order system does not influence the steady-state error. Thus, the steady-state error for PD-control of a second-order system is the same as that obtained with P-control for such a system. This result is not repeated here. Figure 10.27 shows the step response under closed-loop PD-control of the transfer function given in Equation (10.24) without and with constant disturbance ($K_p = 0.2$, $K_d = 0.15$, and $D = 0.05$). Notice how the addition of the D-action almost eliminated the overshoot (compare with Figure 10.23 which uses the same K_p gain), so the system can reach the final value faster. Notice also how the response when the disturbance is present shows a large steady-state error, similar to that of P-control, but with almost no overshoot.

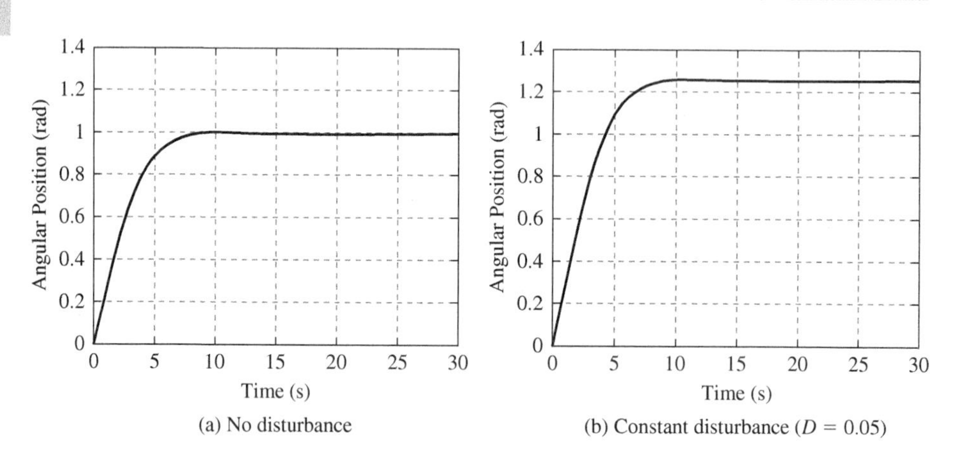

Figure 10.27

Position response of the system in Equation (10.24) with PD-control without and with constant disturbance ($K_p = 0.2$, $K_d = 0.15$, $D = 0.05$)

Example 10.6 illustrates the selection of the gains for a PD controller.

Example 10.6 Selection of Gains for PD-Control of a Second-Order System

The model given by Equation (10.24) is controlled with a PD position controller. Determine the proportional and derivative gains of the PD controller such that the controlled system has two repeated poles at the location of -1 along the negative real axis.

Solution:

For the model given by Equation (10.24) and using Equation (10.28) with $a = J = 0.7$, $b = B = 0.5$, and $c = 0$, the characteristic equation of the system with PD-control is:

$$0.7s^2 + (0.5 + K_d)s + K_p = 0 \quad (1)$$

A characteristic equation with poles $p_1 = -1$ and $p_2 = -1$ is given by:

$$(s + 1)(s + 1) = s^2 + 2s + 1 = 0 \quad (2)$$

To determine the gains K_p and K_d, we need to match the coefficients of the characteristic equations given by Equations (1) and (2). Since the coefficient of the s^2 term is different in both equations, we first multiply each term in Equation (2) by 0.7. This gives:

$$0.7s^2 + 1.4s + 0.7 = 0 \quad (3)$$

Then matching the corresponding terms between Equations (1) and (3), we get $0.5 + K_d = 1.4$, or $K_d = 0.9$, and $K_p = 0.7$. A plot of the response of this system using these gains under a unit step reference input is shown in Figure 10.28. The desired pole locations imply a critically damped response. However, due to the presence of the $K_d s + K_p$ term in the numerator, the response has a very slight overshoot. Also, since the gains in this example are larger than those used in Figure 10.27, the response is faster here.

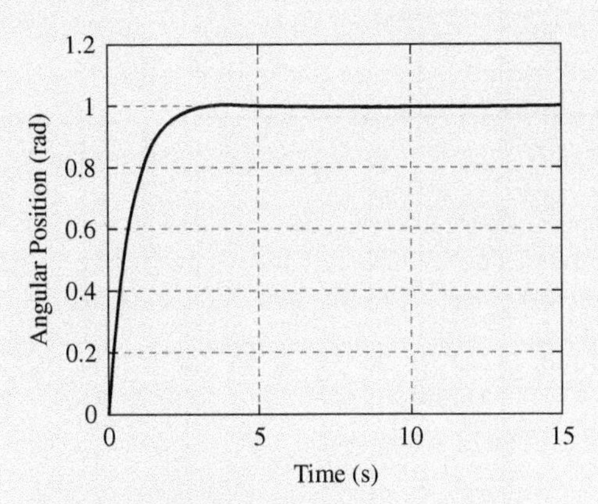

Figure 10.28 Position response of the system in Equation (10.24) with PD-control using the gains $K_p = 0.7$ and $K_d = 0.9$

10.8.4 PID-Control of a Second Order System

Figure 10.29 shows a block diagram of the second-order system that we have been discussing with PID-control.

Figure 10.29

Block diagram of a second-order system with PID-control

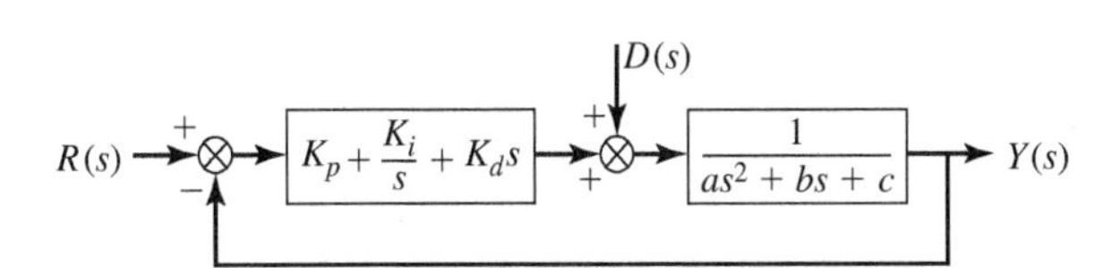

The closed-loop transfer function for the overall system with PID-control can be determined using Equation (10.3), or

$$G_{\text{closed}}(s) = \frac{Y(s)}{R(s)} = \frac{\dfrac{K_p s + K_i + K_d s^2}{s}\,\dfrac{1}{as^2 + bs + c}}{1 + \dfrac{K_p s + K_i + K_d s^2}{s}\,\dfrac{1}{as^2 + bs + c}}$$

$$= \frac{K_d s^2 + K_p s + K_i}{s(as^2 + bs + c) + K_p s + K_i + K_d s^2} = \frac{K_d s^2 + K_p s + K_i}{as^3 + (b + K_d)s^2 + (c + K_p)s + K_i} \tag{10.29}$$

The transfer function between the output $Y(s)$ and the disturbance $D(s)$ can be determined similarly to the previous control cases and is not shown here. Similar to PI-control, the closed-loop transfer function is of third-order in this case. A PID-control combines the best aspects of PI-control (zero steady error) and PD-control (reduced oscillation). Figure 10.30 shows the step response of the system given by Equation (10.24) using larger gains ($K_p = 3$, $K_i = 1$) than those used for the PI-control case (see Figure 10.25), but with D-action added ($K_d = 3$). Note how the response has zero steady-state error, and also a faster response with much less overshoot due to the addition of the D-action.

Figure 10.30

Position response of the system in Equation (10.24) with PID-control without and with disturbance ($K_p = 3$, $K_i = 1$, $K_d = 3$, and $D = 0.05$).

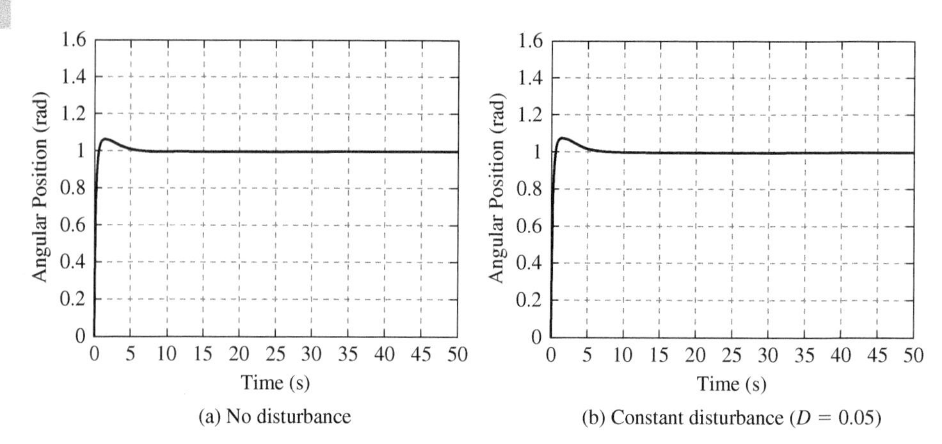

The gains used in Figure 10.30 were determined by trial and error. Example 10.7 discusses a systematic way of determining the gains of a PID controller.

Example 10.7 Selection of Gains for PID-Control of a Second-Order System

The model given by Equation (10.24) is controlled using a PID-controller. Determine the control gains of the PID controller such that the dominant two poles, or those closest to the imaginary axis, of the controlled system exhibit a critically damped response ratio and a desired time constant of 1 s.

Solution:

For the model given by Equation (10.24) and using Equation (10.29) with $a = J = 0.7$, $b = B = 0.5$, and $c = 0$, the characteristic equation of the system under PID-control is:

$$0.7s^3 + (0.5 + K_d)s^2 + K_p s + K_i = 0 \tag{1}$$

In Example 10.5, we solved a similar problem, but the characteristic equation of the system was second-order. Since a third-order characteristic equation has three poles or three roots and we are given only two specifications (damping ratio and time constant), we need to assume a location for the third root to enable us to solve this example. For the third root, we will set the location at $s = p_3 = -5$ or a time constant for that root of $1/5 = 0.2$ s. From Figure 7.16, we know that the further the location of the real part of the root on the real axis is away from the imaginary axis, the faster the response of the system due to that root (the time constant is the inverse of the real axis root distance). By setting the third root at -5, the response is then dominated by the two other roots for whom the specification is setting a time constant of 1 second. By referring to Table 7.1 for the characteristic equation of a second-order system, we can write the characteristic equation for this third-order system as

$$(s + 5)(s^2 + 2\zeta\omega_n s + \omega_n^2) = 0 \tag{2}$$

For a critically damped response, $\zeta = 1$. From Table 7.1, the time constant τ is given as $1/\zeta\omega_n$ when $\zeta \leq 1$. Substituting 1 for τ gives ω_n as 1. Substituting these values for ζ and ω_n into Equation (2) gives

$$(s + 5)(s^2 + 2s + 1) = s^3 + 7s^2 + 11s + 5 = 0 \tag{3}$$

To determine the gains K_p, K_i, and K_d, we need to match the coefficients of the characteristic equations given by Equations (1) and (3). Since the coefficient of the s^3 term is different in both equations, we first multiply each term in Equation (3) by 0.7. This gives

$$0.7s^3 + 4.9s^2 + 7.7s + 3.5 = 0 \tag{4}$$

Then matching the corresponding terms between Equations (1) and (4), we get $0.5 + K_d = 4.9$, or $K_d = 4.4$, $K_p = 7.7$, and $K_i = 3.5$. A plot of the response of this system using these gains under a unit step reference input is shown in Figure 10.31. The response is faster than that shown in Figure 10.30.

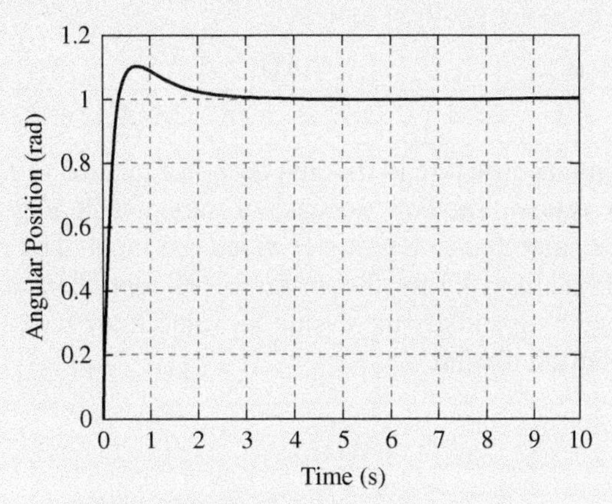

Figure 10.31 Position response with PID-control

The PID controller discussed in this section is commonly referred to as a single-loop PID controller because the feedback control system consists of only one closed-loop path. Such a controller can achieve zero steady-state error and can also reduce the overshoot in the system response, especially when applied to second-order systems. However, when controlling the position of a system, the single-loop PID controller presents several disadvantages. A significant concern is the derivative term (D-action) in the PID controller, which can amplify high-frequency noise from position measurements. This amplification can result in erratic and aggressive control actions that might not only be detrimental to the system but could also reduce the lifespan of actuators due to increased wear. Furthermore, tuning such a controller poses challenges. While Example 10.7 provided one tuning method and Section 10.9.3 will introduce another, tuning a PID controller to achieve a rapid response without overshoot in positioning systems is not easy. To address these limitations, the following section discusses the cascaded control loop structure, which finds extensive use in the servo control of positioning systems.

10.8.5 Cascaded Control Loop Structure

Figure 10.32 shows a **cascaded** or **nested control loop** structure for a positioning system consisting of an inner PI-control velocity loop and an outer P-control position loop. Utilizing cascaded control loops allows for addressing different dynamics of the system separately. This often simplifies the tuning process since it can be easier to tune two simpler loops independently than a single complex one. The inner loop, typically having a higher bandwidth or faster response, is tuned to control velocity and can respond quickly to disturbances, ensuring the desired velocity is achieved. The outer loop is more focused on ensuring the accuracy and precision of the position. Such a structure also improves disturbance rejection in the inner velocity loop, potentially enhancing system robustness and stability.

Figure 10.32

Cascaded or nested control loop structure for a positioning system

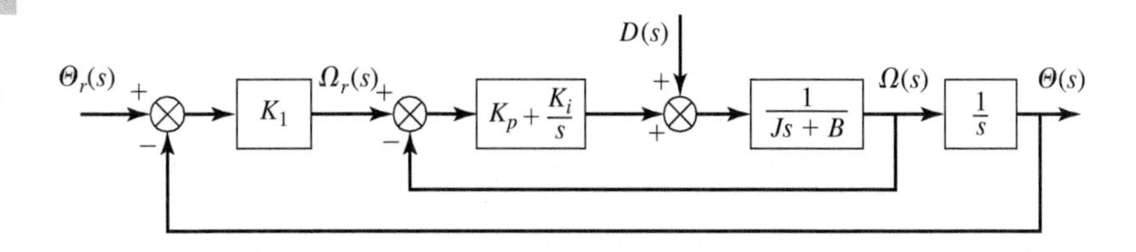

Using an approach similar to that done in Example 10.2, the transfer function for the inner velocity loop is

(10.30)
$$\frac{\Omega(s)}{\Omega_r(s)} = \frac{K_p s + K_i}{Js^2 + Bs + K_p s + K_i}$$

and the transfer function for the outer position loop is

(10.31)
$$\frac{\Theta(s)}{\Theta_r(s)} = \frac{K_1 K_p s + K_1 K_i}{Js^3 + (B + K_p)s^2 + (K_i + K_1 K_p)s + K_1 K_i}$$

Equation (10.31) has a similar structure to the first term in Equation (10.25) which was obtained with PI-control of a second-order system. To show the ease of tuning such a system, Figure 10.33 shows the response of this system when subjected to a unit step reference input. The gains K_p and K_i are the same ones that were obtained in Example 10.5 for the design of PI-control for a first-order system. The gain K_1 ($K_1 = 1.3$) was determined by gradually increasing its value from zero until the system response had the fastest response but without oscillation.

Figure 10.33

Position response of the cascaded system

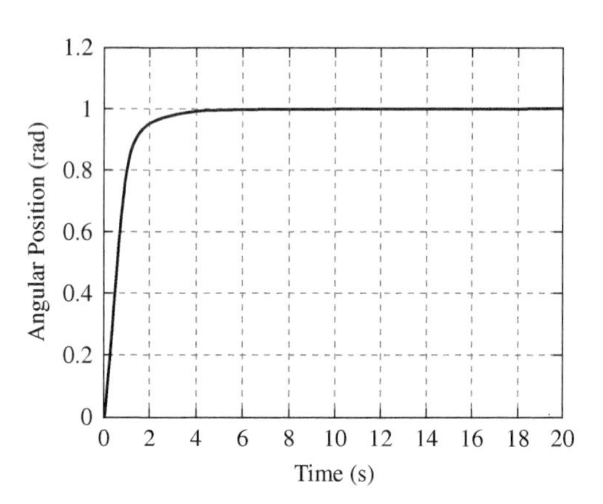

To illustrate the superior rejection capability of such a loop, Figure 10.34 shows the response of this system using the previously mentioned gains when subjected to disturbances of different magnitudes acting on the inner loop. The disturbances shown in this plot are of much greater magnitude than used before.

Figure 10.34

Position response of the cascaded system with disturbance applied to the velocity loop

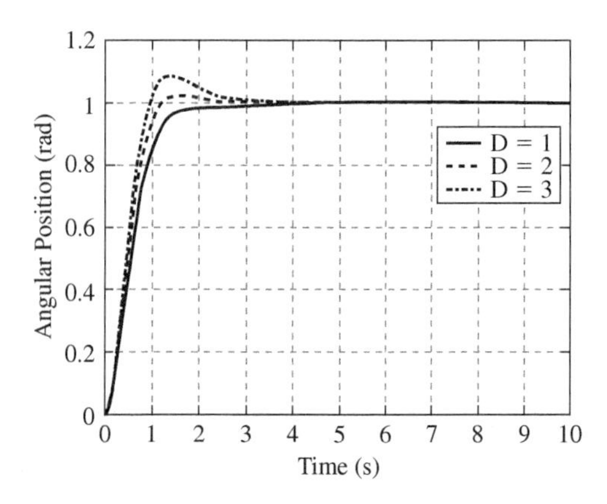

The cascaded control loop structure for a positioning system requires the availability of both the speed and position signals, which increases the cost of the system. Integrated Case Study II.D discusses the use of this cascaded control loop structure in the control of a linear motion slide system.

Integrated Case Study II.D: Closed-Loop Position Controller

In the Integrated Case Study II, which focuses on a DC Motor-Driven Linear Motion Slide (as detailed in Section 11.3), we employed a cascaded control loop structure similar to the one depicted in Figure 10.32 for managing the slide's movement. We derived a first-order dynamic model from the step response data of the actual system. This model, which is given in Equation (11.1), establishes a relationship between the motor speed and the input voltage applied to the servo drive. We used this model to calculate the gains for the inner-loop PI controller, aiming for a response that featured a damping ratio of 1 and a time constant of 0.1 second, resulting in gains of $K_p = 3.1 \times 10^{-4}$ and $K_i = 1.9 \times 10^{-3}$. Following this, we simulated the designed controller using Simulink, with the developed model illustrated in Figure 11.10. To achieve a reasonably quick response, we adjusted the outer loop gain, K_1, by incrementally increasing its value in the model. A trapezoidal profile was used for the reference position trajectory (see Figure 10.35). This profile, after being amplified by a scaling factor of 6000, directed the slide to travel a set distance at a consistent speed, pause momentarily, and then return to its starting position.

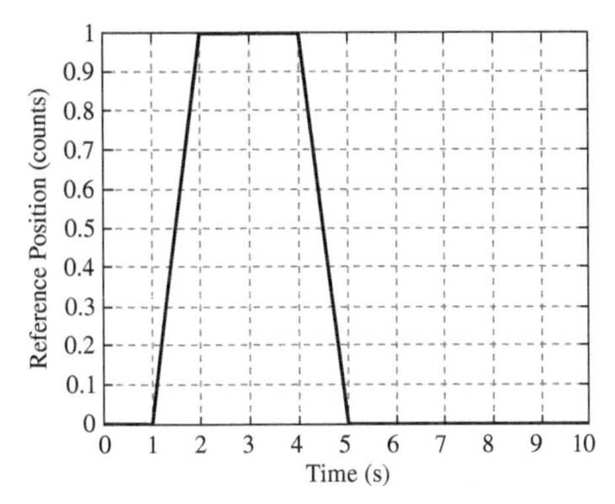

Figure 10.35 Reference position profile

The response of the simulated model with this reference trajectory and $K_1 = 10$ is shown in Figure 10.36. The plot shows a fast response with a very slight overshoot. This controller was implemented on the real system (see Section 11.3.5) using the same K_p and K_1 gains but using half the value of the K_i gain to eliminate the slight position overshoot, and the real system response (see Figure 11.12) is very similar to that of the simulated model.

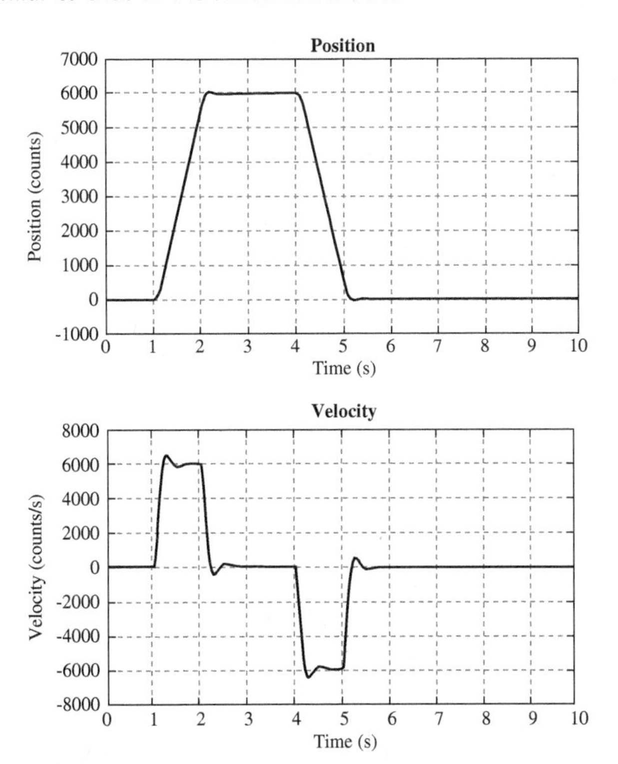

Figure 10.36 Simulated response of the slide system

10.9 Controller Design Considerations

10.9.1 Controller Stability

One important issue in the design of a feedback controller is the **stability** of the closed-loop control system. Ruth and Hurwitz [22] have established a quick method to determine the stability of a control system based on the values of the coefficients of the closed-loop characteristic equation of the system instead of solving for the roots of the characteristic equation. Table 10.1 lists these stability conditions for first-, second-, and third-order systems.

Table 10.1

Ruth-Hurwitz stability conditions

System Order	Characteristic Equation	Stability Conditions
First	$a_1 s + a_0 = 0$	Stable if and only if a_1 and a_0 have the same sign
Second	$a_2 s^2 + a_1 s + a_0 = 0$	Stable if and only if a_2, a_1, and a_0 all have the same sign
Third	$a_3 s^3 + a_2 s^2 + a_1 s + a_0 = 0$	For $a_3 > 0$, stable if and only if a_2, a_1, and a_0 all have the same sign and $a_2\, a_1 > a_3\, a_0$

As an example, let us apply these stability conditions for the characteristic equation of the transfer function of Equation (10.25) with $a = J$, $b = B$, and $c = 0$. Since the inertia and viscous damping coefficients are always positive for this third-order system, the system is stable if both K_p and K_i are positive and if $K_p/K_i > J/B$. This is the case for the parameters used to obtain the plots of Figure 10.25.

10.9.2 Controller Bandwidth

In Chapter 7, we discussed the concept of bandwidth, defined for a transfer function as the frequency at which the magnitude of the system's frequency response drops to -3 dB (or about 70.7% of its peak value). This concept plays a critical role not only in sensor selection but also in control system design. If we have the closed-loop transfer function, we can determine the system's bandwidth using its Bode plot. Figure 10.37 displays the Bode plot for the system discussed in Example 10.5. Based on the definition of bandwidth (see Section 7.5.3), this system has a bandwidth of 5.95 rad/s or 0.947 Hz. But what does this mean for system performance? Generally, a larger bandwidth can yield faster response times but might make a system more sensitive to high-frequency disturbances. On the other hand, a smaller bandwidth can slow system response but improve noise rejection.

Figure 10.37 Bode plot for the system considered in Example 10.5

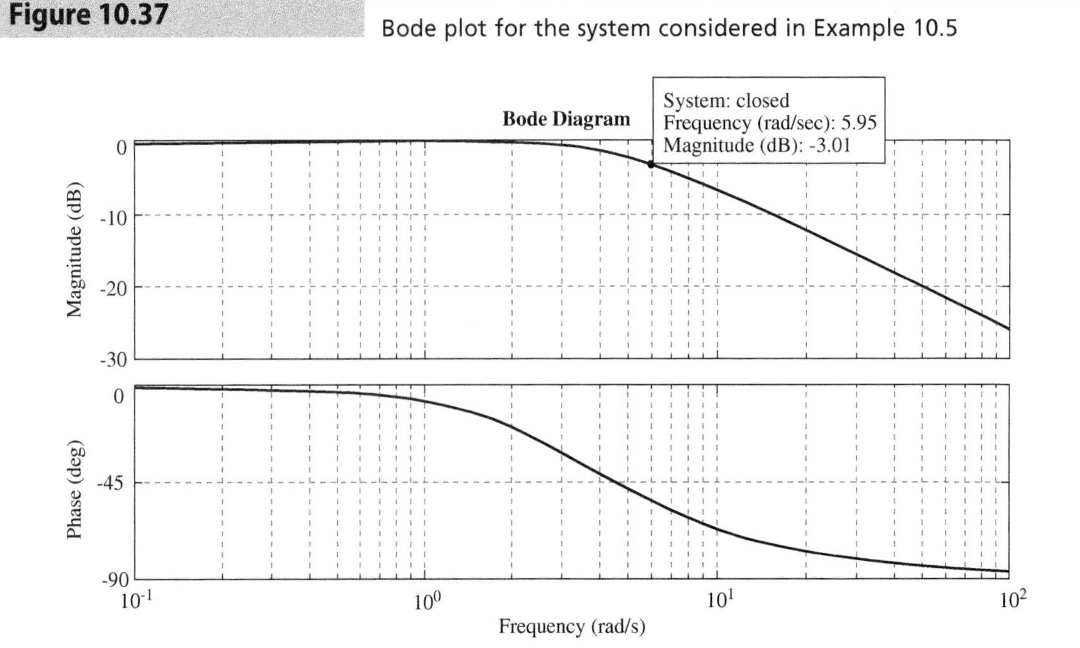

10.9.3 Controller Tuning

Tuning a feedback controller (▶ available—V10.8) involves adjusting its parameters or gains to achieve desired system performance and stability. We previously discussed techniques for adjusting these gains based on either pole placement, matching certain performance parameters such as time constant and damping ratio, or trial and error. For industrial process control applications, analytical models for the process are difficult to obtain. For these cases, it is desirable if one can tune the controller based on obtaining some experimental data from the process. The **Ziegler-Nichols method** [47], proposed by John G. Ziegler and Nathaniel B. Nichols in the 1940s, is a popular empirical technique for tuning feedback controllers, especially the Proportional-Integral-Derivative (PID) controllers. The method has formulas that provide initial settings for the proportional, integral, and derivative gain terms. After obtaining these initial settings, engineers often fine-tune the controller gains using trial-and-error, making iterative adjustments while observing the system's response.

The method offers two primary tuning rules: the Process Reaction Method and the Ultimate Sensitivity Method. The **Process Reaction Method** is an open-loop method in which a step input is applied to the system, and the time response is observed. The key parameters derived from this response are the delay time L and the slope R of a line tangent to the steepest part of the response curve, as shown in Figure 10.38. It is not possible to obtain a positive value for L for first- and second-order systems without a time delay, so this method cannot be applied to them. However, it can be applied to well-damped third-order or higher systems. For these cases, the controller gains are given in Table 10.2. When using this table, an alternative form of the PID controller is used. This is shown in Equation (10.32) where T_i is the reset time and T_d is the derivative time.

$$G_c(s) = \left(K_p + \frac{K_i}{s} + K_d s\right) = K_p\left(1 + \frac{1}{T_i s} + T_d s\right) \tag{10.32}$$

In this form, $K_i = K_p/T_i$, and $K_d = K_p T_d$.

Figure 10.38

Open-loop step response

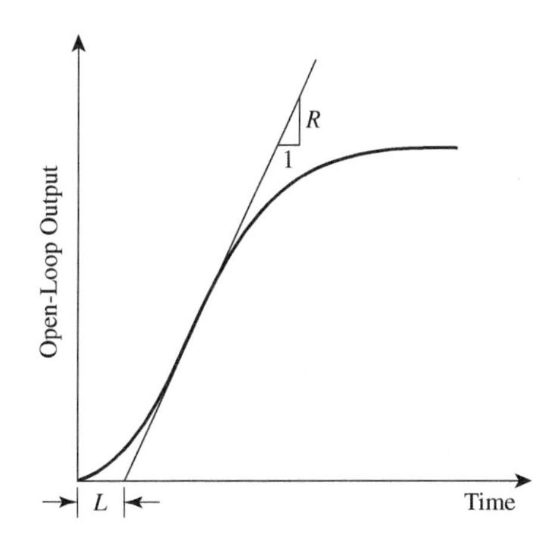

In the **Ultimate Sensitivity Method**, which is a closed-loop method, the controller is initially set to proportional-only control mode, thereby turning off the integral and derivative actions. The proportional gain, K_p, is then progressively increased until the system's output exhibits sustained oscillations. The gain at which these sustained oscillations occur is known as the system's ultimate gain, K_{Pu}. The period of these oscillations is referred to as P_u, or the ultimate period. Using K_{Pu} and P_u, prescribed formulas that provide initial tuning settings for the PID controller are shown in Table 10.2.

Table 10.2

PID controller parameters using the Ziegler-Nichols tuning method

Controller Transfer Function $G_c(s) = \left(K_p + \frac{K_i}{s} + K_d s\right) = K_p\left(1 + \frac{1}{T_i s} + T_d s\right), K_i = \frac{K_p}{T_i}, K_d = K_p T_d$		
Controller Type	**Process Reaction Method**	**Ultimate Sensitivity Method**
P-control	$K_p = 1/RL$	$K_p = 0.5K_{Pu}$
PI-control	$K_p = 0.9/RL$ $T_i = 3.3L$	$K_p = 0.45K_{Pu}$ $T_i = 0.83P_u$
PID-control	$K_p = 1.2/RL$ $T_i = 2L$ $T_d = 0.5L$	$K_p = 0.6K_{Pu}$ $T_i = 0.5P_u$ $T_d = 0.125P_u$

Example 10.8 illustrates the use of the Ultimate Sensitivity Method.

Example 10.8 Determining PID Gains Using the Ultimate Sensitivity Method

Determine the PID controller gains using the Ultimate Sensitivity Method for the system given by Equation (1) for the case $a = 2$ and $b = 4$.

$$G(s) = \frac{1}{s(s + a)(s + b)} \tag{1}$$

Solution:

To use the Ultimate Sensitivity Method, it is required to have the ultimate gain, K_{Pu}, and the oscillation period P_u. While in practice these are obtained experimentally, we can analytically determine them here from the closed-loop transfer function of the system with P-control. The closed-loop transfer function of the system given in Equation (1) with P-control is given by Equation (2).

$$G_{\text{closed}}(s) = \frac{K_p}{s(s + a)(s + b) + K_p} \tag{2}$$

A system undergoes sustained oscillation if some of its poles are located on the imaginary axis. By substituting $s = j\omega$ and $K_p = K_{Pu}$ in the denominator polynomial in Equation (2), and then separating the resulting expression into its real and imaginary parts and setting each to zero, we get the following two expressions:

$$\omega^2 = ab \tag{3}$$

$$K_{Pu} = (a + b)\omega^2 \tag{4}$$

For $a = 2$ and $b = 4$, this gives $\omega^2 = 8$ and $K_{Pu} = 48$. The period P_u is given as $2\pi/\omega$ or $\pi/\sqrt{2}$.

Substituting these values into the expressions given in Table 10.2 gives the gains of the PID controller.

For P-control, $K_p = 0.5K_{Pu} = 24$.

For PI-control, $K_p = 0.45K_{Pu} = 21.6$, $T_i = 0.83P_u = 1.84$, and $K_i = K_p/T_i = 11.7$.

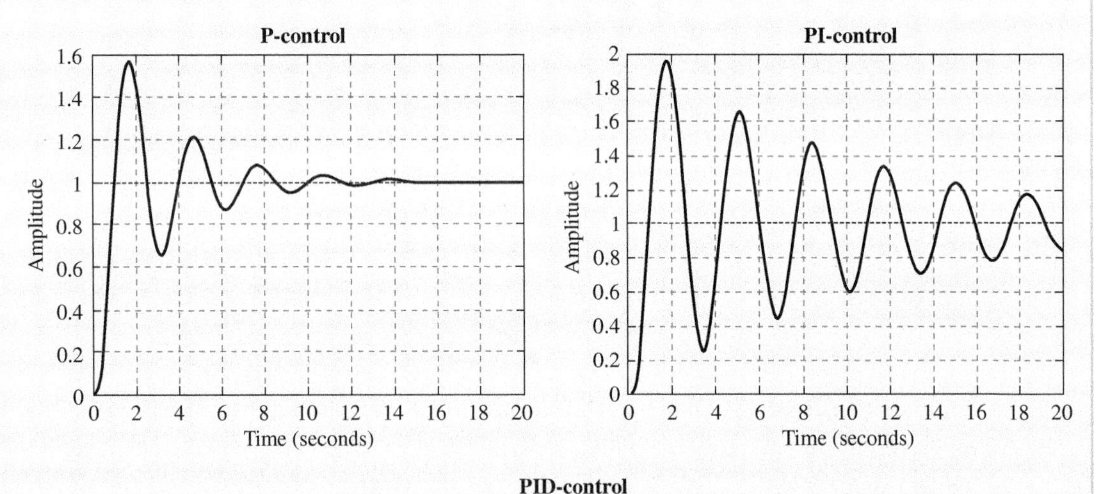

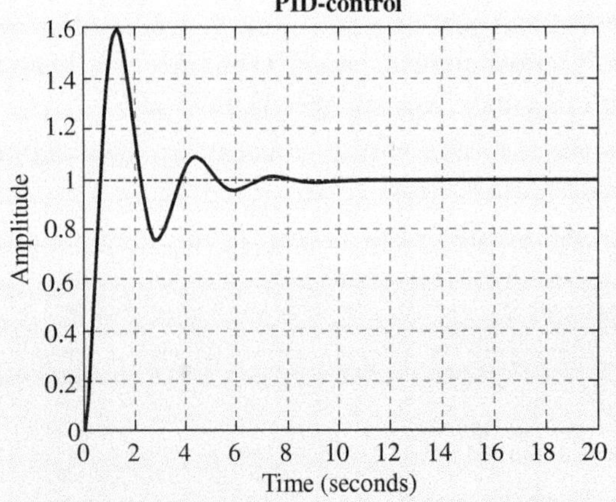

Figure 10.39 Response of the system given in Equation (1) with P-, PI-, and PID-control

For PID-control, $K_p = 0.6K_{Pu} = 28.8$, $T_i = 0.5P_u = 1.11$, $T_d = 0.125P_u = 0.278$, $K_i = K_p/T_i = 25.9$, and $K_d = K_p T_d = 8.01$.

Figure 10.39 shows the response of the system with P-, PI-, and PID-control using these gains. Note how the PI-control makes the system have more oscillations than the P-control, and how the PID-control produces the least oscillation. These gains can be further modified to adjust the response. For example, to reduce the overshoot and decrease the oscillation for the PID-control case, Figure 10.40 shows the response plot with K_i decreased by a factor of one-half and K_d increased by a factor of two.

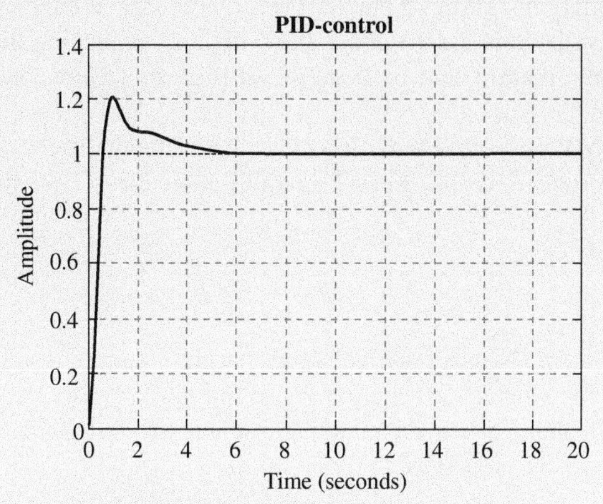

Figure 10.40 Response of the system given in Equation (1) with PID-control with gain adjustment

While the Ziegler-Nichols method provides a systematic approach to initial controller tuning, it is often just a starting point. Real-world systems, with their intricacies and nonlinearities, often require further fine-tuning to achieve the desired performance. Modern control systems might also employ optimization algorithms or software tools to automate the tuning process, ensuring optimal performance across a range of operating conditions.

10.9.4 Controller Robustness

The robustness of a feedback controller refers to its ability to maintain stability and desired performance in the presence of uncertainties in the system parameters and external disturbances. Evaluating robustness typically involves various methods. Two common techniques, especially in classical control theory, are determining the stability margins and estimating the steady-state response to a specific disturbance input. Both methods are elaborated upon below. However, in modern control theory and more complex systems, there are additional methods and tools used to assess robustness. These include singular value analysis, H-infinity control, and Monte Carlo Simulations [48], but they are not discussed in this text.

Stability Margins: Stability margins, including gain and phase margins (▶ available—V10.9), provide insight into the system's ability to maintain stability amidst changes in system parameters. They provide information on how much the gain and phase of an open-loop system can change before the closed-loop system becomes unstable if the open-loop system is placed in a unity feedback closed-loop system, as shown in Figure 10.41.

Figure 10.41

Open-loop system with unity feedback

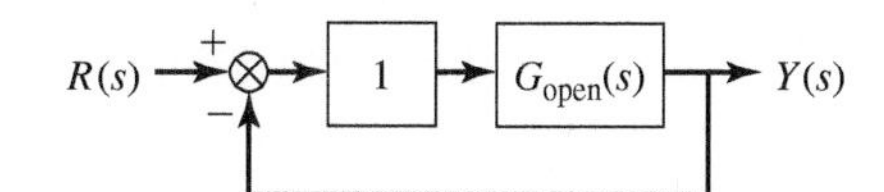

The **gain margin (GM)**, which is determined from the Bode plot of the open-loop system (see Figure 10.42), indicates how much the open-loop gain can be raised before the closed-loop system becomes unstable. Specifically, at the frequency where the phase is -180 degrees (known as the **phase crossover frequency** or **PCF**), if the gain is below the 0 dB line then the system can tolerate an increase in gain before becoming unstable. The gain margin in dB is defined by

(10.33)
$$GM = 0 - G$$

where G is the open-loop gain at the PCF. If the gain is below the 0 dB line at PCF, then the gain margin is positive. A positive gain margin signifies a stable system, with larger values pointing to greater stability and robustness. Alternatively, the gain can be above the 0 dB line, indicating the system needs a reduction in gain to be stable. The gain margin will be negative in this case. Conversely, a negative gain margin indicates an unstable system.

The **phase margin (PM)** represents the allowable phase change at the **gain crossover frequency** or **GCF** (where the magnitude is 0 dB) before instability ensues. The phase margin in degrees is defined by

(10.34)
$$PM = \phi - (-180)$$

where ϕ is the phase angle at the GCF. A positive phase margin indicates a stable system with some resilience against phase variations, whereas a negative phase margin suggests that the system is unstable. A control system designed with ample positive gain and phase margins will be more resilient against changes in the plant model.

Figure 10.42

Gain and phase margins for stable and unstable systems

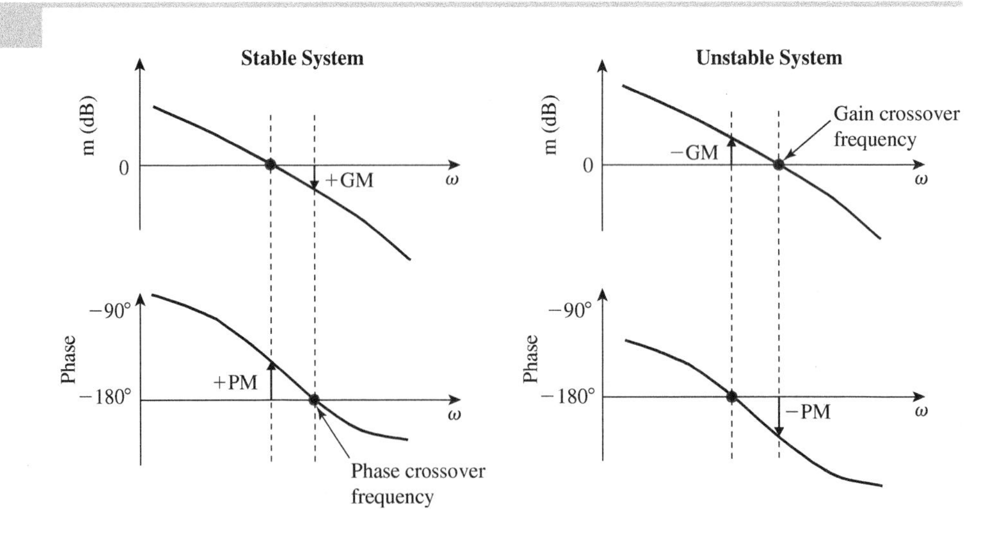

Despite their utility, stability margins have limitations. They are predicated on the system being linear and time-invariant, which can be a significant constraint since many real-world systems exhibit nonlinear or time-varying behavior. Additionally, gain and phase margins offer a conservative measure of stability, often leading to less optimal performance due to overdesign. They also focus solely on stability and do not necessarily reflect the overall performance of the system, such as how accurately or quickly the system responds to changes in input or disturbances. Furthermore, for complex or high-order systems, gain and phase margins may not provide a complete stability assessment, requiring engineers to use these tools alongside other methods to ensure both robustness and performance are adequately addressed. MATLAB offers a function to obtain the phase and gain margins. The function is called *margin*. Example 10.9 illustrates computing the gain and phase margins for a second-order system using this function.

Example 10.9 Gain and Phase Margins

Obtain the gain and phase margins for the following transfer function using the *margin* function in MATLAB.

$$G(s) = \frac{1}{s(0.7s + 0.5)}$$

Solution:

The bode plot with the gain and phase margins identified on the plot by the *margin* function is shown in Figure 10.43. This transfer function, which is the same one as that shown in Equation (10.24), has an infinite gain margin and a phase margin of 33.1 degrees. This means the system will be stable if placed in a unity feedback closed-loop control system. An infinite GM is typically the case for systems where the phase never drops to −180 degrees at any frequency (such as this second-order system), meaning there is no frequency at which the system, with unity feedback, would oscillate on the edge of stability. Hence, theoretically, the gain could be increased indefinitely without the system becoming unstable. However, in practical scenarios, there are always physical limits and nonlinearities not captured in a simple linear model that would prevent infinite gain. A positive phase margin of 33.1 degrees means that the system can tolerate additional phase delay up to this margin before it reaches the critical phase of −180 degrees at the gain crossover frequency.

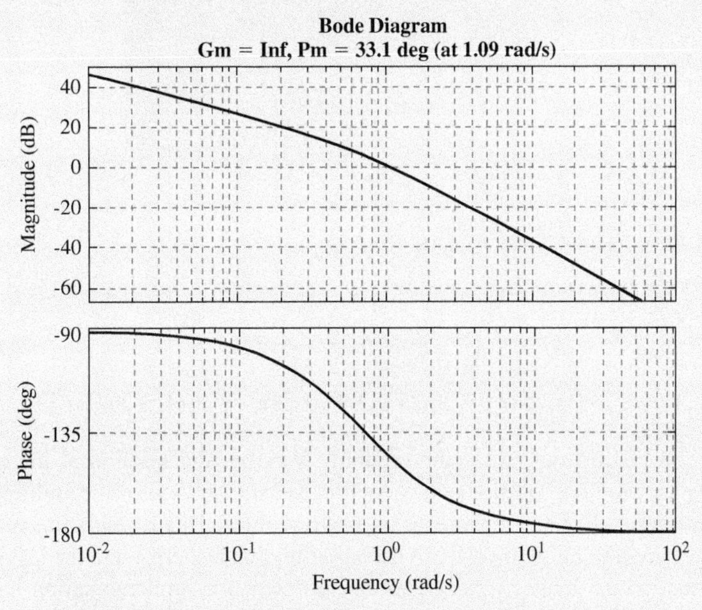

Figure 10.43 Gain and phase margin for Example 10.9

Estimating the Steady-State Response to a Specific Disturbance Input: For this technique, a predetermined disturbance, such as a step input, is introduced. By deriving the closed-loop transfer function between the disturbance and the system output, the Final Value Theorem from the Laplace Transform can be employed to estimate the system's steady-state output due to the disturbance. This is valid for systems with poles in the left half of the s-plane (stable systems). If the steady-state value is zero for a given disturbance, it implies strong disturbance rejection for that specific input, but not necessarily global robustness to all disturbances or uncertainties. This technique was illustrated in several of the examples throughout this chapter.

10.9.5 Controller Digital Implementation

Due to the discrete nature of digital control, when an analog controller such as a PID controller is implemented on a PC or a microcontroller, the controller is approximated by

$$y(kT) = K_p e(kT) + K_i T \sum_{j=0}^{k-1} e(jT) + K_d((e(kT) - e(k-1)T))/T \tag{10.35}$$

where T is the sampling interval, and k ($k = 0, 1, 2, \ldots$) is an index that represents the number of instances at which the feedback control is done. Notice how the integral for the I-action term has been replaced by a summation and the derivative for the D-action has been replaced by a difference. This summation expression comes from approximating the area under the error vs. time plot. While there are several ways to do the approximation (such as backward approximation, and trapezoidal), Equation (10.35) is based on the **forward rectangular approximation** scheme where the error value at the beginning of the sampling period is used for the height of each rectangle. This is illustrated in Figure 10.44 for example, where at $k = 3$, the sum of all the errors to that point is given by the sum of the three shaded areas.

Figure 10.44

Forward rectangular approximation

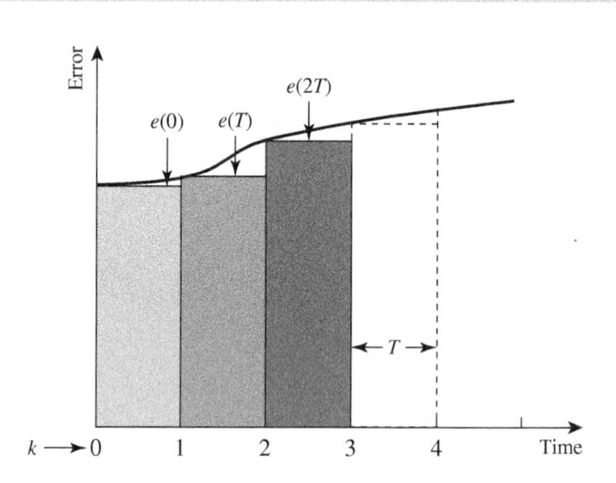

Equation (10.35) can be written in a recursive version as shown in Equation (10.36)

$$u_i(kT) = u_i((k-1)T) + K_iTe((k-1)T)$$

(10.36)
$$y(kT) = K_pe(kT) + u_i(kT) + K_d(e(kT) - e((k-1)T))/T$$

where u_i is the I-action control output. In the recursive version, the I-action at the current time step is obtained by summing the previous I-action control output and the previous error multiplied by the sampling interval. This method avoids the need to store all past error values. This formulation is easier to implement in code than the form shown in Equation (10.35) because only the last error and the last value of the integral term need to be stored, which makes it very memory-efficient for implementation in digital controllers. Figure 11.17 provides a code listing for the implementation of a slightly different version of the recursive formula for a digital PI controller using an MCU where the code continuously updates a single variable that accumulates the sum of errors over time, with the latest error being added after the current iteration. In Section 6.5.2, we discussed a control software structure for the implementation of a discrete feedback control system.

10.10 Nonlinearities

Real systems have nonlinearities (▶ available—V10.10) such as saturation and Coulomb friction. These nonlinearities cause a deviation from the ideal linear system behavior discussed before. This section will discuss the effect of these nonlinearities.

10.10.1 Saturation

In real-world systems, controller outputs have inherent limits. This limitation often arises because the amplifiers, which amplify the control signals sent from computing devices like computers or micro-controllers to the actuators, have power constraints. This phenomenon, where the system cannot exceed a specific output level, is termed **saturation,** and a plot of the **saturation nonlinearity** is shown in Figure 10.45, where u is the controller output and m is the amplifier output. When the controller's output goes beyond $u+$ or drops below $u-$, the amplifier's output saturates at its maximum m_{max} or minimum m_{min}, respectively. For certain systems, like heating systems, m_{min} is zero as heaters cannot produce negative heat.

Figure 10.45

Illustration of the saturation nonlinearity

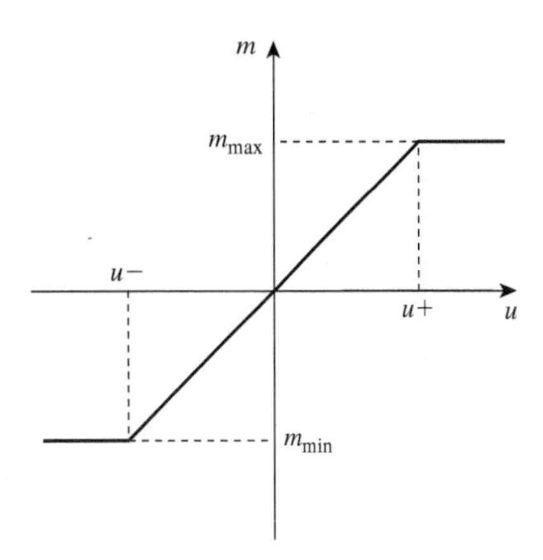

The saturation of the controller output causes the PID controller to overshoot and delay the response of the system. To illustrate this point, consider Figure 10.46 which shows a PI simulation ($K_p = 3$ and $K_i = 3$) of the rotating mass system considered earlier with no saturation limits placed on the controller output. Notice how the speed overshoot is very limited, and the speed error goes to zero just before $t = 3.4$ s. However, the controller output is initially high and stays above 1.0 N · m for approximately the first 0.4 s. The Simulink model (see Section 7.7.4) that is used to obtain this plot is shown in Figure 10.47. The model uses the continuous time *PID Controller* block in Simulink.

Figure 10.46

PI simulation with no controller output limits

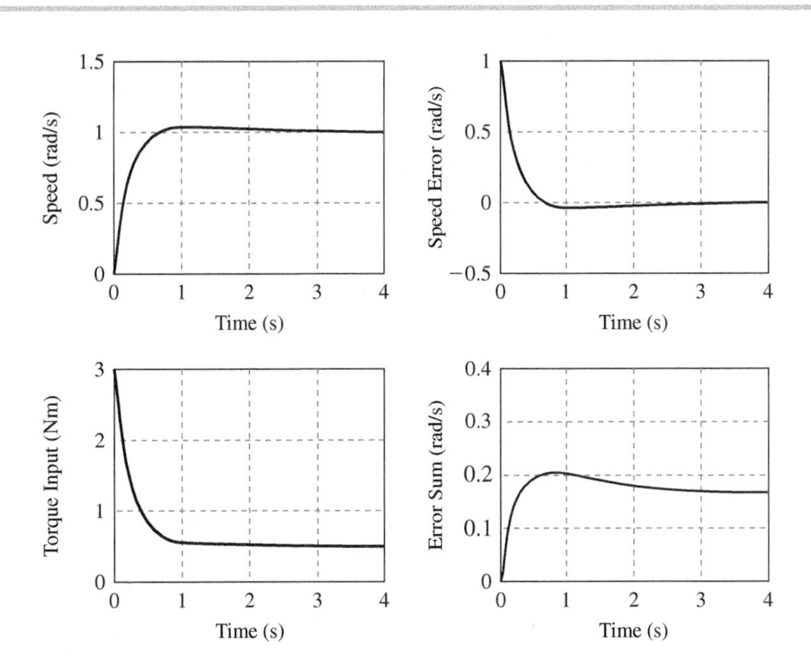

Now the PI simulation is repeated but with a limit of + /−1 N · m placed on the controller output. This simulation is obtained by specifying the *lower saturation limit* and the *upper saturation limit* in the *PID advanced* section of the block parameters for the *PID controller* block. The simulation results are shown in Figure 10.48. The controller saturation is shown in the plot of the torque input to the system

Figure 10.47

Simulink model for simulating a PI controller of a first-order system

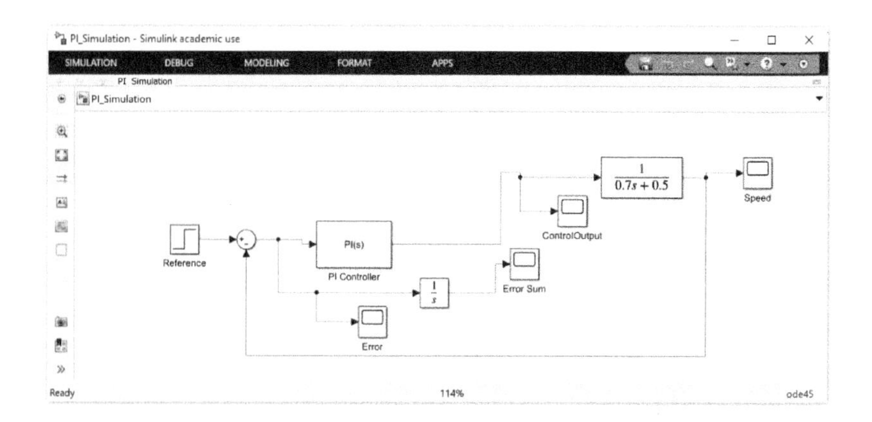

where the torque stays at the limit value for approximately the first 1.1 s. Notice the significant overshoot in the speed response plot. This overshoot is a direct result of the I-action and is explained below. Due to P-action alone, the controller output is close to 3 N·m at the beginning of the simulation, assuming zero initial conditions. Having I-action in this time interval is not useful since the additional control output from the I-action will not be utilized due to saturation of the control output. Furthermore, due to the summing nature of the I-action, the I-action term keeps increasing in value until the error switches signs and will supply a nonzero input to the system even if the error is zero. Thus, for systems that have saturation, it is better to completely shut off the I-action while the contribution from the P-action alone (or from the P- and D-action if a PID controller is implemented) exceeds the controller limit. This behavior of the PI controller is called **reset windup** or **integrator buildup** and can occur with any controller with integrator action and saturation.

Figure 10.48

PI simulation with a limit of +/−1 N·m placed on the controller output

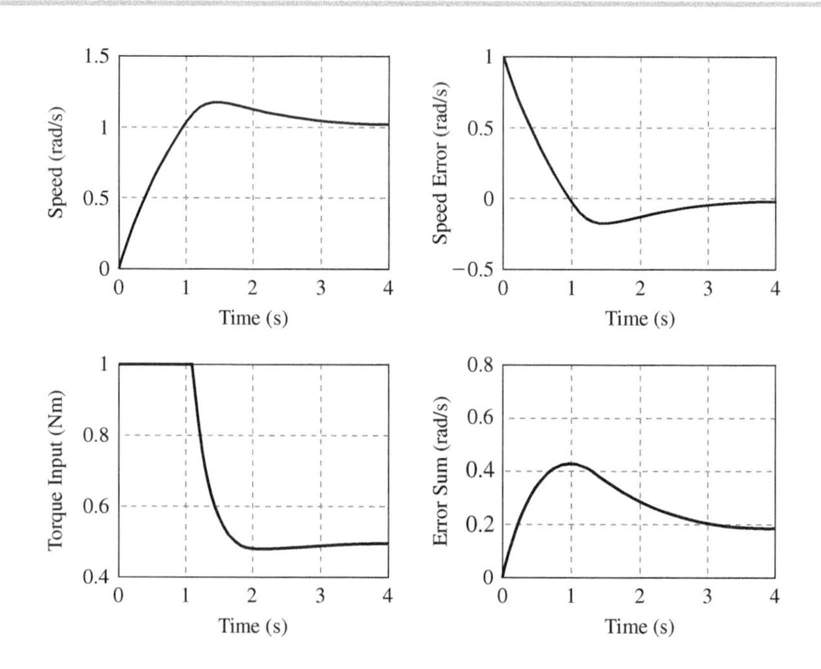

With the use of digital controllers, it is possible to implement the PID controller in software to avoid the above-mentioned problem. Such implementation is called a PID controller with anti-windup or **no reset windup.** The turning off or adjustment of the I-action contribution helps to prevent the reset windup of the system. In Simulink, no reset windup simulation is obtained by selecting an anti-windup method in the *PID advanced* section of the block parameters for the *PID controller* block. Figure 10.49 shows the simulation results with the *clamping* method chosen as the anti-windup method. In this method, Simulink stops the integration action within the PID block when the sum of the PID block components exceeds the output limits and the integrator output and block input signal have the same sign. The I-action integration is resumed when the sum of the block components exceeds the output limits and the integrator output and PID block input signal have opposite signs. Here the speed response is slower than the response shown in Figure 10.48, but there is no overshoot.

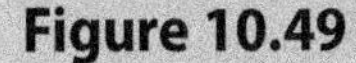

Figure 10.49

Simulation of PI controller with no reset windup

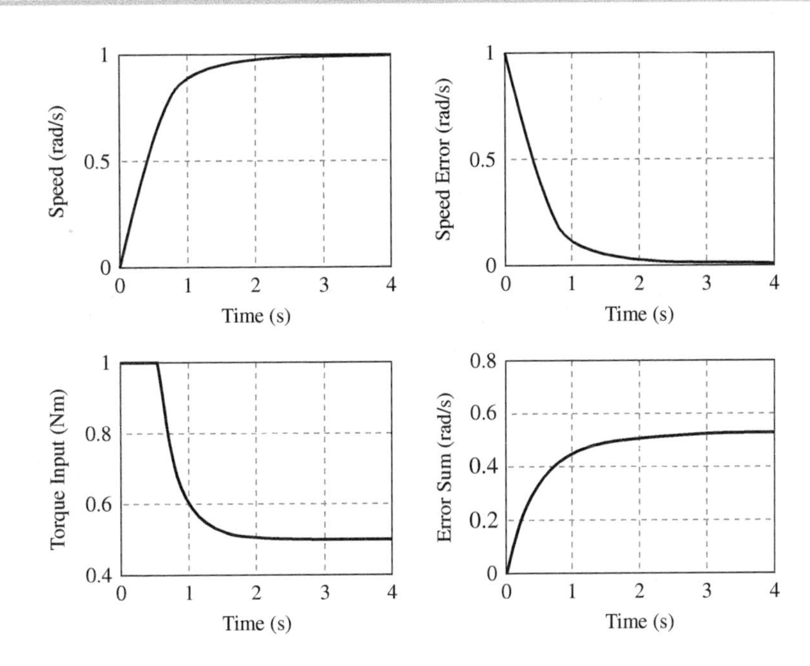

10.10.2 Nonlinear Friction

Mechanical systems subjected to dry or Coulomb friction exhibit a nonlinear response behavior. This point is illustrated in Figure 10.50, which shows the open-loop speed step response of a small PM brush DC motor subjected to different input voltages. At a 3.0 V input, the final speed steady-state value as measured by the tachometer that is attached to the motor is less than 1 V. At double the input voltage value or 6 V, the final steady-state value is about 3 V, which is more than three times the response at 3.0 V input. This behavior is characteristic of **dry** or **Coulomb friction** where the output speed is not proportional to the input voltage. In addition, the input voltage (or torque) has to exceed a certain value before motion occurs. For example, in this particular system, the motor does not rotate unless the input voltage exceeds 2.25 V.

Figure 10.50

Open-loop step response of a small DC motor system with a tachometer

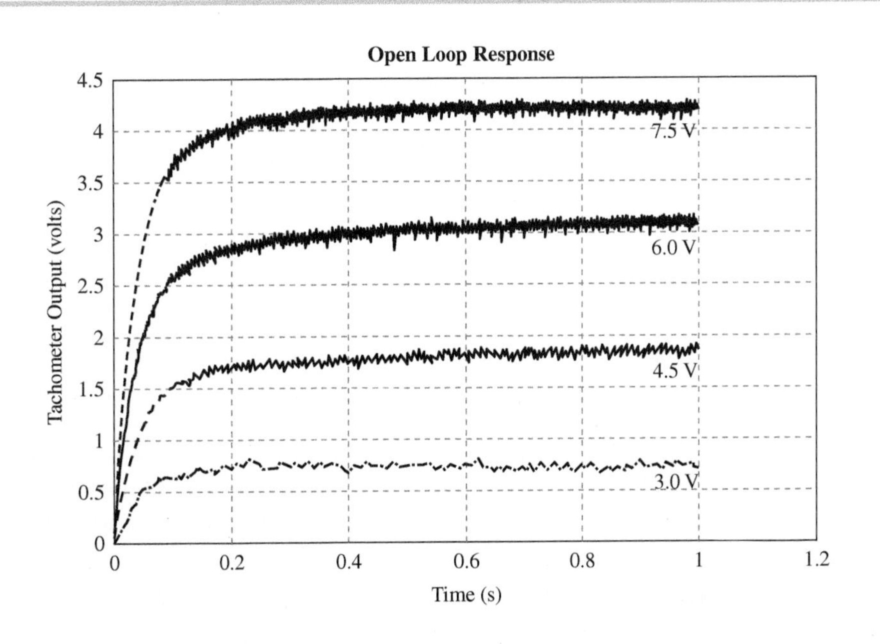

A linear model for such a system (such as Equation (10.8)) will only reproduce the actual behavior at input conditions similar to those used in obtaining the model parameters. At other input values, the model prediction would deviate from real behavior. Fortunately, in many cases with closed-loop PI- or PID-control action, the nonlinear friction behavior becomes a disturbance to the system, and the feedback control system can produce a response with zero steady-state error for such a system. Figure 10.51 shows a step response of this motor with PI-control. The reference speed was set at 4 V (the same units as the tachometer feedback units), and the response showed no steady-state error with no overshoot. It should be noted that there are numerous nonlinear behaviors (e.g., backlash, dead zones, or saturation) that cannot be easily compensated for with simple feedback.

Figure 10.51

Step response of a small DC motor system with a tachometer with PI-control and a reference speed of 4 V

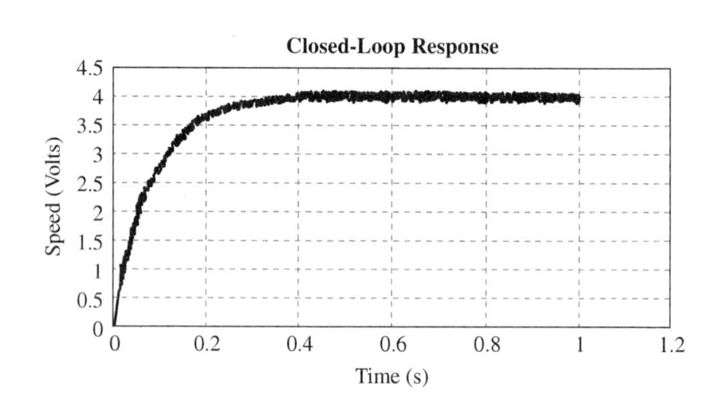

10.11 State Feedback Controller

While PID controllers offer a range of control possibilities, they have a limitation: the closed-loop poles' locations are inherently tied to the selected control gains and cannot be freely chosen. For more precise control over the closed-loop poles' locations—and consequently, a specific dynamic response behavior—a **state feedback controller** (▶ available—V10.11) is typically required. The controller is named so because it uses the states or the variables that describe the system. These states are typically the position, velocity, and so on in a mechanical system, or the voltage, current, and so on in an electrical system. However, this approach assumes the ability to measure or estimate all the states of the system, which may pose challenges.

For a linear system represented in the form (see Section 7.7.1)

(10.37)
$$\dot{x} = Ax + Bu$$

the state feedback controller takes the form

(10.38)
$$u = -Kx$$

where K is the $1 \times n$ state feedback gains matrix. Substituting the controller expression into Equation (10.37), a system under state feedback control has its dynamics given by

(10.39)
$$\dot{x} = (A - BK)x$$

The gain matrix K is determined by matching the poles (or eigenvalues) of the system matrix A-BK to a user-specified set of pole locations. In MATLAB, this operation is performed using the *PLACE* function. For arbitrary placement of the closed-loop poles, the given system needs to be controllable [25].

The controller form given by Equation (10.38) performs a regulation of the system around the origin (i.e., if the system is subjected to disturbances, it will bring the state vector back to the origin). If we need the state feedback controller to **track a particular reference signal** $\boldsymbol{r}$, then the state feedback controller is given in the form

(10.40)
$$u = -Kx + \widetilde{N}r$$

where $\widetilde{N}$ is a $1 \times m$ gain matrix to produce zero steady-state error for a constant reference r. The determination of the $\widetilde{N}$ matrix is discussed in many control texts (see for example [25]). For a single input, single output system, $\widetilde{N}$ is given by:

$$\widetilde{N} = -1/(C(A - BK)^{-1}B) \tag{10.41}$$

Note that this choice of $\widetilde{N}$ is not robust to changes in the plant model. To make the state-space controller robust to changes in the plant model and to constant disturbances, one needs to incorporate integral control into the state-space tracking system. This approach is not covered in this text, and the reader can refer to reference [25] for details about it. Example 10.10 illustrates the design of a state feedback controller for a second-order system, while Example 10.11 illustrates the design of a state feedback controller for a more complex non-rigid gear drive system.

Example 10.10 State Feedback Controller for a Second-Order System

For a rotating mass system represented by the model $T = J\ddot{\theta} + b\dot{\theta}$ with $J = 0.7$ and $b = 0.5$, and using the angular velocity and angular position as state variables, determine:

(a) State space model matrices (A, B, C, and D) for the system, assuming that the angular position (x_1) is used as the output
(b) The state feedback gains for desired closed-loop poles of $-5 + 3i$ and $-5 - 3i$

Solution:

(a) The system model is given by $T = J\ddot{\theta} + b\dot{\theta}$. Setting $x_1 = \theta$ and $x_2 = \dot{\theta}$, we get

$$\dot{x}_1 = x_2$$

$$\dot{x}_2 = 1/J(T - bx_2)$$

$$y = x_1$$

Writing it in matrix form, we get

$$\begin{bmatrix}\dot{x}_1 \\ \dot{x}_2\end{bmatrix} = \begin{bmatrix}0 & 1 \\ 0 & -b/J\end{bmatrix}\begin{bmatrix}x_1 \\ x_2\end{bmatrix} + \begin{bmatrix}0 \\ 1/J\end{bmatrix} T \text{ or } A = \begin{bmatrix}0 & 1 \\ 0 & -b/J\end{bmatrix} \text{ and } B = \begin{bmatrix}0 \\ 1/J\end{bmatrix}$$

$$y = [1 \;\; 0]\begin{bmatrix}x_1 \\ x_2\end{bmatrix} \text{ or } C = [1 \;\; 0] \text{ and } D = [0]$$

(b) When the system is placed under state feedback control, the dynamic model of the system is given by:

$$\dot{x} = (A - BK)x$$

The matrix K is obtained using the *place* command in MATLAB. This is done in MATLAB as shown below:

```
b = 0.5;
J = 0.7;
A = [0 1;0 -b/J];
B = [0 1/J]';
C = [1 0];
D = [0];
clp = [-5+3.0j -5-3.0j];
K = place(A,B,clp)
```

Which gives K as [23.8000 6.50000]

From Equation (10.41), the $\widetilde{N}$ parameter is 23.8. The state feedback simulation of the above system model for a unit step response of x_1 is shown in Figure 10.52. The model is simulated as the following system

```
sys1 = ss(A-B x K, B, C, D);
```

with the elements of the input vector multiplied by the parameter $\widetilde{N}$. As shown in Figure 10.52, the state feedback controller achieved a zero steady-state error with very little overshoot for this perfect plant model.

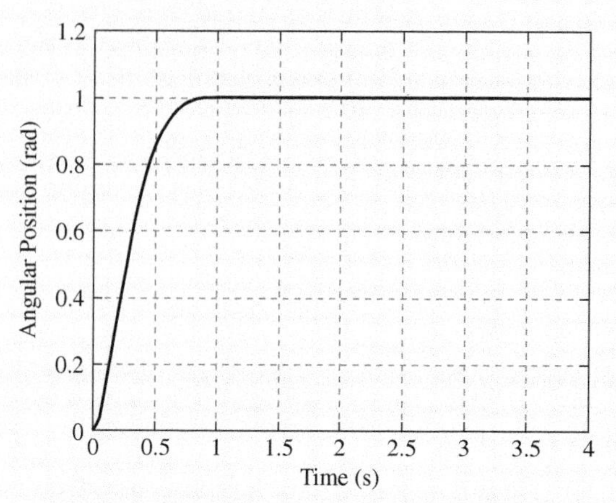

Figure 10.52 System response using a state feedback controller

Example 10.11 State Feedback Controller for Non-Rigid Gear Drive System

Design and simulate in MATLAB a state feedback controller to control the load link displacement in the gear drive system considered in Example 9.3 and reproduced in Figure 10.53. Let the torque output of the motor be the input to the system and take into consideration the compliance of the input and output shafts. Use the following parameter values: $N = 10$, $k_1 = 1200$ N/m, $k_2 = 1900$ N/m, $b_1 = 0.015$ Nm s/rad, $b_2 = 0.030$ Nm s/rad, $I_1 = 0.002$ kg m^2, and $I_2 = 0.925$ kg m^2.

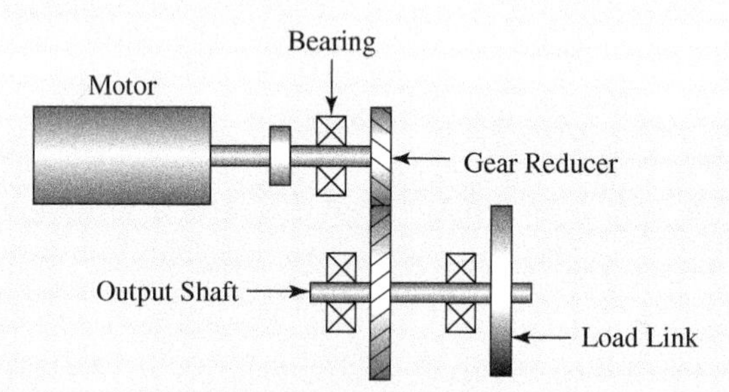

Figure 10.53 Drive system

Solution:

If the shaft compliance is considered, then the motor inertia and the load inertia are connected through compliant members and our model needs to reflect this fact. The system is modeled as shown in Figure 10.54(a). The rotor is connected in series with a spring k_1 that represents the elasticity of the shaft connecting the rotor to the gear. Similarly, spring k_2 represents the elasticity of the shaft connecting the gear to the link. We assume the gears to be rigid since the gear stiffness is normally much higher than the shaft stiffness. The model in Figure 10.54(a) can be represented by an equivalent system based on the input shaft as shown in Figure 10.54(b). In this representation, the gear reduction is eliminated and the parameters that represent the output shaft stiffness k_2, the output inertia I_2, and the friction torque Tf_2 are modified to reflect the effect of the gear reduction (i.e., $k'_2 = k_2/N^2$, $I'_2 = I_2/N^2$, and $Tf'_2 = Tf_2/N$). The springs k_1 and k'_2 are in series and can be combined to have an effective stiffness of k. Let θ_1 be the motor angular position, and let θ_2 be the angular position of the output link measured in the input shaft coordinate system.

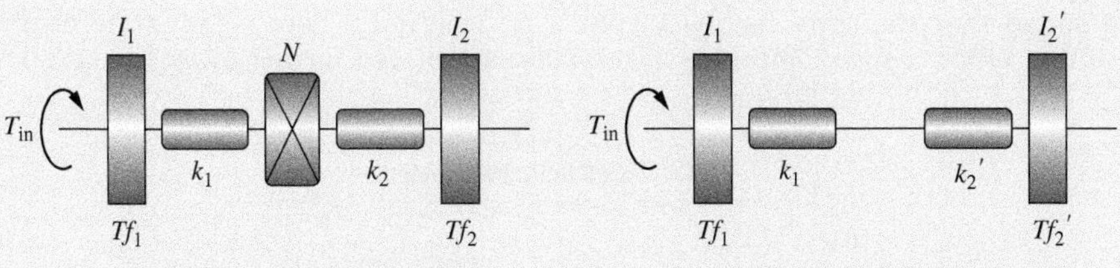

(a) Model with Shaft Compliance (b) Model Incorporating Gear Effects

Figure 10.54 Model of the system

The equations of motion for the system are then obtained as

$$T_{in} = k(\theta_1 - \theta_2) + b_1\dot{\theta}_1 + I_1\ddot{\theta}_1 \quad (1)$$

$$k(\theta_1 - \theta_2) - b'_2\dot{\theta}_2 = I'_2\ddot{\theta}_2 \quad (2)$$

where $1/k = 1/k_1 + 1/k'_2$, $k'_2 = k_2/N^2$, $I'_2 = I_2/N^2$, and $b'_2 = b_2/N^2$, where N is the gear ratio. If we let x_1 be the input shaft (or motor) angular displacement, x_2 be the input shaft angular speed, x_3 be the output link angular displacement measured in the input shaft coordinate system, and x_4 be the output link angular speed measured in the input shaft coordinate system, then the above equations can be represented in state space form with:

$$A = \begin{bmatrix} 0 & 1 & 0 & 0 \\ -k/I_1 & -b_1/I_1 & k/I_1 & 0 \\ 0 & 0 & 0 & 1 \\ k/I'_2 & 0 & -k/I'_2 & -b_2/I'_2 \end{bmatrix}, B = \begin{bmatrix} 0 \\ 1/I_1 \\ 0 \\ 0 \end{bmatrix}, C = [0\ \ 0\ \ 1\ \ 0], \text{ and } D = [0] \quad (3)$$

For the given parameters, the A and B matrices are given in MATLAB as:

$$A = 1.0e+003* \begin{matrix} 0 & 0.0010 & 0 & 0 \\ -9.3519 & -0.0075 & 9.3519 & 0 \\ 0 & 0 & 0 & 0.0010 \\ 2.0220 & 0 & -2.0220 & -0.0000 \end{matrix} \quad \text{and} \quad B = \begin{matrix} 0 \\ 500 \\ 0 \\ 0 \end{matrix}$$

If we let the desired closed-loop poles of the system be clp [−3 −4 −5 −6], then the state feedback gain matrix K is determined from the MATLAB command:

```
K = place (A, B, clp)
```

This gives the K vector as:

$$K = [-22.5111 \quad 0.0209 \quad 22.5115 \quad -0.0355]$$

From Equation (10.41), the $\widetilde{N}$ parameter is 3.5608e-004. The state feedback simulation of the above system model for a unit step response of x_3 is shown in Figure 10.55. The plot also shows the response for two other sets of closed-loop pole locations. The model is simulated as the following system

```
sys1 = ss(A-B x K, B, C, D);
```

with the elements of the input vector multiplied by the parameter $\widetilde{N}$. For a given set of pole locations, the response is obtained using the following code:

```
rd = ones(size(t));

w = [N*rd];

[y,t] = lsim(sys1,w,t);
```

The plot shows that the state feedback controller achieved a zero steady-state error and, as expected, the response speed improves as the poles are placed farther away from the imaginary axis.

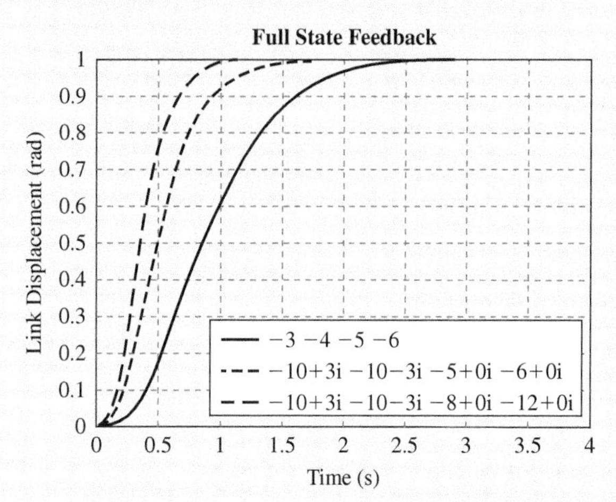

Figure 10.55 State feedback simulation of the drive system

A state feedback controller requires the availability of all the states of the system. In reality, all states may not be available. For example, in many positioning systems, only the position signal is available, and the velocity signal is not measured. The velocity signal can be obtained by differentiating the position signal (if the signal is not noisy) or the unavailable state(s) can be created in software using an observer [25], but this topic is beyond the scope of this textbook. State feedback control can also produce large actuator output if the poles are not selected carefully.

Even if the position and velocity signals are available, for many positioning applications, we have a choice to make to reduce the cost and complexity of the feedback system. In systems such as the one considered in Example 10.11, we can place the position/velocity sensors on either the motor shaft or the output link shaft, but not on both shafts. To illustrate these choices, consider the system of Example 10.11 with a simplified state feedback controller of the form

(10.42)
$$T_{in} = R K_p - K_p x_1 - K_d x_2$$

where x_1 is the motor shaft angular displacement, x_2 is the motor shaft angular speed, and R is the reference input signal. The unit step response of the link position for different combinations of K_p and K_d gains is shown in Figure 10.56(a). The figure shows that the motor shaft–based state feedback controller does not become unstable as the K_d value increases. Let us now replace the controller in Equation (10.42) with one that is based on the output link shaft. The simplified state feedback controller in this case is

(10.43)
$$T_{in} = R K_p - K_p x_3 - K_d x_4$$

where x_3 is the output link angular displacement measured in the input shaft coordinate system and x_4 is the output link angular speed measured in the input shaft coordinate system. The response of the system under such a controller is shown in Figure 10.56(b). Notice here how the system becomes unstable with increasing values of K_d. The second control configuration is an example of a **non-collocated** actuator-sensor system [49]. The instability arises from the compliance of the drive elements between the actuator and the sensor. To avoid instability problems, the controller should be based on the input shaft at the expense of less accurate positioning of the link, as a motor shaft controller does not compensate for any hysteresis or compliance in the gear train.

Figure 10.56

Response of the system in Example 10.11 to (a) a simplified state feedback controller based on the input (motor) shaft and (b) a simplified state feedback controller based on the output (link) shaft

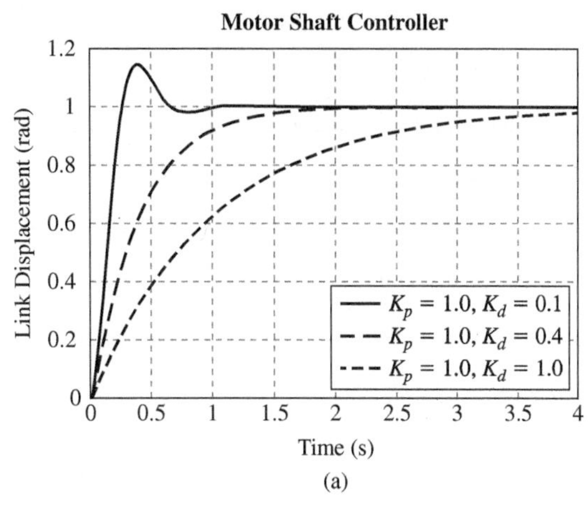

(a)

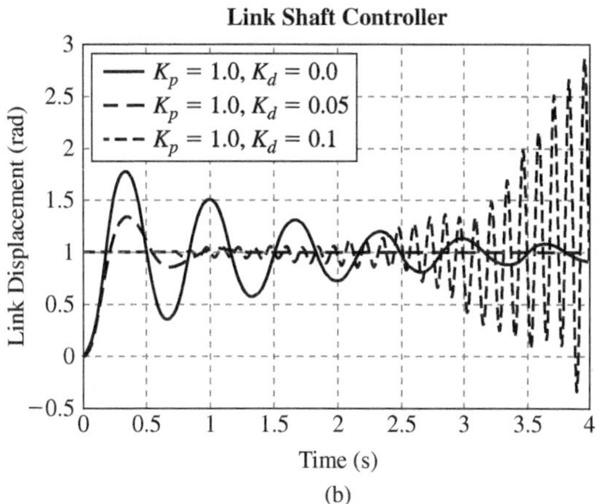

(b)

10.12 Chapter Summary

This chapter gave an overview of feedback control systems. A feedback control system is one in which the input to the system is a function of *both* the actual output and the reference input, unlike an open-loop system in which the input is only a function of the reference input. The chapter focused primarily on the PID-control algorithm, which is one of the most widely used feedback control laws. Through analysis, it was shown that a P or PD controller would not achieve a zero steady-state error in controlling a first-order system under a step input. Meanwhile, a PI or PID controller would achieve a zero steady-state error for the same system with or without constant disturbances acting on the system. For a second-order system, a P controller would not achieve zero steady-state error for a step input unless the system has integral action in its dynamics and is not subjected to disturbances. In contrast, a PI or PID controller would achieve zero steady-state error in a second-order system, with or without the presence of constant disturbances. The cascaded control loop structure, which is widely used in positioning control applications, was also discussed. Controller design considerations including stability, bandwidth, tuning, robustness, and digital implementation were also covered. The effect of saturation nonlinearity on the behavior of a feedback control system with integrator action was considered. A simulation of the behavior of a PI controller with no reset windup was presented in this chapter. The behavior of a feedback control system with nonlinear friction present in the system was also considered. We also discussed other control schemes, such as the on-off controller and the state feedback controller.

Questions

10.1 List several common applications of closed-loop control.

10.2 List two limitations of open-loop control.

10.3 How are analog controllers typically implemented?

10.4 What is the advantage of a digital controller?

10.5 How can one determine the order of a closed-loop transfer function?

10.6 What do the poles and zeros of a transfer function mean?

10.7 What are some of the limitations of the on-off controller?

10.8 Does a P-action closed-loop control of a first-order system achieve a zero steady-state error?

10.9 Which variation of the PID controller results in a zero steady-state error regardless of whether a disturbance is present or not?

10.10 Which variation of the PID controller does not result in an increase in the order of the system under control?

10.11 What is the advantage of the PD controller?

10.12 List several techniques for selecting the gains of a PI or PD controller.

10.13 What condition assures the stability of a closed-loop control system?

10.14 What is the advantage of using the Ruth-Hurwitz stability conditions?

10.15 Which Ziegler-Nichols tuning method uses open-loop data to select the control gains?

10.16 What is meant by phase and gain margins?

10.17 Is it desirable for a system to have a high gain margin, and why?

10.18 What causes reset-windup problems when using a PID controller?

10.19 List two nonlinear effects that are encountered in the control of real systems.

10.20 What limits the application of the state feedback controller in some cases?

Problems

Section 10.2 Open- and Closed-Loop Control

P10.1 Draw a block diagram representation of an open-loop control system to control a window air-conditioner functioning without thermostat control.

P10.2 Draw a block diagram representation of a closed-loop control system to control the speed of an electric motor.

Section 10.4 Control Basics

P10.3 The controller transfer function and the plant transfer function are given below. Determine the overall closed-loop transfer function for each of the cases listed below.

a. $G_c(s) = 3$ $\qquad G_p(s) = \dfrac{5}{s+4}$

b. $G_c(s) = \dfrac{s+3}{s+6}$ $\qquad G_p(s) = \dfrac{5}{s+5}$

c. $G_c(s) = \dfrac{3s+3}{s}$ $\qquad G_p(s) = \dfrac{5}{s^2+s+5}$

P10.4 Determine the closed-loop transfer function $Y(s)/R(s)$ for each of the systems shown in Figure P10.4.

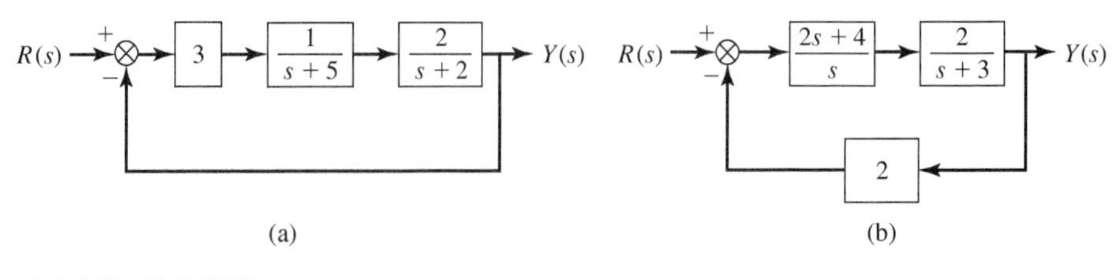

Figure P10.4

P10.5 Determine the closed-loop transfer function $Y(s)/R(s)$ for each of the systems shown in Figure P10.5.

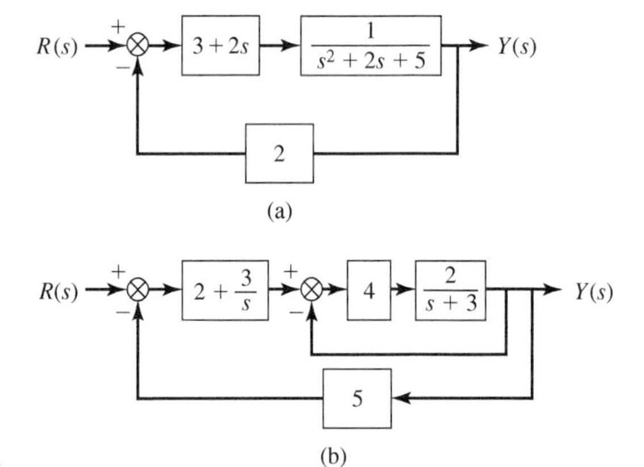

Figure P10.5

Section 10.5 On-Off Controller

P10.6 An on-off controller for an air-conditioning (A/C) system needs to operate as follows. The A/C should be on when the room temperature is one degree or more above the set or desired temperature and be off when the temperature is one degree or more below the set or desired temperature. If the temperature is between these limits, then the A/C should maintain the last controller input setting. Write a mathematical expression for the operation of this on-off controller.

Section 10.7 PID-Control of a First-Order Model

P10.7 Determine the K_p gain required for a system with a transfer function of $1/(s+2)$ so that the time constant of the closed-loop system with proportional feedback is 0.2 second.

P10.8 A first-order system with the transfer function $G(s) = 1/(0.5s+1)$ was placed under P-control. Determine the value of the control gain K_p so that the closed-loop control system has a time constant of 0.1 second. What is the value of the steady-state error for a unit-step input applied to the closed-loop system when using this K_p gain?

P10.9 A system whose transfer function is $2/(0.4s+2)$ was placed under P-control. Determine the steady-state error if the closed-loop control system was subjected to a step input of magnitude 3 and the gain K_p is 4.

P10.10 A first-order system whose transfer function is $1/(0.5s+1)$ was placed under PI-control. The control gains are $K_p = 2$ and $K_i = 4$. Determine the poles of the closed-loop system.

P10.11 A first-order system whose transfer function is $1/(0.5s+1)$ was placed under PI control. Determine if the system will have an overshoot for a step reference input if the gains are:

a. $K_p = 2$ and $K_i = 1$

b. $K_p = 1$ and $K_i = 3$

P10.12 A first-order system whose transfer function is $1/(2s+1)$ was placed under PI-control. Determine the gains K_p and K_i such that the controlled system has a damping ratio of 0.8 and a time constant of 1 s.

P10.13 A first-order system whose transfer function is $1/(0.5s+1)$ was placed under PI-control. Determine the gains K_p and K_i such that the controlled system has a natural frequency of 5 rad/s and a damping ratio of 1.

P10.14 A first-order system whose transfer function is $1/(0.5s+1)$ was placed under PD-control. The control gains are $K_p = 5$ and $K_d = 3$. If the closed-loop system was subjected to a unit step reference input and a constant disturbance of magnitude 0.1, what is the error at steady-state?

Section 10.8 PID-Control of a Second-Order Model

P10.15 A second-order system whose transfer function is $1/(0.5s^2+2s)$ was placed under P-control. Determine if the system will have an overshoot for a step input for the following cases

a. $K_p = 2$

b. $K_p = 4$

P10.16 A second-order system whose transfer function is $1/(0.5s^2+2s)$ was placed under PD-control. Determine the gains K_p and K_d so the closed-loop system has poles at $s = -2$ and $s = -3$.

P10.17 A second-order system whose transfer function is $1/(s^2+2s)$ was placed under PD-control. Determine the gains K_p and K_d so the closed-loop system has poles at $s = -3$ and $s = -5$.

P10.18 A second-order system whose transfer function is $1/(0.5s^2+2s)$ was placed under PD-control with $K_p = 4$ and $K_d = 1$. Determine the steady-state error if the closed-loop system was subjected to a step reference input of magnitude 2 and a constant disturbance of magnitude 0.1.

P10.19 A second-order system whose transfer function is $1/(0.5s^2+2s)$ was placed under PID-control. Select gains for the PID controller such that the closed-loop poles of the controller are located at $s = -2$, $s = -3$, and $s = -5$.

P10.20 A second-order system whose transfer function is $1/(s^2+2s)$ was controlled using PID-control. Determine the control gains of the PID controller such that the dominant two poles (those closest to the imaginary axis) of the controlled system exhibit a critically damped response and a desired time constant of 0.5 s.

P10.21 For the system shown in Figure P10.21, determine the following expressions.

a. $Y(s)$ as a function of $D(s)$ and $R(s)$.

b. $E(s)$ as a function of $R(s)$ and $D(s)$.

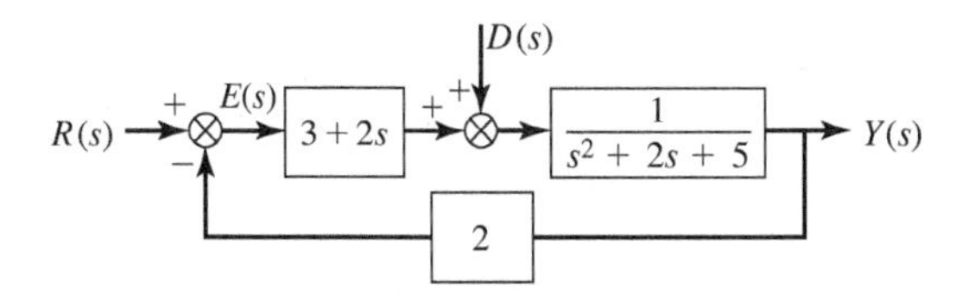

Figure P10.21

P10.22 For the system shown in Figure P10.21, determine the steady-state error if the system was subjected to a unit step reference input and a constant disturbance that has a magnitude of 0.1.

P10.23 For the block diagram shown in Figure 10.7, determine the transfer function between $\Theta(s)$ and a disturbance $D(s)$ applied to the diagram between the K_v and the $K/(\tau s+1)$ blocks. What is the final steady-state output if the disturbance is constant?

P10.24 For the cascade controller shown in Figure P10.24, determine the transfer function between $\Theta(s)$ and a disturbance $D(s)$ applied to the diagram.

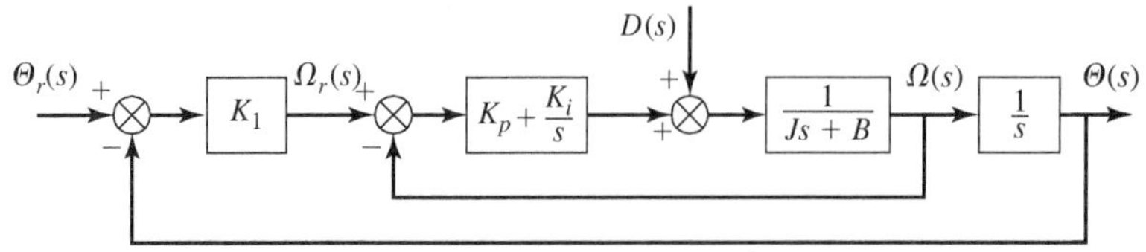

Figure P10.24

Section 10.9 Controller Design Considerations

P10.25 For the following closed-loop transfer functions, determine the limits on the gains based on applying the Routh-Horwitz stability criterion.

a. $\dfrac{K_p + K_d s}{as^2 + (b + K_d)s + c + K_p}$

b. $\dfrac{K_p s + K_i + K_d s^2}{as^3 + (b + K_d)s^2 + (c + K_p)s + K_i}$

P10.26 If the oscillation period $P_u = 2$ s and the ultimate gain $K_{Pu} = 30$, determine the initial gains for a PI controller based on using the Ziegler-Nichols tuning rule.

P10.27 For the transfer function $K_p/(s(s + a)(s + b) + K_p)$, show how the ultimate gain, K_{Pu}, and the oscillation period P_u can be obtained.

P10.28 Use the MATLAB *margin* function to obtain the phase and gain margins for the following transfer functions.

a. $G(s) = \dfrac{1}{s(s + 2)(s + 4)}$

b. $G(s) = \dfrac{2}{s^3 + 2s^2 + 2}$

P10.29 The transfer function $G_{open}(s)$ shown in Figure P10.29 has a gain margin of 27.6 dB at 2.83 rad/s and a phase margin of 79.4 degrees at 0.248 rad/s. Determine the maximum value of K that the closed-loop system can tolerate before it becomes unstable.

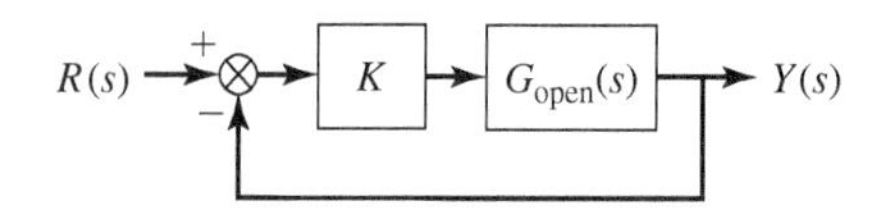

Figure P10.29

Section 10.10 Nonlinearities

P10.30 Research how the back-calculation anti-windup method in Simulink is different from the clamping method.

Section 10.11 State Feedback Controller

P10.31 For the rotating mass system represented by Equation (10.7) with $J = 0.2$ and $B = 0.2$, determine the state feedback gains for desired closed-loop poles of -3 and -5. Use the velocity and position as state variables, and assume that the position (x_1) is used as the output.

Laboratory/Programming Exercises

L/P10.1 Use the *relay* block in Simulink to simulate the On/Off controller discussed in Section 10.2. Use an error band of 0.5 degrees.

L/P10.2 Using the following transfer function $G(s) = 1/(0.5s + 0.1)$, model a PI controller for this system in Simulink and perform the following.

a. Obtain the unit step response for $K_p = 1$ and $K_i = 0$.

b. Repeat the step response using $K_p = 3$ and $K_i = 0.5$.

c. Repeat the step response using $K_p = 1$ and $K_i = 1$.

d. Repeat the step responses in parts a through c but consider the effect of saturation on system response by adding a saturation block after the controller block with output limits of +/– 0.5.

L/P10.3 Using the following transfer function $G(s) = 1/(s(0.5s + 0.1))$, model a PID controller for this system in Simulink and perform the following:

a. Obtain the unit step response for $K_p = 0.2$ and $K_d = 1$.

b. Repeat the step response but use $K_p = 1$, $K_i = 0.1$, and $K_d = 5$.

L/P10.4 Consider the following transfer function $1/(s + 0.2)$ and use MATLAB to obtain the closed-loop poles of the system under PI-control for the following cases.

a. $K_i = 1$ and K_p ranging from 1 to 5 in increments of 1.

b. $K_p = 1$ and K_i ranging from 1 to 5 in increments of 1.

Comment on what happens to the poles for each of the above cases.

L/P10.5 Simulate in MATLAB the step response of the system described by the transfer function $G(s) = 1/(s(s + a)(s + b))$ for the case $a = 1$ and $b = 3$ with P-, PI-, and PID-control. Obtain the controller gains using the Ultimate Sensitivity Method discussed in Section 10.9.3.

L/P10.6 Implement a closed-loop PI controller using an Arduino Uno or similar board for a first-order system such as a motor-tachometer system. Use the A/D converter on the board to get the speed of the system and the PWM feature to send the control output to the system. Use an H-bridge board such as L298 as a driver to the motor, and use the Serial Monitor to display to the user information about the performance of the system.

L/P10.7 Use Simulink to model the Coulomb friction nonlinearity. Use the data shown in Figure 10.50 to obtain the parameters of the model. Test your model with different inputs and compare the model results to the actual data in Figure 10.50.

L/P10.8 Use the PID function in Simulink to model a PI controller with no reset windup for a first-order system with saturation. Use the data in Section 10.10.1 to obtain the parameters of the controller and the system. Compare the Simulink results to that shown in Figure 10.49.

L/P10.9 Design and implement in MATLAB a state feedback controller for controlling the position of the system given by Equation (10.7). Use the model parameters given in Example 10.5. Try at least two different locations for the closed-loop system poles. Plot the response of the system for each of these cases.

11

Chapter

Mechatronics Case Studies

Chapter Objectives:

When you have finished this chapter, you should be able to:

- Apply state-transition diagram concepts to the operation and control of different mechatronic systems
- Apply circuit design for the construction of circuits to interface AVR microcontrollers with physical systems
- Develop software for control of mechatronic systems
- Apply modeling techniques to develop a dynamic model of a mechatronic system
- Apply MATLAB to simulate the response of mechatronic systems
- Explain the integration of the different components of a mechatronic system such as sensors, actuators, amplifiers, interface circuits, and control software that were covered in the book

11.1 Introduction

This chapter provides detailed setup information about the four integrated case studies discussed throughout the text. It also provides code listings, additional modeling, simulation, experimental results, and ways to extend or modify these case studies. Furthermore, this chapter provides a list of the main parts needed for each case study so instructors can duplicate the hardware at their institutions. The purpose of these case studies is to illustrate the integration of the many topics covered in this book in the form of experimental systems, as well as to provide practical information on the design and control of the mechatronic systems used in these case studies. The case studies are I. Stepper Motor-Driven Rotary Table, II. DC Motor-Driven Linear Motion Slide, III. Temperature-Controlled Heating System, and IV. Mobile Robot. All four case studies combine various mechatronic elements, and they include software design, hardware interfacing, data acquisition, timing, and control software. The second and third case studies also include dynamic modeling. Video demonstrations, additional pictures, and code listings for the four case studies are provided on the website for the book.

11.2 Case Study I: Stepper Motor-Driven Rotary Table

Stepper motors are widely available and offer a low-cost actuation system that can operate in an open-loop fashion without the need for feedback sensors. Stepper motors require digital signals for actuation, and these can be conveniently supplied from a PC or microcontroller. This project uses an AVR MCU as the controller.

11.2.1 Case Study Objectives

This case study focuses on the open-loop control of a stepper motor-driven rotary table. The case study objectives are to illustrate:

- Interfacing the microcontroller (Arduino Uno) and the stepper motor system
- The use of a photo-interrupter as a homing sensor
- The use of PWM signals for driving the stepper motor
- The selection of a stepper driver for the system
- The design of a state-transition diagram for the operation of this discrete-event control system

11.2.2 Setup Description

The setup consists of a two-phase hybrid stepper motor (see Section 9.5) with its rotation axis oriented vertically and mounted on a small aluminum support base, as shown in Figure 11.1(a). An optical CD with a notch is mounted on the stepper motor shaft, as shown in Figure 11.1(b). The notch location on the disk is detected by a photo interrupter optical sensor. An Arduino Uno board is used as the controller, and it transmits the step and pulse information to a stepper motor driver.

Figure 11.1

(a) Rotary table system and (b) mounting details of the CD

(Jouaneh, University of Rhode Island)

(a)

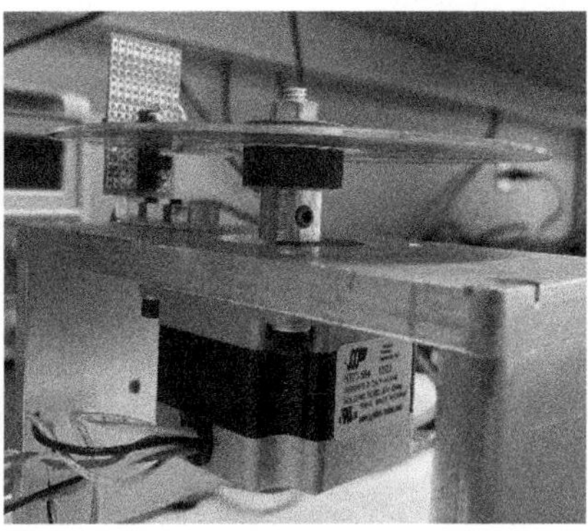

(b)

11.2.3 Sensor and Motor Wiring

The table uses a photo interrupter as a sensor to perform homing, in which the notch on the optical CD is brought to a known position. The wiring of this homing sensor was discussed in Integrated Case Study I.A (see Section 3.4.4), and a wiring diagram was shown in Figure 3.24. As discussed in Section 3.4.4, a photo interrupter is a combination of an LED and a phototransistor. When the optical disk obstructs the infrared light sent to the phototransistor, the phototransistor does not conduct and the sensor output will be the same as the voltage connected to the collector, or 5 VDC in this case. When the notch in the CD is aligned with the opening in the sensor or the table is at the 'home position,' the voltage output will drop to a low value which is read as a low logic in the Arduino. The output from this sensor is connected to one of the digital inputs on the Arduino Uno board.

The Arduino Uno is interfaced with the stepper motor using a commercial stepper motor driver (STR2 from Applied Motion Products—see Section 9.5.2). Integrated Case Study I.C (see Section 9.5.2) discussed the selection and interfacing of the stepper driver for this system, and the interface circuit between the Arduino Uno and the driver is shown in Figure 9.45. The stepper motor used in this project is a two-phase, eight-lead motor (see Figure 9.37), and the motor leads were connected in series bipolar configuration to the driver's A+, A−, B+, and B− connector. The MCU sends pulse and step signals to the driver, and the driver in turn sends the appropriate signals to the phases of the stepper motor. The direction signal is supplied from the digital I/O pin #4, while the pulses are supplied from pin #11. Full, half, and microstepping modes, as well as the current setting per phase, are set using dip switches on the driver.

11.2.4 Operation Commands and Control Software

As discussed in Integrated Case Study I.B (see Section 6.5.3), the motion of the table is controlled by three commands issued by the user. These commands are rotate clockwise, rotate counterclockwise, and stop motion. For clockwise or counterclockwise rotation, these commands should cause the table to make one complete rotation in the selected direction. The desired operation commands were translated into seven states as shown in the state-transition diagram in Figure 6.17. The state-transition diagram serves as a blueprint for the code that controls the operation of the table, which was implemented on an Arduino Uno board. The code is shown in multiple figures. The variable definitions, the *setup()* and the *loop()* functions, are shown in Figure 11.2, the code listing for the *Table_Task()* function is shown in Figures 11.3 and 11.4, and the code for the *MoveTable()* and the *GetTimeNow()* functions are shown in Figure 11.5. The *GetTimeNow()* function uses the *millis()* function for timekeeping. The user commands are entered using a single character interface scheme (*c* for clockwise motion (CW), *w* for counterclockwise motion (CCW), and *s* for stop) in which the commands are sent from the Serial Monitor on the Arduino IDE. The monitoring for these commands is done in the *loop()* function.

Figure 11.2

Variable definitions and *setup()* and *loop()* functions for the rotary table project

```
//////////////////////////////////////////////////////////////////////
///                    RotaryTable                                 ///
//////////////////////////////////////////////////////////////////////
#define Start 1                               // Define the states
#define AtHome 2
#define InitialCWMotion 3
#define InitialCCWMotion 4
#define CWMotion 5
#define CCWMotion 6
#define Stopped 7

#define HomeSensor 6                          // Define input line for home sensor
#define DirectionLine 4                       // Define output line for direction signal
#define PWM_Line 11                           // Define PWM output line
#define CW 1                                  // CW rotation direction constant
#define CCW 2                                 // CCW rotation direction constant
#define STOP 3                                // Stop constant
byte EntryStartState = 0;                     // Start state entry flag
byte EntryInitialCWMotionState = 0;           // InitialCWotion state entry flag
byte EntryInitialCCWMotionState = 0;          // InitialCCWotion state entry flag
byte EntryCWMotionState = 0;                  // CWMotion state entry flag
byte EntryCCWMotionState = 0;                 // CCWMotion state entry flag
byte EntryStoppedState = 0;                   // Stopped state entry flag

unsigned long StartTime;                      // Variable used for interval timing
int State, NextState;                         // State and NextState of transition diagram
char Command;                                 // User command: CW - c , CCW - w or STOP - s
void setup()
{
  Serial.begin(38400);                        // Initialize serial communication at a baud rate of 38400
  pinMode(DirectionLine, OUTPUT);             // Set the direction pin as an output pin
  pinMode(PWM_Line, OUTPUT);                  // Set the PWM pin as an output pin
  NextState = Start;
}
void loop()                                   // put your main code here, to run repeatedly:
{
  if (Serial.available() > 0) {
    Command = Serial.read();                          // Read command
    Serial.print(" The command received is: ");       // Print back the command to the user
    Serial.println(Command);
  }
    Table_Task();                                     // Call Table_Task repeatedly
}
```

Figure 11.3 Code listing for *TableTask()*

```
void Table_Task(void) {
 State = NextState;                              // Update State variable
 switch (State)                                  // Go to active state
 {
  case Start:
   if (digitalRead(HomeSensor) == 0) {           // Is disk at home position?
    NextState = AtHome;
    MoveTable(STOP);
    EntryStartState = 0; }                       // Reset entry flag on transition to another state
   else {
   if (EntryStartState == 0) {
     MoveTable(CW);
     EntryStartState = 1; }                      // Set entry flag on first entry to state
   }
   break;
  case AtHome:
   if (Command == 'c') {                         // Check if user selected CW motion
    NextState = InitialCWMotion;
    Command = " ";}
   else if (Command == 'w'){                     // Check if user selected CCW motion
    NextState = InitialCCWMotion;
    Command = " ";}
   break;
  case InitialCWMotion:
   if (EntryInitialCWMotionState == 0) {
    StartTime = GetTimeNow();                    // Record the start time
    MoveTable(CW);
    EntryInitialCWMotionState = 1; }             // Set entry Flag on first entry to state
   else if ((GetTimeNow() - StartTime) >= 100){  // Check if 0.1 second has elapsed
     NextState = CWMotion;
     EntryInitialCWMotionState = 0;}             // Reset entry flag on transition to another state
   break;
  case InitialCCWMotion:
   if (EntryInitialCCWMotionState == 0) {
    StartTime = GetTimeNow();                    // Record the start time
    MoveTable(CCW);
    EntryInitialCCWMotionState = 1;}             // Set entry flag on first entry to state
   else if ((GetTimeNow() - StartTime) >= 100){  // Check if 0.1 second has elapsed
    NextState = CCWMotion;
    EntryInitialCCWMotionState = 0;}             // Reset entry flag on transition to another state
   break;
```

The *TableTask()* function implements the state-transition diagram that is shown in Figure 6.17 and is called repeatedly inside the *loop()* function. The coding for this function follows the material discussed in Section 6.6. Note the code division in some of the states into an entry section, action section, and test and exit section, as was previously discussed.

The *MoveTable()* function controls the motion of the table by controlling the step signal, which is sent as a PWM signal from Pin #11 to the stepper driver using the *analogWrite()* function in Arduino. If the table is rotating either CW or CCW, the function also sets the value of the direction bit (pin #4) to cause the proper motion. If the table is commanded to stop, then the PWM duty cycle is set to 0. Pin #11 on the Arduino Uno board is controlled by *Timer2* and operates at a default PWM frequency of 490.2 Hz (see Table 4.7). The stepper motor used in this project has a specification of 200 pulses per revolution, so this gives a disk rotation speed of 2.45 rev/s. The user can change the rotation speed of the table by changing the prescaler or divisor value for *Timer2* (see Section 4.6.2).

11.2.5 Summary and Modifications

The material presented in this section and the text related to this integrated case study show how to design, control, and interface a sensor-driven, stepper motor-actuated rotary table that uses an MCU as a

Figure 11.4

Continuation of code listing for *TableTask()*

```
case CWMotion:
  if (EntryCWMotionState == 0) {
    MoveTable(CW);
    EntryCWMotionState = 1;}                    // Set entry flag on first entry to state
  else if (digitalRead(HomeSensor) == 0){      // Is disk at home position?
    NextState = AtHome;
    MoveTable(STOP);
    EntryCWMotionState = 0;}                    // Reset entry flag on transition to another state
  else if (Command == 'w'){                    // Check if user selected CCW motion
    NextState = CCWMotion;
    Command = ' ';
    EntryCWMotionState = 0;}                    // Reset entry flag on transition to another state
  else if (Command == 's') {
    NextState = Stopped;
    EntryCWMotionState = 0;}                    // Reset entry flag on transition to another state
  break;
 case CCWMotion:
  if (EntryCCWMotionState == 0) {
    MoveTable(CCW);
    EntryCCWMotionState = 1;}                   // Set entry Flag on first entry to state
  else if (digitalRead(HomeSensor) == 0){      // Is disk at home position?
    NextState = AtHome;
    MoveTable(STOP);
    EntryCCWMotionState = 0;}                   // Reset entry flag on transition to another state
  else if (Command == 'c'){                    // Check if user selected CW motion
    NextState = CWMotion;
    Command = ' ';
    EntryCCWMotionState = 0;}                   // Reset entry flag on transition to another state
  else if (Command == 's') {
    NextState = Stopped;
    EntryCCWMotionState = 0;}                   // Reset entry flag on transition to another state
  break;
 case Stopped:
  if (EntryStoppedState == 0){
    MoveTable(STOP);
    EntryStoppedState = 1;}                     // Set entry Flag on first entry to state
  else if (Command == 'c'){                    // Check if user selected CW motion
    NextState = CWMotion;
    Command = ' ';
    EntryStoppedState = 0;}                     // Reset entry flag on transition to another state
  else if (Command == 'w'){                    // Check if user selected CCW motion
    NextState = CCWMotion;
    Command = ' ';
    EntryStoppedState = 0;}                     // Reset entry flag on transition to another state
  break;
 }
}
```

standalone controller for this discrete-event system. Emphasis is placed on the sensor wiring, on matching the motor specifications with the stepper driver, and on the design of a state-transition diagram to guide the code development for this case study. The development of a state-transition diagram for the operation of a physical system is a particularly important step. One needs to make sure that the state-transition diagram operates correctly on paper before coding it to prevent wasting time on debugging the code. The code presented for this system was divided into modular pieces for ease of readability and possible reuse in other projects.

This case study can be modified in several ways. The code, for example, can be modified to allow the disk to rotate multiple times instead of a single rotation as was done here. Also, the rotation speed can be made variable. One or more photo-interrupter sensors can be added to allow the disk to stop at user-selected locations. The rotary table can also be replaced by a linear slide with limit switches at its end of travel, where one of the limit switches can be used as a homing sensor.

Figure 11.5

Code listing for *MoveTable()* and *GetTimeNow()* functions

```
void MoveTable(byte value)                    // Send step and direction information to stepper driver
{
 switch (value) {
  case CW:
   digitalWrite(DirectionLine, LOW);          // Set direction bit for CW motion
   analogWrite(PWM_Line, 128);                // Send pulses to stepper driver
   break;

  case CCW:
   digitalWrite(DirectionLine, HIGH);         // Set direction bit for CCW motion
   analogWrite(PWM_Line, 128);                // Send pulses to stepper driver
   break;

  case STOP:
   analogWrite(PWM_Line, 0);                  // Shut of pulses when table is stopped
   break;
 }
}
unsigned long GetTimeNow(void)                // Returns time in units of msec
{
 return (millis());
}
```

11.2.6 List of Parts Needed

Table 11.1 shows a list of the main components needed to fabricate this setup. As mentioned in the table, any two-phase or four-phase stepper motor with comparable specifications can be used. Also, the Arduino board and the stepper driver can be replaced by comparable ones.

Table 11.1

Main components for the stepper motor-driven rotary table setup

Component	Manufacturer/Part #	Comments
Stepper motor	Applied Motion Products HT23-594 two-phase hybrid stepper motor	Any two-phase or four-phase stepper motor with comparable specifications can be used
Sensor	Fairchild H21A1 Photo-interrupter	A similar sensor made by Isocom Components can be used
Microcontroller	Arduino Uno board	Any other Arduino board will work
Stepper motor driver	Applied Motion Products STR2 DC Powered Advanced Microstep Drive	DM542T driver from STEPPERONLINE can be used
Disk	Any CD computer disk	The notch is created by sawing a small slit in the disk

11.3 Case Study II: DC Motor-Driven Linear Motion Slide

11.3.1 Case Study Objectives

This case study focuses on the control of a DC motor-driven linear motion slide using MATLAB Simulink as the control medium. The case study objectives are to illustrate:

- Wiring of the power supply for a mechatronic system
- Interfacing of a Brushless DC motor and a servo drive
- Data acquisition using MATLAB
- Model identification from experimental data
- Simulation of a feedback controller
- Implementation of a feedback controller

11.3.2 Setup Description

The DC motor-driven linear motion slide is shown in Figure 11.6 and consists of a carriage, made up of an aluminum plate that is attached to a sliding ball-bearing, that moves linearly on a guided precision rail. A NEMA 23 BLDC servo motor drives the carriage through a timing belt which is tensioned by an idler sprocket. The sprockets are 50 mm in diameter and run parallel–above and below–the 30 mm in-line *t*-slot bar. The timing belt is fastened directly to the carriage via two aluminum timing belt end plates of compatible pitch. These plates are mounted to the underside of the plate on the carriage. As a precaution for over-travel of the carriage, mechanical stops made up of two-layer polyurethane foam mounted on the top of 3-D printed seat pads were implemented and designed to shear when impacted by the carriage at a high enough speed.

Figure 11.6

DC motor-driven linear motion slide

(Jouaneh, University of Rhode Island)

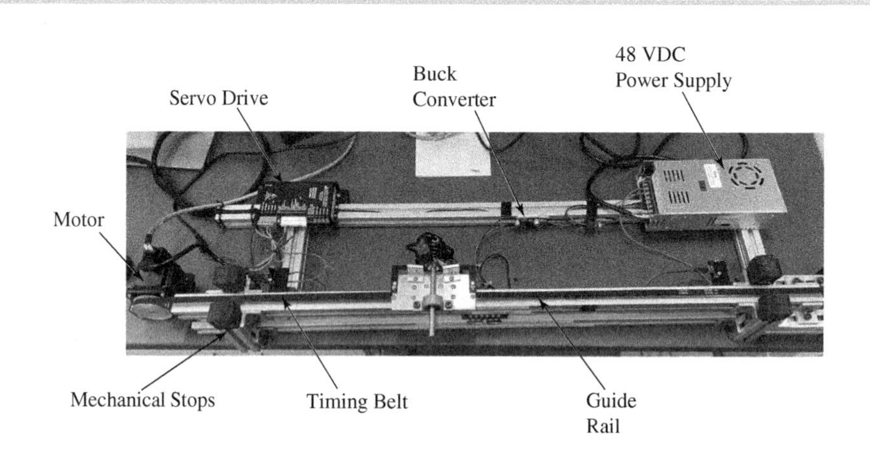

A 0–48 VDC adjustable switching type power supply is used to provide power to all subcomponents of the system. Integrated Case Study II.A (see Section 2.11) discusses how this power supply, in conjunction with a 24 to 5 VDC buck converter, was wired to provide power to the main components of the system. A BE12A6 analog servo drive manufactured by Advanced Motion Control was used to drive the brushless DC motor used in this project, and the servo drive was configured to operate in the current control mode. Integrated Case Study II.C (see Section 9.2.3) discusses the motor and servo drive interface. An incremental encoder attached to the BLDC motor was used to provide feedback information. The signals from the 1000 lines per revolution incremental encoder are fed to a 24-bit hardware counter board that is part of the Measurement Computing PCI-QUAD04 four-channel quadrature encoder board. The encoder operates in quadrature mode (see Section 8.3.3), resulting in an angular resolution of 0.09 degree per count or 0.039 mm per count. Note that the 24-bit counter will overflow once every 16.77 million counts. This will allow the linear slide to travel a total distance of about 659 m before it overflows.

The system was controlled through the MATLAB Simulink Desktop Real-Time toolbox, and the software/hardware interfacing details are detailed in Integrated Case Study II.B (see Section 5.5.2). The control signal from the controller is sent to the BE12A6 servo drive (see Figure 9.22) as a ± 10 V command signal using a 16-bit D/A converter, resulting in approximately 0.3 mV output resolution. A block diagram of the components of the control system for this setup is shown in Figure 11.7.

Figure 11.7

Block diagram of system components

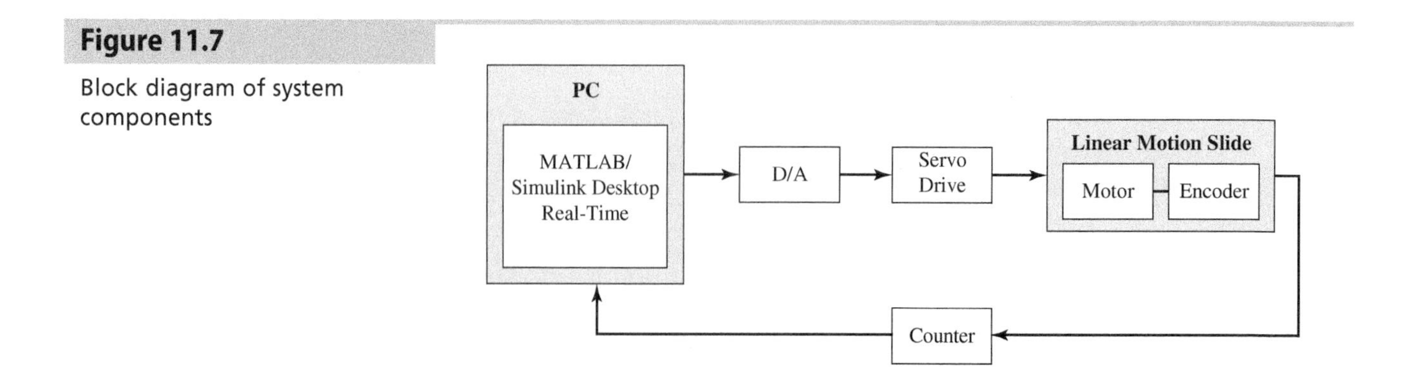

11.3.3 Modeling of the System

To identify the dynamics of the system, a series of step-input voltages are applied to the BLDC motor through the servo drive, and the angular displacement of the motor is recorded using MATLAB Simulink as discussed in Integrated Case Study II.B (see Section 5.5.2). The magnitude of the voltage applied to the system and the test duration time were limited to prevent the slide from exceeding its travel limits. An example of a voltage signal that is sent to the servo drive is shown in Figure 11.8. This signal causes the slide to move in one direction, pause, and then move back in the opposite direction.

Figure 11.8

Input voltage profile applied to the linear slide

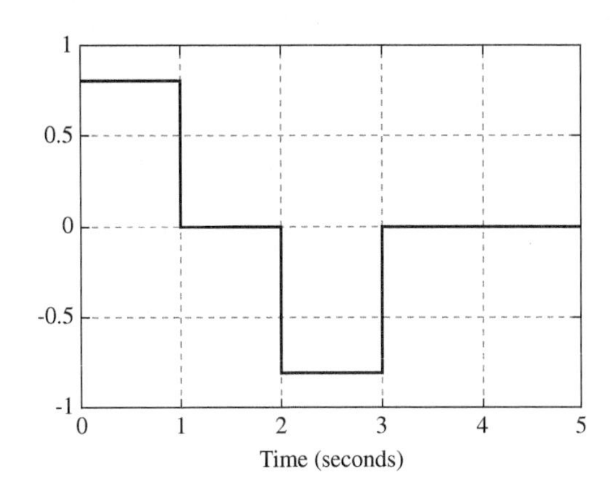

The angular displacement data is digitally differentiated in MATLAB Simulink (see Figure 5.12) to obtain the speed response. A plot of the open-loop position and velocity step responses are shown in Figure 11.9 for input voltages of 0.7 and 0.8 volt. The results show that the system is nonlinear since the final steady state speed reached for each input voltage is not proportional to the input voltage. Also, note that the final position of the slide after $t = 3$ seconds is different from its zero initial position. This nonlinearity is caused primarily by nonlinear friction, which is typical of many positioning systems.

Figure 11.9

Measured open-loop position response of the BLDC motor and measured (differentiated) open-loop velocity response of the BLDC motor to an input profile similar to that shown in Figure 11.8: (a) input voltage = 0.7 volt and (b) input voltage = 0.8 volt

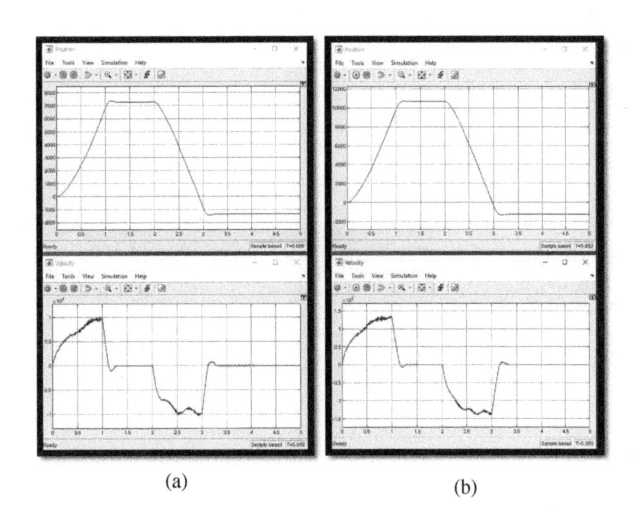

Using the speed data for the 0.8 V case, the parameters of a linear first-order model can be identified. The model, which relates the output speed of the motor (counts/s) to the input voltage sent to the servo drive, is given by Equation (11.1), and it will be utilized in the design of a feedback controller to control the motion of the slide.

$$\frac{\Omega(s)}{V_{\text{In}}(s)} = \frac{k_s}{\tau_s s + 1} = \frac{16250}{0.3s + 1} \tag{11.1}$$

In the model, k_s is the open-loop gain of the system, which is obtained by dividing the steady state speed (13,000 counts/s) by the input voltage (0.8 V), and τ_s is the time constant of the system which is obtained from the definition of the time constant (see Section 7.2). Note that due to the non-linear behavior of this system, this model will not replicate the open-loop response of the slide for voltages other than 0.8 V. However, it will be sufficient for designing a feedback controller for this system. In this design, the non-linear friction will be considered as a disturbance that the feedback control system will compensate for.

11.3.4 Feedback Controller Simulation in MATLAB

A model of a cascaded position controller (see Section 10.8.5) was created in Simulink to simulate the response of the system using this control scheme. The model is shown in Figure 11.10. The motor dynamics are represented by the linear model given in Equation (11.1).

Figure 11.10 Simulink model for simulating the cascaded controller for the linear slide system

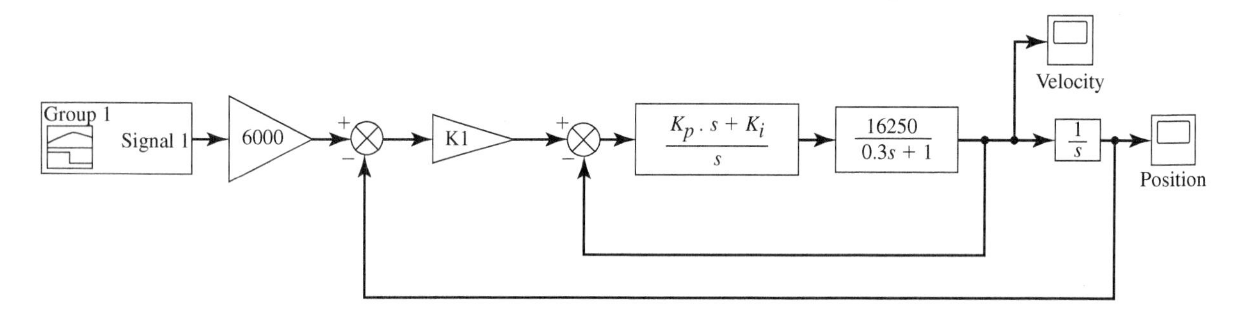

Integrated Case Study II.D (see Section 10.8.5) discusses the selection of the control gains for this controller and the performance of this controller. Figure 10.36 shows the simulated closed-loop response of this system using the desired position profile shown in Figure 10.35. The simulation shows the control system produces a fast response with a very slight overshoot.

11.3.5 Experimental Results

The cascaded control loop was tested on the actual slide using the Simulink Desktop Real-Time toolbox. Figure 11.11 shows the Simulink Desktop Real-Time model to perform the actual testing. Note how the output of the inner velocity loop is sent to the servo drive through the D/A on the data acquisition card. Note also how the measured position of the motor, which is obtained from the encoder, is used in the outer position loop, and how the differentiated position signal is used to provide the velocity feedback signal to

Figure 11.11

Simulink Desktop Real-Time model for controlling the slide

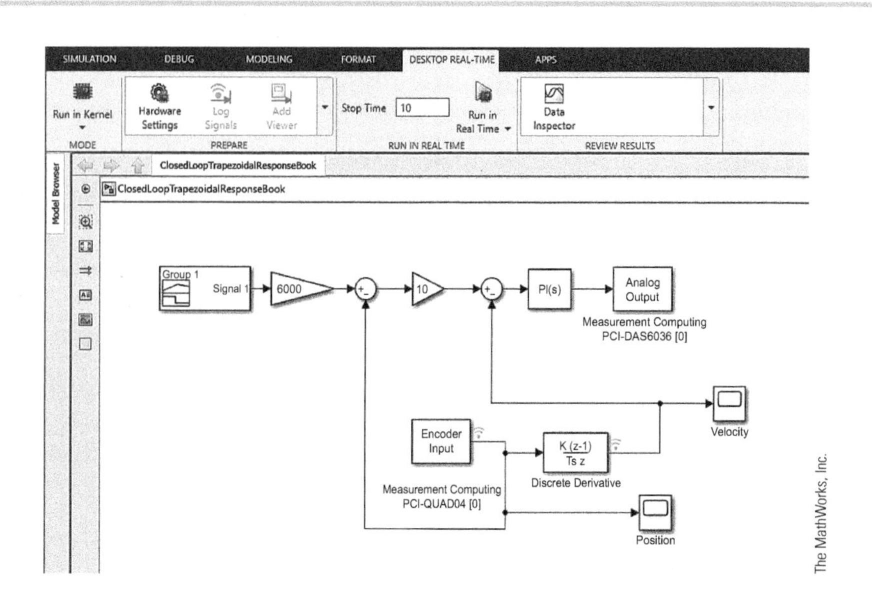

the inner velocity loop. Initially, the controller gains that were determined by simulation were used in the real system, but they resulted in more overshoot in the system response than desired, so the K_i gain was reduced by a factor of 0.5, but the outer loop gain and the K_p gains remained the same. Note that in the implementation of control systems, controller gains determined from an approximate model of the real system are usually modified when implemented on the real system.

Figure 11.12 shows the real response of the system using the modified gains, showing that the cascaded control system achieves a good job of controlling the motion of the motor despite the nonlinearities present in the system.

Figure 11.12

Actual response of the system

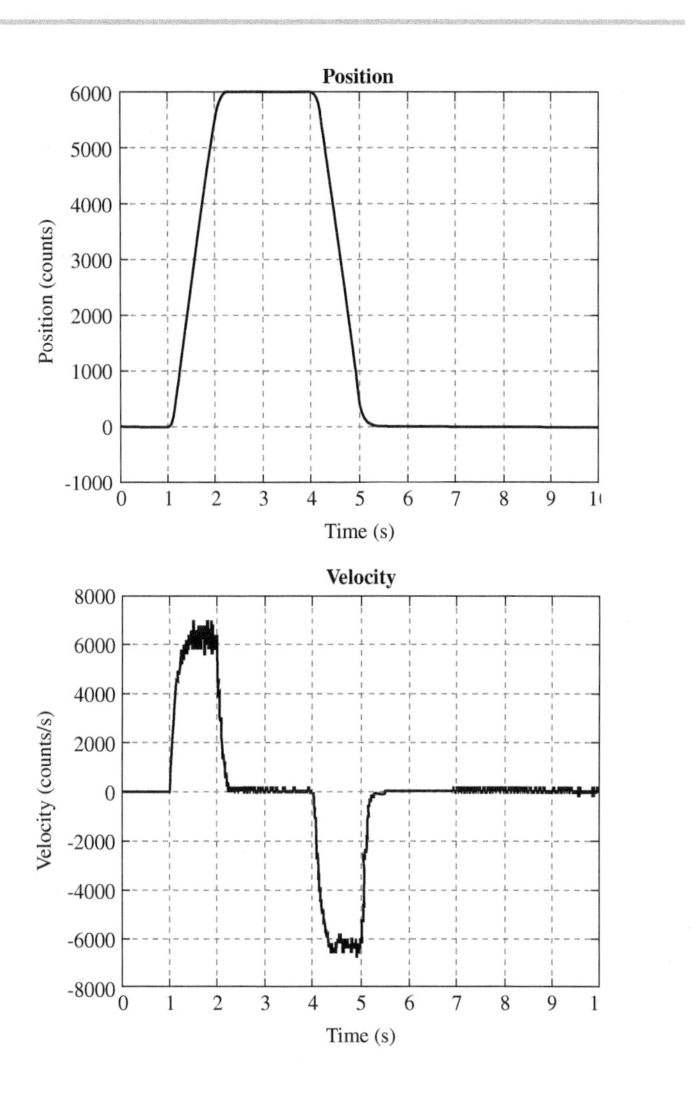

11.3.6 Summary and Modifications

This section and the accompanying integrated case study demonstrate the interfacing, modeling, simulation, and control of a linear motion positioning system. This system utilizes a brushless DC motor with an integrated encoder as the actuator and includes a servo drive for amplifying the control signal sent to the motor. It also features a data-acquisition system for interfacing and uses Simulink Desktop Real-Time as the control software. The case study highlights how a feedback controller can be integrated with a physical system, providing a practical example applicable to the control of various mechatronic systems with similar components.

This case study offers several modification opportunities. First, a more detailed model of the slide, accounting for the timing belt's compliance, can enhance the control design. Second, varying the load on the slide allows for testing the controller's robustness against load changes. Third, the current trapezoidal position trajectory, serving as the reference, can be substituted with alternative trajectory types. Finally, experimenting with different controller types, beyond the one used in this case study, can also be explored for controlling the linear slide.

11.3.7 List of Parts Needed

A list of the main parts (excluding the support frame and sprockets) needed for this case study is given in Table 11.2. Although this study specifically uses a brushless DC motor to control a linear motion system, offering a realistic system control experience, most tasks can be achieved with just the motor, the servo drive, and the controller, without needing the slide component.

Table 11.2

Main components needed for the DC motor-driven linear motion slide case study

Component	Manufacturer/Part #	Comments
BLDC motor with incremental encoder	Lin Engineering BL23E22-01D-06RO (available at Digi-Key)	Any brush or brushless DC motor with an encoder can be used
Timing belt	XL Series: 3/8 in wide timing belt - 9 feet (McMaster 7959K25)	
Rail and bearing	WON M12 series mini rail, stainless steel, 8 × 12 × 695 mm and WON M12 series mini bearing block	Available from AutomationDirect.com
Data acquisition card	Measurement Computing (now Digilent) PCIM-DAS1602/16	This PCI-type board can be replaced by the newer PCIe-DAS1602/16 board (or a USB-type DAQ board if it is compatible with Simulink Desktop Real-Time)
Counter board	Measurement Computing (now Digilent) PCI-QUAD04 four-channel quadrature encoder board	
Servo drive	BE12A6 analog servo drive from Advanced Motion Controls	This drive can be replaced by a newer model such as the AB15A100

11.4 Case Study III: Temperature-Controlled Heating System

11.4.1 Case Study Objectives

This case study focuses on the implementation of a feedback controller for a heating system in a microcontroller. The case study objectives are to illustrate:

- Identification of the dynamics of the heating system
- Determination of feedback-control gains
- Simulation of a control system in MATLAB
- Interfacing of system components to the microcontroller
- Implementation of a feedback controller in an MCU

11.4.2 Setup Description

The experimental hardware (see Figure 11.13) consists of a small rectangular (50.8 mm × 38.1 mm × 12.7 mm) copper plate heated by a 10 W flexible silicone-rubber heat strip that is glued to the bottom of the plate. The plate is mounted horizontally on a 76.2 mm × 102 mm polycarbonate base that acts as an insulator. A small hole is drilled into one side of the plate, and a thermo-transistor temperature sensor (LM35C plastic package from Texas Instruments) is inserted into the plate to read the temperature of the plate. The temperature sensor has a sensitivity of 10 mV/°C and a measurement range of −40 to 110°C.

Figure 11.13

Plate and heater experimental setup

(Jouaneh, University of Rhode Island)

A block diagram of the components of the system is shown in Figure 11.14. The ATmega328P MCU implements a feedback controller to control the plate temperature. A PC acts as a user interface for this control system and uses the RS-232 serial line to communicate with the MCU. The control input to the heater is supplied from the PWM output of the microcontroller through a transistor. The temperature is measured using the 10-bit A/D converter on the microcontroller. With a voltage reference of 5.0 V for the A/D, the temperature measurement resolution is 0.488°C. The heat output rate q from the heater is directly proportional to the heater voltage v as $q = Kv$, where $K = 10/12$ W/V.

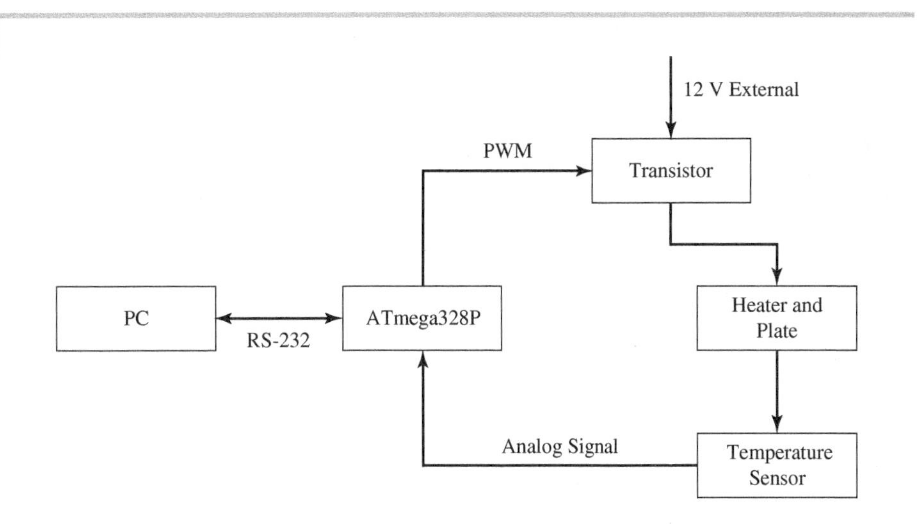

Figure 11.14

A block diagram of the control system components

The interface circuit between the microcontroller and the heater was discussed in Integrated Case Study III.A (see Section 3.5), in which an IRFZ14 power transistor was used as the switching element to modulate the heat energy applied to the system.

11.4.3 PC User Interface

Instead of using a **graphical user interface** (GUI), a simple user interface is used to control the operation of this system. The user simply sends a single lowercase letter command with a numerical argument to configure and control the system. The used commands are:

p pvalue: Sets the proportional gain K_p to the numerical value *pvalue*

i ivalue: Sets the integral gain K_i to the numerical value *ivalue d* interval: Sets the control duration interval in seconds to *interval*

v desired: Sets the desired temperature in °C for closed-loop operation or the open-loop step input voltage

m mode: Sets the control mode (*mode* = 0 means open-loop, and *mode* = 1 means closed-loop operation)

s num: Start the control action if *num* = 1 and stop it if *num* = 0

The commands can be entered in any order, but the start command (or *s*) should be the last one to be entered if one wants to obtain a run with no change in its parameters during its duration. Due to the limited RAM space in the Arduino UNO, arrays are not used to store the output of the program during its operation. The output (time and actual temperature) is simply printed to the Serial Monitor in Arduino. The user can then copy this output and plot it in programs such as Excel or Google Sheets. Note that a Graphical User Interface (GUI) developed using either Python or Visual Basic Express can also be developed to control the operation of this project.

11.4.4 Microcontroller Code

As discussed in Integrated Case Study III.B (see Section 6.5.3), the control software for this case study is structured as two control tasks that are continuously scanned in a loop. The control software was implemented on an Arduino Uno microcontroller, and the variables used in the code along with the *setup()* and the *loop()* functions are shown in Figure 11.15. The task that handles the user interface is shown in Figure 11.16, and it uses the *Serial.available()* function in Arduino to check for the presence of any characters in the serial input buffer. If the buffer has any characters in it, then its contents are read using the *readString()* function. The user-sent command and its associated value are then obtained using functions that extract characters from strings and the string value is converted to float.

Figure 11.15

Variables used in the code along with the *setup()* and *loop()* functions

```
//////////////////////////////////////////////////////////////////////////////
///                          Heated Plate                                  ///
//////////////////////////////////////////////////////////////////////////////
#define TempLine 0              // A/D channel to read temperature
int duty;                       // Control output expressed as duty cycle
int NumSamples;                 // Number of control samples to perform
float Tsamp = 1;                // Sampling time in seconds
float Tsampm;                   // Sampling time in msec
float desired, kp, ki;          // desired value, Kp, and Ki gains - sent from PC
float sumerror = 0.0;           // Sum of errors
byte Start;                     // Start command 1-start 0-Stop
long Counter;                   // Counter for number of Tsamp
char command;                   // Command sent by the user
float value;                    // Value associated with a command
String InputString;             // Command string entered by the user
byte ClosedLoop = 0;            // Flag to indicate open (0) or closed loop (1) operation
float error;                    // Temperature error
float controlout;               // Control output
float ActTemp;                  // Plate temperature in degree Celsius
unsigned long StartTime;        // Time at start of control action

void setup() {
  Serial.begin(38400);          // Initialize serial communication at a baud rate of 38400
}

void loop() {
  ProcessCommandsTask();
  ControlTask();
}
```

Figure 11.16

Code for the control task that handles the user interface

```
void ProcessCommandsTask(void) {
 if (Serial.available() > 0) {                    // Check if the input buffer has any characters
  InputString = Serial.readString();              // Read input string
  command = InputString.charAt(0);                // Extract the 1st character
  String Values = InputString.substring(1, 10);   // Get the remainder of the string
  value = Values.toFloat();                       // Convert string to float
  Serial.print(" The command received is: ");     // Print the command to the user
  Serial.println(command);
  Serial.print(" The value received is: ");       // Print the value to the user
  Serial.println(value);
  switch (command) {                              // Process the different commands
   case 'p':                                      // Proportional gain
    kp = value;
    break;
   case 'i':                                      // Integral gain
    ki = value;
    break;
   case 'd':                                      // Duration in seconds
    if (value >= 1800)
     Tsamp = 2;
    NumSamples = value / Tsamp;                   // Compute the number of samples to do
    Tsampm = Tsamp * 1000;                        // Convert sampling time into msec units
    break;
   case 'v':                        // Desired value (volts in open loop, and temp in closed loop)
    desired = value;
    if (ClosedLoop == 0) {
     duty = ((desired / 12.0) * 255);   // For open loop, convert voltage to duty cycle
    } else {
     desired = desired / 100.0;
    }
    break;
   case 'm':                        // Control mode: 1 -closed loop, 0-open loop
    ClosedLoop = int(value);
    if (ClosedLoop == 0) {
     duty = ((desired / 12.0) * 255);   // For open loop, convert voltage to duty cycle
    } else {
     desired = desired / 100.0;
    }
    break;
   case 's':                        // Start/Stop:  1-start 0-stop
    Start = int(value);
    if (Start == 1) {
     Counter = 0;
     StartTime = GetTimeNow();
    }
    analogWrite(3, 0);              // Shut off the voltage to the heater
    break;
  }
 }
}
```

The code for the other task that handles the control of the plate temperature in either open or closed-loop fashion, along with the PI controller code and the *GetTimeNow()* function, is shown in Figure 11.17. The control code will be skipped unless the user issues a start command. If the start command is active, the code will call a PI-control (see Chapter 10) function to perform closed-loop control of the system at every sampling interval for the desired control duration. In the case of open-loop operation, the PI-control function is skipped, and the system simply sends a voltage signal at the beginning of the control duration interval and reads the temperature of the plate at each sampling interval. The control output from the controller sets the duty cycle of the voltage signal applied to the heater (see Integrated Case Study III.A in Section 3.5). Since the system has no cooling function, a negative controller output simply sets the duty cycle to zero. The time and temperature data are printed to the Serial Monitor at each sampling instant. At the end of the control interval, the user can simply copy these values and plot them in software such as Excel.

Figure 11.17 Code for the functions that perform feedback control and reading time

```
void ControlTask(void) {
  if (Start > 0) {
    if ((GetTimeNow() - StartTime) >= (Counter * Tsampm)) {
      DoControl();                              // Call control function
      Serial.println("Time " + String(Counter * Tsamp) + " Temp " + String(ActTemp));
      Counter++;                                // Increment counter for number of control samples
      if (Counter == (NumSamples + 1)) {
        Start = 0;
        analogWrite(3, 0);                      // Shut off control output at the end of the control action
      }
    }
  }
}

void DoControl(void) {
  int A2D_Data;
  A2D_Data = analogRead(TempLine);
  ActTemp = A2D_Data * 0.0048828;               // Convert A/D value to volt units
  if (ClosedLoop == 1) {                        // Closed loop mode
    error = desired - ActTemp;                  // Compute the error
    controlout = PIcontrol(error);
    duty = (controlout / 12.0) * 255;           // Convert control output to duty cycle
    if (duty >= 0) {
      if (duty >= 255)                          // Check if duty cycle exceeds limit
      {
        duty = 255;
      }
    } else {
      duty = 0;                                 // Set duty to 0 for negative control output
    }
  }
  analogWrite(3, duty);                         // Send control output as a PWM signal
}

float PIcontrol(float error) {
  float out;
  out = kp * error + ki * sumerror * Tsamp;     // P and I terms of PI controller
  sumerror = sumerror + error;                  // Sum of errors
  return (out);
}

unsigned long GetTimeNow(void) {
  return (millis());
}
```

11.4.5 System Modeling

The modeling of this system and the determination of model parameters from step response data were discussed in Integrated Case Study III.C (see Section 7.2). The model was represented by Equation (11.2), where R was determined to be 7.9 °C/W and $RC = 1.1 \times 10^3$ s.

(11.2)
$$RC\frac{dT}{dt} = T_a - T + Rq$$

Letting $\Delta T = T - T_a$ and $q = Kv$, the plate model transfer function with the heater voltage $V(s)$ as the input is

(11.3)
$$\frac{\Delta T(s)}{V(s)} = \frac{RK}{RCs + 1}$$

Using PI control action, the closed-loop transfer function (see Section 10.7.2) is

(11.4)
$$\frac{\Delta T(s)}{\Delta T_R(s)} = \frac{RK(K_p s + K_i)}{RCs^2 + (RKK_p + 1)s + RKK_i}$$

where ΔT_R is the desired value of ΔT.

For a damping ratio of $\zeta = 1$ and a desired closed-loop time constant τ_d, the PI gains can be calculated as

$$K_p = \frac{2\tau/\tau_d - 1}{RK} \tag{11.5}$$

and

$$K_i = \frac{(RKK_P + 1)^2}{4\tau RK} \tag{11.6}$$

Note that, in implementing the PI controller in an AVR MCU, the control gains obtained from Equations (11.5) and (11.6) need to be multiplied by a factor of 100 because the temperature sensor has an output sensitivity of 0.01 V/°C.

11.4.6 Controller Simulation in MATLAB

A Simulink model of the heater control system is shown in Figure 11.18. The model incorporates a saturation limit of 12 V to simulate the maximum voltage that can be sent to the heater for a 100% duty cycle.

Figure 11.18

Heater model in MATLAB

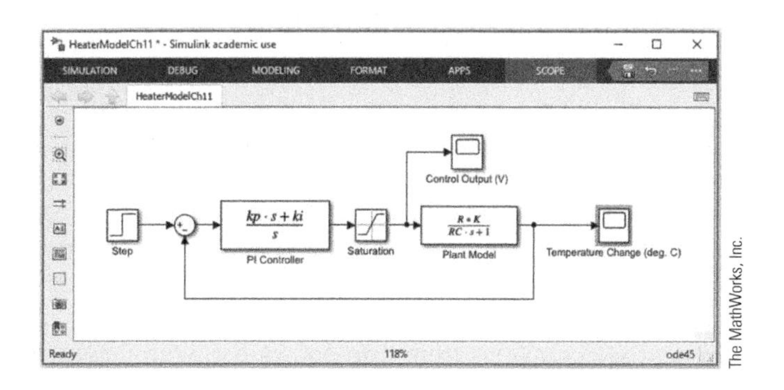

A plot of the simulated closed-loop temperature response of the system using the gains $K_p = 0.4141$ and $K_i = 4.824 \times 10^{-4}$ for a 30° step change in temperature is shown in Figure 11.19. Since the heater voltage is limited to 12 V, if the value of τ_d is selected too small, the heater will saturate. The Simulink model allows us to investigate how small τ_d can be made without causing saturation. It was found that τ_d close to 585 s is the smallest possible value.

Figure 11.19

Simulated response of heater control system in MATLAB: (a) control output (volts) and (b) temperature change (°C) with the horizontal axis indicating time in seconds

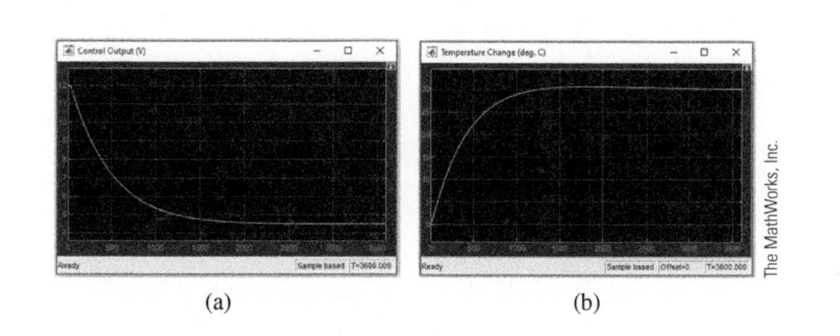

(a) (b)

11.4.7 Experimental Results

A plot of the actual closed-loop control system performance for a one-hour test with a desired temperature of 50°C that uses the same gains used in the MATLAB simulation is shown in Figure 11.20. The result shows a good agreement between the simulated and the real system behavior.

Figure 11.20

Experimental and simulated data for the plate setup

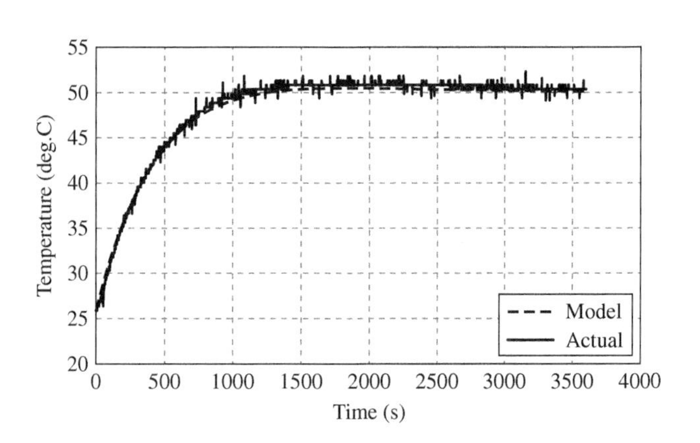

The measured temperature data is noisy. A main contribution of noise is the coarse resolution of the temperature sensor. The measurement resolution can be improved by using an external supply voltage to act as a reference for the A/D. For example, if we have connected a 2.5 V signal to the AREF line on the MCU, then the temperature measurement resolution is 0.244°C instead of 0.488°C.

11.4.8 Summary and Modifications

This section and the accompanying integrated case study demonstrate the interfacing, modeling, simulation, and control of a heating system. The case study delves into the use of state-transition diagrams for structuring the control code. It also explains the selection process for controller gains and details how the system was simulated in MATLAB before being controlled with an MCU.

This case study offers various avenues for modification. First, different sensors, like a thermistor or another digital IC temperature sensor, can be employed to measure temperature. Second, a Graphical User Interface (GUI) can be developed using either Python or Visual Basic Express, enabling users to control the system's operation. This interface can allow users to input numerical values through textboxes or select options from a dropdown list, with these commands being sent to an Arduino upon clicking a command button, thereby configuring or initiating the program. Additionally, the heated plate system in this study can be replaced with different systems, such as a small motor-tachometer system. However, this would require adjustments due to the different time constants; the motor-tachometer system has a much smaller time constant (tens of milliseconds) compared to the heater system (several minutes), necessitating a shorter sampling interval (1 ms versus 1 second) and correspondingly briefer control times.

11.4.9 List of Parts Needed

A list of the main components needed to implement the heater control system is shown in Table 11.3.

Table 11.3

Main components of heater control system

Component	Manufacturer/Part #	Comments
Heater	McMaster-Carr #7945T42 DC Volt Flexible Silicone-Rubber Heat Strip Adhesive Backed, 1″ × 2″, 10 Watts	This heater has a 10 W output power value; using a heater with a different power output will change the heating time
Temperature sensor	Texas Instruments LM35C	
Arduino Uno board	Arduino Uno Rev3 board	Any Arduino-compatible board can be used
Transistor	IRFZ14 transistor	Any transistor with a power rating of 10 W or higher and a current limit of 1 A can be used

11.5 Case Study IV: Mobile Robot

Mobile robots play a crucial role in mechatronic projects, serving as a versatile platform that integrates various components such as sensors, actuators, and controllers. These robots are widely employed across diverse industries and applications, contributing to the advancement of automation and robotics technology. This case study uses an educational robot (Romi from Pololu) that is available as a kit. In the context of mechatronics education, the use of an educational mobile robot such as Romi provides an accessible and affordable solution for educators and students to explore and understand key concepts related to mechatronics.

11.5.1 Case Study Objectives

This case study's objectives are to illustrate the integration of the different topics covered in this text in the operation and control of an educational-type mobile robot. Specifically, this case study will illustrate:

- Motor actuation through an H-bridge driver
- Use of an MCU as the control medium
- Feedback using encoders
- Inertial measurement sensors
- State-transition diagram for system control
- Implementation of a closed-loop motion controller

11.5.2 Setup Description

A picture of the Romi mobile robot is shown in Figure 11.21. The Romi robot is available from Pololu and comes as a kit that can be assembled without any soldering. This mobile robot is used in the For Inspiration and Recognition of Science and Technology (FIRST) robotic competition and can be controlled with a Raspberry Pi computer (not covered in this case study). The Romi robot control board is built around a USB-enabled ATmega32U4 AVR MCU, and it comes preloaded with an Arduino-compatible bootloader. The **ATmega32U4** MCU is slightly different from the ATmega328P MCU that is provided with the Arduino Uno board, and Integrated Case Study IV.B (see Section 4.4.5) discusses this microcontroller.

Figure 11.21

A picture of the Romi robot

(Jouaneh, University of Rhode Island)

The robot is powered by six AA batteries, and it includes a 5 V switching step-down regulator that can supply up to 2 A continuously. The control board is equipped with several sensors including a three-axis accelerometer and gyro which enable the Romi robot to make inertial measurements, estimate its orientation, and detect external forces. In addition, three onboard pushbuttons offer a convenient interface for user input, while indicator LEDs, a buzzer, and a connector for an optional LCD allow the robot to provide feedback. The AVR microcontroller uses an external 16 MHz crystal as the clock. One nice feature of this robot is that its microcontroller can be programmed using the standard Arduino IDE. The Romi32U4 Arduino library needs to be downloaded, and the board is set as the Arduino Leonardo in the Arduino IDE.

11.5.3 Motors and Encoders

The Romi control board has two Texas Instruments **DRV8838 motor drivers** that are used to power the two mini plastic gear motors. Refer to Integrated Case Study IV.A (see Section 3.10), which discusses this driver chip. These motors are PM brush-type DC motors with a rated voltage of 4.5, a gear ratio of 120:1, a stall current of 1.25 A, and a no-load current of 0.13 A at 4.5 volts. The DRV8838 provides a maximum current of about 1.7 A, so it is sufficient to drive these motors. Each of the two motors on this robot is equipped with a Hall-effect type (see Section 8.4.1) incremental encoder sensor that has a resolution of 12 counts per revolution of the motor shaft when operating in quadrature mode (see Section 8.3.3). While this encoder alone does not have good resolution, the use of a gearbox on each motor with a gear reduction ratio of 120:1 gives a reasonable resolution of 1440 counts per revolution of the Romi's wheel.

The Romi robot is an example of a **differential-type wheeled robot**, which is a mobile robot whose movement is based on two separately driven wheels placed on either side of the robot body. If both motors are commanded at the same speed in the same direction (i.e., clockwise or counterclockwise), then the robot moves forward or backward. If the speed of each motor is in the same direction but has a different magnitude, then the robot's motion is circular. If the speed of each motor has a different sign but the same magnitude, then the robot's motion will be rotation in place, where the robot rotates about its center and its center does not undergo any net displacement. While Romi's software provides a function (the function *setmotor()*) to set the speed of the motor, one can use the standard Arduino *digitalWrite()* function to set the direction and the *analogWrite()* function to set the magnitude of the PWM signal to accomplish the same thing. Figures 11.22 and 11.23 provide an example code to drive the robot in an open-loop fashion in different ways, including forward or backward motion, rotation in place, circular motion, or spiral motion. The program uses simple single-letter commands that can be entered through the Serial Monitor in the Arduino IDE to select the motion type. The *loop()* function continuously scans for user input as well as calls the *SpiralMove()* function. This function runs only if the user has selected a spiral move. In the spiral motion mode, the speed of one of the motors keeps increasing, and the radius of the spiral also keeps increasing with time. Since the robot is operating in an open loop, the number of completed spirals is dependent on the charge level of the batteries, the choice of initial speeds, and the time for each round of the spiral.

11.5.4 Inertial Sensor

The Romi control board is equipped with an LSM6DS33 gyro sensor, which provides acceleration values along the x, y, and z axes as well as rotation rates about these axes. Refer to Integrated Case IV.D in Section 8.10.3, which discusses this sensor in detail. Figure 11.24 (see page 375) shows the *loop()* function of an Arduino program that uses this sensor. In this code, the Romi robot will stop if it hits a rigid obstacle while it is moving, and this application is written to simulate the operation of an airbag system in a vehicle. A typical vehicle airbag system uses more than one sensor to arm the airbag to prevent false triggering (see Section 1.2), while only one sensor is used in this application. A simple method is used in this code to detect collision that is based on computing the magnitude of the resultant acceleration along the x and y axes. To prevent false triggering that could occur due to robot acceleration at the start of a motion, an averaging scheme is used, where the last five values are added and compared to a selected threshold. The code makes use of the LSM6 library that is provided by Pololu to access the LSM sensor.

11.5.5 Closed-Loop Control of the Robot Position

In a typical application, the motion of a mobile robot incorporates external sensors to guide and navigate the robot. These sensors include non contact proximity sensors, global positioning systems (GPS), and

Figure 11.22 Variable *declarations*, *setup()*, and *SpiralMove()* functions

```
///////////////////////////////////////////////////////////
///                MotorDrive                           ///
///////////////////////////////////////////////////////////
#define RightMotDir 15        // Define direction pin for right motor
#define LeftMotDir 16         // Define direction pin for left motor
#define RightMotPWM 9         // Define right motor PWM pin
#define LeftMotPWM 10         // Define left motor PWM pin
#define Speed 60              // Motor speed as PWM duty cycle (0-255)
int SpiralOn = 0;             // Flag to indicate if spiral motion is ON or OFF
int EntryState1 = 0;          // Entry state variable for the spiral motion code
unsigned long StartTime;      // Variable to store the start time for each round of the spiral motion
unsigned long RoundTime;      // Time that the spiral spends in each round
int NumberofSpirals;          // Number of spirals
int MinimumSpeed;             // Minimum speed
int SpiralCount = 0;          // Counter for number of spirals
void setup()                  // This function executes only once
{
 Serial.begin(38400);                        // Initialize serial communication at a baud rate of 38400
 pinMode(RightMotDir, OUTPUT);               // Set the right motor direction pin as an output pin
 pinMode(LeftMotDir, OUTPUT);                // Set the left motor direction pin as an output pin
 pinMode(RightMotPWM, OUTPUT);               // Set the right motor PWM pin as an output pin
 pinMode(LeftMotPWM, OUTPUT);                // Set the left motor PWM pin as an output pin
 Serial.println(" ");
 Serial.println("Enter Command");            // Print a message to the user
}
void SpiralMove(void) {                      // Function to implement spiral motion
 if (SpiralOn == 1) {                        // Execute if spiral motion was selected
  if (EntryState1 == 0) {
   MinimumSpeed = 20;                        // Starting speed for left motor
   digitalWrite(LeftMotDir, LOW);
   digitalWrite(RightMotDir, LOW);
   analogWrite(RightMotPWM, Speed);          // Send PWM duty cycle to the right motor
   analogWrite(LeftMotPWM, MinimumSpeed);    // Send PWM duty cycle to the left motor
   NumberofSpirals = 5;                      // Number of spirals to do
   RoundTime = 3300;                         // Time for each circle of the spiral in ms
   StartTime = millis();                     // Record the starting time at the beginning of the motion
   EntryState1 = 1;
  }
  if (((millis() - StartTime) >= RoundTime) && (EntryState1 == 1)) {
   RoundTime = RoundTime + 2000;             // Increase motion time for the next circle in the spiral
   MinimumSpeed = MinimumSpeed + 5;          // Increase the speed to increase the spiral radius
   analogWrite(LeftMotPWM, MinimumSpeed);    // Send PWM duty cycle to the left motor
   StartTime = millis();
   SpiralCount++;                            // Increment the number of spirals
   if (SpiralCount >= NumberofSpirals) {
    SpiralOn = 0;
    SpiralCount = 0;
    analogWrite(LeftMotPWM, 0);              // Send PWM duty cycle to the left motor
    analogWrite(RightMotPWM, 0);             // Send PWM duty cycle to the right motor
    EntryState1 = 0;                         // Reset Entry variable on end of spiral motion
   }
  }
 }
}
```

light detection and ranging (LIDAR) sensors. The Romi robot is not equipped with any of these sensors. While the Romi robot is equipped with incremental encoders (typically called **wheel odometry sensors**), due to wheel slippage and gear backlash, it is not possible to accurately control a mobile robot position just based on wheel odometry, especially for long travel. Furthermore, since a mobile robot moves in two directions, and can also rotate about its axis, it is difficult to obtain an accurate location of the robot without using sophisticated methods such as Simultaneous Localization and Mapping or SLAM (see reference [50] for more detail). SLAM is a technique that allows a robot or a vehicle to create a map of its environment

Figure 11.23

The *loop()* function to drive the robot in different types of motion

```
void loop() // This function executes repeatedly
{
 if (Serial.available() > 0) {
  char command = Serial.read();                          // Read command
  Serial.print(" The command received is: ");            // Print back the command to the user
  Serial.println(command);
  switch (command) {
   case 'f': // Move Forward
    Serial.println("Moving the robot forward");
    digitalWrite(LeftMotDir, LOW);
    analogWrite(LeftMotPWM, Speed);                      // Send PWM duty cycle to the left motor
    digitalWrite(RightMotDir, LOW);
    analogWrite(RightMotPWM, Speed);                     // Send PWM duty cycle to the right motor
    break;
   case 'b': // Move Backward
    Serial.println("Moving the robot backward");
    digitalWrite(LeftMotDir, HIGH);
    analogWrite(LeftMotPWM, Speed);                      // Send PWM duty cycle to the left motor
    digitalWrite(RightMotDir, HIGH);
    analogWrite(RightMotPWM, Speed);                     // Send PWM duty cycle to the right motor
    break;
   case 'c': // Circular Motion
    Serial.println("Moving the robot in circular motion");
    digitalWrite(LeftMotDir, LOW);
    analogWrite(LeftMotPWM, Speed * 0.5);                // Send PWM duty cycle to the right motor
    digitalWrite(RightMotDir, LOW);
    analogWrite(RightMotPWM, Speed);                     // Send PWM duty cycle to the left motor
    break;
   case 'r': // Rotate in place
    Serial.println("Rotating the motor in place");
    digitalWrite(LeftMotDir, HIGH);
    analogWrite(LeftMotPWM, Speed);                      // Send PWM duty cycle to the left motor
    digitalWrite(RightMotDir, LOW);
    analogWrite(RightMotPWM, Speed);                     // Send PWM duty cycle to the right motor
    break;
   case 'x': // Stop
    SpiralOn = 0;
    EntryState1 = 0;
    SpiralCount = 0;
    Serial.println("Stopping the robot");
    analogWrite(LeftMotPWM, 0);                          // Send PWM duty cycle to the left motor
    analogWrite(RightMotPWM, 0);                         // Send PWM duty cycle to the right motor
    break;
   case 's': // Spiral motion
    Serial.println("Spiral Motion");
    SpiralOn = 1;
    break;
   default:
    Serial.println("Stopping the robot");
    analogWrite(LeftMotPWM, 0);                          // Send PWM duty cycle to the left motor
    analogWrite(RightMotPWM, 0);                         // Send PWM duty cycle to the right motor
    break;
  }
 }
 SpiralMove();
}
```

while simultaneously estimating its position within that map. By using various sensors such as LIDAR, cameras, and odometry sensors, the SLAM algorithm processes the sensor data to construct a map of the surroundings. It also incorporates the robot's movement and sensor measurements to estimate its position within that map.

Without the use of any external sensors or SLAM, we will illustrate in this section how to control the position of a mobile robot in a closed-loop fashion to allow it to track a combination of linear and rotational

Figure 11.24

The *loop()* function that detects collision of the Romi robot with a rigid obstacle

```
void loop()
{
  imu.read();                                  // Read the IMU sensor
  int acc = sqrt(imu.a.x ^ 2 + imu.a.y ^ 2);   // Determine the acceleration magnitude in the xy plane
  acc_sum = acc_sum + acc;                     // Update the sum of acceleration values
  count = count + 1;                           // Increment counter
  if (count >= Num)                            // Check if the counter equal or exceed the set number of counts
  {
    if (acc_sum >= 125)                        // Check if the sum of acceleration values exceeds the threshold
    {
      Move_Forward(0);                         // Stop the robot
    }
    count = 0;                                 // Reset the counter
    acc_sum = 0;                               // Reset the sum of the acceleration values
    delay(5);                                  // Use a delay of 5 ms between measurements
  }
}

void Move_Forward(byte speed)                  // A function to move the robot forward
{
  digitalWrite(LeftMotDir, LOW);
  analogWrite(LeftMotPWM, speed);              // Send PWM duty cycle to the left motor
  digitalWrite(RightMotDir, LOW);
  analogWrite(RightMotPWM, speed);             // Send PWM duty cycle to the right motor
}
```

motion segments that produce polygon shapes in 2-D such as a square or a hexagon with reasonable accuracy. The approach is an implementation of the state-transition diagram that is discussed in Integrated Case Study IV.C (See Section 6.5.3). Rather than just specifying the desired position of each wheel at the end of each motion segment (which is not very desirable since it results in uncontrolled acceleration of the wheels and wheel slippage), the desired position for each wheel at every *Tupdate* interval is obtained from integrating a trapezoidal velocity profile (see Figure 11.25) or a triangular velocity profile that uses a pre-specified acceleration and deceleration time intervals.

Figure 11.25

Trapezoidal velocity profile used for each wheel motion

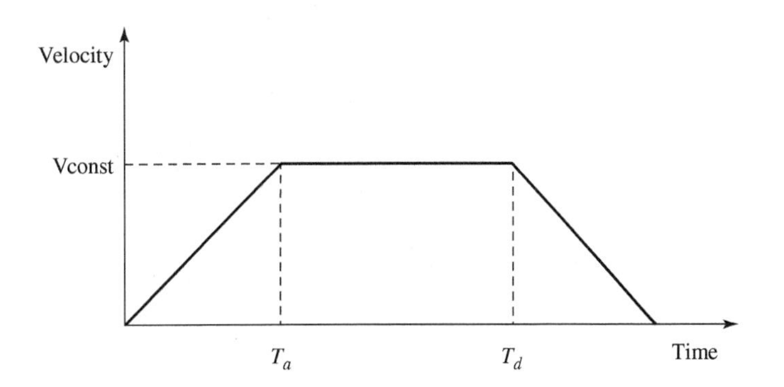

The code for generating the position trajectory is shown in Figure 11.26. This position trajectory generator requires the start and end positions of the segment (in units of encoder counts), the constant speed value (in encoder counts per second), the update interval in seconds, as well as the array name to store the trajectory. Depending on the motion segment length, as well as the speed value, the trajectory may not have enough time to reach constant speed. In that case, the function will generate a triangular velocity profile instead of a trapezoidal one. Figure 11.27 shows two examples of position trajectories generated by this

Figure 11.26 Code for generating the position trajectory for each motion segment

```
int GeneratePosTraj(int StartPos, int FinalPos, int Vconst, float Tupdate, int Traj[]) {
  int NumAccel = 20;                    // Number of points during the acceleration and deceleration phases
  int NumConst;                         // Number of points during the constant speed phase
  int m;
  int TravelDist = abs(FinalPos - StartPos);       // Total travel distance
  int DistAcc = 0.5 * NumAccel * Tupdate * abs(Vconst);        // Travel distance during acceleration phase
  int Ntotal;
  if ((2 * DistAcc) >= TravelDist) {    // Check if the travel distance during the acceleration and deceleration phases
                                        // exceeds the total distance
    DistAcc = 0.5 * (FinalPos - StartPos);  // Compute trajectory parameters for a triangular velocity profile
    NumConst = 0;
    Vconst = (2 * DistAcc) / (NumAccel * Tupdate);
  } else {
    NumConst = int((TravelDist - 2 * DistAcc) / (abs(Vconst) * Tupdate));  // Compute # of points during constant speed
                                                                           // phase
  }
  Ntotal = 2 * NumAccel + NumConst;     // Get total number points for the position profile

  for (int i = 1; i <= NumAccel; i++) {     // Generate trajectory during acceleration phase
    Traj[i] = StartPos + 0.5 * i * Tupdate * (i / (NumAccel * 1.0)) * Vconst;
  }
  if (NumConst > 0) {                   // Generate trajectory during constant speed phase
    for (int j = NumAccel + 1; j <= (NumAccel + NumConst); j++) {
      Traj[j] = Traj[NumAccel] + Vconst * ((j - NumAccel) * Tupdate);
    }
  }                                     // Generate trajectory during the deceleration phase
  for (int k = NumAccel + NumConst + 1; k <= (2 * NumAccel + NumConst); k++) {
    m = k - (NumAccel + NumConst);
    Traj[k] = Traj[NumAccel + NumConst] + Traj[NumAccel] - StartPos - 0.5 * (NumAccel - m) * Tupdate * (Vconst *
((NumAccel - m) / (NumAccel * 1.0)));
  }
  return (Ntotal);
}
```

function. Figure 11.27(a) corresponds to a trapezoidal velocity profile, while Figure 11.27(b) corresponds to a triangular velocity profile.

A PI controller was designed to allow each wheel to follow the trajectories supplied by the trajectory generator. The trajectory update interval was set at five times the sampling interval (5 ms) to allow the wheel some time to reach the position specified by the trajectory generator. The proportional and integral gains were selected by experimentation, where the integral gain was first set to zero, and the proportional gain was increased until a reasonable fast-time response was obtained. Then the integral gain was increased until zero state error was obtained with no overshoot. The determined gains are $K_p = 0.01$ and $K_i = 0.0025$. A listing for the code to perform PI control of the right wheel is shown in Figure 11.28. An identical code was used for the left wheel.

The structure of the code for this project was designed to allow the mobile robot to track any n-sided, equal sides polygon with a user-specified polygon side length (see Figure 6.19). Due to the limited RAM storage capacity of the microcontroller on this robot, the trajectory for the entire polygon shape cannot be generated before the motion begins, and each segment trajectory is generated just before the motion along that segment starts. A flag (the variable *ControlOnFlag*) was used to enable/disable the PI-feedback controller based on the timing of the trajectory generation and usage in the code. Figures 11.29 and 11.30 (see pages 378 and 379) show the code that implements the state-transition diagram shown in Figure 6.20. The code for each state follows the structure shown in that figure.

Due to the use of a USB cable that is connected to the robot, short distances such as 1000–2000 encoder counts per linear segment are recommended to be used. In addition, due to the accumulation of positioning errors after a few polygon moves which may cause a jolt in the first segment motion, it is recommended to reset the MCU after every few polygon moves.

Figure 11.27

Examples of position trajectories generated by the code shown in Figure 11.26: (a) trapezoidal velocity profile and (b) triangular velocity profile

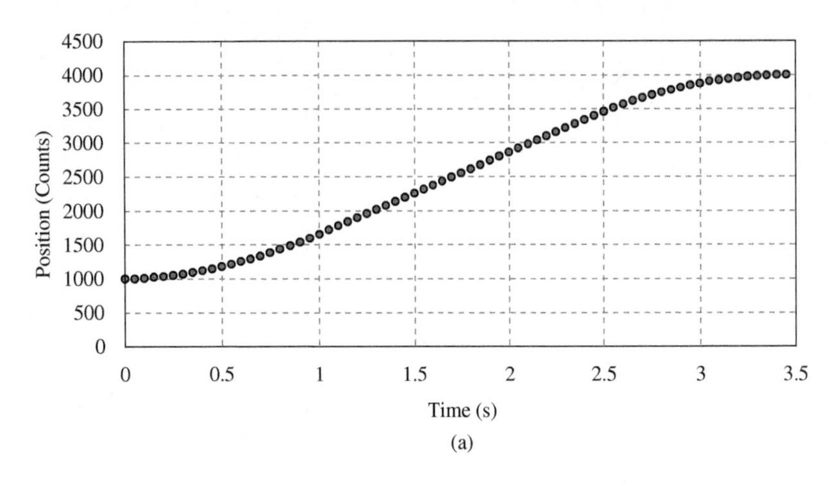

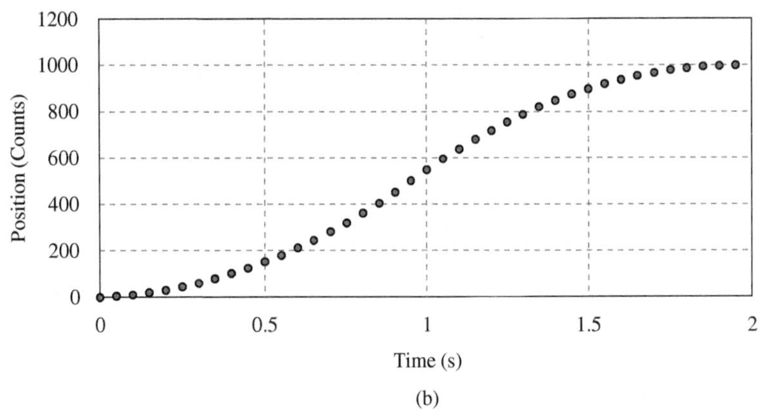

Figure 11.28

PI controller for the right wheel

```
void DoControlRightMotor(int DesiredRight, int CountsRight) {
  int duty;
  float ControlOutput;                                  // Output of PI-controller
  static float SumErrorR;                               // Sum of errors
  float error = DesiredRight - CountsRight;             // Compute current error
  ControlOutput = kp * error + ki * SumErrorR * Tsamp_s; // Compute control output
                                                        // P and I terms of PI controller. Current error is
                                                        //  not used in the computation of I-term
  SumErrorR = SumErrorR + error;                        // Update sum of errors expression

  duty = (ControlOutput / 9.0) * 255;                   // Convert control output to duty cycle
                                                        // 9 is the output voltage available on the robot
  if (duty >= 0)                                        // Positive control output case
  {
    if (duty >= 255)                                    // Check if the duty cycle exceeds the limit
    {
      duty = 255;
    }
    digitalWrite(RightMotDir, LOW);
    analogWrite(RightMotPWM, duty);                     // Send PWM signal to the H-bridge driver
  } else {                                              // Negative control output case

    if (duty <= -255)                                   // Check if duty cycle exceeds limit
    {
      duty = -255;
    }
    digitalWrite(RightMotDir, HIGH);
    analogWrite(RightMotPWM, -1 * duty);                // Send PWM signal to the H-bridge driver
  }
}
```

Figure 11.29 Code that implements that state-transition diagram shown in Figure 6.20

```
void MotionTask(void) {
 State = NextState;                                   // Update the state
 int RightError, LeftError;
 switch (State) {
  case 1:                                             // Start state
   if (Command > 1 && Command < 11) {                 // Check if number of segments is between 2 to 10
    Command = 15;  // Reset Command                   // Reset Command variable
    RotationCount = CountsPerTurn / NumofSegments;    // Compute number of counts for the rotation
                                                      //  segment that follows each linear segment
    SegmentsCount = 0;                                // Initialize segment counter
    NextState = 2;
   }
   break;
  case 2:                                             // Linear motion state
   if (EntryState2 == false) {                        // Check if it is the first time in this state
    SegmentsCount++;                                  // Increment segment counter
                          // Generate the right and left wheel position trajectory for the current segment
    Npoints = GeneratePosTraj(GoalRight, GoalRight + SegmentLength, 1200, Tup / 1000.0, TrajR);
    Serial.println("Right wheel # points " + String(Npoints));
    Npoints = GeneratePosTraj(GoalLeft, GoalLeft + SegmentLength, 1200, Tup / 1000.0, TrajL);
    Serial.println("Left wheel # points " + String(Npoints));
    GoalRight = GoalRight + SegmentLength;            // Set the desired position at the end of the segment
    GoalLeft = GoalLeft + SegmentLength;
    LastTime = millis();                              // Record the time
    NSamp = 0;                                        // Initialize the samples count
    NTraj = 0;                                        // Initialize the trajectory points counter
    ControlOnFlag = 1;                                // Set the ControlOnFlag to 1
    EntryState2 = true;                               // Set the state entry variable to true
    Serial.println("Linear Motion Segment " + String(SegmentsCount));
   }
   RightError = GoalRight - encoders.getCountsRight(); // Determine deviation from segment end point
   LeftError = GoalLeft - encoders.getCountsLeft();
         // Check if the robot has completed the motion segment within the specified error band
   if ((abs(RightError) <= PositionError) && (abs(LeftError) <= PositionError)) {
    NextState = 3;                                    // Transition to state 3
    ControlOnFlag = 0;                                // Reset the ControlOnFlag to 0
    EntryState2 = false;                              // Reset the state entry flag to false
   }
   if (SegmentsCount > NumofSegments || Command == 0) { //check if path completed or user issued a stop command
    ControlOnFlag = 0;                                // Reset the ControlOnFlag to 0
    NextState = 4;                                    // Transition to state 4
    EntryState2 = 0;                                  // Reset the state entry flag to false
   }
   break;
```

11.5.6 Summary and Modifications

This section, along with the integrated case study, delves into the operation of various components in an educational mobile robot. It highlights the use of the ATmega32U4 microcontroller for controlling the robot and explores the DRV8838 driver chip that powers the robot's brush motors. The study also examines the inertial measurement sensor on the robot's control board. Furthermore, it includes code examples for different operating modes of the robot and the implementation of a closed-loop feedback controller to precisely manage its motion

The case study presented offers multiple avenues for customization and enhancement. Users have the option to add new components, like an ultrasonic sensor or a display board, to the existing setup. Furthermore, the project can be expanded by integrating a Graphical User Interface (GUI). This GUI could enable users to command the robot, plot its current location, or display status information. In terms of software, different

Figure 11.30 Continuation of the code that implements that state-transition diagram shown in Figure 6.20

```
case 3:                                              // Rotational motion state
  if (EntryState3 == false) {                        // Check if it is the first time in this state
    // Generate the right and left wheel position trajectory for the current segment
    Npoints = GeneratePosTraj(GoalRight, GoalRight + RotationCount, 1200, Tup / 1000.0, TrajR);
    Serial.println("Right wheel # points rotation" + String(Npoints));
    Npoints = GeneratePosTraj(GoalLeft, GoalLeft - RotationCount, -1200, Tup / 1000.0, TrajL);
    Serial.println("Left wheel # points rotation" + String(Npoints));
    GoalRight = GoalRight + RotationCount;
    GoalLeft = GoalLeft - RotationCount;
    LastTime = millis();
    NSamp = 0;                                       // Initialize the samples count
    NTraj = 0;                                       // Initialize the trajectory points counter
    ControlOnFlag = 1;                               // Set the ControlOnFlag to 1
    EntryState3 = true;                              // Set the state entry variable to true
    Serial.println("Rotational Motion State " + String(SegmentsCount));
  }
  RightError = GoalRight - encoders.getCountsRight();     // Determine deviation from segment end point
  LeftError = GoalLeft - encoders.getCountsLeft();
  if ((abs(RightError) <= PositionError) && (abs(LeftError) <= PositionError)) {
    NextState = 2;                                   // Transition to state 2
    EntryState3 = false;                             // Reset the state entry flag to false
    ControlOnFlag = 0;                               // Reset the ControlOnFlag to 0
  }

  if (Command == 0) {
    ControlOnFlag = 0;                               // Reset the ControlOnFlag to 0
    NextState = 4;                                   // Transition to state 4
    EntryState3 = 0;                                 // Reset the state entry flag to false
  }
  break;
 case 4:                                             // Stop State
  if (EntryState4 == false) {
    ControlOnFlag = 0;                               // Reset the ControlOnFlag to 0
    NSamp = 0;                                       // Initialize the samples count
    NTraj = 0;                                       // Initialize the trajectory points counter
    analogWrite(RightMotPWM, 0);                     // Shut off both motors
    analogWrite(LeftMotPWM, 0);
    Serial.println("Stop Motion State");
    EntryState4 = true;                              // Set the state entry variable to true
  }
  if (Command == 0) {
    NextState = 1;                                   // Transition to state 1
    EntryState4 = 0;                                 // Reset the state entry flag to false
  }
 }
}
```

averaging methods can be applied to simulate the operation of the airbag sensor. The feedback controller, initially developed for polygon motion, can be adapted for various driving modes of the robot. Additionally, programming diverse path types, such as a mix of linear and circular segments, is another area for exploration.

11.5.7 List of Parts Needed

The only component that is required for this project is the Romi robot which is available from Pololu. The Romi control board includes an ATmega32U4 MCU which can be conveniently programmed from the Arduino IDE as an Arduino Leonardo board.

11.6 Chapter Summary

This chapter provided detailed setup information about the four integrated case studies that were discussed throughout the text. It also provided code listings, additional modeling, simulation, experimental results, a list of the parts needed for each case study, and ways to extend or modify these case studies. The purpose of these case studies is to illustrate the integration of the many topics covered in this book, as well as to provide practical information on the design and control of the mechatronic systems used in these case studies.

The first case study illustrated the matching of a stepper motor with a stepper driver and the use of an Arduino MCU to perform open-loop control of the motion of a stepper-driven rotary table that uses a photo interrupter as a homing sensor. A state-transition diagram was created to handle the commands that were specified for the operation of the table, and the control code was coded in the Arduino Programming Language. The second case study considered the closed-loop position control of a custom-built DC motor-driven linear slide. The dynamics of the motor were identified using step-response tests, and a cascaded type controller was designed and simulated in MATLAB before it was implemented on the actual system using the Simulink Desktop Real-Time toolbox. The third case study considered the temperature control of a small copper plate heated by a flexible heater. The dynamics of the heated plate were identified from open-loop step-response tests, and the developed model was used to design a PI controller. The PI controller was implemented in an Arduino microcontroller. The last case study considered the motion and control of an educational-type robot. It discussed feedback using encoders and the inertial measurement sensor on the robot. It also illustrated code for driving this robot in different ways, as well as implementing a closed-loop position feedback controller to control the motion of the robot.

Bibliography

[1] F. Harashima, M. Tomizuka, and T. Fukuda. "Mechatronics—What Is It, Why and How," *IEEE/ASME Trans. on Mechatronics*, Vol. 1, No. 1, pp. 1–3, 1996.

[2] D. Auslander, J. Ridgely, and J. Ringgenberg. *Control Software for Mechanical Systems: Object-Oriented Design in a Real-Time World*, Prentice Hall PTR, Upper Saddle River, NJ, 2002.

[3] R. Boylestad. *Introductory Circuit Analysis*, 3rd Edition, Charles E. Merrill Publishing Company, Columbus, OH, 1977.

[4] J. Irwin and R. Nelms. *Basic Engineering Circuit Analysis*, 9th Edition, Wiley, Hoboken, NJ, 2008.

[5] G. Rizzoni and J. Kearns. *Principles and Applications of Electrical Engineering*, 6th Edition, McGraw-Hill, 2015.

[6] R. Coughlin and F. Driscoll. *Operational Amplifiers and Linear Integrated Circuits*, 6th Edition, Prentice Hall, 2000.

[7] P. Scherz and S. Monk. *Practical Electronics for Inventors*, 4th Edition, McGraw-Hill TAB, 2016.

[8] P. Horowitz and W. Hill. *The Art of Electronics*, Cambridge University Press, Cambridge, UK, 1980.

[9] C. Roth. *Fundamentals of Logic Design*, 3rd Edition, West Publishing Co., St. Paul, MN, 1985.

[10] M. Rafiquzzaman. *Fundamentals of Digital Logic and Microcomputer Design*, 5th Edition, John Wiley & Sons, 2005.

[11] M. Rafiquzzaman. *Microcontroller Theory and Applications with the PIC18F*, 2nd Edition, John Wiley & Sons, 2018.

[12] S. Barrett and D. Pack. *Atmel AVR Microcontroller Primer: Programming and Interfacing*, 2nd Edition, Morgan & Claypool Publishers, 2012.

[13] C. Maxfield. *The Design Warrior's Guide to FPGAs: Devices, Tools and Flows*, Elsevier, 2004.

[14] *ATmega48A/PA/88A/PA/168A/PA/328/P megaAVR Data Sheet*, Microchip Technology Inc., 2020.

[15] Y. Elder. *Sampling Theory: Beyond Bandlimited Systems*, Cambridge University Press, 2015.

[16] K. Astrom and B. Wittenmark. *Computer-Controlled Systems: Theory and Design*, 3rd Edition, Prentice-Hall, 1997.

[17] S. Smith. *The Scientist & Engineer's Guide to Digital Signal Processing*, California Technical Publishing, San Diego, CA, 1997.

[18] J. Seams. "R/2R Ladder Networks," *Application Note IRC/AFD006*, International Resistive Company, Inc., 1988.

[19] S. Lyshevski. "Mechatronic Curriculum—Retrospect and Prospect," *Mechatronics*, Vol. 12, pp. 195–205, 2002.

[20] P. Laplante. *Real-Time Systems Design and Analysis*, 3rd Edition, Wiley, 2004.

[21] D. Auslander, A. Huang, and M. Lemkin. "A Design and Implementation Methodology for Real Time Control of Mechanical Systems," *Mechatronics*, Vol. 5, No. 7, pp. 811–832, 1995.

[22] W. Palm. *System Dynamics*, 4th Edition, McGraw-Hill, New York, NY, 2021.

[23] K. Ogata. *System Dynamics*, 4th Edition, Pearson, 2003.

[24] D. Karnopp, D. Margolis, and R. Rosenberg. *System Dynamics: Modeling, Simulation, and Control of Mechatronic Systems*, 5th Edition, John Wiley & Sons, 2012.

[25] G. Franklin, J. Powell, and A. Emami-Naeini. *Feedback Control of Dynamic Systems*, 7th Edition, Prentice-Hall, 2014.

[26] W. Palm III. *Introduction to MATLAB for Engineers*, 3rd Edition, McGraw-Hill, New York, NY, 2011.

[27] A. Gilat. *MATLAB: An Introduction with Applications*, 3rd Edition, Wiley, Hoboken, NJ, 2008.

[28] T. Beckwith, R. Marangoni, and J. Lienhard V. *Mechanical Measurements*, 6th Edition, Prentice-Hall, 2007.

[29] J. Webster (Editor-in-Chief). *The Measurement, Instrumentation and Sensors Handbook*, CRC Press, 1999.

[30] J. Holman. *Experimental Methods for Engineers*, 7th Edition, McGraw-Hill, 2001.

[31] NIST ITS-90 *Thermocouple Database*, https://srdata.nist.gov/its90/main/ [Last accessed: December 3, 2023]

[32] S. Rao. *Mechanical Vibrations*, 3rd Edition, Addison-Wesley, Reading, MA, 1995.

[33] P. Schumacher and M. Jouaneh. "A Force Sensing Tool for Disassembly Operations," *Robotics and Computer-Integrated Manufacturing*, Vol. 30, No. 2, pp. 206–217, 2014.

[34] A. Hughes. *Electric Motors and Drives: Fundamentals, Types and Applications*, 3rd Edition, Newness, 2006.

[35] I. Gottlieb. *Electric Motors and Control Techniques*, 2nd Edition, TAB Books, McGraw-Hill, 2004.

[36] T. Kenjo. *Electric Motors and Their Controls: An Introduction*, Oxford University Press, 1991.

[37] A. Parr. *Hydraulics and Pneumatics: A Technician's and Engineer's Guide*, Elsevier, 2011.

[38] W. Brown. "Brushless DC Motor Control Made Easy," *Application Note AN857*, Microchip Technology, Inc., 2002.

[39] L. Elevich. "3-Phase BLDC Motor Control with Hall Sensors Using 56800/E Digital Signal Controllers," *Application Note AN1916*, Rev. 2.0, Freescale Semiconductor, 2005.

[40] ISO 1219-1:2012—*Fluid power systems and components*, https://www.iso.org/standard/60184.html

[41] K. Uchino. *Piezoelectric Actuators and Ultrasonic Motors*, Vol. 1., Springer Science & Business Media, 1996.

[42] A. Rao, A. Srinivasa, and J. Reddy. *Design of Shape Memory Alloy (SMA) Actuators*, Springer International Publishing, Cham, 2015.

[43] S. Wu, G. Baker, J. Yin, and Y. Zhu. "Fast Thermal Actuators for Soft Robotics," *Soft Robotics*, Vol. 9, No. 6, pp. 1031–1039, 2022.

[44] Y. Bar-Cohen (Editor). "Electroactive Polymer (EAP) Actuators as Artificial Muscles: Reality, Potential, and Challenges," Vol. 136, SPIE Press, 2004.

[45] K. Ogata. *Modern Control Engineering*, 4th Edition, Prentice-Hall, 2002.

[46] W. Levine (Editor). *The Control Handbook*, Chapter 10, CRC Press, 1996.

[47] J. Ziegler and N. Nichols. "Optimum Settings for Automatic Controllers," *Transactions of the American Society of Mechanical Engineers*, Vol. 64, No. 8, pp. 759–65, 1942.

[48] G. Balas, R. Chiang, A. Packard, and M. Safonov. *Robust Control Toolbox 3*, The Mathworks, Inc., Natick, MA, 2005.

[49] A. Preumont. *Vibration Control of Active Structures*, 2nd Edition, Kluwer Academic Publishers, Dordrecht, The Netherlands, 2002.

[50] S. Thrun, W. Burgard, and D. Fox. *Probabilistic Robotics* (*Intelligent Robotics and Autonomous Agents*), The MIT Press, 2005.

Appendix A

Standard Electronic Decade Value Tables

Standard Decade Resistance Values

The following table lists four established number series which are used as preferred values in electronic design. Each series is shown under an associated value of tolerance %. The number series under the ±10% column is known as the E12 Series because there are 12 standard values within a decade range. E24 series is utilized by ±2% and ±5%, ± 1% uses E96, and ± 0.1%, ± 0.25% and ± 0.5% use E192. Successive values within a decade series are related (approximately) by a factor of $\sqrt[12]{10}$ for the E12 Series, $\sqrt[24]{10}$ for the E24 Series, $\sqrt[96]{10}$ for the E96 Series, and $\sqrt[192]{10}$ for the E192 Series.

Use of standard values is encouraged because stocking programs are designed around them.

±0.1 % ±0.25 % ±0.5 %	±1 %	±0.1 % ±0.25 % ±0.5 %	±1 %	±0.1 % ±0.25 % ±0.5 %	±1 %	±0.1 % ±0.25 % ±0.5 %	±1 %	±0.1 % ±0.25 % ±0.5 %	±1 %	±0.1 % ±0.25 % ±0.5 %	±1 %	±2 % ±5 %	±10 %
10.0	10.0	14.7	14.7	21.5	21.5	31.6	31.6	46.4	46.4	68.1	68.1	10	10
10.1		14.9		21.8		32.0		47.0		69.0		11	–
10.2	10.2	15.0	15.0	22.1	22.1	32.4	32.4	47.5	47.5	69.8	69.8	12	12
10.4		15.2		22.3		32.8		48.1		70.6		13	–
10.5	10.5	15.4	15.4	22.6	22.6	33.2	33.2	48.7	48.7	71.5	71.5	15	15
10.6		15.6		22.9		33.6		49.3		72.3		16	–
10.7	10.7	15.8	15.8	23.2	23.2	34.0	34.0	49.9	49.9	73.2	73.2	18	18
10.9		16.0		23.4		34.4		50.5		74.1		20	–
11.0	11.0	16.2	16.2	23.7	23.7	34.8	34.8	51.1	51.1	75.0	75.0	22	22
11.1		16.4		24.0		35.2		51.7		75.9		24	–
11.3	11.3	16.5	16.5	24.3	24.3	35.7	35.7	52.3	52.3	76.8	76.8	27	27
11.4		16.7		24.6		36.1		53.0		77.7		30	–
11.5	11.5	16.9	16.9	24.9	24.9	36.5	36.5	53.6	53.6	78.7	78.7	33	33
11.7		17.2		25.2		37.0		54.2		79.6		36	–
11.8	11.8	17.4	17.4	25.5	25.5	37.4	37.4	54.9	54.9	80.6	80.6	39	39
12.0		17.6		25.8		37.9		55.6		81.6		43	–
12.1	12.1	17.8	17.8	26.1	26.1	38.3	38.3	56.2	56.2	82.5	82.5	47	47
12.3		18.0		26.4		38.8		56.9		83.5		51	–
12.4	12.4	18.2	18.2	26.7	26.7	39.2	39.2	57.6	57.6	84.5	84.5	56	56
12.6		18.4		27.1		39.7		58.3		85.6		62	–
12.7	12.7	18.7	18.7	27.4	27.4	40.2	40.2	59.0	59.0	86.6	86.6	68	68
12.9		18.9		27.7		40.7		59.7		87.6		75	–
13.0	13.0	19.1	19.1	28.0	28.0	41.2	41.2	60.4	60.4	88.7	88.7	82	82
13.2		19.3		28.4		41.7		61.2		89.8		91	–
13.3	13.3	19.6	19.6	28.7	28.7	42.2	42.2	61.9	61.9	90.9	90.9		
13.5		19.8		29.1		42.7		62.6		92.0			
13.7	13.7	20.0	20.0	29.4	29.4	43.2	43.2	63.4	63.4	93.1	93.1		
13.8		20.3		29.8		43.7		64.2		94.2			
14.0	14.0	20.5	20.5	30.1	30.1	44.2	44.2	64.9	64.9	95.3	95.3		
14.2		20.8		30.5		44.8		65.7		96.5			
14.3	14.3	21.0	21.0	30.9	30.9	45.3	45.3	66.5	66.5	97.6	97.6		
14.5		21.3		31.2		45.9		67.3		98.8			

Standard resistance values are obtained from the decade table by multiplying by powers of 10. As an example, 13.3 can represent Ω, 133 Ω, 1.33 kΩ, 13.3 kΩ, 133 kΩ, and 1.33 MΩ.

Source: Standard Electronic Decade Value Tables, Decade Table, Vishay Intertechnology, Inc.

Appendix B

7-Bit ASCII Code

ASCII Codes

Dec	Hex	Code	Dec	Hex	Code	Dec	Hex	Code	Dec	Hex	Code
0	00	NUL	32	20	space	64	40	@	96	60	`
1	01	SOH	33	21	!	65	41	A	97	61	a
2	02	STX	34	22	"	66	42	B	98	62	b
3	03	ETX	35	23	#	67	43	C	99	63	c
4	04	EOT	36	24	&	68	44	D	100	64	d
5	05	ENQ	37	25	%	69	45	E	101	65	e
6	06	ACK	38	26	$	70	46	F	102	66	f
7	07	BEL	39	27	'	71	47	G	103	67	g
8	08	BS	40	28	(	72	48	H	104	68	h
9	09	HT	41	29	)	73	49	I	105	69	i
10	0A	LF	42	2A	*	74	4A	J	106	6A	j
11	0B	VT	43	2B	+	75	4B	K	107	6B	k
12	0C	FF	44	2C	,	76	4C	L	108	6C	l
13	0D	CR	45	2D	–	77	4D	M	109	6D	m
14	0E	SO	46	2E	.	78	4E	N	110	6E	n
15	0F	SI	47	2F	/	79	4F	O	111	6F	o
16	10	DLE	48	30	0	80	50	P	112	70	p
17	11	DC1	49	31	1	81	51	Q	113	71	q
18	12	DC2	50	32	2	82	52	R	114	72	r
19	13	DC3	51	33	3	83	53	S	115	73	s
20	14	DC4	52	34	4	84	54	T	116	74	t
21	15	NAK	53	35	5	85	55	U	117	75	u
22	16	SYN	54	36	6	86	56	V	118	76	v
23	17	ETB	55	37	7	87	57	W	119	77	w
24	18	CAN	56	38	8	88	58	X	120	78	x
25	19	EM	57	39	9	89	59	Y	121	79	y
26	1A	SUB	58	3A	:	90	5A	Z	122	7A	z
27	1B	ESC	59	3B	;	91	5B	[	123	7B	{
28	1C	FS	60	3C	<	92	5C	\	124	7C	|
29	1D	GS	61	3D	=	93	5D	]	125	7D	}
30	1E	RS	62	3E	>	94	5E	^	126	7E	~
31	1F	US	63	3F	?	95	5F	_	127	7F	DEL

Appendix C

Introduction to the Laplace Transform

This Appendix briefly reviews the Laplace Transform that is used in engineering and mathematics for solving differential equations and analyzing linear time-invariant systems such as electrical circuits, mechanical systems, and control systems. It also provides a table of useful properties for the Laplace Transform as well as a listing of the Laplace Transform of some common time functions. The Laplace Transform is an integral transform named after Pierre-Simon Laplace. It transforms complex differential equations into simpler algebraic forms. **The Laplace Transform of a function $f(t)$** is defined as:

$$F(s) = \mathcal{L}(f(t)) = \int_0^{\infty} f(t)\,e^{-st}\,dt \qquad \text{(C.1)}$$

where s is a complex number. The above definition can be used to obtain the Laplace Transform of a given time function. For example, the Laplace Transform of the unit step function $u(t)$ is obtained as:

$$F(s) = \int_0^{\infty} f(t)\,e^{-st}\,dt = \int_0^{\infty} 1e^{-st}\,dt = \left[-\frac{1}{s}e^{-st}\right]_0^{\infty} = \frac{1}{s} \qquad \text{(C.2)}$$

As an illustration of how the Laplace Transform can be used to solve differential equations, let us consider the first-order differential equation given by Equation (C.3) with an initial condition $x(0) = 2$.

$$2\dot{x} + 3x = 5 \qquad \text{(C.3)}$$

Applying the Laplace Transform to both sides of Equation (C.3) and using Property 7 in Table C.1 for the $\dot{x}$ term and Case 3 in Table C.2 for the right term gives

$$2(sX(s) - x(0)) + 3X(s) = \frac{5}{s} \qquad \text{(C.4)}$$

Substituting the value for the initial condition $x(0)$ in Equation (C.4) and solving for $X(s)$ gives

$$X(s) = \frac{4}{2s+3} + \frac{5}{s(2s+3)} \qquad \text{(C.5)}$$

To obtain the solution of the differential equation or $x(t)$, we need to obtain the inverse Laplace Transform for each of the two expressions on the right side of Equation (C.5). The term $\frac{4}{2s+3}$ is obtained from Case 5 in Table C.2, while the term $\frac{5}{s(2s+3)}$ is obtained from Case 11 in Table C.2. To use Table C.2, we express each of these two terms in a form that matches the format of the corresponding expressions in Table C.2, or

$$X(s) = 2\left(\frac{1}{s+3/2}\right) + 5/3\left(\frac{3/2}{s(s+3/2)}\right) \qquad \text{(C.6)}$$

Hence, $x(t)$ is then

(C.7)
$$x(t) = 2e^{-\frac{3}{2}t} + \frac{5}{3}(1 - e^{-\frac{3}{2}t}) = \frac{5}{3} + \frac{1}{3}e^{-\frac{3}{2}t}$$

Table C.1 has a list of useful Laplace Transform properties, and Table C.2 lists the Laplace Transform of common time functions.

Table C.1

Useful properties of the Laplace Transform

1. **Linearity:** Used for simplifying the transform of a linear combination of functions $$\mathcal{L}(af(t) + bg(t)) = a\mathcal{L}(f(t)) + b\mathcal{L}(g(t))$$
2. **Time Shifting:** Applies to shifted functions in time, useful in solving delayed systems $$\mathcal{L}(u_s(t - a)f(t - a)) = e^{-as}F(s)$$
3. **Exponential Shifting:** Used for functions multiplied by an exponential term, often in damping systems $$\mathcal{L}(e^{at}f(t)) = F(s - a)$$
4. **Convolution Formula:** Converts convolution in time domain to multiplication in the s-domain, useful in system analysis $$\mathcal{L}(f(t) * g(t)) = F(s)G(s)$$
5. **Initial Value Theorem:** Determines the initial behavior of a function without performing inverse transform $$\underset{t\to 0+}{f(t)} = \lim_{s\to\infty} sF(s)$$
6. **Final Value Theorem:** Predicts the long-term behavior of a function, especially in stable systems $$\underset{t\to\infty}{f(t)} = \lim_{s\to 0} sF(s)$$
7. **First Derivative in Time Domain:** Transforms the first derivative of a function $$\mathcal{L}\left(\frac{df}{dt}\right) = sF(s) - f(0)$$
8. **Second Derivative in Time Domain:** Transforms the second derivative of a function $$\mathcal{L}\left(\frac{d^2f}{dt^2}\right) = s^2F(s) - sf(0) - \dot{f}(0)$$
9. **Integration in Time Domain:** Used for transforming the integral of a function $$\mathcal{L}\left(\int_0^t f(t)\,dt\right) = \frac{F(s)}{s}$$

Table C.2

Table of Laplace Transform of common time functions

	Time Function $f(t)$, $t \geq 0$	Laplace Transform $F(s)$
1.	Unit impulse, $\delta(t)$	1
2.	Unit step function, $u_s(t)$	$\frac{1}{s}$
3.	Constant, c	$\frac{c}{s}$
4.	Delayed unit step, $u_s(t - D)$	$\frac{e^{-Ds}}{s}$
5.	e^{-at}	$\frac{1}{s + a}$
6.	t^n	$\frac{n!}{s^{n+1}}$
7.	$\sin(bt)$	$\frac{b}{s^2 + b^2}$
8.	$\cos(bt)$	$\frac{s}{s^2 + b^2}$
9.	$e^{-at}\sin(bt)$	$\frac{b}{(s + a)^2 + b^2}$
10.	$e^{-at}\cos(bt)$	$\frac{s + a}{(s + a)^2 + b^2}$
11.	$1 - e^{-at}$	$\frac{a}{s(s + a)}$
12.	$\frac{1}{b - a}(e^{-at} - e^{-bt})$	$\frac{1}{(s + a)(s + b)}$
13.	$\frac{1}{b - a}[(p - a)e^{-at} - (p - b)e^{-bt}]$	$\frac{s + p}{(s + a)(s + b)}$
14.	$at - 1 + e^{-at}$	$\frac{a^2}{s^2(s + a)}$
15.	$1 - (at + 1)e^{-at}$	$\frac{a^2}{s(s + a)^2}$
16.	$t\sin(bt)$	$\frac{2bs}{(s^2 + b^2)^2}$
17.	$t\cos(bt)$	$\frac{s^2 - b^2}{(s^2 + b^2)^2}$
18.	$1 - \cos(bt)$	$\frac{b^2}{s(s^2 + b^2)}$
19.	$\sinh(bt)$	$\frac{b}{s^2 - b^2}$
20.	$\cosh(bt)$	$\frac{s}{s^2 - b^2}$
21.	$\frac{1}{a^2 + b^2}\left[1 - \left(\frac{a}{b}\sin(bt) + \cos(bt)\right)e^{-at}\right]$	$\frac{1}{s[(s + a)^2 + b^2]}$

Appendix D

Modeling of Mechanical Systems

D.1 Introduction

This Appendix reviews the dynamic modeling of mechanical systems that include rigid-body and compliant elements. These models can be combined with dynamic models of actuators and sensors to produce a complete model of a mechatronic system. With the availability of such a model, system characteristics such as time constant and bandwidth can be obtained. Also, the response of the system under a variety of input conditions can be simulated.

A schematic of two typical systems that involve solid mechanical elements is shown in Figure D.1. In Figure D.1(a), a motor drives a disk (an inertia load) that is mounted on a shaft through a gear pair. This is an example of a rotary motion system where the torque supplied by the motor causes the load to rotate. The rotational speed of the load is a function of the torque supplied by the motor.

Figure D.1

(a) Rotational motion and (b) linear motion system

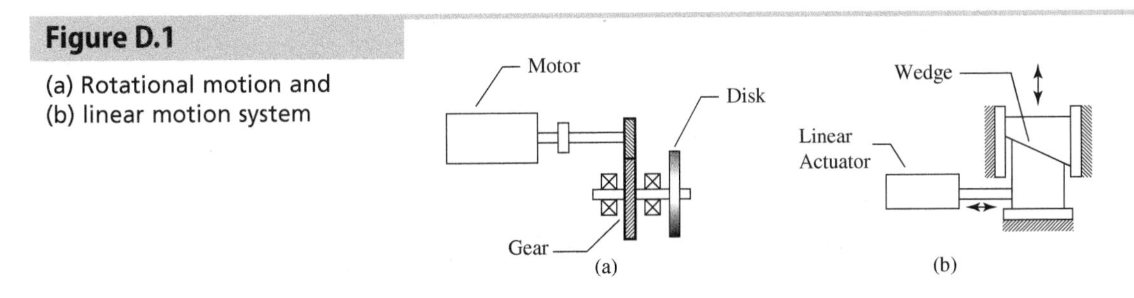

Figure D.1(b) shows a linear motion system where the force applied by the linear actuator causes the upper wedge to move up as the lower wedge is moved to the right. In general, many mechanical systems involve a combination of translational and rotational elements. In modeling such systems, we are interested in obtaining a mathematical model that relates the input(s) to the system such as the force or torque applied by an actuator to the output(s) of the systems such as the position or velocity of one or more of the mass elements in the system. Such a mathematical model can be obtained by either applying Newton's Law or by using an energy approach. We will first consider mechanical systems without spring elements.

D.2 Modeling of Mechanical Systems Without Spring Elements

While all objects deform under the action of forces or torques, we can simplify the modeling if we consider all objects to be rigid. Such an approximation is valid if the deflection of an object due to an applied force or torque is small in comparison to the length scale of the object along the direction of the applied force or torque. Rigid bodies undergo three distinct types of motion. These are translation, rotation about a fixed axis, and general plane motion which involve both translation and rotation. We will consider each type of motion separately.

In **translation**, any line on the object remains parallel to its original direction during the motion. Examples of translational motion include the motion of elevator cars, the motion of feed axes in machine tools, and the guided transport of packages in a straight line. Figure D.2(a) shows an example of translational motion in which a winch pulls up a cart on an inclined plane.

Figure D.2

(a) Example of a translational motion system and (b) FBD of this system

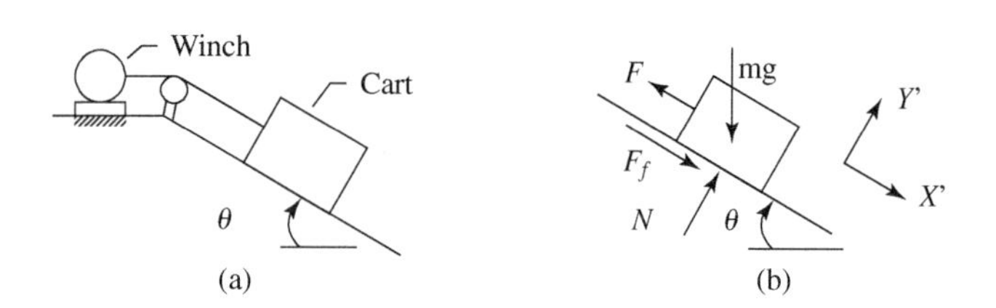

The equation of motion of the cart can be obtained by either applying Newton's 2nd Law of Motion or by using the Work-Energy Principle. We will drive the equation of motion using Newton's 2nd Law in this section, and we will rework this problem again in Section D.4 after we cover the Work-Energy Principle. **Newton's 2nd Law of Motion** states that if an unbalanced system of forces acts on an object that is at rest or moving with a constant velocity in a straight line, the object will change its state and will accelerate with an acceleration that is a function of the mass of the object. In equation form, the law is stated as

(D.1)
$$\Sigma \mathbf{F} = m\mathbf{a}$$

The bold font is to indicate that both the force and acceleration are vector quantities. In applying Newton's 2nd Law, it helps to draw a free-body diagram (FBD) of the object in question. In an FBD, the object is isolated from its surroundings, and all the forces acting on the object are shown. The forces include:

- Forces due to the action of gravity
- Reaction forces from supports
- External forces
- Friction forces that act to resist motion

A free-body diagram of the cart in Figure D.2(a) is shown in Figure D.2(b). The FBD has four forces. These are the gravity force (mg), the normal reaction force (N) from the inclined plate, the pulling force in the cable (F), and the friction force (F_f) at the interface between the block and the inclined plane. Note how the friction force is drawn to oppose the direction of motion. The combined action of these four forces causes the block to accelerate up the plane. This resultant action is represented by the vector m**a** that is drawn on another diagram of the object called the **kinetic diagram** (see Example D.1). From equating the FBD and the kinetic diagrams, we obtain the equations of motion. Instead of applying Newton's 2nd Law in vector form, we will use the scalar form of the law in which Newton's 2nd Law is applied in two orthogonal directions such as the X' and Y' directions that are shown in the figure. Applying Newton's 2nd Law along the X'- direction, we obtain

(D.2)
$$F - mg\sin\theta - F_f = m\ddot{x}$$

And along the Y'- direction

(D.3)
$$N - mg\cos\theta = 0$$

To solve the above two equations, we need a model of the friction force. The particular model of friction depends on the type of lubrication present between the contact surfaces. For un-lubricated surfaces, the **Coulomb** or **dry friction model** is commonly used. Here the frictional force is proportional to the product of the normal force and a coefficient of friction that is dependent on the materials that contact each other and the relative velocity between them. The model takes the following form

(D.4)
$$F_f(v) = \begin{cases} \mu_s N, & v = 0 \\ \mu_k N, & v > 0 \end{cases}$$

where v is the velocity of the object, μ_s is the static coefficient of friction, and μ_k is the kinetic coefficient of friction. Note the Coulomb friction model is nonlinear. For lubricated surfaces, such as when bearings are present, the frictional force is assumed to be linearly proportional to the velocity of the object and is given by the following model, called the **viscous friction model**

$$F_f(v) = bv \tag{D.5}$$

where b is the viscous friction coefficient. In certain situations, the frictional force could be a combination of the above two models. For the cart problem, we will assume an un-lubricated surface. Since we are interested in determining the motion of the cart, the frictional force in this case is

$$F_f(v) = \mu_k N \tag{D.6}$$

Substituting Equation (D.6) into (D.2) and solving Equations (D.2) and (D.3), we obtain the following expression for the acceleration of the cart.

$$\ddot{x} = \frac{1}{m}[F - mg\sin\theta - \mu_k\, mg\cos\theta] \tag{D.7}$$

Equation (D.7) is a dynamic model of the motion of the cart that relates the tension in the cable to the acceleration of the cart. The cable tension is a function of the torque supplied by the motor which is itself a function of the control voltage supplied to the motor. Knowing the initial position and velocity conditions of the cart and the time history of the motor torque, this model can be solved to obtain the position and velocity of the cart as a function of time.

The above procedure can be applied to systems that involve more than one mass. In this case, a separate FBD needs to be drawn for each mass of the system. Newton's 2nd Law is then applied for each mass, and geometric constraints are used to relate the motion of one mass to the motion of the other masses in the system. The resulting equations are solved simultaneously to obtain the dynamic model. Example D.1 illustrates this approach for a system made up of two masses.

The second motion type of a rigid body is **rotation about a fixed axis**. In this mode, all lines on the rigid body undergo the same amount of angular rotation. For planer rotation about a fixed axis, Newton's 2nd Law of Motion takes the form shown below

$$\Sigma \boldsymbol{F} = m\boldsymbol{a}_G \tag{D.8a}$$

$$\Sigma \boldsymbol{M}_G = I_G\boldsymbol{\alpha} \tag{D.8b}$$

where a_G is the acceleration of the center of mass, I_G is the moment of inertia about the centroidal axis of the object, and α is the angular acceleration. If the axis of rotation happens to be a noncentroidal axis, then Equations (D.8a) and (D.8b) can be replaced by a *single* equation

$$\Sigma \boldsymbol{M}_O = I_O\boldsymbol{\alpha} \tag{D.9}$$

where I_O is the mass moment of inertia of the object about axis O-O. The mass moment of inertia is a property of the rigid body which is a function of the distribution of the mass of the rigid body around the axis. I_O is defined as

$$I_O = \int r_o^2\, dm \tag{D.10}$$

where r_o is the radial distance of the point mass dm from the rotation axis. In most cases, we do not need to perform the integration operation in Equation (D.10), but we can determine the mass moment of inertia from tables that list that property for common shapes. Such tables are provided in engineering mechanics books. When the mass moment of inertia for an object is computed about an axis other than its mass center, then we need to apply a theorem called the Parallel Axis Theorem. The **Parallel Axis Theorem** is stated as

$$I_O = I_G + md^2 \tag{D.11}$$

where d is the distance between the two parallel axes and m is the mass of the object. Example D.2 illustrates the application of the Parallel Axis Theorem, and Example D.3 illustrates Newton's 2nd Law for rotation about a fixed axis.

Example D.1 Two-Mass System

Figure D.3(a) shows a simplified model of an elevator system driven by a motor. The elevator car has a mass of m_c and the counterweight has a mass of m_w. Neglect the mass of the pulleys and the cable and derive a model that relates the force supplied by the motor to the acceleration of the car.

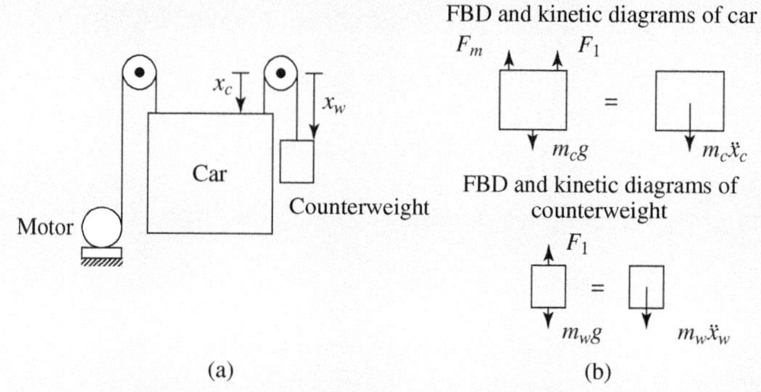

Figure D.3 (a) Simplified model of an elevator and (b) free-body and kinetic diagrams

Solution:

Since we have two masses here, we need to draw two separate FBDs. FBDs and kinetic diagrams for the car and the counterweight are shown in Figure D.3(b). Note how the direction of the kinetic force in each diagram is drawn along the assumed positive coordinate direction for each mass. From the FBD and the kinetic diagram of the car, we obtain

$$m_c g - F_m - F_1 = m_c\ddot{x}_c \tag{1}$$

Similarly for the counterweight, we obtain

$$m_w g - F_1 = m_w\ddot{x}_w \tag{2}$$

The relationship between x_w and x_c can be obtained from the geometric constraint that the length of the cable that connects the counterweight to the car is constant. Mathematically this is stated as

$$x_c + x_w = constant \tag{3}$$

Differentiating Equation (3) twice gives $\ddot{x}_w = -\ddot{x}_c$. Substituting this relationship into Equation (2), and substituting the expression for F_1 from Equation (1) gives

$$\ddot{x}_c(m_c + m_w) = [(m_c - m_w)g - F_m] \tag{4}$$

Equation (4) gives the model that relates the motor force to the car acceleration.

Example D.2 Mass Moment of Inertia

Determine the mass moment of inertia of the compound pendulum shown in Figure D.4 about an axis perpendicular to the page and passing through O. The pendulum consists of a uniform slender rod of mass m_1 and length l_1 attached to a rectangular object of mass m_2.

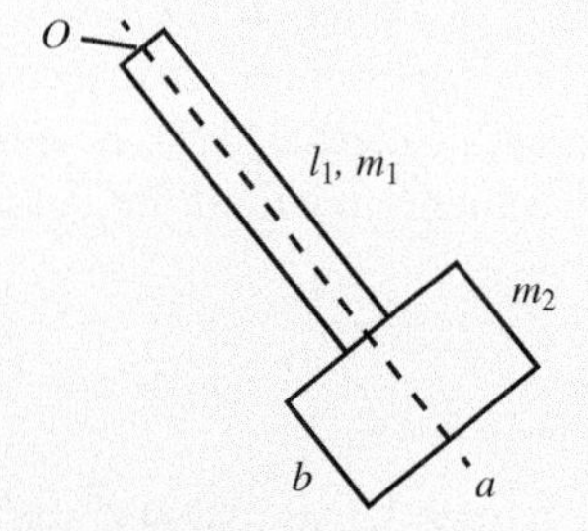

Figure D.4 Compound pendulum

Solution:
The mass moment of inertia of the slender rod about its mass center is $\frac{1}{12}m_1 l_1^2$. Using the Parallel Axis Theorem, the moment of inertia of the rod about axis O is

$$Irod_O = \frac{1}{12}m_1 l_1^2 + m_1\left(\frac{l_1}{2}\right)^2 = \frac{1}{3}m_1 l_1^2 \tag{1}$$

The mass moment of inertia of the rectangular object about its mass center is $\frac{1}{12}m_2(a^2 + b^2)$. Using the Parallel Axis Theorem, the moment of inertia of the rectangular object about axis O is

$$Isolid_O = \frac{1}{12}m_2(a^2 + b^2) + m_2\left(l_1 + \left(\frac{b}{2}\right)\right)^2 \tag{2}$$

Combining the expressions in (1) and (2) gives the mass moment of inertia of the compound pendulum or

$$I_O = Irod_O + Isolid_O$$

Example D.3 Equation of Motion for Rotation about a Fixed Axis

Derive the equation of motion for the system considered in Example D.2. The pendulum was released from rest. What is the natural frequency of oscillation of the pendulum?

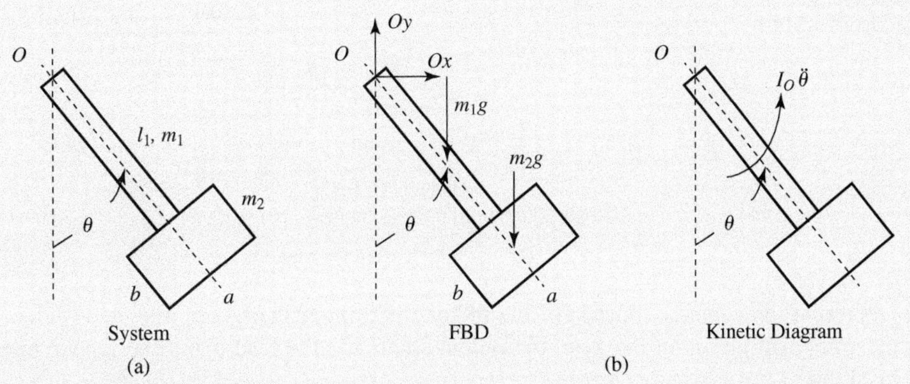

Figure D.5 (a) Compound pendulum and (b) FBD and kinetic diagrams

Solution:
The FBD and the kinetic diagrams for the pendulum are shown in Figure D.5(b). Note how the kinetic moment $I_O\ddot{\theta}$ is shown along the positive direction for θ. Summing moments about point O (with CW direction treated as positive) gives

$$m_1 g\left(\frac{l_1}{2}\right)\sin\theta + m_2 g\left(l_1 + \frac{b}{2}\right)\sin\theta = -I_O\ddot{\theta} \tag{1}$$

where I_O is as determined in Example D.2. This is a nonlinear differential equation that is depended on θ. For small θ, $\sin\theta \sim \theta$, and the above equation becomes

$$I_O\ddot{\theta} + \left(m_1 g\left(\frac{l_1}{2}\right) + m_2 g\left(l_1 + \frac{b}{2}\right)\right)\theta = 0 \tag{2}$$

Equation (2) is of the form

$$I_O\ddot{\theta} + k\theta = 0$$

and the natural frequency (see Chapter 7) is then given as $\sqrt{k/I_O}$.

Many mechanical actuation systems consist of several elements that are interconnected by components such as gears and chains. An example of such a system is shown in Figure D.6 where the load disk is rotated by a motor through a gear drive system. In deriving the equations of motion of such systems, it simplifies the derivation if we can combine the inertia of all the moving elements into a single effective inertia. This is possible if the drive system is rigid enough so that its elasticity does not affect the dynamic response. The effective inertia is calculated by summing the kinetic energy of all moving elements in the system and equating it to the kinetic energy of the effective inertia. The system in Figure D.6 has two speeds: the motor input speed and the load output speed. Hence the effective inertia can be referenced to either the motor shaft or the load shaft. Note that the **kinetic energy** of an object that undergoes translation is

(D.12)
$$K.E. - Translation = \frac{1}{2} m v_G^2$$

where v_G is the velocity of the center of mass. For an object that undergoes rotation, the kinetic energy is

(D.13)
$$K.E. - Rotation = \frac{1}{2} I_G\ \omega^2$$

For fixed-axis rotation about a non-centroidal axis O-O, the total kinetic (translational and rotational) is given by

(D.14)
$$K.E. - Rotation\ about\ axis\ O - O = \frac{1}{2} I_O \omega^2$$

Figure D.6

A simple drive system

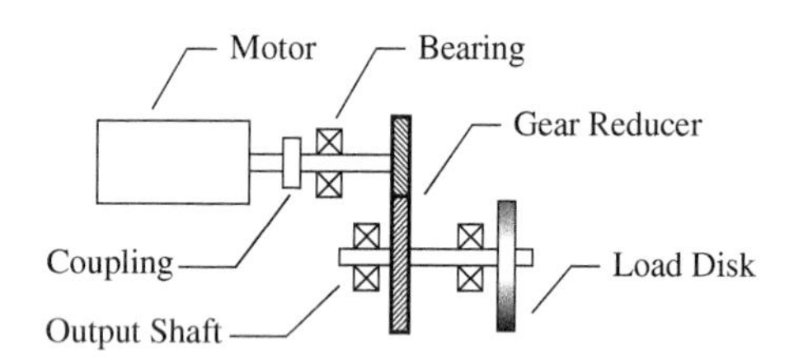

Let I_1 represent the combined inertia of the motor shaft, input shaft, coupling, and pinion, and let I_2 represent the combined inertia of the gear, the output shaft, and the load disk. The kinetic energy of the system is given by

(D.15)
$$K.E. = \frac{1}{2} I_1 \omega_1^2 + \frac{1}{2} I_2\ \omega_2^2$$

where ω_1 is the motor speed and ω_2 is the load rotational speed. Since $\omega_1/\omega_2 = N$, where N is the gear ratio, we can write the kinetic energy of the system using either the motor (input) speed or the load (output) speed. Using the input speed, the kinetic energy is then

(D.16)
$$K.E. = \frac{1}{2} I_1 \omega_1^2 + \frac{1}{2} I_2 \left(\frac{\omega_1}{N}\right)^2 = \frac{1}{2}\left(I_1 + \frac{I_2}{N^2}\right)\omega_1^2 = \frac{1}{2} Ie_1 \omega_1^2$$

where Ie_1 is the effective inertia of the system referenced to the input shaft. Similarly, using the output speed, the kinetic energy of the system can be expressed as

(D.17)
$$K.E. = \frac{1}{2} I_1 (N\omega_2)^2 + \frac{1}{2} I_2 \omega_2^2 = \frac{1}{2}(I_1 N^2 + I_2)\omega_2^2 = \frac{1}{2} Ie_2 \omega_2^2$$

where Ie_2 is the effective inertia of the system referenced to the output shaft. Note how the output inertia is reduced by the factor N^2 when referenced to the input shaft and is multiplied by the factor N^2 when referenced to the output shaft. Thus, a system with speed reduction has the advantage that load inertia variation has minimal impact on the effective inertia as seen by the motor that drives that system. This simplifies the design of a control system for the actuator. The effective inertia as seen from the input side and as seen from the output side are always related by

$$Ie_1 = Ie_2/N^2 \tag{D.18}$$

Thus if we know the effective inertia as seen from the output side, we can compute the effective inertia as seen from the input side by simply using Equation (D.18). Example 7.3 illustrates the derivation of the dynamic model for the system in Figure D.6. Examples D.4 and D.5 illustrate the computation of the effective inertia for typical drive systems.

The third type of rigid body motion is **general plane motion**. It involves both translation and rotation. Examples of this kind of motion are the motion of a rolling wheel and the motion of the connecting link in a crank and slider mechanism. For general plane motion, Newton's Law of Motion is similar to rotation about a fixed axis and takes the form shown below

$$\Sigma \boldsymbol{F} = m\boldsymbol{a}_G \tag{D.19a}$$

$$\Sigma \boldsymbol{M}_G = \boldsymbol{I}_G\boldsymbol{\alpha} \tag{D.19b}$$

In general plane motion, unlike rotation about a fixed axis, the linear acceleration and the angular acceleration of the center of mass are not necessarily related unless the object undergoes special motion such as rolling without slipping. Example D.6 illustrates this case.

Example D.4 Effective Inertia for a Positioning Table System

Determine the effective inertia of the drive system shown in Figure D.7 as seen from the motor side. The lead screw is double threaded and has a pitch of p. The table has a mass m, and the combined inertia of the motor shaft, the coupling, and the lead screw is I.

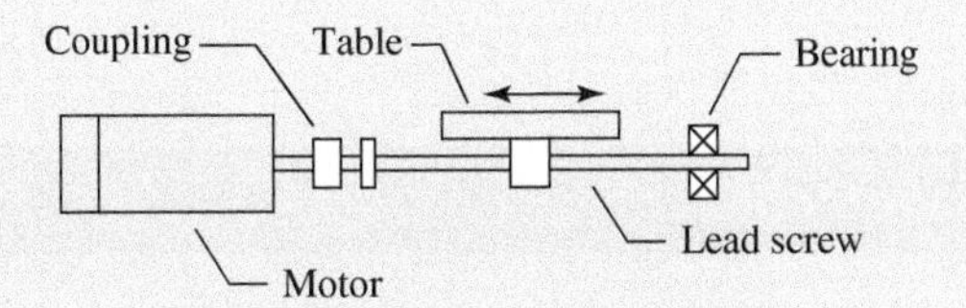

Figure D.7 Positioning table

Solution:

The combined kinetic energy of the translating table and the rotating screw is

$$K.\,E. = \frac{1}{2}mv^2 + \frac{1}{2}I\omega^2 \tag{1}$$

The linear speed of the table and the rotational speed of the screw are related by

$$v = \frac{\omega}{2\pi}l \tag{2}$$

where l is the lead or the linear displacement of the table per one revolution. The 2π factor in the denominator of Equation (2) is to convert the rotational speed ω from rad/s to rev/s. Note that the lead for this screw is twice the thread pitch since the screw is double threaded. Substituting the expression for v into Equation (1), we obtain

$$K.\,E. = \frac{1}{2}m\left(\frac{\omega l}{2\pi}\right)^2 + \frac{1}{2}I\omega^2 = \frac{1}{2}\left(m\frac{l^2}{4\pi^2} + I\right)\omega^2 \tag{3}$$

Thus, the effective inertia as seen by the motor (or input) side is

$$I_{eff} = m\frac{l^2}{4\pi^2} + I \tag{4}$$

Example D.5 Effective Inertia for a Geared System

Determine the effective inertia of the drive system shown in Figure D.8 as seen from the motor side. The system has two stages of gear reduction with the first stage having a gear ratio of N_1 and the second stage having a gear ratio of N_2. The inertia of the rotating elements on the input shaft are represented by I_1, the rotating elements on the intermediate shaft by I_2, and the rotating elements on the output shaft by I_3. The load link inertia is given as I_L.

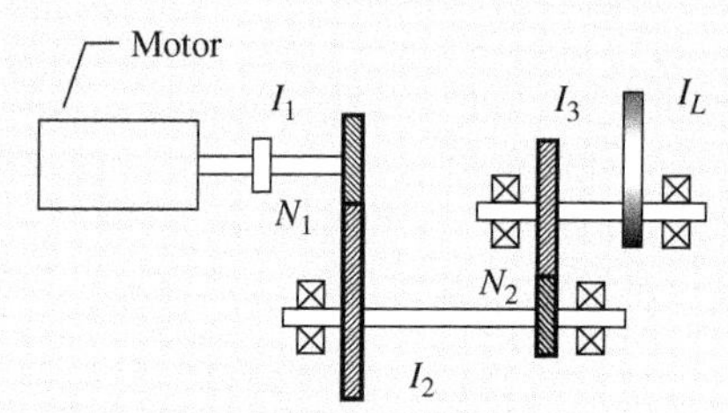

Figure D.8 Geared drive system

Solution:

The load link inertia and the output shaft inertia can be simply added together since they are rotating at the same speed. Reflecting these elements to the intermediate shaft and adding I_2, we obtain

$$I_2' = \frac{1}{N_2^2}(I_3 + I_L) + I_2$$

Reflecting the above expression to the input shaft and adding I_1, the effective inertia of the system as seen from the motor side is then

$$I_{eff} = I_1 + \frac{I_2'}{N_1^2} = I_1 + \frac{1}{N_1^2 N_2^2}(I_3 + I_L) + \frac{I_2}{N_1^2}$$

Note how the load and output inertia are reduced by the factor $(N_1N_2)^2$.

Example D.6 General Plane Motion

A disk is subjected to a force P as shown on Figure D.9(a). Determine the angular acceleration of the disk. The disk has a radius r, a mass m, and a moment of inertia about its center of mass I_G. The coefficient of static and dynamic friction between the disk and the support surface are μ_s and μ_k, respectively.

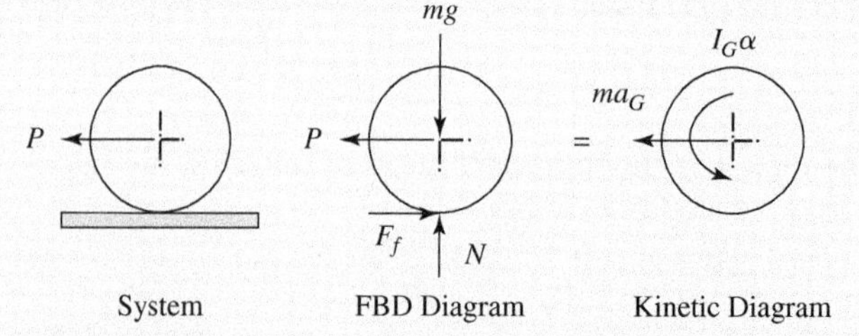

Figure D.9 (a) Disk and (b) FBD and kinetic diagrams

Solution:

The FBD and the kinetic diagrams of the disk are shown in Figure D.9(b). Applying Equation (D.19a), we obtain

$$P - F_f = ma_G \tag{1}$$

$$N - mg = 0 \tag{2}$$

And from Equation (D.19b), we get

$$F_f r = I_G\alpha \tag{3}$$

The above three equations have 4 unknowns. These are F_f, a_G, N, and α. We need a fourth relationship to solve this problem. The fourth relationship will be obtained from kinematics. We will assume that the disk does not slip. Under such an assumption, a_G and α are related by

$$a_G = \alpha r \tag{4}$$

Solving Equations (1) – (4), we obtain

$$\alpha = \frac{Pr}{I_G + mr^2} \tag{5}$$

We need to check if our assumption is correct. To do this, we solve for F_f and check if $F_f \leq \mu_s N$. If this is not correct, then our assumption is not valid, and the disk is slipping. In this case, a_G and α are not related, F_f should be set equal to $\mu_k N$, and Equations (1)–(3) are solved again to obtain α.

D.3 Modeling of Mechanical Systems with Spring Elements

All objects deform under the application of forces or torques on them. If the magnitude of the deformation is large, then we need to include in our model spring elements to represent the compliance of these objects. The spring elements can represent actual springs such as coil or leaf springs that are explicitly present in the system, or the compliance of system members such as gears, bars, shafts, or beams. An important property of a spring element is its stiffness. For linear (rotary) motion members, the stiffness is defined as the ratio of the applied force (torque) to the object displacement (angular displacement) due to this force (torque). The stiffness of common structural members is shown in Table D.1.

Table D.1

Stiffness of common structural members

Member	Shape	Stiffness
Bar under axial tension/compression	P, P	$k = \frac{AE}{L}$ A: Cross-sectional area E: Modulus of elasticity L: Member length
Circular rod subjected to torsion	T, T	$k_t = \frac{JG}{L}$ J: Polar moment of inertia G: Modulus of rigidity L: Member length
Cantilever beam	P, L	$k = \frac{3EI}{L^3}$ I: Area moment of inertia E: Modulus of elasticity L: Member length
Simply supported beam	P, L	$k = \frac{48EI}{L^3}$ I: Area moment of inertia E: Modulus of elasticity L: Member length

A passive elastic element, such as a spring, typically opposes displacement away from its static equilibrium position. Thus, if we have a vertically oriented spring as shown in Figure D.10(a), and we press down on it through an object, the spring force will be pushing up against the object. Similarly, if we try to stretch the spring, the spring force will act down on the object as shown in Figure D.10(b).

Figure D.10

Direction of spring force:
(a) compression and (b) tension

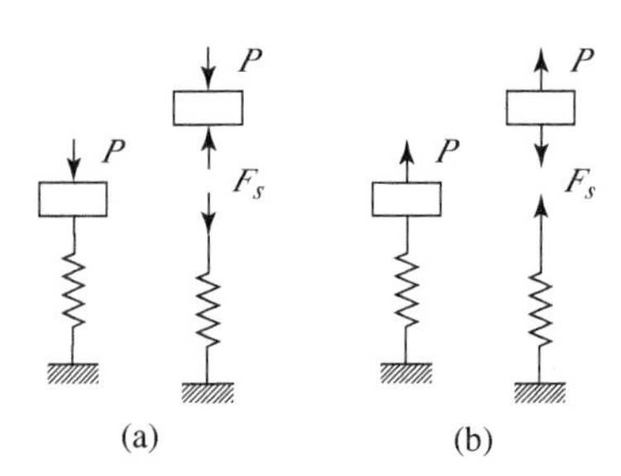

In deriving the equation of motion for systems with spring elements, we assume that the displacement of the object is made from the **static equilibrium position** which is defined as the position at which the spring stretches/compresses to just balance the weight of the object attached to it. In this way, we only account for forces that result from stretching/compressing the spring beyond the static equilibrium position. This is not an issue for planar systems in which gravity does not cause the spring to stretch or compress. Furthermore, with systems that involve more than one mass, we need to make an assumption about the displacement of the masses relative to each other to obtain the correct equations of motion. Such an approach is shown in Example D.7 which illustrates the derivation of the equation of motion for a system with two masses.

Example D.7 Equation of Motion for a Spring-Mass-Damper System

Derive the equation of motion for the system shown in Figure D.11(a).

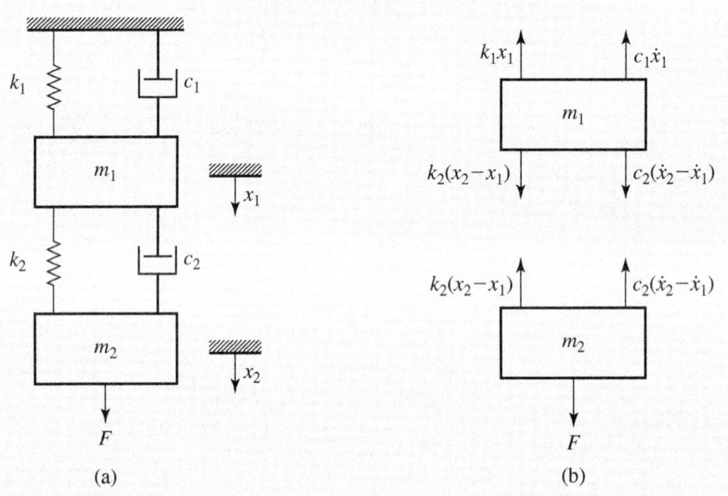

Figure D.11 (a) Mass-spring-damper system and (b) FBD diagrams

Solution:

Let us assume that $x_2 > x_1$ and $\dot{x}_2 > \dot{x}_1$. The free-body diagram for this assumption is shown in Figure D.11(b). Because x_2 was assumed to be larger than x_1, spring k_2 is stretched in this case and the force in that spring acts upward on mass m_2 and downward on mass m_1. Similarly, because $\dot{x}_2 > \dot{x}_1$, the force in the linear damper c_2 acts upward on mass m_2 and downward on mass m_1.

For m_1, summing forces in the vertical direction and equating to $m_1\ddot{x}_1$, the kinetic force, gives

$$k_2(x_2 - x_1) + c_2(\dot{x}_2 - \dot{x}_1) - k_1x_1 - c_1\dot{x}_1 = m_1\ddot{x}_1 \tag{1}$$

Similarly, we obtain the following equation for the mass m_2.

$$F - k_2(x_2 - x_1) - c_2(\dot{x}_2 - \dot{x}_1) = m_2\ddot{x}_2 \tag{2}$$

Note that an identical set of equations will be obtained if we had assumed that $x_2 < x_1$ and $\dot{x}_2 < \dot{x}_1$. The direction of the force in spring k_2 will be opposite to that shown in Figure D.11(b) since the spring will be compressed in this case, and the spring force will also be different and given by $k_2(x_1 - x_2)$, but the resulting equation of motion will be the same. The reader can verify that the same equations of motion will be obtained for *any* assumption regarding x_1 and x_2 and $\dot{x}_1$ and $\dot{x}_2$.

We previously presented the concept of effective inertia for a system made up of several rigid bodies. For compliant members, a similar concept is the effective mass. The **effective mass** of a compliant element, such as a spring, is the portion of the mass element that has the same kinetic energy as the entire element. Thus, the effective mass of an axial spring under tension/compression loads is equal to 1/3 of the mass of the spring, and for a cantilever beam is equal to 0.23 of the mass of the beam. A table of the effective mass of common shapes is normally included in system dynamics and vibration textbooks. Example D.8 illustrates the use of effective mass in modeling an object mounted at the end of a flexible platform.

Example D.8 Object Mounted on a Flexible Platform

Figure D.12(a) shows a winch of mass m_w mounted at the end of a flexible platform of mass m_p and a stiffness k_p. The winch is lifting a load of mass m_l. Obtain a dynamic model for this system, and an expression for the natural frequency of oscillation.

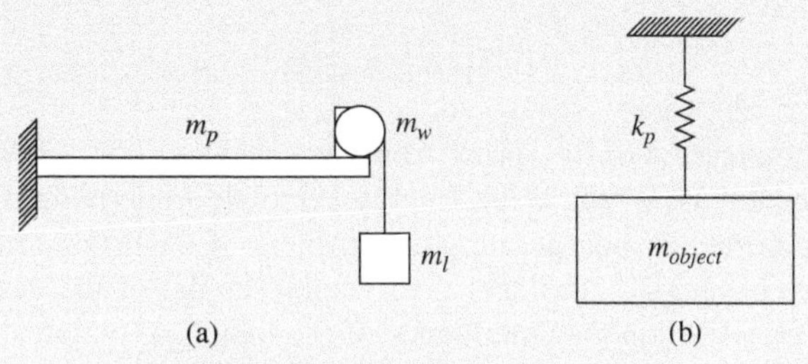

Figure D.12 (a) System and (b) model

Solution:

The given system can be modeled as a spring with an object attached at its end as shown in Figure D.12(b) if we neglect the compliance of the winch cable that is attached to the load. The spring stiffness is the platform stiffness k_p, while the mass of the object is the sum of the effective mass of the platform, the mass of the winch and the mass of the load or

$$m_{object} = 0.23 * m_p + m_w + m_l \quad (1)$$

The dynamic model is thus

$$m_{object}\ddot{x} + k_p x = 0 \quad (2)$$

And the natural frequency of oscillation is

$$\omega_n = \sqrt{k_p/m_{object}} \quad (3)$$

D.4 Work-Energy Principle

The Work-Energy equation provides another method of deriving the equation of motion for mechanical systems. The **Work-Energy Principle** states that the change in kinetic energy from state 1 to state 2 is equal to the work done on the system. In mathematical form, it is stated as

$$KineticEnergy1 + Work1\text{–}2 = KineticEnergy2 \quad \textbf{(D.20)}$$

Expressions for the kinetic energy of a rigid body were given in Equations (D.12) and (D.13). The work done by a force (torque) on a rigid body is the product of the force (torque) and the displacement (angular displacement) of the rigid body due to that force (torque). The work is considered positive if the force and the displacement caused by it are in the same direction and negative otherwise. Thus, the work done by gravity in lifting (lowering) an object is negative (positive). The work done by a spring force is given by

$$Work_{spring} = -\frac{1}{2}k\left(x_2^2 - x_1^2\right) \quad \textbf{(D.21)}$$

where x_1 and x_2 are the initial and final displacements, respectively. If the system is subjected to conservative forces, then the law of Conservation of Energy applies. A conservative force is one in which the work done by that force on an object is independent of the path followed by the object. Gravity and spring forces are examples of conservative forces, while friction is an example of a non-conservative force. The law of **Conservation of Energy** states that the sum of the kinetic and potential energy is constant at any point in time. In equation form, it is stated as

$$KineticEnergy1 + PotentialEnergy1 = KineticEnergy2 + PotentialEnergy2 \quad \textbf{(D.22)}$$

where 1 and 2 are two different states of the system. The potential energy (P.E.) of a rigid body due to gravity is given by

$$\text{P.E. Gravity} = mgh \quad \textbf{(D.23)}$$

where h is the height of the body above the reference datum. The potential energy due to spring extension/compression is given by

$$\text{P.E. Spring} = \frac{1}{2}kx^2 \quad \textbf{(D.24)}$$

where x is the extension/compression of the spring from the static equilibrium position.

Example D.9 illustrates the use of the Work-Energy Principle in deriving the equation of motion. In many situations, we are interested in incorporating the stiffness of the driving members into the dynamic model. Example 10.11 (see page 346) considers the dynamic model of the gear drive system shown in Figure D.6 but takes into consideration the compliance of the input and output shafts.

Example D.9 Equation of Motion Using Work-Energy Principle

Reconsider the system shown in Figure D.2(a) and reproduced in Figure D.13(a). Using the Work-Energy Principle, determine the speed of the cart after it had travelled up a distance of *d* starting from rest.

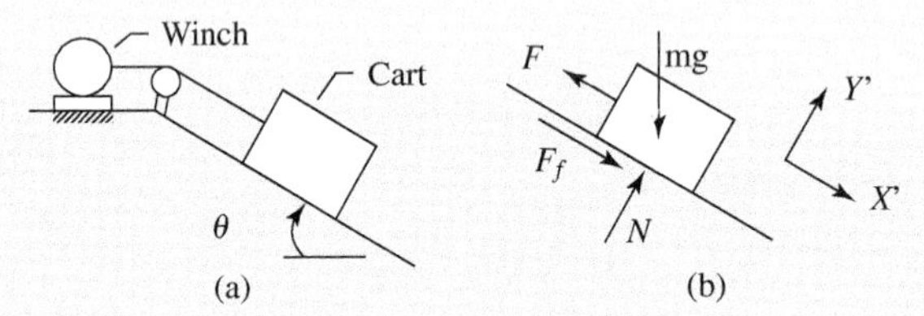

Figure D.13 (a) System and (b) FBD

Solution:

Since friction is present, we cannot use the law of Conservation of Energy. To apply the Work-Energy Principle, we need to evaluate the work done by all the forces on the cart. From the FBD shown in Figure D.13(b), the gravity force, the tension in the cable, and the friction force will do work on the cart. The normal force will not do any work since it is perpendicular to the cart displacement. The work done by the tension in the cable is

$$Work_{tension} = F\,d \quad (1)$$

The work done by gravity is equal to

$$Work_{gravity} = -mg\,\sin\theta\,d \quad (2)$$

And the work done by friction is

$$Work_{friction} = -\mu_k\,mg\,\cos\theta\,d \quad (3)$$

Using Equation (D.20), we obtain

$$Fd - mg\sin\theta\, d - \mu_k\, mg\cos\theta\, d = \frac{1}{2}mv_f^2 \tag{4}$$

Equation (4) can be written as

$$\frac{1}{2}\frac{v_f^2}{d} = 1/m(F - mg\sin\theta - \mu_k\, mg\cos\theta) \tag{5}$$

The left term of Equation (5) is the acceleration in terms of displacement and final speed. We obtained a similar expression (Equation (D.7)) for the acceleration of the cart when we used Newton's Law.

Index

Note: Page number followed by f and t indicates figure and table.

C

D

E

H

I

J

K

L

N

No reset windup, 342

O

Q

R

S

T

U

V

W

X

Z